SOTHIS PRESS

Sothis Press aims to completely reimagine the public domain book with an emphasis on universal access to captivatingly designed print editions. We believe that technology will enable every single book ever published to be perpetually accessible in beautiful print form. Sothis is beauty in ubiquity.

www.sothis.press
1 Waterhouse Square, London, ECIN 2ST

Asia

With Ethnological Appendix

Augustus H. Keane

SOTHIS PRESS

ISBN 978-1-0365-2487-6

Asia is in the public domain.

This edition is published by
Sothis Press © 2023. The company
is registered in England and Wales.
The Company Number is 14900389.

STANFORD'S

COMPENDIUM OF GEOGRAPHY AND TRAVEL

BASED ON HELLWALD'S 'DIE ERDE UND IHRE VÖLKER'

ASIA

WITH ETHNOLOGICAL APPENDIX

BY

AUGUSTUS H. KEANE

VICE-PRESIDENT OF THE ANTHROPOLOGICAL INSTITUTE

EDITED BY

SIR RICHARD TEMPLE, BART., M.P.

G.C.S.I., C.I.E., D.C.L., LL.D.

MAPS AND ILLUSTRATIONS

SECOND EDITION

LONDON: EDWARD STANFORD

26 & 27 COCKSPUR STREET, CHARING CROSS, S.W.

1886

PREFACE TO THE FIRST EDITION.

THIS book, relating to Asia, forms part of a compendium of geography and travel for all quarters of the world. The compendium was originally based on Von Hellwald's comprehensive work, *Die Erde und ihre Völker*.

But when the task of adapting the foreign work, regarding Asia, to the requirements of English readers was undertaken, it was found necessary to enlarge the treatment of the whole subject, and to rearrange the topics comprised therein. This was found to involve the composing of a new book, recasting the contents of Hellwald's work, for which due acknowledgment is made—but including also much additional matter as the result of recent research, and bringing the information up to the latest dates.

The composition was, as will be seen from the title-page, entrusted to Mr. A. H. Keane, who possesses special qualifications, having translated from the German the whole of Hellwald's work, and having prepared ethnological appendices to several numbers of the series contained in this general compendium. Indeed Mr. Keane's ethnological acquirements, which are of the first rank, and in which he is equalled by few, mark him as singularly fitted for this task. His wide acquaintance with physical geography, and his literary aptitude generally, will be apparent from the book itself.

As Author, then, he is responsible for the contents of the book, and for the verification of the facts. On the other hand, I have, as Editor, carefully revised the whole composition throughout all the chapters of the work itself, and to that extent I fully acknowledge my responsibility.

But for the Ethnological Appendix Mr. Keane is solely responsible, and it will be found instructive as well as interesting. I regret that his alphabetical list of the races and languages of Asia has been unavoidably sacrificed to the exigencies of space. The list will be very full, containing 3000 entries with copious references. For the information of those who take a special interest in ethnography, however, it may be mentioned that this list, together with much additional matter, will probably soon be issued under a different form.

In the general arrangement of the chapters and of the heads in each chapter, I have taken a more particular part. It is hoped that clearness and facility of reference will result from the pains which have been bestowed on this arrangement. As Asia contains many diverse nations and countries, simplicity and uniformity of arrangement become peculiarly important. The aim has been to present the same kind of information for each of these many nations and countries, arranged under the same heads. Hereby the student will find his studies easier than they would otherwise be, and will be enabled to compare readily the state of the several nations and countries under their different forms of civilisation. He will also have the means of immediately measuring the stage which the available information regarding them has reached under their various types of administration.

Thus Asia is divided, for the purposes of this work, into four main sections :—

A. Western Asia: Muhammadan States.
B. Southern Asia: British Political System.
C. Northern Asia: Russian Political System.
D. Eastern Asia: Buddhist States.

The chapters are numbered in one series consecutively from the beginning to the end of the book. Thus Chapter I., the introduction, presents a general survey of Asia, Chapters II. to VI. are comprised in Section A, Chapters VII. and VIII. in Section B, Chapters IX. to XI. in Section C, and Chapters XII. to XIV. in Section D.

Then in each chapter there are the same heads as follows:—1. Area—Extent—Boundaries; 2. Relief of the Land; 3. Hydrography: Rivers and Lakes; 4. Natural and Political Divisions; 5. Climate; 6. Fauna and Flora; 7. Inhabitants; 8. Topography: Chief Towns; 9. Highways of Communication; 10. Administration; 11. Statistics.

Thus, although the work deals with the utmost variety of conditions and of circumstances, there is symmetry in its arrangement as a volume of reference.

To the English reader a description of Asia may prove interesting by reason of the interests which England possesses in that quarter of the globe. The Asiatic dominions or dependencies of the British Crown contain more than 2½ millions of square miles. The people, under direct administration or political control of England, consists of 270 millions of souls, speaking at least twenty languages. Her European troops in Asia are maintained at a strength of about 78,000 men, while her native troops are 140,000 in number. Her ships of war stationed in Asiatic waters are about forty. She possesses 12,000 miles of railway in Asia, and 25,000 miles of electric telegraph on land, besides 8000 miles of submarine cable. The capital invested by the British people in these dominions or dependencies, either in State loans or in railways under State supervision, is not less than 250 millions sterling altogether. Besides this, there is a vast sum invested in private enterprises—agricultural, commercial, industrial—and amounting to scores of millions sterling, whereof the aggregate cannot be exactly computed. The foreign trade of these dominions or dependencies amounts to 150 millions sterling in value annually, of which about half is with the United Kingdom. Apart from this, the trade of the United Kingdom with other Asiatic countries has a yearly value of 20 millions sterling.

The Asiatic dominions or dependencies of Russia are even more vast in area, containing 7 millions of square miles; but they have a small population, 18 millions of souls, and in other economic respects they are not comparable to the British Empire in Asia, however interesting and important they may be. Still the progress of Russia from the Altai Mountains, which form the southern boundary of Siberia, towards the Himalayas, which are the northern boundary of India, has been remarkable within the present generation. During the most recent years she has established a base of progress, starting from the east shore of the Caspian, towards Central Asia. With the Caspian she now possesses complete communication by rail. As an inland sea the Caspian is now a Russian lake, the maritime share which Persia once owned in it having dwindled to insignificance.

The Asiatic interests of France are growing fast, and her possessions in Cambodia and Annam now contain 56,000 square miles, with a population of 750,000 souls. Besides this, she has a protectorate over the neighbouring provinces of Tongkin and Cochin-China, with an area of 200,000 square miles and a population of about 20,000,000. Her dominions in this quarter, though as yet undeveloped, are capable of indefinite expansion—

the soil is splendidly fertile, and the population may multiply under a civilised administration. This French jurisdiction is separated from British India by the intervening kingdom of Siam.

Among the comparatively novel points elucidated by this work, the following may be mentioned:—

The circumstances of Central Arabia, the decay of the Wahhabi and the rise of the Shammar State;

The results of the late Russian campaign in Turkestán, the new Russo-Persian frontier, and the present condition of Merv;

The orographic and lacustrine systems of Zungaria, with the lines of approach between the Chinese and Russian empires in that direction;

The recent exploration in Palestine and beyond the Jordan;

The geographical results of the late Afghan war, the approaches to the Iranian tableland from the Indus valley and from Turkestán;

The investigations by Prejevalsky in the basins of the Lob-nor and the Kuku-nor;

The travels by Gill, Baber, and Desgodins, on the frontiers between Tibet, China, and Burma;

The determination of the farthest source of the Irtish;

The political and social changes in Japan.

Although the information afforded by this book may be of general assistance in mastering or elucidating some among the political problems of the time, yet the utmost care has been taken to keep the matter free from controversial elements.

It is hoped that the physical geography of Asia will be presented to the imagination of the reader more vividly than it ever has been before. The prominent features are drawn with as much precision as possible. Such is the great Central Plateau, containing the most elevated areas to be found anywhere, walled in by mountains some of which are the loftiest yet discovered, giving birth to all the greatest rivers which flow towards the Arctic, the Pacific, the Indian Oceans, and affording a home to the nomad tribes that once ravaged the gardens of Central Asia and overran parts of Europe. Such, also, is the wondrous drainage which finds no vent towards the ocean, and is collected in inland seas, as the Caspian, the Aral, and the Siberian Balkhash, or in the lakes of Yarkand and Tibet, the swamps of Sístán, the saline depressions of Persia, the inner waters of Asia Minor and of Palestine.

The climatic conditions are set forth and their causes indicated—the heat and moisture of the south, the prevailing drought of the west, the temperate zone in the east, the amazing rainfall, measured not by inches but by feet, on the uplands which confront the vapours surging from the Indian Ocean, the burning winds hot as the blast from a furnace, the alternations of temperature in the centre and the north ranging between the extremes of heat and cold. In fact, the main Asiatic continent presents remarkable instances of what has been termed the Continental climate, which, at one season, is marked by burning heat, at another by intense frost, and which is so strange to English experience.

Some attempt is made to depict the scenery—the summits of Caucasus and Ararat, the brilliant colouring of Arabian rocks, the glittering snows and cloud-piercing peaks of the Himalayas, the groves and woods of Southern India or Ceylon, the teak forests of Burma, the tropical vegetation of Siam and Malacca, the majestic rivers of China, the sylvan and architectural beauties of Japan, the countless boats in Indian estuaries and Chinese waters, the steamers riding at anchor at Calcutta and Bombay, the fires in Mongolian prairies, the desolate lakes at alpine altitudes in mid-Asian ranges, the atmospheric gloom of some regions in Siberia, the frozen plains bordering on the Arctic Ocean.

Mention is made of the noble structures still preserved, as attesting the Asiatic genius of the past—the marble mausoleum at Agra, the Jama mosque of Delhi, the tomb of Tamerlane at Samarkand, the gilt pagodas of Burma and Siam, the Hindu temples in Southern India, the Lama monasteries of Tibet, the Buddhist fanes of China and Japan.

Conspicuous in the Indian landscape are the works of British engineering in railways and in canals for irrigation on the largest scale in the world. To these are now added the tall chimneys of the factories worked by European machinery and appliances in the manufacturing centres of India. The steam-engine has crossed most of the classic Indian rivers, penetrated parts of the Himalayas, and invaded Baluchistán.

Nor are the art-industries to be disregarded—the carpets of Persia, the shawls of Kashmir, the lacquer-work of Japan, the silks of China, the enamelling and embroidery of India. While the principles of European art are freely communicated to Asiatics, we should take care that the fresh springs of native genius are not quenched. Though perhaps inferior to Europeans in accuracy of drawing, Asiatics have a strong perception respecting vigour in outline, and a fine feeling for harmonised richness in colour.

The brute creation also deserves notice—the elephant of Ceylon, the tiger of Bengal, the ibex and ovis ammon of the Himalayas, the deer of India, the goat of Angora, the wild ass of Persia, the powerful horse of Turkestán, the fleet and gentle steed of Nejd, the Bactrian camel, and the swift-footed dromedary—truly named the ships of the desert—in the basin of the Oxus and the heart of Arabia.

An account of Asia summons up the memory of the remotest past known to human history, and presents to the mind's eye the antiquities with which most parts of the Continent are bestrewn, the remnants of a bygone civilisation pertaining to ancient times or the monuments of mediæval greatness. Such are the remains of Iconium and Balbec, of Persepolis, of Susiana, and of Ctesiphon—the rocks forming the ancient gates of Syria, and bearing memorial tablets of invaders from the Pharaohs to Napoleon III.—the Scythian cromlechs—the colossal *bas-reliefs* on the scarped side of the pass over the Indian Caucasus,—the sculptured caves, the rock-cut temples, the monumental mounds,—the ruins of Isfahán, of old Delhi, of Ayuthia in Siam,—the vestiges of an otherwise unknown Hindu dynasty in the upper valley of the Cambodia, and of the national deities in Japan. Such, too, are the scarcely distinguishable sites of places which were once mighty cities—of Balkh, of Merv, of Ayodhya in Northern India, and of a nameless capital near Kandahar.

The various phases of civilisation are displayed—the Muhammadan, whether in Asia Minor, Mesopotamia, Persia, or Turkestán, withering the minds and energies of the people with a blight; the Arabian, fostering an isolated independence within the limits of an arid and inaccessible peninsula; the British, enriching a vast population materially, and striving to elevate the upper classes morally and intellectually by enlightened legislation, by honest administration, and by national education; the Russian, enforcing among semi-barbarous tribes that degree of order which must ultimately soften rude violence; the Chinese, compelling or encouraging the most exact and assiduous study, yet implacably hostile to real progress; the Japanese, sweeping away ancient institutions, and imitating with sudden zeal the material improvements of Europe; the French, endeavouring by missionary effort and by scientific exploration to sow the seeds of improvement.

Then the present condition is described of many races whose social state hardly attains even a humble degree of civilisation — the Bedawin plundering the travellers in the desert; the squalid and inert Mongol, once the invader of fair domains, and still among the first of equestrians; the aborigines dwelling amidst the malarious forests of India: the solitary hunter in Siberian wilds.

The characteristics are sketched of the nations which make up the Asiatic population of 835 millions of souls—the modern Turk, patient, much-enduring, self-sacrificing, but stolid and incapable of mental effort; the Armenian, quick-witted and skilful in business, heretofore wanting in the stamina which constitute national character, but now beginning to assert a claim to nationality; the Arab, devout, sedate, resolute, and independent in isolation; the Persian, gay, refined, fluent in speech, but unstable in action; the Turkoman, ruthlessly dragging his victims into bondage; the fanatical priesthood of Islám; the high-caste Hindu, cherishing the traditions of centuries, and clinging to caste, despite the inroads of new learning; the Indian, trained by the Western education, abandoning the ancestral faith and fairly facing the moral or spiritual problems of the age; the Annamese, described as merry-hearted by sympathetic Frenchmen; the Chinese, diligent and practical in trade and agriculture, assiduous in severe study, but still hating modern progress; the Japanese, eagerly opening the windows of his mind to the social ideas of Europe.

The labours of some among the most distinguished of

modern travellers should be gratefully remembered: the Englishmen, Rawlinson, Ney Elias, Baber, Gill, Delmar Morgan, Freshfield, Bryce, Lansdell, Seebohm, Burnaby, Palgrave; the Russians, Prejevalsky, Kropotkin, Radde, Khanikoff; the Germans, Richthofen, Schlagintweit, Hüber, Wallin; the Frenchmen, Huc, Garnier, Mouhot, Pallegoix, Desgodins; the Swedish, Nordenskjöld; the Italian, Manzoni; the Dutchman, Siebold; the Hungarian, Vambery; the Americans, Schuyler, Collins, Kennan, Bush. In this difficult work, involving severe privations and distress, European ladies have borne an honourable part—Lady Anne Blunt, Miss Bird, Mrs. Atkinson, Madame Ujfalvy Bourdon. The pioneers of geographical discovery in Asia have to encounter not only the savagery of Nature but also the fierceness of Man, and need the sternest resolution besides the stiffest physical strength. Many of them have, in the service of geography, injured their health, and some of them have sacrificed their lives. But all have been supported under their load of labour by a love of natural beauty and an enthusiasm in the cause of knowledge.

Prominent among geographical explorers have been the Christian missionaries, in whom intellectual gifts and courageous resolution have been added to spiritual fervour.

Geography has ofttimes proved to be a foster-mother to the physical sciences in Asia. She has also been the handmaid to those who conduct the trigonometrical surveys, the geological investigations, the antiquarian explorations, who make the botanical collections, who manage the meteorological observatories, who follow the ethnological and philological studies.

Notwithstanding the matchless interest attaching to it as "the birthplace of mankind" and "the cradle of civilisation," the Asia of to-day hardly presents a happy appearance. Her three widespread creeds—Buddhism, Brahmanism, Muhammadanism, by their effects as now prevailing—obscure the reason, damp the aspirations, and deaden the energies of the people. Weighed in the scales of modern civilisation she is found practically wanting; viewed in the light of religion and reason she seems incapable of self-elevation. She is, in short, unable to attain moral or spiritual enlightenment by any strivings of her own, or to propel herself onward in the path of progress by spontaneous energy. Decrepitude has long been stealing over her, and old age has supervened without any future in hopeful prospect, unless she shall be rendered amenable to external influences. One-third of her population is already subject to one or other of the European powers, and of the remainder much is dominated by European influence in many essential respects. It almost seems to be decreed that Europe shall, in the immediate future, mould the destinies of Asia. Some Asiatic nations enjoy, indeed, a certain independence, but in practice it often degenerates into a liberty to wage internecine conflicts, to maintain an intellectual bondage, and to hold themselves aloof from foreign culture. In two of her recent expeditions, also, China has acted without deference either to England or to Russia. Nevertheless Europe is gradually becoming the mistress of Asia—is connected with the Asiatic continent by growing interests, and is deriving material advantage from this connexion. It is only by the means of such a connexion that rejuvenescence seems possible for Asiatic races. One immediate consequence is that the Christian religion is unreservedly preached in many parts of the continent. The progress of Christianity at first sight seems slow because of the vastness of the field in which its operations are conducted. Actually, however, it is considerable, and in some instances rapid.

The advantages springing from European influence in Asia may be aptly illustrated by a summary comparison between India and China. Of the two countries China is not only the larger, but also the finer in respect to the conditions of its soil and climate, as well as the qualities of its people. Yet it has neither a railway nor an electric telegraph, nor any enlightened system of legislation and education. Its grand canals are ill repaired and imperfectly managed. Its mineral resources are undeveloped, and its foreign trade is relatively small. Whereas India is permeated by the railway and the telegraph, has scientifically framed laws, and a liberal system of State education. Its canals are magnificently managed, its coal-mines are worked, and its foreign trade is threefold that of China, though its population is less than that of the Chinese by one third.

After all, Asiatic progress is supremely important to the interests of mankind at large, for Asia—despite all the devastation and depopulation to which she has been subjected — still contains more than half of the human race. Those regions of hers which, lying on her southern and eastern shores, have been the most readily approached by foreigners, are the most densely peopled parts of the earth. Though some Asiatic races are unimpressionable, and others almost brutish, yet many possess a lively intellect, a studious disposition, an expansive imagination, a sensitive conscience. But none of them have the mental stamina and the moral fibre which are found in Europe. For the present they all need guidance from without; with that guidance most of them are susceptible of such improvement and capable of such advancement as may give rise to boundless aspirations among all who feel the enthusiasm of humanity. They ought to receive from Europe the benefit of all that knowledge which has raised European civilisation to its present height. In short, Europe is incurring much responsibility in respect of Asia, and will doubtless discharge the duties arising therefrom until the Asiatic races shall, under Providence, be able to dispense with guidance and to soar aloft with their own wings.

R. T.

THE NASH, KEMPSEY, NEAR WORCESTER,
April 1882.

EDITOR'S PREFACE TO THE SECOND EDITION.

IN this new edition Professor Keane has judiciously made many alterations in detail, so as to embody the results of geographical exploration, and of territorial or political changes, effected since the first edition in 1882. The principal statistics, too, have been brought as nearly up to date as practicable.

Four years have passed since the first edition was issued. During this time there has been a marked growth in the prevailing tendency, as noticed by me in the former Preface. That tendency was described as causing the destinies of Asia to be moulded by Europe, and as gradually rendering European influence the main factor in Asiatic conditions. Within the short time that has ensued, the three advancing Powers in Asia, England, Russia, and France, have made strides onwards.

England has moved from the delta and the mid-region of the Irawaddy to the upper valley of that great river. The last remnant of the Burmese kingdom has fallen under British administration. The latest reigning successor of Alompra is now a royal exile in British India. The whole of the Burmese race and several mountaineer tribes will thus come more or less directly under that sway which has wrought such marvels in the Indian continent. A contact will be established with South-western China, and a trade route, long existing and imperfectly opened, will be so developed as to be virtually new. Probably a series of railways will be begun between the coast of the Bay of Bengal and Yunnan, by the valleys of the Irawaddy and the Mekhong.

Russia has now completed the subjugation of Turkomania, and is the undisputed mistress of all the Turkoman tribes. The question of Merv, so long pending, has been finally closed in favour of the Muscovite empire. The Russian arms, advancing from the new Turkoman base, have made good their movement along two main lines touching the Afghan borders, one near Panjdeh, the other near Zulfikar. These events will, on the one hand, check the marauding or slave-hunting disposition of the Turkomans, and, on the other hand, will for the present draw closer the relations between England and Afghanistan.

France, already suzerain of Annam, has now obtained a protectorate over Tonquin (or Tonking), the largest by far of all her new acquisitions. The extent, fertility, and populousness of this region are such that French administrators will, if they make the best use of their opportunities, have a wonderful scope before them. Thus the older settlement of Saigon in Cochin-China, the later dominion in Annam, and the new protectorate over Tonking, may perhaps constitute a French empire in South-eastern Asia.

The position of Siam, hemmed in by England on one side and France on the other, will be a matter for anxiety in the future.

R. T.

April 1886.

SYNOPTICAL TABLE.

CONTENTS.

CHAPTER I.

General Survey.

SECTION A.

WESTERN ASIA: MUHAMMADAN STATES.

(TURKEY IN ASIA, ARABIA, AND PERSIA.)

CHAPTER II.

Asia Minor.

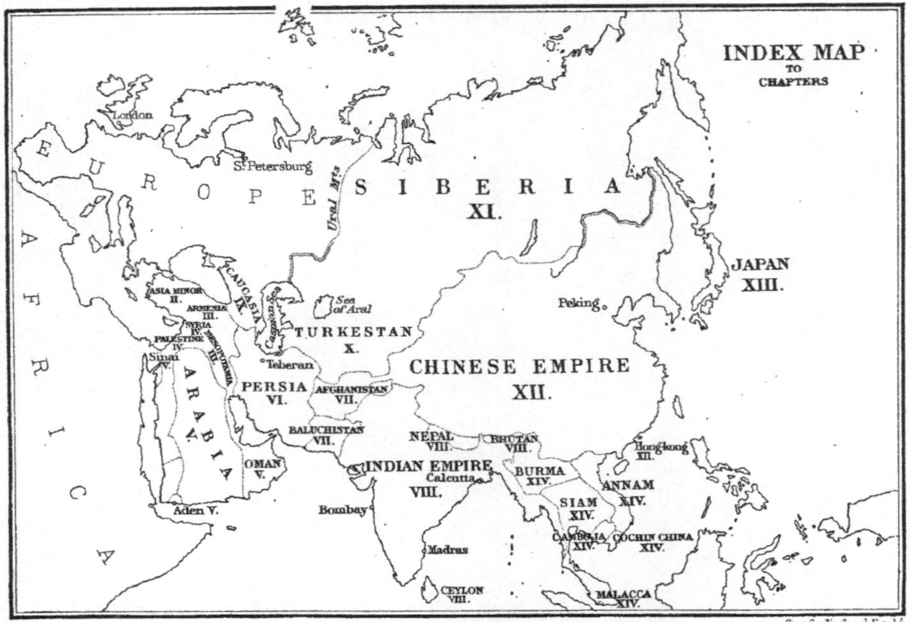

London: Edward Stanford, 26 & 27, Cockspur St, Charing Cross, S.W.

CHAPTER III.

The Euphrates and Tigris Basin.

CHAPTER IV.

Syria and Palestine.

CHAPTER V.

Arabia.

CHAPTER VI.

Persia.

SECTION B.

SOUTHERN ASIA: BRITISH POLITICAL SYSTEM.

CHAPTER VII.

AFGHANISTÁN AND BALUCHISTÁN (KÁBUL AND KELÁT).

CHAPTER VIII.

THE INDIAN EMPIRE.

SECTION C.

NORTHERN ASIA: RUSSIAN POLITICAL SYSTEM.

CHAPTER IX.

CAUCASIA.

CHAPTER X.

Russian Turkestán.

CHAPTER XI.

Siberia.

SECTION D.

EASTERN ASIA : BUDDHIST STATES.

CHAPTER XII.

CHINESE EMPIRE.

CHAPTER XIII.

JAPAN.

CHAPTER XIV.

INDO-CHINA AND MALACCA.

APPENDIX.

ETHNOLOGY AND PHILOLOGY OF THE ASIATIC RACES.

LIST OF ILLUSTRATIONS.

LIST OF MAPS.

PHYSICAL MAP OF ASIA

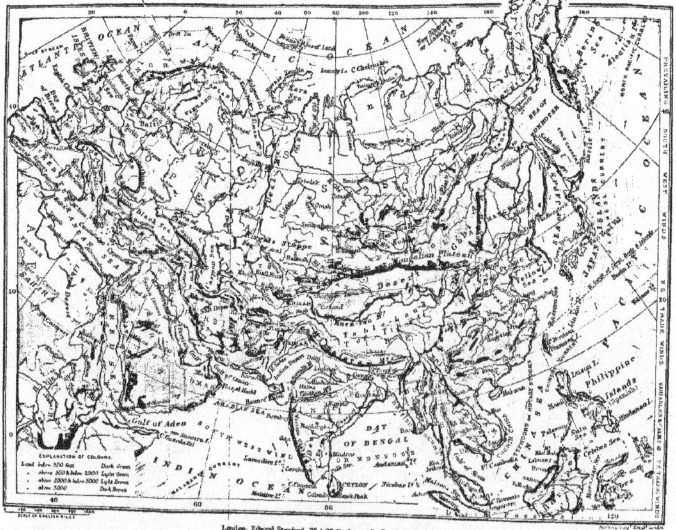

London: Edward Stanford, 26 & 27, Cockspur St., Charing Cross, S.W.

ASIA.

CHAPTER I.

INTRODUCTION: GENERAL SURVEY.

1. *Extent—Area—Boundaries.*

ASIA is not only the largest, but in many respects the grandest and most interesting among the main divisions of the globe. In size it exceeds by perhaps one million square miles the New World, while falling to about the same extent short of Europe, Africa, and Australasia, that is, of the remaining divisions of the Old World, taken collectively. Indeed, two of these—Europe and Africa—might be regarded geographically as appendages, or western peninsulas, of the Asiatic mainland, and geology has already determined beyond doubt the former connection of the austral-insular world with the south-eastern seaboard. Here the seas separating Trans-Gangetic India and China from Sumatra, Borneo, and the Philippines, rarely exceed 600 feet in depth, while the shallow waters are continued almost uninterruptedly from the Philippines south-eastwards to New Guinea and the north coast of Australia.

The principal features of this vast region everywhere present the same majestic proportions, or are drawn upon the same colossal scale. On three sides oceans form its natural boundaries, the Arctic on the north, the Pacific on the east, and the Indian on the south. Towards the west the frontier-line is extremely irregular, and at some points almost arbitrary, running in one place along the 60th, projecting in another westwards to the 30th degree of the meridian, and elsewhere presenting no conspicuous landmarks. Nevertheless, even here extensive mountain ranges and inland seas form on the whole a natural frontier sufficiently well defined between Europe and Asia. Proceeding southwards we have the Ural Mountains and the Ural River forming the line of demarcation from the 70th to the 50th parallel, beyond which the separation is even more strongly marked by the great barrier of the Caucasus, the Black, Mediterranean, and Red Seas. In the Isthmus of Suez, Asia is cut off from its vast African peninsula only by the narrow canal now connecting the Mediterranean with the Red Sea, and offering a continuous water highway from Great Britain to her remotest eastern possessions. From this point the mainland stretches in a compact body for about 6700 miles east and west to East Cape, where Bering Strait,[1] here scarcely 36 miles wide, separates it from the New World. Its greatest breadth north and south lies between Cape Chelyuskin in the Arctic Ocean, and Cape Romania, the southern extremity of Malacca; and these two points, which are some 5300 miles apart, might almost be connected by a straight line passing along the 104° east longitude, and dividing the continent into two unequal parts.

Within these limits Asia presents a compact mass of land, of a somewhat quadrangular shape, with its four sides facing towards the four points of the compass. But the line is broken on the south by three great projections, the Arabian, Indian, and Indo-Chinese peninsulas,

[1] Usually written with an *h*, *Behring*, according to German orthography. But the illustrious navigator was a Dane, and always spelled his name *Bering*.

presenting many striking points of analogy with Spain, Italy, and Greece, the three corresponding peninsulas of South Europe. Arabia, like Spain, forms a vast tableland with a monotonous coast-line, unvaried by any deep inlets. Like Italy, India is sheltered from the north by a great alpine region, is traversed by a mountain range running north and south, and terminates at its southern extremity with a large and fertile island. In the same way the Eastern Archipelago, continuing the Indo-Chinese peninsula towards Australasia, answers to that of the Ægean Sea, serving to connect the Hellenic peninsula with Asia Minor. Asia Minor, itself the westernmost projecting peninsula of Asia, may be compared with Brittany, the westernmost promontory of Central Europe, and the analogy is completed by the peninsulas of Corea and the Crimea, both projecting into narrow inland seas, and by the great archipelagoes of Japan and the British Isles, nearly equal in extent and even in population, but with their positions towards the mainland necessarily reversed.

Along the whole northern section of the continent there stretches a boundless lowland region, which, for hundreds of miles inland, is covered with the so-called *tundra*—dreary and almost uninhabitable wastes, exposed to the full fury of the fierce Arctic gales, ice-bound for nine months in the year, and in many places permanently frozen to a considerable depth. Further south the land ascends gradually to the south Siberian highlands, whence flow the Ob, Yenisei, Lena, and other great streams, which during the short open season roll their sluggish waters northwards to the Arctic basin. This great polar sea washes the whole of the flat and low-lying North Asiatic seaboard, the exploration of which has been but recently completed by the Swedish navigator, Nordenskjöld, who for the first time made the north-east passage in 1878-9, and determined the northernmost point of the continent, at Cape Severo, close to Cape Chelyuskin, in 78° 20′ N. lat. and 104° E. long. These bleak Northern shores, facing the neighbouring archipelago of New Siberia, and the recently-visited Wrangel Land, form a true Arctic region, lying entirely within the Arctic Circle, and sparsely inhabited by a few nomad Samoyede, Yakut, Yukaghir, and Chukchi tribes. Its low level and exposed northern aspect, combined with its high latitude and enormous extension southwards, are the chief reasons which cause the climate of this region to be the most "continental"—as it is technically termed, that is, subject to the greatest extremes of cold and heat—of any other on the globe. "Siberian" winters have become proverbial, but the summers are almost equally intense; and while the mercury becomes frozen to a hard malleable mass during the clear Arctic nights in mid-winter, it will occasionally rise to above 100° F. at mid-day in June. The most unfavourably situated tracts are undoubtedly those which stretch along the Frozen Ocean, from the Taimur peninsula to the River Kolima, which in many places are permanently frozen for some distance below the surface. But further east also, and generally speaking throughout the whole of the north-east, the aspect of the land is extremely dreary, especially in the Chukchi country, which reaches quite to Bering Strait. Here is developed the great peninsula of Kamchatka, which stretches southwards, and is continued across the Kurile archipelago as far as the northern extremity of the large Japanese island of Yesso. Igneous agencies, elsewhere all but exhausted or long quiescent on the Asiatic mainland, are still active in Kamchatka, whose eastern seaboard is traversed by an imposing line of burning mountains. These volcanoes are continued across the barren Kurile group, which, with the peninsula, encloses a deep inlet of the Pacific Ocean known as the Sea of Okhotsk, along whose desolate shores dwell a few scattered Lamut and other tribes of Tungus stock. More inviting and more favourable for agricultural life becomes the region where the mighty Amur rolls its waters to the sea over against the long and narrow island of Sakhalin. This metalliferous island, which is at one point almost connected with the continent, marks the extreme eastern limit of the Czar's authority, but since its cession to Russia by Japan it has been chiefly used as a convict station for political offenders.

Sakhalin is separated by the narrow Strait of La Pérouse from the Japanese group of islands which stretch, in a slightly curved arch, southwards to the Korean peninsula. Thus is formed the almost land-locked Sea of Japan, which communicates, through the Gulf of Tartary northwards with the Sea of Okhotsk, and through the Strait of Korea southwards, with the Yellow and Eastern Seas. East of these waters, and along the east coast of Japan, flows the Kuro Siwo, or "Black Stream," which is situated nearly under the same parallels of latitude as the Gulf Stream, and which plays almost a more important part in the Pacific than that remarkable current does in the Atlantic. Favoured by the boundless extent of the Pacific Ocean, which is here encumbered by but few island groups, the Kuro Siwo finds far fewer obstacles to its full development than its Western rival, and is thus enabled to pursue a more decided course, attended by correspondingly greater influences on the climate and vegetation of the lands lying in its course. Its effects are especially visible in Japan itself, where everything reminds us that we have entered a mild, and, in some places, even a sub-tropical zone. Here a delightful climate, combined with a lavish display of grand natural scenery, unites all the conditions required for the development of that peculiar civilisation which cannot fail to excite the admiration of the Western world, and create a deep sympathetic feeling for the Japanese people, with their varied industrial pursuits; their populous cities lying at the foot of threatening volcanoes; their well-tilled lands; their many ingenious social and political institutions.

A southern continuation of the Japanese archipelago is formed by the much smaller group of Liu-kiu Islands, which have long constituted a subject of contention between the governments of China and Japan. This group forms a link in the chain of islands, which are developed in a series of successive festoons, as it were, along the east Asiatic seaboard, between the Bering and China Seas. Another, and a still more important link, is formed by the extensive but little known island of Formosa, whence the transition is easily effected through the Batanes and Babuyan groups to the Philippines. Formosa occupies an important position, both physically and ethnically, for it is crossed, nearly in its centre, by the Tropic of Cancer. It thus stands on the verge of the torrid and temperate zones, marking the extreme northern extension of the Malay race, which here meets the Chinese on common ground. Beyond this point we pass with the Philippines into Australasia proper, and the great Archipelago of Malaysia, through which the south-eastern extremity of Asia merges imperceptibly with the continent of Australia.

Notwithstanding the labours of Ney Elias, Montgomerie, Forsyth, Margery, Gill, Prejevalsky, Krapotkin, Kostenko, Richthofen, Vambery, Schlagintweit, Desgodins, and many other illustrious modern explorers, a vast amount of geographical work still remains to be done in almost every part of the continent. British India, West Siberia, Palestine, and the Caucasus, alone can be said to have been thoroughly surveyed, and India especially is certainly one of the best-known countries in the world. But with these exceptions, and although most of the great geographical problems have been solved, our knowledge of most of the mainland is still far from complete, and often extremely inadequate.

2. *Relief of the Land: Plateaux and Highlands.*

The bold lines on which Asia has been framed are especially conspicuous, no less in its main political and social, than in its physical features. The heart of the continent consists of a vast tableland, by far the most elevated and extensive on the globe, with a mean altitude of from 6000 to 15,000 feet, above which tower the mighty Himalayan, Kuen-lun, Tian-shan, and Altai ranges. This tableland broadens out eastwards, and converges westwards in the nucleus of the Great Pamir, or "Roof of the World." A western extension of the same tableland is formed by the Iranian plateau, which stretches from the Hindu-Kush and Suliman Mountains, across Afghanistán, Baluchistán, and Persia, to the Persian Gulf and Mesopotamian lowlands. Culminating towards the north-west in the Kurdish and Armenian highlands, Irania merges westwards in the tableland of Asia Minor and the snowy crests of Lebanon, but falls abruptly northwards to the valley of the Kur. Beyond this historic stream, the land again rises to the mighty barrier of the Caucasus, which is continued north-westwards through the Taman peninsula into the Crimea, and south-eastwards across the Caspian to the highlands separating Irania from the Turkestán lowlands.

The vast central plateau itself is enclosed on the south by the mighty barrier of the Himalayas, sweeping round from Afghanistán to Burma in a graceful curve, which presents its convex side towards the Indian Ocean. On the north the tableland is hemmed in by the Altai, with its eastern projections, the Sayan, Yablonovoi, and other Siberian ranges; on the east by the less continuous

Yung-Ling, Inshan, and other Chinese ranges; on the west the Himalayas and Altai, through the Karakorum, Hindu-Kush, Tian-shan, and Alai, close round the Great Pamir, here interlacing in the focus of the whole continental mountain system.

But within these stupendous rocky walls the central tableland, occupying an area of perhaps 3,000,000 square miles altogether, presents several clearly-defined divisions, differing greatly in their relief, and even in their physical aspect, one from the other. The great Tibetan plateau maintains, between the Himalayas and the Kuen-lun, a mean elevation of 14,000 to 15,000 feet. The Pamir steppe in the west, and the Koko-nor basin in the east, fall to 12,000 and 9000 feet respectively. But beyond the Kuen-lun and its possible eastern extensions, there is almost an abrupt descent to the vast region of the Gobi desert, which is scarcely more than 4000, and which sinks westwards in the Tarim or Lob-nor depression as low as 2000 feet above sea-level. Yet, notwithstanding these deviations, the enormous extent and great mean elevation of the whole region are sufficient to give to the entire continent an average altitude of no less than 1600 feet, or about 600 feet more than Europe, and 500 more than the estimate made by Humboldt, on insufficient data, early in the present century.

The mountain ranges intersecting the plateaux, mainly in the direction from the north-west to the south-east, but occasionally running nearly due west and east, consist chiefly of crystalline rocks, old schists, palæozoic and other primitive formations, in the Siberian, Kuen-lun, and Karakorum sections. But the Himalayas, although resting on granite masses, which crop out in many of the highest peaks, are, to a very large extent, of comparatively more recent formation, having been upheaved during the secondary and tertiary epochs, when the eocene strata in Ladak were raised to an elevation of nearly 12,000 feet.

Simultaneously with the tendency towards greater dryness in the interior of the continent, there is clear evidence to show that a process of slow upheaval has been going on, at least around most of the seaboard, throughout the present geological epoch. On the north coast, islands, which a hundred years ago stood at some distance from the land, are now connected with it by rocky isthmuses. The upheaved coral reefs skirting the west coast of Arabia show that here also the land is rising, and similar tendencies have been observed in the Euxine and Ægean in the extreme west, about the Amur delta, Kamchatka, and China, in the extreme east; along the shores of Burma, Ceylon, Malabar, and Baluchistán, in the extreme south. On the other hand, symptoms of subsidence have been detected at a few points on the coast of Syria, near the Indus delta, on the shores of Annam and Fo-kien over against Formosa, and especially in the Laccadive and Maldive islands, where the atolls or round coral reefs are disappearing, and where the Chagos bank has already vanished.

3. *Hydrography: Rivers and Lakes—Inland and Seaward Drainage.*

Several distinct systems of inland drainage are formed by deep depressions, partly within the tablelands themselves, partly in the plains by which they are nearly everywhere surrounded. Such is the depression of Eastern Turkestán, 2000 feet above sea-level, through which the Tarim and its tributaries drain eastwards to Lake Lob, recently explored by Prejevalsky; that of Lake Sístán, which receives, through the Helmand and other streams, a great part of the Afghanistán drainage; and the remarkable trough of the Dead Sea, the deepest on the surface of the earth, fed mainly by the Jordan from the north. But by far the most important system of inland drainage is that of the Aral Sea, which comprises the whole of the western Turkestán lowlands, and which was formerly even still more extensive. At present it drains the Great Pamir, Alai, and western Tian-shan highlands alone, through the twin rivers Oxus (Amu-darya) and Jaxartes (Sir-darya). But it seems to have at one time stretched eastwards to Lake Balkhash, westwards to the Caspian, southwards to the north Iranian highlands, and northwards to the low range of hills forming the water-parting between the Ob and Aral basins. Altogether, the area of all the lands, which have no present outflow seawards, is estimated at about 4,000,000 square miles, or nearly one-fourth of the whole continent. The significance of this fact will be best realised when it is added that both Europe and America are almost destitute of an inland drainage, while that of Africa seems limited mainly to the Chad and Ngami basins.

The seaward drainage of the continent is determined only to a very small extent by the lofty ranges enclosing the great central and western tablelands. These ranges form scarcely anywhere true water-partings; for, except where they converge about the Pamir, they are everywhere pierced by the great continental rivers, which rise, not in their outer flanks, but within the plateaux, and which have thus to force their rocky barriers to reach the surrounding oceans. Thus, of the three great Siberian rivers flowing north to the Arctic, both the Ob and Yenisei have their farthest head-streams south of the mountains fringing the Kobdo and Mongolian plateaux; and even the Lena, now rising on the outer slopes, seems to have formerly been connected with the Angara

(Yenisei basin), in the neighbourhood of Irkutsk. So also the Amur, Hoang-ho, and Yang-tse-kiang, the three main streams flowing each to the Pacific, rise all of them far beyond the encircling ranges of the Mongolian, Koko-nor, and Tibetan tablelands. The great southern rivers, Mekhong, Salwen, Irawady, Brahmaputra (San-po), and Indus, have also their sources behind the Himalayas on the Tibetan steppe. Here a solitary but important exception is the Ganges system, of which both the head-streams, the Ganges and the Jamna, rise on the outer or southern flanks of the Himalayas. The same remarkable phenomenon is presented in the extreme west of the continent, where the Tigris and Euphrates flow to the Persian Gulf and the Araxis to the Caspian, from the very heart of the Armenian and Kurdistán highlands. Here also the Kizil-Irmak has to force its way from the Anatolian tableland through the Anti-Taurus to the Euxine, while the Orontes reaches the Mediterranean from the Bekáa (Cœle-Syria), behind the Lebanon and Nusarieh coast ranges. The list of great Asiatic rivers is almost completed by those of Southern India, where the Nerbadda flows from the furthest extremity of the Vindhya hills westwards to the Arabian Sea, and where the Godavari and the Kistna, rising also on the plateau of the Deccan, find more easy access through the low and broken line of the Eastern Ghats to the Bay of Bengal.

Compared with the other divisions of the globe, Asia is singularly deficient in large fresh-water lakes. Apart from the intensely salt Dead Sea and the salt or brackish Caspian, Aral, and Balkhash,—apparently remnants of a vast Asiatic Mediterranean communicating on the one hand with the Euxine, on the other through the Ob basin with the Arctic Ocean,—the only sheet of fresh water worthy of mention, by the side of the great inland seas of equatorial Africa and North America, is Lake Baikal, which discharges its overflow through the River Angara to the Yenisei. The so-called lakes of the West Siberian steppes are little better than swamps, and no large bodies of water occur in West Asia except Gokchai, Van, Urumiah, and the marshy Hamún or Sístán, none of which have any outflow seawards. In the whole of India and Indo-China the Tonlé-sap of Cambodia is the only lake of any size; in China, the Tong-ting and Po-yang alone deserve mention, and even on the Tibetan uplands, although lakes are very numerous in some places, none appear to be of large size except the Tengri, Koko, Buka, Palti, and Ike-Namur. The Kos-gol, Ubsa, and Kulon of North Mongolia, the Kenka of Manchuria, the Zaizan and Ulyungur of the Upper Irtish, the Issik-kul of the Tian-shan highlands, and the Lob-nor of Eastern Turkestán almost complete the list of large Asiatic lakes.

4. *Main Political Divisions.*

While Europe may geographically be described as a dependency of Asia, yet politically Asia may almost be regarded as a dependency of Europe. Notwithstanding its vast extent and enormous population, this continent has comparatively few independent States, and of these not one can be said to be entirely independent of European influences. In fact all these States are grouped like so many satellites around a few central suns, forming altogether four great political systems, of which two are directly controlled and two indirectly affected by European powers. The whole of the northern division, comprising nearly one-third of the mainland, may be regarded as practically a mere extension of European Russia eastwards to the Pacific seaboard. In the south the British Queen and Empress of India ("Kaisar-i-Hind") is either the absolute sovereign or the suzerain of the Indian peninsula, together with a large portion of Burma in Further India, of the Malay Peninsula, of Ceylon, and most of the islands scattered over the Indian Ocean, besides possessing either treaty rights or a political status in Baluchistán and Afghanistán, which bring a large portion of the Iranian tableland within the British political system. In the west, the Muhammadan world, embracing the rest of Irania, Anatolia, Syria, and Arabia, is mainly ruled over by the Turkish Sultan and the Shah of Persia. The Sultan has, by the recent Anglo-Turkish Convention, practically accepted the protectorate of England for his Asiatic possessions, while the Shah remains much under the influence of England and of Russia. Lastly, in the east the Buddhist world is divided between China, Japan, Siam, and Burma. But even here European influences are in many respects predominant. The "Middle Kingdom" has opened its ports to the trade of the world in virtue of treaties concluded after the close of military operations. Japan also has adopted the culture of the west, perhaps with the view of preserving its political independence. Lastly, in Further India, Annam and Cambodia are rapidly becoming French territory; while Siam, Burma, and the petty Moslem States of Malacca, naturally gravitate towards the power whose meteor flag sweeps the southern waters from Aden to Singapore and Hong-Kong, and whose beneficent voice is ever heard in the cause of freedom and humanity.

Thus we behold the Asiatic world mapped out into four political regions, which roughly correspond to four main natural divisions, and even to the four religious systems predominant in this quarter of the globe. The Russian possessions in the north, comprising Siberia, Caucasia, Western Turkestán, and part of Manchuria, have mainly an Arctic and inland drainage through the Ob, Yenisei, Lena, Jaxartes, and Oxus, and here is the original home of Shamanism. In the west, still held by the two great Moslem powers of Turkey and Persia, the drainage is chiefly through the Shat-el-Arab, the Orontes, the Kizil-Irmak, and other Anatolian streams, to the south-western land-locked basins of the Euxine, Mediterranean, and Persian Gulf. The southern, or British division, drains almost exclusively to the Indian Ocean through the Indus, Ganges, Brahmaputra, Nerbadda, Godavari, Kistna, and other streams of the Deccan, and here Brahmanism is the prevailing form of belief. Lastly, the Buddhist world, occupying the whole of the east, and comprising the Chinese Empire, Japan, and most of Further India, drains partly through the Tarim to the inland basin of the Lob-nor, but mainly through the Hoang-ho, Yang-tse-kiang, Mekhong, and Menam, to the Pacific Ocean. These remarks may suffice to render intelligible the tabulated scheme of the plan and subject matter of the present volume.

5. *Climate: Diminished Moisture—Rainfall.*

Although the great bulk of the land lies within the temperate zone, the climate is essentially continental, that is, characterised by the extremes of heat and cold, and by great dryness. Excluding the three southern peninsulas, which are mainly tropical, and China proper and Japan in the east, and parts of Persia, Syria, and Australia in the west, which are mainly temperate, the general climatic conditions are remarkably uniform, notwithstanding the great differences in the relief of the land. Thus, the Aral basin, which in many places is scarcely 200 feet above sea-level, and the Tibetan tableland, which is nowhere less than 12,000 feet, and occasionally attains an altitude of 18,000 feet, are both subject to the same intense heat and long droughts in summer, followed in winter by almost equally intense cold. On the whole, the climate may be said to depend rather on the aspect, elevation, and configuration of the land, than on its distance from the equator or the pole. It is affected especially by the great elevation of the tablelands with their excessively rarefied atmosphere, and by the vast extent of the continent, which is thus far less exposed to oceanic influences, and receives a correspondingly less amount of moisture than Europe or even America. The central regions, mostly enclosed by lofty ranges, which intercept the course of the humid sea-breezes, have necessarily a slighter annual rainfall than the surrounding lowlands. Yet, notwithstanding the different elevations and latitudes, great uniformity is produced in the central regions by the prevailing aridity of the soil, the sudden changes of temperature, and the dryness of the atmosphere. Even the abrupt transition from the uplands to the encircling plains is attended by far less change than is elsewhere caused by a slight difference of latitude. The elevated steppes of the great Pamir, 11,000 to 12,000 feet above the sea, the Mongolian desert of Gobi, the bare and barren plateaux of Tibet, and the dried-up bed of the great inland sea, jointly covering a space of over 1200 miles north and south, present almost everywhere the same monotonous aspect, varied only with a few green oases in the more favoured spots. But even here the native vegetation is scanty, and the running waters are lined chiefly with the poplar and the willow.

A careful survey of these regions seems to show that moisture was formerly far more abundant in Central Asia than at present, and that even within the historic period the climate has become much drier throughout most of the continent. Formerly, the Tarim basin was flooded by the Si-hai, or "Western Sea," a vast mediterranean communicating with the still more extensive Han-hai, but now represented only by the shallow and sedgy Lob-nor. The Han-hai itself covered an area in the great central depression nearly as long as the present Mediterranean Sea, stretching eastwards, through an island-studded strait to the Shamo basin, and developing between the Tian-shan and the Altai a large inlet which occupied the whole of the present basin of Zungaria. Through the so-called "Zungarian Strait" it seems to have even communicated with the vast depression of Western Turkestán, so that there was probably a time when a great water highway extended from the Atlantic through the Mediterranean, Euxine, and Caspian, eastwards to the Gobi desert, and through the Ob basin northwards to the Frozen Ocean. And of all these inland waters little now remains except the Aral, Caspian, Balkhash, and some smaller saline lakes and marshes—

> "Mere sluggish leagues of peat and black morass,
> Without a shrub or tree or blade of grass."

In many places the waters were succeeded by fertile diluvial plains, which in their turn have been swallowed up by the sands of the desert. This process is still going on, not only in the Tarim basin, in Bokhara, and other parts of Turkestán, where flourishing States and many populous cities have already disappeared, but even on the Iranian plateau, where Colonel MacGregor recently saw the sands in the very act of surging up above the walls and overflowing into the streets of Yazd and other Persian towns. These sandy wastes, formed by the weathering of crystalline, siliceous, and other old rocks, have already covered the greater part of Arabia, beyond which they stretch almost uninterruptedly across the Libyan Desert and the Sahara to the Atlantic seaboard. Rivers, which formerly had an outlet, if not seawards at least to the great land-locked basins, are now lost in the desert. Thus it is that the Zarafshan, Murgh-ab, and Hari-rúd no longer reach the Oxus or the Caspian, and the map of East Persia is scored with many watercourses which seem to run nowhere, but which formerly combined to fertilise the now arid wastes of Kerman and Khorasán.

But while the inland and south-western plateaux are amongst the driest, the great southern and south-eastern peninsulas are perhaps the wettest regions on the globe. Over one-half of the total annual rainfall is said to be absorbed by India, Indo-China, and the neighbouring archipelagoes of the Philippines and Malaysia. The coasts of Malabar and British Burma are deluged by the summer monsoons, which also discharge tremendous downpours on the advanced ramparts of the Himalayas. At the head of the Bay of Bengal the moisture-charged clouds from the Indian Ocean are almost completely arrested by the lofty ranges enclosing the lower Brahmaputra basin, and the annual rainfall, varying in the Indian peninsula from 240 to 480 inches, amounts on the Assam highlands in some years to no less than 600 inches. Hence arise the striking contrasts everywhere presented by the climate, flora, and fauna of the north Indian lowlands to those of the neighbouring Tibetan tablelands. On one side of the dividing range we have tropical heats, a magnificent southern vegetation, varied animal life, flourishing cities, and teeming populations ; on the other Arctic winters, bleak and almost uninhabited steppes, stunted vegetable growths, a fauna restricted to a few hardy upland species. Such a contrast scarcely occurs elsewhere in regions separated from each other by forty or fifty degrees of latitude.

6. *Flora and Fauna.*

Within the vast limits of the Asiatic mainland, which almost touches the equator at Cape Romania, and advances to within twelve degrees of the North Pole at

Cape Chelyuskin, every variety of animal and vegetable life finds a congenial home. While the southern peninsulas abound in tropical and aromatic products, the northern tundras are almost destitute of vegetation. Cereals cease to be cultivated beyond the 62d parallel; but, on the other hand, a tropical and sub-tropical flora prevails, even in the temperate zone on the eastern seaboard. Here, as well as in India and Indo-China, rice forms the staple food of many hundred millions of human beings, whereas the nomad Kirghiz and Kalmuk tribes of the Mongolian and West Siberian steppes are limited almost exclusively to an animal diet. The tea plant flourishes in Japan, China, Annam, and has in recent years been successfully cultivated in Assam, and along the southern slopes of the Himalayas. Coffee, indigenous in Arabia, is now also successfully cultivated in Ceylon and the uplands of Southern India. Opium is largely grown in India, and the area of its cultivation is yearly increasing in China. Cotton, indigo, and sugar, flourish in the two eastern peninsulas; cinnamon in Annam and Ceylon; aromatic plants in Arabia. Forest trees occupy, on the whole, a relatively limited area, being restricted mainly to the north coast of the Euxine, Caucasia, the southern shores of the Caspian, India, Indo-China, and the South Siberian uplands. The most useful species are the oak, walnut, pine, cedar, box, poplar, teak, bamboo, cocoa-nut, date, palm, apricot, peach, and other stone-fruit trees.

Of the larger animals the elephant, tiger, buffalo, and bear, abound chiefly in India and Further India, the yak in Tibet, the horse, wild ass, camel, and dromedary in Turkestán, the West Siberian steppes, Irania and Arabia, the reindeer in the extreme north and north-east. But the tiger has penetrated north as far as the Altai highlands, and the buffalo is indigenous in China. Characteristic of the central tablelands and Mongolia, are the argali, ovis poli, and other large-sized wild sheep and goats, and the hair of the Kashmir and Angora breeds is unequalled for its delicate texture. The sable, civet, marten, blue and silver fox, and other valuable fur-bearing animals, are widely diffused throughout Siberia and Manchuria, but are almost everywhere rapidly disappearing before the Russian, Ostiak, Tungus, and other trappers.

7. *Inhabitants: Social Culture—Religions.*

Asia, which is believed to be the cradle of the human race, is still the home of perhaps two-thirds of the inhabitants of the globe. But these teeming multitudes are far from being evenly distributed over its surface. While the bleak plateau of the Great Pamir, the frozen tundra stretching along the Arctic Ocean, the deserts of Gobi and Turkestán, are almost uninhabited, and the greater part of Siberia, Tibet, Persia, and Arabia, occupied only by a scanty nomad population; the rich and well-watered alluvial plains of the Ganges, Yang-tse-kiang, and Hoang-ho, are amongst the most densely-peopled regions in the world. On the whole the density of the population is in direct ratio to the abundance of the rainfall, and in the southern and eastern lands—India, Indo-China, China, and Japan—which are directly exposed to the moist winds from the Indian and Pacific Oceans, are concentrated over half of the human race. The popular views long entertained regarding the enormous masses of people occupying these countries had often been suspected of exaggeration, but they have been more than confirmed by the results of the official enumerations which are now regularly taken in India and Japan. The estimates for China are still matter of conjecture; but when the latest census for India reveals a total population of considerably over 252,000,000, strength is added to the generally received opinion that the inhabitants of the "Flowery Land" may number from 250,000,000 to 300,000,000.

Almost every variety of physical types, of speech, social culture, and religion, finds representatives amongst the Asiatic peoples. A few specimens of the dark, woolly-haired Negrito stock occur not only in the Andaman Islands, but in the interior of Malacca, and possibly also amongst the low caste hill tribes of Southern India. In the extreme north-east certain affinities have been traced between the Chukchis and the Eskimo of the North American seaboard. The relations of the neighbouring Yukaghirs, Kamchadales, and Koriaks, as well as of the Ainos and Giliaks of Yesso, Sakhalin, and the Amur delta, have not yet been satisfactorily determined. But, with the exception of these few outlying communities, all the inhabitants of the continent belong to the two great fair and yellow stocks, conventionally known as the Caucasic and Mongolic. The physical characteristics, main subdivisions, and linguistic families of these races will be found comprehensively treated in the Appendix to this volume. Here it will suffice to observe that throughout the historic period the Caucasic peoples have been mainly confined to the south-western region of Caucasia, which gives its name to the type, to Asia Minor, Syria, Arabia, Irania, and Northern India. They are roughly cut off by the Himalayas, the Hindu-Kush, and its western extensions to the Caspian, from the Mongolian, sometimes collectively grouped as the "Turanian" races, which occupy all the rest of the mainland. At the same time there have at all times been frequent crossings and overlappings, especially in Turkestán, Persia, and Asia Minor, which have presented many complicated problems to the student of ethnology in this division of the globe. Mongol and Caucasic tribes seem to be intermingled in the Cochin-Chinese and Yun-nan highlands; many of the low-caste tribes of the Deccan, if not all the Dravidian and Kolarian peoples, must be classed rather with the Mongol than with the Caucasic races; the Akkads, a people apparently of Turanian origin, were the founders of the earliest civilisation in Babylonia, and numerous members of the Mongol family have been so long settled in Irania and Anatolia that they have become almost entirely assimilated in physique to the surrounding Caucasic peoples. On the other hand the Caucasic Tajiks have from the remotest times been settled amid the nomad Mongols of the Aral basin, and, according to Prejevalsky, have penetrated eastwards as far as the dreary shores of the Lob-nor.

The various grades of human culture, broadly described as the hunting, pastoral, and agricultural states, depend in Asia rather on soil and climate than on race. Thus the Mongoloid Chinese, and Japanese, have for ages been settled agricultural peoples, while the Caucasic Arab tribes still remain mostly in the pastoral condition. The Tunguses, a large north-eastern branch of the yellow stock, follow the chase, tend their herds, or till the land, according to their position on the shores of the Arctic, in the Siberian steppes, or along the fertile banks of the Amur. Some of the Túrki races also, such as the Usbegs and Osmanli, have formed settled communities in Bokhara, Khiva, and Asia Minor, while the kindred Kirghiz and Kara-Kirghiz hordes of the West Siberian steppes and the Tian-shan still dwell in tents, and migrate with the seasons between the lowlands and the uplands of Central Asia. But, speaking generally, the hunting and fishing state is confined to a northern zone, reaching from the Frozen Ocean southwards to about the 60th parallel. The nomad pastoral tribes occupy the heart of the continent as far south as the 35th parallel, besides the arid plains of Irania and Arabia. Elsewhere, and especially in Japan, China, India, Indo-China, and Anatolia, the populations have long formed settled and more or less civilised communities on an agricultural basis.

But if social culture is chiefly conditioned by the outward surroundings, religion, on the other hand, is still largely determined by race and nationality. Asia, the original home of monotheism, is also still the land of paganism in some of its crudest aspects, while the originally pure doctrines of Brahma and Buddha alike have almost everywhere degenerated to the grossest polytheism and superstition. Judaism has almost vanished from Palestine, but there are scattered Jewish communities in many parts. Christianity is spreading very gradually through missionary effort, but is at present professed only by the Hellenes of Anatolia, the Maronites of Mount Lebanon, the Armenians, the Georgians, and kindred peoples of Caucasia, the so-called "Nestorians" or "Chaldeans" of the Upper Tigris and Lake Urumiyah, the Eurasians and the more recently converted communities in India, some 500,000 converts in China, Further India, and Japan; lastly, the Russians of Siberia and Caucasia, more numerous than all the rest put together. Most of the hill tribes in India, in Kafiristán, on the Indo-Tibetan frontier, in Further India, and in the western and southern highlands of China, are still pagans. A few survivors of the old Iranian fire-worshippers still linger on in Persia; and the flourishing tribe of Parsis in India follow the religion of Zoroaster. But, with these exceptions, the whole of the Asiatic populations belong either to the Shamanist, the Buddhist, the Brahmanic, or the Muhammadan religious world. Shamanism, open or thinly disguised, is diffused throughout all the Finno-Tatar tribes of Siberia and Manchuria. Buddhism is the religion of fully one-fourth of mankind, nearly all of Mongol stock, and concentrated mainly in Japan, the Chinese Empire, Further India, and Ceylon. Hinduism, or, as it should be more exactly termed, Brahmanism, is professed by 180 of the 250 millions of inhabitants of India. Muhammadanism, divided into the two great sects of the Sunnis and the Shiahs, prevails amongst the Tatar peoples of Turkestán, and South-West Siberia, and West China, in Irania, Anatolia, Syria, and Arabia, and has 50,000,000 adherents in different parts of India. The tenacity with which the Asiatic peoples almost everywhere adhere to their particular forms of belief is curiously illustrated by the Kalmuk and Kirghiz nomad tribes settled side by side in the steppe lands of the Lower Volga. The Mongolian Kalmuks, like their remote kinsmen of Zungaria and Mongolia, are all still Buddhists. But the Kirghiz, like all the other Kirghiz hordes of the Siberian steppes and uplands, are all Muhammadans. In the same way, all the Iranians of pure Persian blood belong to the Shiah sect, while the mixed Tajik communities of Western Turkestán and Afghanistán are invariably Sunnis. An exception, however, to this rule is presented by the Aimaks and Hazarahs of the North Afghan highlands, between Herat and Kábul, both of Mongol origin, but the former of whom, like those of Persia, are Sunnis, while the Hazarahs are of the Shiah sect. Besides these two great divisions, Muhammadanism embraces some other communities which are addicted to mysterious rites, and are consequently looked on with suspicion by their neighbours. Some of these, such as the Kizil-Bashis of Anatolia, Persia, and Afghanistán, are doubtless recent developments. But others, like the Druses and Nusarieh of Syria, and the Yezides, or so-called "Devil-worshippers" of Kurdistán, seem to date back from pre-Moslem and even pre-Christian times.

8. *Topography: Chief Towns.*

In reference to the aggregate of its population, Asia is not remarkable for the number of its large towns. A consideration of the present statistics relating to many Asiatic cities of historic renown, within comparatively recent times—such as Bagdad, Isfahán, Shiraz, Tabríz, Samarkand, Pekin, Ormuz, Goa—would disclose a remarkable decline in population, affording melancholy instances of the instability of material greatness. On the other hand, within the last century an equally striking progress is perceptible in many seaport towns. But that has generally arisen under European auspices, or, as it might be more accurately said, under British influence. Bombay having in round numbers nearly three-quarters of a million of inhabitants; Calcutta and Madras each having, roughly, half a million, may be reckoned in the second rank of the cities of the world—Calcutta, indeed, with its suburbs, has more than three-quarters of a million. The increase of Singapore, Penang, and Hong-Kong in populousness is rapid. The prosperity of Canton, Shanghai, and Tokio—all towns of the second rank in the world—is largely due to British trade. But of towns in the first rank—like London, Paris, New York, and others —there is not now any example in Asia.

9. *Highways of Communication.*

Except in India, where 12,000 miles of railway are open to traffic, there are no railways worth mentioning in Asia. The railway, however, which the Russians have begun from the east shore of the Caspian towards Central Asia may have momentous effects hereafter. Some short lines are being constructed in Japan. China offers a vast field for railways, but the Chinese Government at present sets its face against this mode of communication. India also can show several highways, each many hundred miles in length, which may bear comparison with the roads in Europe. Elsewhere in Asia there are no highways fit to be called such in the European sense of the term. Irrespective of highways, properly so termed, there are very few tracks easily passable in Asia; except in Siberia, there is not one such track traversing the continent from end to end. There is no through road from the British to the Russian dominions in Asia; no road from India to China, or to Tibet, or to Central Asia. The great central plateau, already described, interposes extraordinary obstacles in the way of such communication. The only instance of this nature is in the south. A horseman might, without meeting any real difficulty as regards ground, ride from any part of India through southern Afghanistán to Persia, and thence to the shore of the Euxine. Both China and India have magnificent rivers navigated by small craft. China also has navigable canals of great length. In Mesopotamia the two rivers are the natural highways. But extensive regions in Asia are destitute of water traffic.

Connected with communications is the subject of the electric telegraph. British India is the only Asiatic country which has telegraphic communication between all the principal towns. But some other countries in Asia have one or two through lines. From Constantinople there runs a telegraphic line across Asia Minor, then down Mesopotamia to the head of the Persian Gulf. From Tiflis, in Russian territory, there runs a line to Tehrán, then southwards across Persia to the head of the Gulf. Both these lines are joined to the Indian system by a line along the shore of the Persian Gulf and of Baluchistán. A long line passes from European Russia, near the Ural Mountains, across Siberia to the Pacific Ocean, near the mouth of the Amur. These are all land lines, but there are submarine lines also. One such line runs from Egypt down the Red Sea to Aden, and thence across the Arabian Sea to India. Another line passes from Madras across the Gulf of Bengal to the Straits of Malacca, and thence turning northwards passes near the Chinese coast to join the Japanese and Russian systems. The introduction of telegraphs is entirely due to the British and Russian Governments, and, in some degree, to the French Government. Japan is the only Asiatic country that has adopted the electric telegraph *ex proprio motu*.

10. *Administration*.

In all Asia, British India and Ceylon alone have an administration completely organised, in the European sense of organisation. The administration in Siberia doubtless approaches this standard so far as may be possible in a country thinly peopled and wild in parts. In Central Asia—that is, Khokand, Bokhara, and Khiva—civilised principles are being gradually introduced under Russian auspices. The French have established their rule in Cochin-China and Cambodia. The Chinese management of affairs, while evincing an elaborate culture in some respects, is in other respects semi-barbaric. Japan has been remodelling all its institutions after the European example, but whether these multifarious reforms have really taken root is more than the best-informed authorities seem able to pronounce. Both Persia and Asiatic Turkey are unreformed, and have nothing commendable in their administration. Independent Arabia has scarcely an administration in the strict sense of the term; its political organisation is for the most part tribal.

Respecting geography, the administrative results are in this wise:—

"The greater part of Asia has not yet been touched by scientific operations on a complete scale. In the whole of Asia, only India, Ceylon, Cyprus, Western Palestine, Caucasia, the Caspian basin, part of Western Siberia, and part of Japan, also many points in the Asiatic coast-line, have been subjected to trigonometrical observation. The altitudes of mountains have been determined only in the Himalayas, the Caucasus, and the Urals, by trigonometry. But in many ranges the heights have been approximately ascertained by the barometer. Professional surveys in detail have been completed only in India, Ceylon, Western Palestine, Caucasia, parts of Western and Eastern Siberia, the Tian-shan region, the greater part of Western Turkestán, Cambodia, parts of Cochin-China, parts of Afghanistán, also on certain lines of Persia, Mesopotamia, and Asia Minor. Even in the professionally surveyed territories many defects and imperfections are acknowledged to remain."

"Non-professional surveys have been carried out in Japan, in China proper, in parts of Arabia, on the frontiers of Tibet, China, and Burma, and on certain lines in Afghanistán and Baluchistán. Explorations without any actual survey have been made in Mongolia, Siam, the interior of Arabia, most parts of Persia, the Turkoman country, the Ust Urt plateau between the Aral and the Caspian, Manchuria, and in some parts of Afghanistan and Baluchistán."

"Though the southern coasts of Asia have been surveyed in sufficient detail for geographical purposes, yet according to the demands of a growing traffic and of maritime resort, these surveys need frequently to be amplified in detail. The old surveys by the Indian Navy were good in their day, reflecting honour on Moresby, Ross, and others; still the Government have ordered a new survey to be made for nautical purposes. A fresh survey, like that made by Nares for the Gulf of Suez, may have to be ordered one day for the whole of the Red Sea and the Persian Gulf. The British Admiralty are making yearly additions to the surveys of the Chinese coasts, of which the work done by St. John (R.N.) is an example. Whether the Russians will see fit to attempt a scientific survey throughout the Arctic coast of Siberia remains to be seen."

"Of geological surveys, the largest example is that in India, which, though far advanced, is far from complete. Very much remains to be done in this respect for the Himalayas. Geological surveys have been made in the Caucasus, the Urals, the Tian-shan and Altai ranges, Kamchatka, many parts of China and Japan, Cambodia, Ceylon, some parts of Arabia and Persia, much of Asia Minor and Palestine."[1]

11. *Statistics.*

The size of Asia, in comparison with that of the other main divisions of the globe, may be seen by the following statement of the relative areas of the five continents :—

	Square miles.
Asia, including Malaysia	17,300,000
America	16,000,000
Africa	11,800,000
Europe	3,800,000
Australasia	3,000,000
Total	51,900,000

The relative populousness of Asia will be seen thus. The population of the world, according to Behm and Wagner, was in 1880 :—

[1] Paper read before the British Association for the Advancement of Science, on 2d September 1881, by Sir Richard Temple.

Asia	834,707,000
Europe	315,929,000
Africa	205,679,000
America	95,495,500
Australasia	4,031,000

Of the four great political systems, already described as existing in Asia, the approximate areas and populations are as under :—

		Area in sq. miles.	Population.
I.	Western Asia: Muhammadan States .	2,200,000	32,000,000
II.	Southern Asia: British Political System	2,700,000	270,000,000
III.	Eastern Asia: Buddhist States . .	5,500,000	500,000,000
IV.	Northern Asia: Russian Political System	6,730,000	18,000,000
		17,230,000	[1]820,000,000

The direct and indirect European possessions in Asia are :—

	Area in sq. miles.	Population.
British Political System .	2,700,000	270,000,000
Asiatic Russia	6,730,000	18,000,000
French Territory . . .	50,000	2,750,000
Portuguese ,, ...	7,000	750,000
	9,493,000	291,500,000

[1] This does not quite agree with Behm's total given above; but except in the British and Russian dominions, the figures are either approximate or estimated only.

SECTION A.

WESTERN ASIA: MUHAMMADAN STATES.

(TURKEY IN ASIA, ARABIA, AND PERSIA.)

CHAPTER II.

ASIA MINOR.

1. *Boundaries—Extent—Area.*

THE Asiatic portion of the Turkish Empire comprises with Arabia the whole of the south-western section of the continent west of the Tigris. It is thus conterminous eastwards with Persia and Russia, the three empires converging about Mount Ararat, the culminating point of the Iranian plateau towards the north-west. Elsewhere Asiatic Turkey is, except towards Egypt, everywhere surrounded by water—Persian Gulf on the east, Arabian Sea on the south, Red Sea, Mediterranean, and Ægean on the west, Black Sea on the north. It comprises four well-marked natural and historical divisions—the two peninsulas of Arabia and Asia Minor, the basin of the Euphrates and Tigris, and the upland region of Syria and Palestine. These four main divisions will here be treated under four separate heads, a general survey of the empire being reserved for the end of the chapter.

Asia Minor or the "Lesser Asia," is so named relatively to the Greater Asia of which it forms the westernmost projection. But relatively to Europe it is the "Anatolia" of the Greeks, and the "Levant" of the

Italians—that is, the "Orient," or "Land of the Rising Sun."[1] Projecting far into the Mediterranean, it is washed on three sides by inland seas—the Euxine on the north, Marmora and Ægean with their connecting straits on the west, the eastermost section of the Mediterranean on the south. Eastwards it is limited by a somewhat arbitrary line running from Alexandretta Bay east to the great bend of the Euphrates, and thence follows the course of this river to its source, where it trends northwards to the Euxine, mainly along the valley of the Choruk-su. Anatolia thus lies between the 36° and 42° north latitude, consequently between the same parallels as the southern sections of the three European peninsulas and the northernmost portion of Barbary. Its greatest length from Cape Baba to the Euphrates, west and east, is about 700 miles, and its extreme breadth from Cape Anamur opposite Cyprus to Cape Injeh near Sinope in the Black Sea is rather over 400 miles. Within these limits its area is roughly estimated at about 220,000 square miles, or nearly 20,000 more than France, but with scarcely one-fourth of the population of that country.

2. *Relief of the Land: Taurus, Anti-Taurus, and Amanus —Passes—Plains—Volcanic Agencies—Geological Formation.*

Geographically Asia Minor must be regarded as a western extension of the Armenian and Kurdistán high-

[1] Anatolia, from ανατέλλω; Levant, from *levare*, both of which terms mean "to rise," hence are the exact equivalents of the "Orient," from *oriri*, the corresponding Latin word. But while *Anatolia* is by the Greeks strictly limited to Asia Minor, *Levante* is by the Italians extended to all the lands lying east of the Mediterranean, and *Orient* is applied to the East in general. *Anadoli*, the Turkish form of Anatolia, is more usually restricted to the western and northern provinces of Asia Minor, while the rest of the country is known as Karamania.

lands, from which it can nowhere be separated by any hard-and-fast line. The plateau formation prevails throughout, the interior of the peninsula forming an extensive tableland at a mean elevation of from 3500 to 4000 feet above sea-level, and stretching north-east and south-west for a distance of over 200 miles with an average breadth of about 140 miles. Above this tableland rise several loosely-connected mountain ranges, while over its surface are scattered a number of salt-lakes, morasses, and watercourses, without any visible outflow seawards, besides several streams which find their way mainly northwards to the Euxine and westwards to the Ægean. The plateau is skirted south and north by two broken mountain ranges, which radiate from the Armenian uplands, and to which the terms Taurus and Anti-Taurus were somewhat vaguely applied by the Ancients. The Taurus or southern branch, which forms a continuation of what Kiepert calls the "Armenian Taurus," rises close to the Euphrates, where one of its peaks attains an elevation of 10,000 feet. From this point it pursues a very irregular course under the more specific name of the Amanus down to Karamania, and thence along the Mediterranean coast to the Ægean, and with ramifications projecting northwards and southwards at various points. These branches, like the several sections of the main range itself, bear special names, such as the Ala-dagh, the Karmez-dagh, the Bulgar-dagh, the Sultan-dagh, the Jebel-kúm, and others, ranging from 7000 to 10,000 and even 13,000 feet high. The large island of Cyprus here lies, like a detached fragment of the mountain mass, opposite the angle formed by the Anatolian and Syrian coast-lines, while the south-western extremity of the peninsula is continued seawards by the lofty island of Rhodes, facing which the Massacitus spur terminates and culminates with Mount Takhtalu, 7820 feet high. But elsewhere the escarpments of the tableland fall westwards down to the Ægean, whose southern islands may be regarded as their advanced terminal peaks. Between the hills and the coast space is left in many places for lower valleys, and even for alluvial plains, varying in width, but mostly of great fertility, and sloping gently in all directions seawards.

The Anti-Taurus,[1] now perhaps better known as the Agha-dagh, forms a western extension of the Lazistán highlands, running in two and occasionally three nearly parallel chains from the neighbourhood of Batúm along the coast of the Euxine, and at no great distance from the sea, as far as the Bosphorus. Here it throws off a southern branch to the great western network, culminating with the Keshish-dagh (Olympus), the Morad-dagh, and the Kas-dagh (Ida), which rises 5700 feet above the plains of Troy at the head of the Gulf of Edremid. The Anti-Taurus forms a water-parting for the streams rising on the southern slopes of the Armenian uplands, and flowing some westwards and others towards the Euphrates. Like the Taurus, it also throws off several side ridges seawards and to the interior. Here a number of smaller and more isolated chains run in various directions, and often attain considerable elevations, culminating with the volcanic Ergish-dagh (Argaios), which is 13,100 feet high, and apparently the culminating point of the peninsula. This cone, which is nearly isolated from the Taurus, forms a striking landmark on the plains of Kaisarieh (Cæsarea), which here attain an elevation of over 3000 feet. It consists altogether of igneous matter, and its summit terminates with two craters, through which in former times the underground forces found an outlet. In its central section the Taurus itself varies in height from

[1] This term is by some geographers applied to the Amanus or north-eastern section of the Taurus between the Armenian highlands and Adana.

2700 to 5500 feet, while the Asi-Kur (Niphates), one of its loftiest summits, rises above the snow-line.

Both the Taurus and Anti-Taurus are crossed at various points by passes generally at low elevations and of moderately easy access. Of these the most important strategically and commercially is the Gölek-Boghaz, or "Cilician Gates," a deep gorge, 3300 feet above sea-level, running about 30 miles north of Tarsus over the Taurus, and connecting Anatolia with North Syria and the Euphrates valley. This famous defile has been followed in all ages by migrating peoples, traders, and conquering hosts. Through it Alexander marched to the overthrow of the Persian Empire, and through it Mehemet Ali in recent times twice penetrated into Anatolia on his march to Constantinople. About 100 miles west of this point the Taurus is crossed by a second pass leading from Karaman southwards to the Gok-su valley, and by a third, 150 miles still farther west, connecting Isbarta southwards with Adalia. The chief openings giving access from the Euxine through the Anti-Taurus to the central plateau are those leading from Ineboli to Kastamuni and Angora, from Sinope to Amasia, from Samsún to the same place, and from Trebizond over the Kolat-dagh to Erzerum.

In the higher regions of the peninsula the chief geological formations seem to be serpentines, granites, and schists, while limestones prevail lower down and almost everywhere in the western provinces. The trachytic formations, which abound in the east, are overlaid towards the centre of the plateau by black volcanic breccia, interspersed with blocks of trachyte. Altogether, igneous formations may be regarded as the dominant feature in the geology of Asia Minor.

3. *Hydrography: Rivers, The Kizil-Irmak, Sakaria, Choruk, Khoja-chai, and others—Lake Tuz-gol.*

The chief Anatolian rivers flow in a north-easterly direction to the Black Sea. Of these rivers, which have not all yet been thoroughly explored, the largest is the Kizil-Irmak (Halys), formed by the junction of two head-streams, one rising in the hills south-west of Tokat and flowing westwards, the other rising farther south on the slopes of Taurus and thence flowing first in a westerly and then in a northerly direction. After pursuing a very winding course of about 800 miles, the Kizil-Irmak discharges its waters through two principal channels into the Euxine below Bafra, and a little to the east of the Gulf of Sinope. Some 50 miles farther east the Yeshil-Irmak (Iris) enters the Black Sea, about 16 miles to the east of Samsún, after flowing by Tokat and Amasia during a tortuous course of nearly 240 miles.

But next in importance to the Halys is the Sakaria (Sangarius), which rises near Angora on the tableland and reaches the Black Sea at a point some 80 miles east of the Bosphorus. The Choruk or Joruk (Bathys), the north-eastern frontier river, crosses Armenian territory and falls into the Euxine just south of Batúm. The affluents of the Ægean Sea are important historically rather than geographically. While all are of small size, most of them are renowned in song and legend. Especially famous are the Gediz-chai (Hermus), flowing to the Gulf of Smyrna, and formerly noted for its auriferous sands; the Bakir-chai (Caïcus?), reaching the coast below Pergamus; the Khoja-chai (Granicus), flowing from the slopes of Ida, and the scene of Alexander's first victory over Darius in 354 B.C.; the Meinder-su, "called Xanthus by the gods and Scamander by men," which with its tributary, the equally famous Simois, traverses the Troas and joins the Ægean at the mouth of the Dardanelles; lastly, the Buyuk Meinder (Mæander), which flows for 250 miles through wild mountain gorges and rich alluvial plains to the coast near Miletus, and whose remarkable windings have given a familiar word to the English tongue. Most of these streams bring down much alluvial matter, which has during the historic period choked up many of the old harbours of the Ionian seaboard. Of less consequence are the rivers running south to the Mediterranean, two only of which, the Jihún-chai (Pyramus) and Sihún-chai (Carus), are of any considerable size. The "Silver Cydnus," associated with the names of Antony and Cleopatra, reaches this coast close to the mouth of the Sihún.

A prominent feature of the plateau consists of its numerous fresh and salt water lakes, of which the largest is the Tuz-gol, or "Salt Lake" (Tatta Palus), lying about 60 miles north of Konia (Iconium). It is nearly 50 miles long by 10 to 12 wide; its waters are very brackish, and the saline incrustations on its banks are rich enough to supply the surrounding districts with salt. It is very shallow, and its area is much diminished by evaporation during the summer months. Of the fresh-water lakes the largest is the Egerdir, which lies 2800 feet above the sea, between the Sultan-dagh and the northern spurs of Taurus, and which is 30 miles long by 9 to 10 broad at its widest point. In the north-west the Isnik-gol, near Brusa, is 50 miles round, and drains to the Sea of Marmora.

4. *Natural and Political Divisions—Islands.*

The Anatolian peninsula forms in reality as well as in name a miniature of the whole continent. Both consist mainly of extensive central plateaux, with an inland and seaward drainage, and both are skirted by lofty ranges, behind which most of the streams have their source, which find their way to the coast. But in Asia Minor the alluvial plains developed by these rivers cannot compare in relative extent with those of the greater Asia. The escarpments of the plateau approach everywhere so near to the sea that no space is left for great lowland plains such as those of Siberia and China. There are a few low-lying and somewhat marshy tracts about the lower course of the Yeshil-Irmak, Kizil-Irmak, and Sakaria on the Black Sea, along the banks of the Meinder below Smyrna, and about Adalia and Mersina on the south coast. But with these and a few other unimportant exceptions, the whole peninsula may be broadly divided into two main natural divisions—the central plateau, here and there intersected by transverse ridges, and the encircling ranges. This disposition of the surface has largely determined the limits of the eight great vilayets or provinces into which Anatolia is divided for administrative purposes. Two or three—Angora and Sivas—comprise the greater part of the tableland. Of the six others, Adana, Aydin, Kastamuni, and Trebizond coincide with so many distinct sections of the coast ranges, while Brusa and Konia alone include portions both of the plateau and of the seaboard.

The old historical divisions, which fluctuated considerably with the many political and ethnical vicissitudes of this region, have been almost entirely effaced by the modern administrative changes of the Ottoman rule. Nevertheless the names of these ancient states have never quite died out of history, and such memorable geographical terms as Phrygia, Lydia, Pamphylia, Paphlagonia, Cilicia, Cappadocia, are still familiar to the ordinary reader. How far all the old divisions correspond with the present administrative departments may be seen in the subjoined comparative table :—

Turkish Vilayets.	Ancient Divisions.
Brusa (Khodavendikiar)	Mysia.
Aydin (Smyrna)	Lydia. Caria.
Konia (Iconium)	Lycia. Pisidia. Pamphylia. Isauria. Lycaonia.
Adana	Cilicia.
Sivas	Part of Cappadocia.
Angora	Phrygia. Galatia. Part of Cappadocia.
Trebizond	Part of Pontus.
Kastamuni	Paphlagonia. Bithynia.

All the islands of the Ægean Sea belonging to Turkey, and collectively known as the Sporades, are grouped together in a separate administrative division called the Vilayet Jezairi Bahr-i-Sefid—that is, the "Vilayet of the White Sea Islands."[1] In this division was included the large island of Cyprus till the year 1878, when its administration was transferred to England. Thasos also is attached to the Egyptian Government, while Samos forms since 1832 a semi-independent tributary Christian State, under the suzerainty of the Porte, by whom its prince is appointed. With these exceptions all the Sporades of the White Sea Vilayet are disposed in five Sanjaks, or "Banners," as under :—

[1] The Ægean, for no apparent reason, is always called the "White Sea" by the Turks and Arabs. See *Das Vilayet der Inseln des Weissen Meeres*, by A. Ritter zur Helle. Vienna, 1878.

Sanjaks.	Islands.	Population.
Bigha . .	Tenedos. Lemnos. Samothrace. Imbros.	26,916 houses, of which 10,544 Greek, 10,308 Moslem.
Mytilene .	Mytilene (Lesbos).	19,522 houses, of which 16,594 Greek, 2,818 Moslem.
Sakyss .	Khios (Sakyss). Ipsaria (Psara).	10,428 houses, nearly all Greek.
Kos . . (Istankoi)	Kos. Kalymnos. Patmos. Nisyros. Nicaria (Icaria). Leros.	6,394 houses, of which 6,085 Greek.
Rhodes .	Rhodes. Karpathos. Kharki. Kasos.	16,762 houses, of which 10,270 Greek, 1,172 Moslem.

Most of these islands enjoy a delightful climate, and are fertile in oil, wine, silk, honey, corn, figs, oranges, and other fruits. Physically they may be regarded as a continuation of the mainland, belonging mostly to the same geological formation as the opposite coast of Anatolia, and, like it, still subject to violent earthquakes. By one of these Khios (Scio) was nearly ruined in the spring of the year 1881. In the Sporades the Greeks have always maintained a large numerical superiority, and the Turks, still numerous in Lemnos, Tenedos, and a few others, are retrograding like their fellow-countrymen on the mainland. The ownership of the land is rapidly passing from them into the hands of the Greeks, Armenians, and Jews.

5. *Climate.*

Owing to the great diversity in its relief, the climate

of Anatolia is so varied that a general description becomes very difficult. In some places the transition from winter to summer may be effected by the traveller within the four-and-twenty hours. Along the west coast, at all times famous for its genial temperature, the thermometer varies in summer from 85° to 98° or 100° F., and here the heavy dews partly compensate for the slight rainfall. On the central plateau the winters are often exceedingly severe, the snow lying deep on the ground for about four months. In Karamania these winters are followed by sultry summers, and here also the rainy days are so few between April and November that the people depend nearly altogether on the tanks and reservoirs for their water. In the mountain passes of the Taurus the winters are excessively severe, and the summers correspondingly oppressive. More favourable is the climate of the north coast, thanks to its mild character and abundant rainfall. But while the interior is generally healthy, malaria, produced by the great heat and moisture, prevails, especially in autumn, near Trebizond and at some other points along the shores of the Euxine.

6. *Flora and Fauna.*

The Anatolian flora forms a transition between those of Persia and Syria in the east, and of Southern Europe in the west. On the south coast we are even reminded of the Nile valley, while the western seaboard strongly resembles that of the Morea. Owing to their abundant moisture the northern shores possess a magnificent forest vegetation, including the oak, beech, box, ash, plane, and other leafy trees. Here we meet with dense groves of the walnut, quince, mulberry, pomegranate, peach, apricot, plum, and cherry, while the valleys of the Kizil-Irmak, Sakaria, and other streams, afford excellent pasturage.

Storax and other plants yielding valuable resins flourish on the Karamanian coast, whose flora resembles that of the shores of Syria.

In the Taurus grow several forest trees, especially of the coniferous order. But thousands of stately pines are yearly destroyed by fire, which is recklessly applied to them in order to stimulate the yield of turpentine. In Adana the sugar-cane grows well, but does not ripen sufficiently to cause the sap to crystallise. Large quantities of excellent grapes, olives, and figs are produced in the southern valleys, while the flora in many parts of the west and south rivals that of Spain and Sicily in splendour and luxuriance. In these respects a striking contrast is presented by the bleak upland plateaux of the interior, which produce little more than a stunted growth of brushwood, some saline plants, wormwood, wild sage, a few species of ferns, and in some districts nothing but two kinds of bramble. Amongst the cereals there is a species of bearded wheat; but oats are little cultivated, and barley is used as fodder for horses and other animals.

Like the flora, the fauna is akin to that of Southern Europe, but still more to those of Syria and Mesopotamia. Amongst the beasts of prey, nowhere numerous, are a few bears, wolves, hyenas, birds, several species of the cat, and wild dog. Jackals are met in the more secluded districts, where the gazelle and other varieties of the deer also abound. Of domestic animals the buffalo is most commonly employed in agriculture, and even its milk generally replaces that of the common cow, which is rarely seen in the country. The camel is the chief beast of burden, although the horses are strong and well built, and had once a high reputation. The asses also are active and above the average size. The famous long-haired Angora or shawl goat, formerly peculiar to this region, but now found also in Persia, thrives in Anatolia only in a tract about 11,000 square miles in extent, stretching westwards from the Kizil-Irmak. Elsewhere the breed soon degenerates and loses the fine fleecy texture of its coating. The indigenous sheep belong mostly to the fat-tailed species, common throughout the east from Syria to the Kirghiz steppes.

Amongst birds the most common are the eagle, falcon, bustard, stork, heron, quail, partridge, besides the ordinary European species. Of butterflies the varieties are endless, and many are noted for their rare and gorgeous colours. The coasts teem with all kinds of fish, amongst which are the dye-producing cuttle-fishes. Land tortoises, lizards, frogs, are common, while leeches are exported in considerable quantities to France and Italy.

7. *Inhabitants: Turks, Greeks, Kizil-Bashis.*

Ethnically speaking, Asia Minor is at present the true home of the Turks. It is one of the mainstays of the Ottoman Empire, from which this power continues to draw most of the resources that have hitherto enabled it to preserve its footing in the Balkan peninsula. Hence it is that the true character of this race can best be studied in Anatolia. All the western provinces are inhabited chiefly by Turks, who, however, even here are compelled to maintain the struggle for existence with other nationalities, and especially with the Hellenes. Farther east other races, such as the Armenians, Kurds, and Lazis, take part in the rivalry.

Yet, strange to say, the term "Turk" itself, at one time a proud title from the shores of the Adriatic to the remotest confines of Central Asia, is now carefully eschewed in Anatolia itself, where it has become a byword of reproach, answering somewhat to the English "boor," or "clod-hopper." And the people themselves have become all the more sensitive on the point, inasmuch as the "effendi," or refined "gentleman" from Stambúl, regards the terms "Turk" and "Anatolian" as practically synonymous with "uncouth" or "clownish." The stalwart and sinewy figure of the Anatolian peasant, his rough manners, his harsh dialect, so different in its primitive type from the Arabo-Persian jargon that passes for

ANATOLIAN PLOUGH.

Turkish in the capital, combined with his rude pronunciation, which has been compared to the gobbling of an enraged turkey-cock, afford a constant source of merriment to the dandies from the other side of the Bosphorus.

At the same time the social condition of the people must be regarded as backward and unsatisfactory. Since the days of the Trojan war the cultivation of the land has undergone but little improvement, and even the simple art of maintaining meadow lands is still unknown, so that during the dry summer months the herds must still be driven to the uplands in quest of a sorry pasturage. The fig, the vine, and the olive supply the Turkish peasant with his frugal fare, and enable him to meet his scanty wants. What need, therefore, to trouble himself with refined systems of husbandry ?

The Turkish village presents a far from inviting appearance. The uncleanly hovels built of adobe, or sun-baked bricks, and pierced with one or two holes for windows, usually comprise two compartments, one for the family, the other for the storage of provisions. The fittings of the interior are extremely simple, the furniture consisting mainly of a straw mat on the floor, a trestle bed with woollen mattress and cotton coverlets in the corner, a rude chest for the linen and best clothes, a few copper vessels and stone water-jars.

Dr. Carl Scherzer, a shrewd observer and a competent judge in Oriental matters, paints the present and the future of the Anatolian Turk in a few pregnant touches: —"The Turk, as a rule, understands his own language only, whereas all the other races in the country speak at least two from their infancy. This is due partly to his pride and contempt for all non-Muhammadan peoples, partly to the lack of enterprise and social rivalry. Earnest, reserved, and perhaps somewhat indolent, the Turk is still gifted with a fair share of intelligence. But though a keen observer of character, he lacks the business habits and the calculating spirit which have enabled the rival races to monopolise nearly all the trade of the country. In the rural districts the Turks are occupied mainly with agriculture and stock-breeding; in the towns they either deal in the local products, or else ply such simple trades as suffice to supply the few wants of their existence. Under proper management they make good seamen, and are also well suited for the caravan trade.

They are, generally speaking, honourable in all their dealings, frank, kind-hearted, and hospitable, while in religious matters they are, contrary to the general impression, the most tolerant of all Oriental races. They are deficient in the qualities of industry, perseverance in the acquisition of wealth and the upward tendency towards social improvement, and indolence may be regarded as one of their most salient national failings. The morrow troubles them but little; hence they will often pay an exorbitant interest for the means wherewith to tide over temporary embarrassments, and will freely sell their lands without giving a thought to the consequent decrease of future income.

"In the districts where they are surrounded by Greek and Armenian communities the Turks have fallen greatly behind; but, thanks to the natural resources of the land and their own frugal lives, they are seldom reduced to absolute want. The recruiting system is a heavy burden, to which the Muhammadan populations alone have hitherto been subjected."

The exclusion of the female element from the social life of the Turk helps but to intensify the evil. The continuance of this practice is due mainly to the low state of education, which completely fails to meet the requirements of modern ideas.

It is not perhaps surprising that under such circumstances the energetic, mercurial, and quick-witted Greek should threaten to usurp the inheritance of the Turk even during his lifetime. Occupied with thoughts of gain, a shrewd calculating man of business, a skilful seafarer, and intelligent husbandman, the Greek outrivals his Moslem neighbour in every pursuit of life. The learned professions he almost entirely monopolises, and the doctor, lawyer, teacher, banker, are everywhere sure to be of Hellenic blood. The Greek is invariably the broker who negotiates all business matters for "his Turkish friend," and he has secured the almost exclusive control over the local and export trade. He is at the same time indefatigable in his efforts to promote scientific and literary work, while also fostering a lively sense of Hellenic nationality. Thus Smyrna has already become a Greek city, and Athens has become the centre of an

ARMENIAN LADY OF SMYRNA.

ably-directed movement aiming especially at the improvement of education amongst the Anatolian Hellenes. With his unflagging efforts to better his social and political status the descendant of the old Ionian stock is gradually resuming possession of the western provinces; while close behind him presses the Armenian, intellectually scarcely his inferior, in restless energy fully his equal. Both are alike hostile to the Turk.

Other ethnical elements in Asia Minor are the Jews,

numerous in the large towns, the Gipsies, the Circassians, Abkhasians, Lazis, and the Yuruks, a nomad Túrki race occupying the uplands between Erzerum and the plains of North Syria. Mention should also be made of the Kizil-Bashis, or "Red Heads," a remarkable race, also of Túrki stock, scattered over Anatolia, Persia, and Afghanistán as far east as Kábul. Outwardly devout Muhammadans, the Kizil-Bashis are none the less tenaciously attached to their own peculiar tenets and observances. These they never reveal to strangers, and Mordtmann, who frequently visited Asia Minor, never succeeded in obtaining any trustworthy information regarding them. He, however, agrees with Van Lennep,[1] in looking on them as the last survivors of the old pagan communities. But W. Gifford Palgrave, when British Consul at Trebizond in 1868, described them as "a sort of Eastern Mormonites, with a dash of Persian or Shiah superstition."[2] He adds that they are as distinct from the Osmanli as the Saxons are from the Swedes. They call themselves "Eski-Túrk"—that is, "Old Turks"—a term often applied to the Anatolian Turkoman tribes, to whom they seem to be closely akin in physique and speech. Although reputed Atheists, they are said to be believers in the doctrine of transmigration, are very hospitable, and entirely free from the absurd feelings of jealousy which degrade women to the level of the brute creation in most Muhammadan countries. The fertile plains of Raz Ova and Ard Ova near Tokat, and the villages between Angora and Amasia, and between Kara-Hissar and Tokat, are the central quarters of the Anatolian Red Heads.

The Circassians and Abkhasians who have migrated to Turkey since the reduction of West Caucasia by the Russians have never found suitable homes in Asia

[1] *Travels in Little Known Parts of Asia Minor*, Lond. 1870.

[2] Official Report on prov. Trebizond, in Blue-book for 1868, part ii.

Minor, where they have consequently become a serious disturbing element. Mrs. Scott-Stevenson, and other recent travellers, represent them as a source of constant trouble, hopelessly indolent, given to plundering and hectoring over the people, levying blackmail right and left, and actually laying siege to the provincial towns.

8. *Topography: Chief Towns.*

The interior of Asia Minor is rich in towns whose names have been famous since classic times, a circumstance which is apt to give them far greater importance than they now really possess. Such are Kaisarieh (Cæsarea), at the north foot of Mount Argaios, which, though much fallen from its former greatness, still derives some importance from its position at the junction of several highways of commerce; Sivas (Sebaste) on the Kizil-Irmak, and Tokat on the Yeshil-Irmak, 60 miles north-west of it, both centres of a considerable inland trade. Farther west, Konia (Iconium), on the road between Brusa and Adana, gives its name to a large vilayet; formerly capital of the Seljuk empire, its numerous shrines of "saints" still attract devout Moslem pilgrims. Angora, or Engurieh (Ancyra), in the centre of the Angora plateau, is noted for its silky, long-haired animals—cats, dogs, rabbits, and goats, the wool of the last mentioned forming the staple of its trade. Afiun-Karahissar, midway between Smyrna and Lake Tuz-gol, is the centre of a large opium trade, whence its name, which means "Black Castle of Opium." On the northern route leading thence to the Sea of Marmora stand Kiutayah, or Kutaieh (Cotyæum), near which are some interesting Phrygian remains, and Brusa at the foot of Mount Olympus, whence its classic name of Prusa ad Olympum. It was formerly the capital of Bithynia, and is at present the chief town of the vilayet of Khodavendikiar. A few miles north-east of it are the once famous towns of Isnik (Nicæa) at the east end of Lake Ascanius, and Ismid, now connected by rail with Scutari, the Asiatic suburb of Constantinople. On the coast of the Euxine are the small ports of Sinope, where the destruction of the Turkish fleet

A GREEK OF SMYRNA.

by the Russians precipitated the Crimean war of 1854, and Samsún (Amisus), near the mouth of the Kizil-Irmak. East of it lies the flourishing port of Tarabúzun (Trebizond), the great emporium of the overland trade with Armenia and Persia. Here the Greeks under Xenophon, on their memorable retreat northwards from Cunaxa, first struck the coast and hailed the blue waters of the Euxine with shouts of Thalatta, Thalatta!

But the true emporium of the Levantine trade and the real capital of Asia Minor is Ismir (Smyrna), which is conveniently situated at the head of the gulf of like name, a magnificent inlet of the Ægean, over 40 miles long, forming a vast and well-sheltered harbour with deep water right up to the quays of the city. In Smyrna there are three perfectly distinct populations—the Turks, Greeks, and Franks. The Turks, by far the most numerous element, reside chiefly in narrow, dirty slums, into which it is dangerous to penetrate alone, and which are cut off from easy access to the more open and safer quarters. The Greeks also occupy a district apart, where the brightly-painted wooden houses produce a very pleasant effect. The Greeks take the leading part in all municipal affairs, and they have monopolised nearly all the retail trade of the place.

Nothing gives us a better idea of the varied natural resources of Anatolia than a glance at the export trade of Smyrna. The tables include such diverse commodities as maize, rice, and other cereals, tobacco, silk, cocoons, opium, madder, valonea, gall nuts, yellow berries, mohair, sponges, besides large quantities of dried figs and raisins of prime quality.

9. *Highways of Communication.*

One of the chief impediments to the development of the resources of Anatolia is the lack of good highways of communication. Railways there are none, except three short coast-lines, one connecting Scutari with Ismid, one running from Smyrna along the valley of the Gediz to Ala-shehr, and the third running also from Smyrna southwards to Aidin. But these may be regarded as the first links of that great Asiatic trunk-line, which perhaps may some day connect Constantinople and the West with the Indus valley. Meanwhile trade and intercourse are dependent on four main and a number of secondary highways, none of which except those connecting Trebizond with Erzerum and Samsún with Amasia, would pass for roads in the West. Of the four main routes the longest runs from Scutari through Ismid, Boli, Amasia, and Tokat, right across the northern section of the peninsula to Erzerum and the frontier Russian fortress of Kars. The second starting from the Euxine at Samsún strikes the former at Amasia, and again leaves it at Tokat, running thence nearly due south to Sivas. Here it branches off in two directions, south-westwards to Kaisarieh and through the Cilician Gates over the Taurus to the Mediterranean at Mersina, eastwards through Arabkir and Erzinghan to Erzerum. Another branch connects Kaisarieh with the Tigris at Diarbekr. The third main line runs from Trebizond southwards to Erzerum, where it trends eastwards to Bayazid on the Russian frontier, and thence across the Persian border to Tabríz. This has from time immemorial formed the great highway of communication for Persia with the Euxine and the West. Lastly, the fourth main route runs from the Sea of Marmora south-eastwards through Brusa, Kiutayah, and Koniah, to Erekli, beyond which it crosses the Taurus also by the Cilician passes, winding thence by Adana round Alexandretta Bay to Skanderún (Alexandretta), where it sends off branches eastwards to Aleppo, southwards to Antiochia. This formerly much-frequented route is now little used for through traffic.

CHAPTER III.

THE EUPHRATES AND TIGRIS BASIN.

1. *Boundaries—Extent—Area.*

NEARLY the whole of the eastern provinces, lying between Anatolia and Syria on the west, and the Russian and Persian empires on the east, are drained through the Rivers Euphrates and Tigris to the Persian Gulf. They consist mainly of two great physical divisions — the Armenian and Kurdistán highlands in the north, the Mesopotamian lowlands in the south. But there are nowhere any sharply-defined natural frontiers. The somewhat arbitrary line marking the limits of Turkey in Asia towards Russia and Persia coincides nearly throughout its entire length with the eastern frontier of this basin, which thus stretches from Lazistán to the Persian Gulf. On the west the northern uplands merge almost imperceptibly in the Anatolian plateau, while the southern lowlands rise very gradually towards the Syrian highlands and the Arabian tableland. Even in the north Turkish Armenia is cut off from the Black Sea by the portion of Lazistán which is still left to the Porte, and which is administratively included in the Anatolian vilayet of Trebizond. In the south alone the Persian Gulf gives for some distance a decided natural limit. In most maps a graceful curve, described almost with the regularity of the compass, and stretching across the Syrian desert from near the Dead Sea to the head of the gulf, is supposed to mark off Turkish territory from independent Arabia. But this line has absolutely no significance at all. In official maps it disappears altogether, or is replaced by a straight line drawn much farther south from about the head of the Gulf of Akaba eastwards to the new vilayet of Basra, which now includes all the Shat-el-Arab district and a large slice of North-East Arabia. The extent of this region will therefore vary enormously according as it is made to include or exclude the Syrian desert and portion of the province of Basra. But taking the southern limit at the 30th parallel, which crosses the head of the Persian Gulf, and the northern at the Lazistán coast range under the 41st parallel, the Mesopotamian basin will have a total length of about 770 miles, with an average breadth of 300 from the Russo-Persian frontier to Anatolia and Syria, and an area of over 300,000 square miles.

2. *Relief of the Land: The Armenian and Kurdistán Uplands—The Mesopotamian Lowlands.*

The northern section of this vast region embraces that portion of the Armenian highlands which still remains under the Ottoman rule. It consists mainly of a lofty plateau 4000 to 7000 feet above sea-level, and culminating with Mount Ararat just on the eastern frontier. Its surface is even more mountainous and irregular than that of Anatolia, for within its narrower limits it is crossed by four main ranges, with many secondary branches, forming connecting links between the Caucasian system on the north, the Anatolian on the west, and the Kurdish on the south. But notwithstanding the great mean elevation of the land, only a few of the peaks rise above the line of perpetual snow, and the chains themselves, which are crossed in several directions by accessible passes, are separated from each other by the deep valleys of the Aras, Choruk, and Euphrates, flowing in three opposite directions to the Caspian, Euxine, and Persian Gulf. The surface of the country between the mountain ranges consists of broad and mostly level steppe-like tablelands at various elevations, and forming a series of terraces one above the other. Deep and narrow valleys, gloomy and occasionally imposing mountain masses, broad and bleak plateaux, a severe climate, with rigorous winters, followed by dry and sultry summers, a marked absence of forest trees, but in the valleys an abundant and even luxurious vegetation, such is the general physical aspect of the Armenian highlands.

The Kars district, recently ceded to Russia, forms a rugged tableland, terminating south-westwards with the lofty Soghanli range, from 7000 to 8000 feet high, beyond which stretches the great valley of the upper Aras (Araxes). This valley, which crosses the district of Erzerum from west to east, is everywhere enclosed by high mountains—on the south by the Aghri-dagh (9400 feet), the Bingol-dagh (12,000 feet), and others; on the north by the Shamar-dagh (9227 feet); on the west the Boyun and Palantukan-dagh (7300 feet), close to Erzerum. North of Erzerum the land falls towards the valley of the Choruk, beyond which it again rises to the Lazistán coast range, which attains an elevation of 11,000 feet, and forms the northern frontier of Armenia proper. Eastwards the range is pierced by the Choruk, which here trends northwards through a narrow gorge at Artvin, beyond which it flows through alluvial plains to the Euxine at Batúm. Here the new Russo-Turkish frontier line has been shifted a few miles westwards to the coast village of Khopa, whence it runs southwards over the hills to the Choruk, thus leaving the eastern and richest division of Lazistán to Russia. The rest of this region, as already stated, is included in the vilayet of Trebizond, which thus stretches between Armenia and the coast eastwards to the Russian frontier. But the Choruk forms a geological and ethnical, as well as a political parting-line. While chalk and jurassic formations prevail in the south, igneous rocks everywhere crop out in the north, where they form the higher ridges of the coast range. The range itself is also inhabited by the Lazis, a western branch of the Georgian race, and consequently quite distinct from the Armenians, whose northernmost limit is marked by the middle course of the Choruk.

The great central tableland of Erzerum, which stretches eastwards to Ararat, may be said to be limited southwards by the valley of the Murad, or eastern head-stream of the Euphrates. Here rise the Sunderlyk-dagh, the Ala-dagh, the snowy Sipan-dagh, and other mountains, attaining an elevation of from 10,000 to 12,000 feet, beyond which the plateau maintains a mean altitude of 5000 feet eastwards to the frontier town of Bayazid. But it falls southward to the land-locked basin of Lake Van, which may be taken as the southern limit of Armenia proper. Since 1876 the Van district has been separated from the vilayet of Erzerum, and a line drawn from Músh through the lake eastwards to the frontier town of Kotúr, recently ceded to Persia, will roughly mark off the Armenian from the Kurdistán highlands. But the delimitation is in every sense arbitrary. The term Kurdistán—that is, "Country of the Kurds"—is so far correct that it is mainly occupied by tribes of Kurdish stock. But, on the other hand, these tribes have spread in almost every direction far beyond its present limits, reaching eastwards to the Bakhtiari highlands in Persia, northwards in isolated communities to the parallels of Batúm and Tiflis. Physically, also, the long and rugged mountain range forming the backbone of Kurdistán stretches beyond the frontier northwards between Lakes Van and Urmia to the foot of Ararat. From this point a second chain branches off south-westwards, sweeping round Lake Van and rejoining the eastern range at the Erdosh-dagh. The united chain runs thence with many ramifications south-eastwards to about the 34th parallel. The main eastern axis thus forms a natural frontier line between Turkey and Persia from Bayazid to Karmanshah, and the whole system encloses an area of nearly 50,000 square miles.

In the north the surface is very rugged and mountainous, but one extensive plateau, from 4000 to 7000 feet high, is developed between the Erdosh-dagh and Jebel-Judi,[1] which, running nearly west and east from Jezireh to Persia, rises from about 2000 feet at its western extremity to upwards of 13,000 in the Jawar and Rowandiz peaks near the Persian frontier. Beyond this range the country is generally level, varied only with a few low ridges culminating with the Jebel-Hamrin, about midway between Mosul and Bagdad. Here Kurdistán and Mesopotamia proper may be said to overlap, for, while the former at this point reaches southwards to the 34° N. lat. beyond the Tigris, the latter stretches between the two rivers northwards to the 37° N. lat.

The prevailing geological formation in the north is limestone, with red sandstones and conglomerate. Here the hills generally present bare crests with rugged slopes partly overgrown with dwarf cedars, junipers, and valonea. Limestones and sandstones also prevail along the southern frontier range, but intermingled with schists, quartz, and granites. Here the bleak brown hills present jagged outlines and steep sides, often deeply scored by the action of the mountain torrents which lower down flow through

[1] *Dagh* is the Turkish and *Jebel* the Arabic word for "Mountain."

narrow winding valleys. Copper, lead, and iron ores are said to abound in the west, and in several places amongst the hills of the Euphrates; but the only minerals available for export are salt from Van, sulphur, alum, naphtha, and a little iron.

South of the province of Erzerum and west of Lake Van, the Armenian and Kurdistán highlands slope continually southwards to the plains of Mesopotamia, and westwards to the Euphrates, which here marks the eastern limits of Anatolia. The tract between the Van district and the Euphrates, east and west, and between the Murad and Khabúr rivers, north and south, is often spoken of as "Kurdistán" in a more restricted sense, and on many maps figures as the Turkish province of Kurdistán; but this use of the word cannot be justified. There is no Turkish province of the name, and the country as above limited is mostly comprised in the vilayet of Diarbekr. Most of this vilayet is watered both by the Tigris and the Euphrates, consequently nearly as far north as Diarbekr it belongs geographically to the region commonly designated as Mesopotamia—that is, the Inter-riverain Country, or what in India would be spoken of as a "Doab," or "Land of Two Waters." It also belongs ethnically to two distinct domains, for the Kurdish and Arab nomad tribes, of Iranian and Semitic stock respectively, here meet on common ground. The term El-Jesireh, or "The Island," as Mesopotamia is always called by the Turks and Arabs, was formerly limited to the land strictly lying between the two rivers southwards to the old wall by which they were connected above Bagdad. The tract from this point to the Persian Gulf (that is, the ancient Babylonia) was and is still known as Irak-Arabi (that is, Irak of the Arabs), to distinguish it from the Irak of Persia. But the whole region from Diarbekr to the Gulf and from Syria to the Persian frontier is now commonly spoken of as Mesopotamia, the two divisions being sometimes distinguished as Upper and Lower Mesopotamia. It has a total area of perhaps 180,000 square miles; but it everywhere presents remarkable uniformity in its physical and ethnical conditions.

In the extreme north the land rises towards the Armenian and Kurdish highlands; but even here the mean elevation is little more than 1500 feet above the sea. The upland tract between Jesireh and Mardin is a stony waste, offering a scanty pasturage to the flocks and herds of the nomads in winter and spring. But the plains stretching farther west towards Urfa and Harran, and southwards to the low Sinjar hills, are well watered and very productive. These Sinjar hills form an isolated ridge, 7 miles wide and 40 miles long, midway between the Tigris and Euphrates, about the parallel of Mosul. Farther south the land is nowhere more than 600 feet above sea-level. It may be regarded as a northern extension of the Persian Gulf, which at one time probably reached to within 80 miles of the Mediterranean, but which has been gradually filled in by the alluvia of the great rivers, and by the advancing sands of the desert. Indeed, before the formation of the Syrian coast ranges the Mediterranean and Persian Gulf were possibly connected, thus isolating Arabia from the rest of the continent, and offering a direct water highway from the Atlantic to the Indian Ocean. Owing to this geological origin of Mesopotamia, the soil is found to consist everywhere of a sandy clay, abounding in excellent agricultural properties, and incapable of cultivation only where water fails. Its astounding fertility is sufficiently shown by the fact that it still remains unexhausted after having supported the teeming populations of the Akkadian, Assyrian, Babylonian, and Persian empires, from the dawn of history down to comparatively recent times. The num-

MESOPOTAMIAN DESERT.

ber and vastness of the ruins scattered over this region from Babylon to Nineveh still bear witness to its former flourishing material condition; and since the cuneiform writings abounding in these ruins have yielded up their secret to the ingenious labours of modern science, we now know that the Mesopotamian plains have been the scene of successive cultures, rivalled in splendour and antiquity by those only of the Nile valley.

3. *Hydrography: The Tigris and Euphrates—Lake Van.*

With the exception of a small area in the extreme north, the whole of this region drains through the Euphrates and Tigris to the Persian Gulf. Since the recent rectification of the Russo-Turkish frontier, the valley of the Kúr (Cyrus) belongs entirely to the Russian territory of Transcaucasia. But the Choruk and the Aras (Araxes) still flow for a considerable distance through Turkish Armenia before crossing the frontier on their course to the Euxine and the Caspian. The Choruk, which rises in the uplands west of Erzerum, is joined below Baiburt by a tributary from Trebizond, after which it flows along the southern base of the Lazistán coast range eastwards to the Russian frontier. Here it bends northwards altogether within Russian territory, and reaches the sea close to Batúm, after a precipitous course of about 200 miles. The Aras rises at the north foot of the Bingol-dagh 30 miles south of Erzerum, and flows north-eastwards to the frontier, which it now soon reaches at a point considerably to the west of Kars and Ararat. In Turkish Armenia, of which it drains a very small area, it is little more than a rapid mountain torrent.

All the rest of the Armenian and nearly the whole of the Kurdish highlands belong to the basin of the twin rivers Euphrates and Tigris, which flow mainly in a south-easterly direction across the Mesopotamian plains. Rising on the Armenian terrace lands, they pursue on the whole a parallel course, although often approaching and diverging from each other, until they at last mingle their waters at Kurnah, where the united stream takes the name of the Shat-el-Arab about 120 miles above its delta at the head of the Persian Gulf. Above Kurnah their channels approach nearest to each other at Bagdad, thus nearly separating Irak-Arabi from upper Mesopotamia.

The upper region of the Euphrates resembles that of the Rhine, while its middle course may be compared with that of the Danube, and its lower with the Nile. The Euphrates proper is formed by the junction of two great head-streams—the Kara-su or western branch, and the Murad or eastern branch, whose sources lie over 120 miles apart, in the very heart of the Armenian highlands. The Kara-su—that is, "Black Water"—rises some 20 miles to the north-east of Erzerum, and flows for 270 miles south-eastwards to Keban-Maaden, a few miles west of Kharpút. Here it is joined from the east by the Murad, which flows from the Ala-dagh south of Bayazid, and near the Russian frontier, and has a total course through Armenia and Kurdistán of about 300 miles. Some 60 miles south of the junction the Euphrates pierces the Upper Taurus near Arghana, beyond which it trends southwards through the vilayet of Aleppo, here coming within 80 miles of the sea. But about the 36th parallel it turns somewhat abruptly to the south-east, and henceforth retains this direction to the Gulf. It is navigable for over 1100 miles for small steamers to Bír (Bírejik), near Urfa, the point where it is crossed by the great caravan route from Syria to Bagdad.

The Tigris is also formed by an eastern and a western head-stream, the former rising close to Bitlis, near the west side of Lake Van, the latter flowing from the neighbourhood of Kharpút by Diarbekr to the confluence above Findúk. Beyond this point it pursues a southerly course by Mosul to Bagdad, and between these points is joined on its left bank by the Great and Little Zab, and some other tributaries from the Kurdistán highlands. It is navigable for vessels of light draught to Nimrúd, 20 miles below Mosul, and again for 300 miles by rafts from Mosul to Diarbekr. But owing to the rapidity of the current the traffic is all down stream, and is still carried on mainly by a primitive style of craft, which is broken up at Bagdad, and transported by camels back to Mosul. The journey between these points occupies three or four days during the floods, and from twelve to fourteen at other times.

Below Bagdad the main streams are connected by several channels and intermittent watercourses, of which the chief are the Nahr Isa or Saklawiyah canal and the Shat-el-Hai. Higher up the Euphrates is joined on its left bank by the Belik near Rakkah, and by the Khabúr at Kerkesia. The latter flows intermittently through the desert from the Karijah-dagh hills, 20 miles west of Mardin, round the western extremity of the Sinjar hills. During the floods it is joined by several streams, which at other times run dry in the sands. Below the junction of the Khabúr there stretches a desolate desert tract between the Euphrates and the Tigris, which is overgrown with wormwood, and still haunted, as in the time of Xenophon, by the wild ass, ostrich, and bustard. This region is visited by terrific whirlwinds, such as that which on May 21, 1836, nearly overwhelmed the English Euphrates Expedition under Colonel Chesney.

Below Kurnah the Shat-el-Arab traverses a flat and fertile plain, dotted over with villages, and covered with artificially irrigated meadow-lands and date groves. At Mohamra, 40 miles above its mouth, and 20 miles below Basra, it is joined by the Karún from Persia, and here properly begins the delta, of which one arm only is navigable. For six months in the year this delta is converted into a swampy lacustrine district by the floods, caused by the melting of the snows about the head-streams in spring, and occasionally by the autumn rains. From its mouth to Bagdad the main stream is navigable throughout the year for steamers of considerable size. For some years past an English line plying between Basra and Bagdad has contributed much towards the development of the resources of Mesopotamia.

BANKS OF THE TIGRIS.

In the whole of the Mesopotamian basin there is only one body of still water deserving the name of lake. This is the magnificent Lake Van, by far the largest in Asiatic Turkey, renowned alike for its romantic beauty and historic associations. It occupies an irregular triangular space, 80 miles by 30, over 5000 feet above sea-level, on the border land between the Armenian and Kurdistán highlands. From the snowy Sipan-dagh, towering above its northern shore, it is seen to occupy "the centre of a magnificent valley, surrounded on three sides by densely-wooded mountains, whose forests of firs, chestnut, beech, walnut, and ash merge in the broad belt of gardens and melon-grounds that fringe most of the shore line as far as the eye can reach."[1] Its waters, which are diversified

[1] J. C. M'Coan, *Our New Protectorate*, i. 45.

with several lovely islets, and teem with fish, are very salt, and have no present outflow.

4. *Natural and Political Divisions: Turkish Armenia and Kurdistán—Mesopotamia.*

The Mesopotamian basin comprises two natural divisions only—the Armenian and Kurdistán uplands, where all the rivers have their sources, and the alluvial plains of Mesopotamia proper, which may be regarded as the creation of those rivers. To these two natural divisions correspond the five Turkish administrative divisions of Erzerum, Aziz, Van, Diarbekr, and Bagdad, together with a portion of Basra, distributed as under:—

Natural Divisions.	Turkish Vilayets.	Area in sq. miles.	Population.
Armenian Highlands	Erzerum	27,000	675,000
	Aziz	12,000	338,000
Kurdistán Highlands	Van	19,000	1,015,000
	Diarbekr	30,000	220,000
Mesopotamian Lowlands	Bagdad	100,000	4,748,000
	Part of Basra	10,000 ?	300,000 ?

The Armenian and Kurdistán highlands, which form a border-land between three empires, possess neither physical, ethnical, nor political unity. Thus their drainage is partly to the Euxine through the Choruk-su, partly to the Caspian through the Kur and Aras, partly inland to the closed basins of Lakes Van and Urmia, but mainly through the Euphrates and Tigris to the Persian Gulf. Ethnically, also, they are occupied by peoples of four distinct stocks—the Lazis, a branch of the Georgian race; the Armenians and Kurds, members of the Iranian family; the Turks and Tatars, of Túrki origin; the Arabs, Jews, and so-called "Nestorians," of

Semitic blood. Lastly, these highlands, taken in their widest sense, are politically distributed between the empires of Russia, Persia, and Turkey, which here converge round the base of Ararat, their culminating point. The recent changes that have taken place on the Russo-Turkish and Turko-Persian frontiers have even increased rather than diminished the difficulty of drawing any clear parting-line between the three states, whose boundaries are here almost everywhere purely conventional and even arbitrary. In Armenia the Russo-Turkish frontier-line is now deflected considerably westwards in the direction of Erzerum, thus leaving Ardahan, Olti, and Kars to Russia, to which power the seaport of Batúm on the Black Sea has also been ceded by the Berlin Congress of 1878.

Armenia is inhabited mainly by Christians, Kurdistán by Muhammadans. But both countries suffer alike from the effects of Turkish misrule. At the same time, the Christians are themselves largely to blame for the grievances of which they complain, nor can it be denied that the whole question of the much-needed local reforms has been complicated by questions of a political character. "Ample cause for discontent is afforded by the really wretched system of Turkish administration, the unequal imposition of taxes, persistent denial of justice, and practical disavowal of the Christians' claim to be treated with the same consideration as their equals among Moslems. But the subordinate officers of the local government are aided and abetted in their disgraceful proceedings by the criminal assistance of the Armenian Mejliss members, ostensibly elected by their co-religionists to guard their interests. As the evil thus lies as much with the Christians as the Turks, there is no remedy for it until the local authorities see for themselves that the Porte's orders are really carried out."[1]

[1] Extract from Report of Mr. Taylor, British Consul in Erzerum, 1878.

In Mesopotamia the northern and southern sections of El-Jesireh and Irak-Arabi differ greatly in their main features. The transition from the elevated plateau of Diarbekr to the alluvial plains is effected by the extensive open tract which maintains an elevation of over 1500 feet between the Tigris and the western bend of the Euphrates. Here the hilly wooded districts in the north are succeeded by grassy steppes or arid wastes, which are converted into highly-productive oases where-

NEIGHBOURHOOD OF BAGDAD.

ever water abounds. Such is the fertile district stretching from Urfa southwards to Harran, where splendid crops of maize, tobacco, and cotton are raised. Below Mosul the date-palm begins to make its appearance, and this plant forms the prevailing feature in the landscape throughout the level alluvial plains of Irak from Bagdad to Basra. In the extreme south the numerous backwaters and channels of the two main streams merge imperceptibly in the lagoons and morasses of the Shat-el-Arab delta. But these magnificent lands, so well suited for agriculture, are now little cultivated. The nomads and even the scantily-settled population rely mainly on the produce of their flocks and herds, and the country shows the same signs of misrule, ruin, and decay that are elsewhere visible in Asiatic Turkey. "Except around Bagdad the traveller now sees hardly a trace of the date-groves, the vineyards, and the gardens which excited the admiration of Xenophon" (*M'Coan*).

5. *Climate.*

The Armenian climate, pleasant enough in spring and autumn, is excessively severe in winter and summer. During the long winter months from October to May the ground is mostly covered with snow, while the mid-summer heats are most oppressive. These conditions also prevail in Kurdistán, where, however, the variations of temperature are not so great as farther north. Here, also, the winter is of shorter duration, with correspondingly longer springs and autumns. In Mesopotamia the mild but short winters become the pleasantest part of the year. But they are succeeded by sultry summers, during which the plains become scorched and bare. Here the Samiel, or "poison wind," prevails in the same season; and the disease known as the "Aleppo button," or "Bagdad date-mark," is seldom absent from the towns fringing the desert.

Throughout the Mesopotamian basin the annual rainfall is below the average. Summer is everywhere very dry, but much snow falls on the uplands in winter; and in Upper Mesopotamia abundant rains prevail from December to March. Farther south vegetation and husbandry depend largely on artificial irrigation, which has been practised in this region from the remotest times.

6. *Flora and Fauna.*

In Armenia there is a marked absence of forest trees and so deficient is the supply of wood that in many places cattle-droppings form the staple of fuel. The well-watered valleys abound in fruits and cereals; but the bleak plateaux are generally bare, or covered with a scanty vegetation of grass. Far more varied is the flora of Kurdistán, where the hills are often clothed with forests of oak, ash, walnut, and pine trees. Here also the lower grounds yield rich crops of maize, wheat, pulse, hemp, besides tobacco, cotton, mulberries, grapes, melons, and other Southern fruits. In Mesopotamia the vegetation becomes more decidedly tropical, and the Shat-el-Arab district produces some of the finest dates in the world.

Wild animals have almost disappeared from this region. But the towns are infested by packs of pariah dogs, which, while doing the work of the scavenger, are occasionally dangerous to the people. M'Coan tells us that on one occasion he nearly fell a victim to the half-jackal breed of Erzerum.

Their countless flocks of sheep form the chief wealth both of the Kurdish and Arab nomads, and the latter also possess many camels, and perhaps the purest breed of Arab horses in Asia.

7. *Inhabitants: The Armenians, Kurds, Nestorians, and Bedouins.*

Although the seat of some of the earliest human cultures, the Mesopotamian basin is still largely occupied by a nomad population. Its inhabitants belong to four distinct stocks—the *Iranian*, represented in the northern highlands by the Armenians and Kurds; the *Semitic*, represented in the north by the so-called "Nestorians," or Chaldeans, in the plains by the Arab Bedouins; the *Túrki*, which, besides some Tatar tribes, supplies the ruling element found chiefly in the towns; lastly, the *Caucasian*, of which there are two branches—Lazis in the extreme north, and Circassians, many of whom have migrated in recent years from Russian to Turkish territory.

The centre of gravity of the Armenian nationality, which formerly lay about the basin of Lake Van, has been gradually shifted northwards to the neighbourhood of the Ala-göz and the famous monastery of Echmiadzin, both within the Russian frontier. The race, like the country itself, has long lost its political unity, and is now distributed over the Russian, Turkish, and Persian empires. Nevertheless, over one-third of the people still continue to reside under the Ottoman rule about the head streams of the Euphrates and Aras. They are distinguished as much by their features, dress, and social habits as by their distinct Christianity from the surrounding Kurdish and Turkish Muhammadans, with whom they cannot be said to enjoy much popularity. Their craft and acuteness have become proverbial; and although there may be some exaggeration in the charges brought against them, it cannot be denied that their moral tone has been affected by the political servitude to which they have been long subjected. Like the Jews, the Armenians, after the loss of their independence, have turned to trade, which is now almost entirely in their hands. They accumulate all the capital of the country, so that the money market is ruled by them. The great influence thus ensured to them naturally causes mutual heart-burnings and rivalries amongst themselves, while against the common enemy they combine together and spare no sacrifice for the general weal. Surpassing others in shrewdness, the main object of the Armenian dealers is to purchase cheap wares of attractive appearance, and then retail them advantageously. Thus they often succeed in amassing great wealth, which, however, they

ARMENIAN ENTERTAINMENT.

are always careful to conceal. Thus capital is hoarded up, which they neither invest nor enjoy.

Timid and taciturn, they display at least an outward

obedience to their rulers, whom they inwardly despise. Naturally of a mild disposition, they have scarcely ever sought to recover their independence by force of arms. Satisfied when allowed to pursue the peaceful paths of commerce, they have ever shown themselves submissive to their fierce and warlike neighbours. They might even be said, of all Christian people, to sympathise most with the Turks, whom they resemble in their earnest temper and frugal habits, and whose language they generally speak like a second mother-tongue. They also stand much on the same social level. Among them the women are little better off than among the Moslems, being practically the drudges of the household. But while the sensual Turk often becomes the slave of his handmaiden, the Armenian man of business still remains the head of the family. All menial work is performed by the wife, who waits on her husband at his meals, which she never shares with him. Although unveiled indoors, she is never seen by strangers, even at entertainments withdrawing to a room set apart for the purpose. This is usually raised a few feet above the level of the large central hall, and shut off by means of a wooden lattice, whence, without being seen, the women command a view of the banquet below.

Betrothed from her childhood by parental arrangement, the bride seldom obtains a sight of her future lord before their union.

The Armenian race, whose national name is Haï, Haïk, or Haïkan, formerly numbered some 8,000,000, but is now reduced to little over 2,000,000, distributed as under:—

Caucasia and Russia in Europe . .	850,000
Turkish Armenia	760,000
Persian Armenia	150,000
Turkey in Europe	250,000
Elsewhere	60,000
	2,070,000

While the settled and peaceful Armenians have been constantly losing ground, the nomad and lawless Kurds have long spread far beyond the limits of the region to which their progenitors, the Karduchi, seem to have been confined. In classic times Armenia included the whole of the Van district southwards to the 38th parallel, and Sachan [1] has recently determined the site of Tigranocerta, one of its many capitals, at the village of Tel Ermen, or the "Armenian Hill," a little to the south-west of Mardin, within the limits of Upper Mesopotamia. But all this region is now mainly occupied by the Kurds, some of whose tribes reach far southwards to the vilayet of Diarbekr, while others have encroached upon the Armenian district round about Ararat, and are found as far north as the 41st parallel about the head-waters of the Kur. Others are scattered over parts of Asia Minor, North Syria, West Persia, and the highlands between Khorasán and the Turkoman country. Semi-independent Kurdish tribes still form a dreaded cordon round about the upland town of Van. Still more formidable is the Hormakli branch, occupying the snowy Bingol-dagh south of Erzerum, between the two forks of the Euphrates.

Although not always so chivalrous as they have been described by the few travellers who have occasionally visited them, they still possess the proud and frank address of independent highland tribes. Nor can it be denied that many of their lawless propensities and notorious indifference to the rights of property must be attributed to the maladministration of their Turkish and Persian rulers. Under some of their semi-independent chiefs, a general rising took place on the Turko-Persian frontier in 1880-81, during which the most deplorable

[1] *Ueber die Lage von Tigranocerta.* Berlin, 1881. This place was hitherto supposed to lie much farther north, at or near Diarbekr, on the Upper Tigris.

excesses were committed, and the Urmia district wasted with fire and sword almost up to the very gates of Tabríz.

But the worst qualities of the race have been developed in the Nestorian district of Hakhiari, about the head-waters of the Great Zab. This tract stretches from the Persian border-land westwards to the Jebel-Judi, between the Zab and Tigris. But the Nestorians are also found in the extreme north-west of Persia, about Lake Urmia, and in small communities scattered over Upper Mesopotamia. They may almost be regarded as the last surviving erratic boulders of a formerly powerful Christian sect, at one time widely diffused over the vast region stretching from the Euphrates to Western China. But few travellers have succeeded in penetrating to their present home in the Kurdish highlands, a circumstance probably due as much to the inaccessible nature of this alpine region as to the savage character of its Christian Nestorian and Moslem Kurdish inhabitants. The heart of the country can be reached only by the Zab valley, on either side of which lie the dangerous haunts of the fierce Leihun tribe, the name of whose dreaded chief, Bedr Khan, is still remembered after two generations by the surrounding Christian communities. Feuds and forays are still frequent enough, especially in the Tiyari district, where nestle the stone huts of the Nestorians under the shade of mighty walnut trees in the well-watered valleys, here everywhere encircled by snowy crests.

The Nestorians, who number altogether about 200,000, reject both the name "Nestúri" and the doctrine of Nestorius. The term is probably a corruption of "Nessaráni," from Nazareth, commonly applied in the East to the Christians. But however this be, they call themselves *Kaldani*, or Chaldeans, and claim to be the survivors of the old Christian people of Mesopotamia, who were of Chaldean or Assyrian stock. Those of Mosul and others still speak a corrupt form of Assyrian, which they call modern Chaldean, and which is certainly an Aramaic dialect closely allied to Syriac.

Notwithstanding their lawless and predatory habits, the Kurds have developed a few simple industries. They breed a degenerate species of the Angora goat, from the hair of which are woven rugs and carpets, which have found their way to the European market. They also produce coarse woollen, silken, and cotton stuffs, besides earthenware, hardware, and especially arms. The widely-scattered tribes of Kurdish stock number altogether probably about 3,000,000, of whom 1,250,000 in Turkey.

In Upper Mesopotamia the Kurdish and Arab nomads are intermingled. But farther south the bulk of the population beyond the walls of the towns consists of Bedouin tribes, whose subjection to the Porte is of a very loose character, and who may in some respects be regarded as the true masters of the land. Besides, the Ottoman Government is quite incapable of introducing a practical system of culture even into the arable tracts of Irak-Arabi. For many years past the governors, pressed by the Anazeh, Shammar, Montefik,[1] Beni-Laam, and other powerful Bedouin tribes, have been able to do little more than keep things from tumbling to pieces. Here, as elsewhere, the history of the last fifty years has been nothing more than a constant feud, in which the advantage has frequently been on the side of the foes of Ottoman rule. Could the Arab tribes be induced to combine their forces, the Government would find it no easy matter to hold in check the powerful hordes, which often number from 10,000 to 20,000 mounted warriors. Along the Shat-el-Arab there is little more than an outward show of

[1] In August 1881 the Montefiks, who stretch along the right bank of the Lower Euphrates and Shat-el-Arab, came into collision with the Turkish troops from Bagdad, but were defeated.

authority, which is to some extent rather endured than obeyed.

8. *Topography: Chief Towns—Erzerum, Van, Nineveh, Bagdad, Kerbela, Basra.*

The constant encroachments of Russia have left to Turkish Armenia no towns of any note, except Erzerum, capital of the vilayet of like name. Even this place is important rather for its strategical position, and as the *entrepôt* of the caravan trade between Persia and the Euxine, than for its size or population. It lies in a fertile district some 30 miles north of the Bingol-dagh, and 100 miles south-east of Trebizond on the great commercial highway leading from that town over the plateau to the Persian frontier. But, like most fortified towns, it is irregularly built, its narrow dirty streets, flanked by mean houses, being crowded together in the small space enclosed by its lofty walls. Here the Moslem largely prevails over the Christian element, although Erzerum is the metropolis of the Armenian Church in union with Rome, as Echmiadzin is of the Orthodox or Independent Armenian Christians. Its mosques are very numerous, and it is a chief halting-place for Persian pilgrims on route for Mecca.

A more interesting place is Van, which, though the chief town in East Kurdistán, is inhabited mostly by Armenians. It is picturesquely situated on the east side of the lake, above which rises an isolated rock crowned with its citadel. Van has suffered much both from earthquakes and from the turbulent Kurdish nomads of the surrounding district. Some time back these marauders took advantage of a fire in the bazaar to plunder the Armenian shops and houses, and since then its trade has greatly declined.

In Mesopotamia nearly all the large towns are situated, not on the Euphrates, but on the Tigris. Of these the northernmost is Diarbekr, capital of a vilayet, and lying on the western head-stream of the Tigris in a debatable land, where the Kurdish, Armenian, Syrian, and Arab races meet on common ground. It is the seat of a Chaldean patriarch, and does a considerable trade by river and caravan.

Lower down the river, and in the heart of the ancient Assyria, stands the town of Mosul, once noted for its fine cotton fabrics, which from this place are still known as *muslins.* Here the Tigris breaks through its southern mountain barrier, which forms a natural boundary between the Kurdistán highlands and the Mesopotamian plains. Although a poverty-stricken and decaying place, Mosul must always remain a hallowed spot in the eyes of the antiquarian, thanks to the neighbouring ruins of Nineveh, which have of late years been so successfully explored. Eastwards there stretches an extensive cultivated tract, limited on the north by the steep walls of an irregular limestone range, and extending beyond the horizon southwards to the confluence of the Great Zab, where the right bank of the main stream is already fringed by the Mesopotamian steppe. The small plateau thus circumscribed is broken only by low hills crowned with numerous hamlets, generally associated with those mysterious artificial mounds or barrows which are found scattered over Western Asia, the Balkan peninsula, Russia, and as far west as the Pomeranian and Mecklenburg marsh lands. Close to these countless tumuli stand the villages of the agricultural Kurds, while the whitewashed tombs of Moslem "saints" are dotted over the boundless grassy plains. On this plateau the ruins of Nineveh cover a space about 18 miles in length along the river, and extending nearly 12 miles from its left bank, thus occupying an area of over 200 square miles, or rather more than that of London. The famous mound of Kuyunjik, where the excavations were begun in 1841, faces Mosul, while those of Nimrud occupy the angle formed by the confluence of the Tigris and Great Zab, 18 miles farther south. Here Layard discovered the colossal winged bulls, lions with human heads, and winged sphinxes placed as guardians at the entrances of the royal palaces, and now preserved in the British Museum. Since then all the European collections have been enriched by the artistic treasures brought to light in the intervening space. The arrow-headed writings of the brick libraries, which are now deciphered, show that Nineveh was the centre of an Assyrian or Semitic civilisation of great antiquity, but still modern compared with that of the Akkads of Babylonia, whose ethnical affinities have not yet been determined.

Nearly midway between Mosul and the Persian Gulf is situated the famous city of Bagdad, in what was once one of the richest and most productive regions in the world. This city was formerly the most brilliant capital of the Moslem world. Arriving with the Persian caravan from Mandali, we enter the city by the gate of Sheikh Omer. The archway has long since fallen in, and the soft-hoofed camels struggle painfully over the breaches formed by time in the dilapidated bastions. In the first purlieus we meet with nothing but piles of rubbish, stagnant waters, and cesspools, while a pack of pariah dogs is scattered in all directions by the shrill voice of the leader of the caravan. Over the city swoops the vulture of the wilderness, and at its very gates flocks of carrion crows settle unmolested on the putrid carcasses strewn about.

East of the river is the district of New Bagdad, containing the Government offices and the chief commercial and public buildings. On the right bank is the old town, enclosed by an extensive tract of orange and date groves. Towards the desert this quarter is protected by a wall with two gates, leaving the part facing the river unenclosed. No other large city of Asiatic Turkey is in-

COFFEE-HOUSE, BAGDAD.

fluenced by the desert to the same extent as is Bagdad; no other stands in such direct contact with Central Arabia. The purest Arabic dialect is here current, and here still prevail the Bedouin manners in the social life of the people, and especially in their intercourse with the non-Muhammadan element. Yet, in spite of their religious fanaticism, their general bearing is preferable to that of most other Asiatic Mussulmans, because of the very sincerity of their belief, combined with the natural dignity and frankness peculiar to the Moslem Arab. The population is of a very motley character, being composed, according to some authorities, of 150,000 Muhammadans of various races, some 18,000 Jews, 2000 "Nestorians," nearly the same number of Latin Christians, several hundred Armenians and Syrians, and scarcely more than 20 Europeans. But the estimate of the total population is reduced to about 60,000 by Dr. Albert Socin, one of the best-informed writers on Eastern subjects. Bagdad, though shorn of the greatness for which it was once famed, still possesses importance commercially and politically, which it owes to its situation on the great water highway in a country nearly destitute of land routes.

Up to this point the Tigris is navigable throughout the year for steamers of considerable size, while from the north there daily arrive the so-called "Kelleks," a sort of craft made of inflated goat-skins, boarded over. On these are floated down quantities of lumber from the Kurdistán uplands, the boatmen returning with the empty skins in company with the caravans. But still more characteristic of Bagdad is the "quffe," or coracle, consisting of a round hull 6 to 8 feet broad, with sides curved inwards, constructed mostly of strong reeds and well pitched on the outside. When the bridge of boats becomes broken, the communication is kept open by means of these frail craft.

West of the Euphrates, though at no great distance from Bagdad, lies the village of Kerbela, a spot held in great veneration by the Shiah or Persian Muhammadans. Here is the tomb of Hosein, the Prophet's grandson and son of Ali, whom the Shiahs regard as his true successor in the Caliphate. They believe that by living or dying here they have nothing to fear in the next world, being thereby rendered irresponsible for their conduct in life. So strong is this belief, that many leave instructions in their wills to have their remains brought from great distances and buried in this hallowed place. Hence many thousands of bodies are yearly brought from Persia and elsewhere and laid in the ground at Kerbela. The place is also visited by numerous caravans of Shiah

RUINS OF BABYLON.

pilgrims, all who have performed this pilgrimage henceforth bearing the proud title of "Kerbelai."

A little south of Kerbela, and on the Euphrates, stands the town of Hillah, opposite which are the ruins of Babylon, scattered over a wide tract of country. Still farther south, on the Shat-el-Arab, which affords excellent navigation to ships of considerable draught, and near the head of the delta, lies the important port of Basra (Bassora). Though surrounded by a marshy and malarious district, Basra is favourably situated for trade. It was formerly a very flourishing place, and is still the emporium of Asiatic Turkey for Eastern produce. Ships of 500 tons burden reach this point, and since the establishment of the English line of steamers affording regular communication with Bagdad and the Gulf, its prosperity has considerably revived.

9. *Highways of Communication.*

In the Mesopotamian basin there are scarcely any roads properly so called. The two great arteries of the Tigris and Euphrates still continue to be the chief highways of communication. But the desert is crossed in various directions by caravan tracks, and in the extreme north there is one good road, the already-mentioned route connecting Trebizond through Erzerum and Bayazid with Persia. Erzerum is also connected eastwards by a military road with Kars, and south-eastwards through Yangali with Van. From Van an important route runs southwards through Mosul, and down the Tigris valley to Bagdad, and another westwards through Músh and Kharput to Anatolia. Of the caravan routes across the desert, by far the most important is that which strikes the Euphrates at Bír (Bírejik), here bifurcating through Urfa northwards to Diarbekr, south-eastwards down the Mesopotamian lowlands to Bagdad. Another route runs from Diarbekr along the left bank of the Tigris through Findúk to Mosul, here crossing to the west bank, which it follows to Bagdad. Mosul is also reached from Diarbekr by an alternative route *via* Mardin and Nisibin (Nisibis). But the most direct route between the Mediterranean and the Persian Gulf runs from Alexandretta through Aleppo to Kalaat-Jabar on the

Euphrates, thence following the right bank of that river *via* Annah and Hit to Kalat-Ambar. Here it crosses over and pursues a straight course south-eastwards to Bagdad. This route is not essentially different from the line which has been examined and partly surveyed for the project of the Euphrates Valley Railway. The line is proposed to run from Alexandretta to Bagdad, and thence south-eastwards to Basra. At Bagdad it would form a junction with the great South Asiatic trunk-line, which, starting from Scutari, is intended to connect the Bosphorus with the Persian Gulf through Anatolia and Mesopotamia. There has been speculation also regarding the possibility of carrying a railway from Mesopotamia on to India, through Persia and Afghanistán or Baluchistán.[1] Meantime there are no railways in the Mesopotamian basin, nor is it probable that the projected trunk-line will be undertaken at present.

[1] See the publications on the Euphrates Valley Railway and *India and Her Neighbours*, by Mr. W. P. Andrew, chairman of the Sind, Panjáb, and Delhi Railway Company.

CHAPTER IV.

SYRIA AND PALESTINE.

1. *Boundaries—Extent—Area.*

THE Mesopotamian plains are separated by the great Syrian desert from the Mediterranean coast region, which here stretches nearly in a straight line from the Sinai Peninsula northwards to Anatolia. The desert forms a chalk and limestone tableland gradually rising to an altitude of over 2000 feet above the sea, stretching away southwards into the peninsula of Arabia, but on the west sinking abruptly down to the long, deep, and narrow depression of El-Ghor, which forms the eastern limit of the southern section of the coast region known as Palestine or the Holy Land. Farther north the desert merges imperceptibly in the plains of Damascus and Aleppo; consequently Syria, or the northern section of this region, presents no natural well-defined limits towards the east. Elsewhere the boundaries of the whole land are sufficiently clear—the sea on the west, the Amanus (eastern Taurus) on the north, the Euphrates on the north-east, the little river El-Arish on the south-west, Arabia Petræa on the south. This gives a total length north and south of about 430 miles, with a mean breadth of 100, narrowing in the south to 50, expanding northwards to 150. The area is vaguely estimated at about 120,000 square miles, of which not more than 12,000 are comprised in Palestine, leaving 108,000 to Syria. The distinction between these terms has long ceased to be recognised in the East; but is still retained in the West, by reason of the religious associations and historical reminiscences with which the southern division is inseparably associated. Palestine is cut off by the Lower Orontes and Mount Hermon from Syria proper, measuring from this point to the southern end of the Dead Sea about 160 miles, with an average breadth of 70.

2. *Relief of the Land: Lebanon and Anti-Lebanon.*

While this strip of coast land serves on the one hand to cut off the desert from the sea, it forms on the other a connecting link between the Anatolian and Arabian tablelands. It is everywhere too mountainous to allow the plateau formation to be clearly developed. But the mass of the land has a mean elevation of probably 3000 feet, above which rise two parallel mountain ranges, clearly marked in the centre, less distinctly defined in the north, and southwards breaking into an irregular upland region, where the hills and low ridges still form two systems west and east of the El-Ghor depression, round which they meet and become interlaced in the Arabian uplands.

The coast line, running nearly due north and south, is varied by but few and unimportant headlands and inlets, the section south of Beyrút forming almost a straight line, broken only by the bold promontory of Mount Carmel nearly midway between Beyrút and Jaffa (Joppa). Throughout its entire length the coast is followed by the outer chain of mountains, leaving but a narrow strip of lowlands between their base and the sea. In Palestine this range is little more than the escarpment of the broad and hilly plateau of Judæa, beyond which the plain of Sharon stretches seawards from Cæsarea southwards to

Gaza. Beyond Carmel the hills still recede sufficiently to make room for the less extensive plain of Acre, after which they continue to rise in height and approach constantly nearer to the coast. North of the valley of the Lower Leontes (Nahr-el-Litany) they culminate in the two parallel chains of the Lebanon and Anti-Lebanon, which form the great physical feature of this region. The Anti-Lebanon, or inner range, falls gradually northwards down to the plains of Upper Mesopotamia. But the Lebanon, or Jebel-el-Gharbi—that is, "Western Range"—is continued by the less elevated Jebel-Nusarieh as far as the plain of Antiochia, about the 36th parallel. North of this plain the Jebel-Nusarieh is continued by the Giaour-dagh and Akma-dagh to the Taurus above the Gulf of Alexandretta.

The Lebanon or central coast range runs for about 90 miles south-west, at some points approaching to within 8 or 10 miles of the Mediterranean. Seen from a vessel out at sea it presents the appearance of bare, rocky walls, here and there surmounted by a few snow-clad peaks, of which the highest are the Dhor-el-Khodih (10,200 feet), and the Jebel-Makmel (10,000 feet). From these the range takes the name of the Lebanon or "White Mountains," a name which was already current in the time of Moses (Deut. i. 7), and which has never since dropped out of history. Notwithstanding its rugged aspect seawards, the Lebanon, which is properly limited southwards by the valley of the Lower Leontes, really contains many fertile slopes and valleys, well cultivated and thickly peopled.

Eastwards it is separated by the still more fertile valley of the Bekáa (Cœle-Syria) from the Anti-Lebanon or inner range, whose naked rocky walls present far more varied outlines and wilder gorges than the coast range. Although with a lower mean elevation, the southern extremity of the Anti-Lebanon rises in the Jebel-es-Sheikh (Mount Hermon) to an altitude of 11,000 feet, the culminating point of the Syrian highlands, some 30 miles south-west of Damascus. Beyond this point it throws off two branches, towards the south-west and south-east, thus enclosing the upper sources of the Jordan, and merging eastwards in the rocky uplands of Gilead and Moab.

3. *Hydrography: Jordan—Dead Sea.*

Syria and Palestine are still sometimes represented as being intersected in their entire length by a deep depression called in the north El-Bekáa, in the south El-Ghor. But more accurate recent surveys have shown that this view is entirely erroneous, and that El-Bekáa and El-Ghor are totally distinct formations. Although the term El-Bekáa means a "deep plain," the tract in question, answering to the ancient Cœle-Syria—that is, "Hollow Syria"—is only "deep" or "hollow" relatively to the two lofty ranges of the Lebanon and Anti-Lebanon, between which it lies. In itself the Bekáa is not a depression at all, but a plateau at an average elevation of no less than 2000 feet above the sea. On the other hand the Ghor is not only a true depression, but the very deepest in the earth's crust, falling in the basin of the Dead Sea to a depth of 1292 feet below the Mediterranean, or over 4000 feet below the Bekáa. Nevertheless, these two features of the country are still to some extent connected by its hydrography, which they largely regulate. At the famous ruins of Baalbek, under the 34th parallel and about midway between Antiochia and the Dead Sea, the Bekáa attains its greatest elevation of about 3000 feet above the sea, and here is consequently the chief water-parting of the whole region. Round about Baalbek rise the four main streams—Jordan, Leontes. Orontes, and Abana—which flow in four opposite directions, south to the Dead Sea, south-west and north-west to the Mediterranean, east to the Bahr-el-Ateibeh beyond Damascus. At Lake Merom the Jordan reaches the trough of the Ghor, which it henceforth follows throughout its entire course to the Dead Sea. The Leontes and Orontes traverse the southern and northern sections of the Bekáa respectively, while the Abana pierces through the deep gorges of Anti-Lebanon down to the smiling plains of Damascus. Of the four rivers, the Jordan and Orontes will here claim a more detailed description.

The Orontes (Nahr-el-Asy) rises with two head-streams on the western slopes of Anti-Lebanon, some 10 miles north of Baalbek, flowing thence northwards to the neighbourhood of Homs (Emessa), where it expands into the lakelet of Kades, 6 miles long by 2 broad. Beyond this point it continues its northerly course by Hamah (Epiphania), and through narrow rocky gorges for about 50 miles to the northern extremity of the Nusarieh range, where it trends suddenly westwards and south-westwards through the plains of Antiochia to the coast, which it reaches near Suedia (Seleucia), after a winding course of about 150 miles. At its northern bend it receives on its right bank the Kara-su, its only important tributary, flowing from the Lake of Antioch 4 miles off.

The Jordan (Sheriat-el-Kebir) is formed by three small head-streams, the farthest of which rises between Baalbek and Mount Hermon. The united stream falls thence over seven low terraces southwards to the muddy little Lake Merom (El-Huleh), which lies at the head of the Ghor in a fertile basin, fringed on the north by an almost impenetrable reedy swamp, and enclosed on the south by a spacious elevated plain. This plain sinks southwards sufficiently to afford an outlet for the Jordan, which now pursues its impetuous course through the deep rocky fissure of the Ghor for 10 miles to the Sea of Galilee (Lake Gennesareth or Tiberias). The fall in this short space is nearly 700 feet, and at this point the trough of the Jordan has already descended to 682 feet below the Mediterranean.

Lake Tiberias is a sheet of clear water, now as of old abounding in fish, and encircled on all sides by lofty mountain walls and hills, which in spring are covered with a soft grassy carpet, but which become parched up during the dry summer months. West of the lake stretches the fertile plain of Gennesareth (El-Ghuweir, or "the Little Ghor"); but the Ghor itself continues still to fall for about 200 miles between the Gilead hills and the escarpment of the plateaux of Galilee and Samaria, southwards to the Dead Sea. The total fall in this space amounts to 610 feet; so that at its lowest level the Jordan has descended to a depth of 1292 feet below the Mediterranean through a chasm, which is by far the longest and deepest on the surface of the earth. All further extension of the river southwards is thus rendered impossible, although it will be seen further on that the Ghor itself continues its southerly course into the Arabian peninsula.

The Dead Sea (Asphaltites Lake, or Bahr-Lút, that is, "Sea of Lot") is enclosed within a basin formed by naked limestone cliffs, 2500 feet high on its east and 1500 on its west side. It is nearly 50 miles long north and south, with an average width of 8 miles and a mean depth of 1300 feet, but shoaling southwards to the ford between the Lisan promontory and the west shore, which is scarcely more than 3 feet deep. M'Coan tells us that its water is "nearly as clear and blue as that of the Mediterranean, but salt, slimy, and fœtid beyond description; its taste like a mixture of brine and rancid oil; and its buoyancy so great that, as I can personally vouch, the human body will not sink in it, strive as the bather may. Bitumen bubbles up plentifully from the bottom, and with the sulphur, nitre, and rock-salt that abound along most of the shore-line, sufficiently explains the density and the nauseous taste and smell of the water. The old traveller's tale that the water itself and the evaporation from it are alike fatal to animal life is less than half true. The 26 per cent of saline matter precludes indeed the existence of fish; but though its exhalations under a burning sun are thick and fever-inducing, they are in no worse degree poisonous, and birds fly along its shores and over its surface as lively as in the mountains on either side" (i. 103).

At the southern extremity of the lake lies the lofty rock-salt ridge of the Jebel-Usdum, beyond which extends the desolate salt marsh of Es-Sebkah, fed by the Wady-es-Safieh flowing from the Wady-el-Arabah. This now dried-up watercourse forms a southern continuation of the Ghor depression. But it does not extend, as was long supposed, to the Gulf of Akaba at the head of the Red Sea, but only to a water-parting near the Bedouin camping-ground of Arabah, some 500 or 600 feet above the Mediterranean.

4. *Natural and Political Divisions: Gilead and Moab—Land of Bashan—Trachonitis—Ala District—Plateau of Aleppo—Canaan—The Plains of Sharon—Galilee—Samaria—Judœa.*

Till recently the uplands of Gilead and Moab, whose position beyond the Jordan is indicated by their ancient name of Peræa, were a veritable *terra incognita.* But notwithstanding the lawlessness of their Bedouin inhabitants, their numerous cromlechs, ruins, and other interesting monuments, have of late years tempted several European explorers to penetrate into its most secluded retreats.[1]

Seen from the western shores of the Dead Sea, Moab looks like a mountain range, but is in reality merely the verge of a rocky upland plateau about 2500 feet above the sea, or 4000 feet above the level of the lake. This plateau, which is furrowed by deep valleys, stretches eastwards for about 25 miles to a bare limestone range, conventionally regarded as the limit of the land towards Arabia. Moab was formerly a well-peopled region. But the eye everywhere lights on ruined villages. Even now, badly cultivated as it is, the land is rich and fertile, and large tracts of a fine red sandy loam, needing no manure, still produce heavy wheat crops. All the streams flow westwards through deep rocky beds to the Dead Sea.

The Moabite country is continued northwards by the volcanic plateau of the Land of Bashan, which attains an elevation of from 4000 to 6000 feet eastwards in the Haurán uplands. Including the three districts of the Leja (Western Trachonitis), Nukrah, and El-Jebel, this region runs 60 miles north and south, and nearly 40 east and west. The Leja is mostly a stony plain; but the Nukrah is a rich tract, containing many small towns and villages, unfortunately exposed to the frequent raids of the Anazeh Bedouins, while the Jebel, or "Highlands," marking the extreme eastern limits of Palestine towards the desert, abound in ruined towns still partly peopled by the Druses.

Between the Haurán and the Oasis of Damascus

[1] The Palestine Exploration Fund, having completed the survey of Palestine proper, has now extended its labours to the region beyond the Jordan, where a good beginning was made during the year 1881 by Lieutenants Conder and Mantell. Over 500 square miles have already been surveyed, and more than 200 ruins examined.

there stretches a broad expanse of volcanic "tell," covered with recent tertiary and pliocene craters, which, although seemingly scattered about in wild confusion, really lie in three tolerably parallel lines, inclining slightly north and south. This is the Eastern Trachonitis (Tulul-el-Safá), towards the northern verge of which stand the stupendous ruins of Palmyra (Tadmor), under the 35th parallel, in 38° E. long. and 120 miles north-east of Damascus. The ruins cover a space of about 3 square miles, and conspicuous amongst them are the sixty columns still standing of the magnificent Temple of the Sun. This "City of the Palms," as both names mean, dates back to the time of Solomon, and is for ever associated with the sad fate of the hapless Queen Zenobia.

Still more interesting to archæologists is the Ala region between the vilayets of Damascus and Aleppo. It forms an extensive basaltic upland tract, stretching for many miles east of the Orontes valley. Here are the ruins of many cities, which have evidently been rebuilt over and over again, besides numbers of remarkable tombs and fortified camping-grounds. Few Europeans besides the English travellers Burton and Drake have visited this extraordinary land, within whose limits, though figuring on the maps as a blank space or portion of the Syrian Desert, the Arabs have indicated the sites of no less than 365 ruined cities.

In the extreme north the extensive inland plateaux of Aleppo, Umk, and Aintab occupy all the space between the great bend of the Euphrates and the coast range. Although intersected by several low ridges, they contain many fine and fertile level tracts, thickly peopled by Turkoman and Armenian agriculturists. This region marks the extreme limits of both of these races towards the south-west. In the west of the Umk plateau lies the Bahr-el-Abiad, or Lake of Antioch, a fine sheet of water 8 miles by 6, formed by the junction of several steppe streams, and draining to the Orontes.

Returning southwards and recrossing the Jordan from Moab, we enter the small territory of Canaan, the "Land of Promise," or Palestine proper, ever venerable as the scene of the history of the "Chosen People," and as the Holy Land of Christianity. This region consists of an irregular hilly plateau falling west of the Jordan down to the level coast lands. This narrow low-lying tract, comprising the ancient land of the Philistines, was at one time studded with large towns and thickly inhabited by a restless warlike population. But at present the only noteworthy places are Gaza, Jaffa, and Ascalon, along a coast stretching in an almost unbroken monotonous line northwards to Cape Carmel. This headland, enclosing the Bay of Acre on the south, forms the northern extremity of the Jebel-Mar-Elias (1800 feet), which runs through the old lands of Manasseh and Asher north-westwards between the plains of Sharon and Acre. The rich plain of Sharon, of which only a small part is now under cultivation, stretches some 15 or 20 miles inland, and skirts the coast from above Cæsarea to Gaza, beyond which its loamy soil gradually mingles with the sands of Arabia Petræa.

The tablelands rising immediately behind Sharon comprise in the north the old land of Galilee lying mainly between the Leontes and Carmel, Samaria in the centre, and Judæa in the south. The regions which fall abruptly eastwards to the El-Ghor depression are generally described as of jurassic formation. But Dr. Oscar Fraas has lately shown that they consist rather of chalk deposits with hippurites and other fossil shells. The same formation prevails throughout the land east of Jordan, the Sinai Peninsula north of the zone of primitive rocks, and the Nile valley far beyond Karnak.

Galilee, the northern division of Palestine, is a hilly district from 2000 to 3000 feet above the sea, sinking eastwards abruptly to the Jordan and Lake Gennesareth, and southwards to the rich alluvial plain of Esdraelon (Jezrîl). Here are many pleasant fertile valleys, varied with bold mountains and splendid woodlands stretching northwards to Mount Hermon.

South of the plain of Esdraelon the plateau again rises to the central district of Samaria, where are also many well-watered and cultivated valleys, producing heavy crops and fruits in abundance. Here the prominent landmarks are the rocky Mounts Ebal (3076 feet) and Gerizim (2849 feet), rising close together about 34 miles due north of Jerusalem.

The southern district of Judæa is traversed by a somewhat ill-defined ridge of bare treeless hills, known collectively as the Mountains of Judah. These hills form a small water-parting between Kedron and other brooks flowing east to the Dead Sea, west to the Mediterranean. But although rich beyond any other land in hallowed memories and stirring events, Judæa is on the whole a somewhat bleak, arid country, far less productive than any other part of Palestine.

5. *Climate.*

In this region climate depends entirely not on latitude, but on the relief of the land. Even in small districts the greatest diversity prevails, according to the varying altitudes. Thus on the exposed upland plateau beyond the Jordan the glass falls at night to 22° F., or 10° below freezing point, when it stands at 76° F. on the shores of the neighbouring Dead Sea. In general a cold temperature prevails on the higher slopes of the Lebanon and other ranges rising above the snow-line. Here the winters, almost as severe as on the southern shores of the Baltic, are followed by genial springs, summers scarcely warmer than in England, and fresh autumns. Along the west coast and the Jordan valley the summer heat is very oppressive, the winter mild, and rain falls in both seasons. Malaria is prevalent at certain marshy spots along the coast, especially near Tripoli and Alexandretta. Central and South Palestine and the vilayet of Damascus enjoy a warm, dry climate, with mild winters and slight rainfall. Here the hot desert winds prevail in summer, drying up the rivulets, and reducing the land to an arid waste. At Jerusalem the mercury rises to 79° at sunset in midsummer, sinking to 49° in January—hottest and coldest means.

A remarkable feature of the Bekáa is the violent, almost tornado-like wind which prevails especially in the central districts, where it blows regularly every day for some hours in the afternoon.

6. *Flora and Fauna: The Cedars of Lebanon.*

As a rule, vegetation is much more varied and luxuriant in the north than in the south. In Syria the slopes and many of the coasts are often densely wooded, whereas in Judæa "the hill vegetation is everywhere scanty, and the general aspect of the country east and south of Sharon rugged, desolate, and barren" (*M'Coan*).

The turpentine tree and the *ballud,* the species of oak which produces the gall-nut of commerce, are common features even beyond the Jordan. The vine, olive, orange, and other Southern fruits, besides the mulberry, cereals, and dates of splendid quality, abound in Sharon, the Damascus district, the sheltered Lebanon valleys, and generally throughout Galilee and Samaria. The tobacco especially of the Latakia district facing Cyprus is noted far and wide for its delicate flavour, and the rose of Sharon still remains more than a reminiscence.

On the other hand, the historic cedars of Lebanon have almost become a thing of the past. At a solitary

THE CEDARS OF LEBANON.

spot a few miles below Tripoli, and not far from Cannobin. seat of the Maronite patriarch, there still survives all that remains of what must be regarded as undoubtedly the most venerable tree in the whole world—the tree to which the Psalmist compares the vine "brought out of Egypt," the boughs of which "were like the goodly cedars" (Ps. lxxx. 10). In 1875 Fraas counted altogether 377 plants of all sizes, but there remain five only of the gigantic trees, whose trunks measure upwards of 30 feet round. Burton and Drake, who visited the place a few years ago, were also greatly disappointed at the appearance of these "Christmas trees on a large scale," which from a distance looked like a clump of enclosed pines, and on a closer inspection were found to consist of a few decayed old stems.[1]

Of wild animals the chief are the Syrian bear, the hyæna, jackal, and buffalo. There is a small but hardy breed of horses, but the camel and mule are also employed as beasts of burden, especially for the transit trade between the coast and the interior. Fat-tailed sheep are numerous, but the Angora breed soon degenerates.

7. *Inhabitants : The Syrian Christians — Missionary Work — The Maronites, Druses, Nusarieh, and Fellahîn.*

With the exception of a few wandering Kurdish and Turkoman tribes in the extreme north, and of the Turkish officials in the large towns, all the inhabitants of Syria and Palestine belong to the Semitic stock. The modern Syrian, who represents the Aramæan branch of that stock, is the result of a happy blending of races, in which the Semitic element largely predominates. The natural endowments of the people are displayed in the best light by the Christian section of the community. The Syrian Christians are a highly intelligent people, with a rare capacity for adopting European ideas. The admixture of Greek and Arab blood has evidently in no way impaired the good qualities of their Phœnician and Aramæan

[1] *Unexplored Syria.* London, 1872.

forefathers. And Phœnicians the inhabitants of the coast districts still remain in their enterprising spirit, commercial skill, and love of travel. In Marseilles, Liverpool, and Manchester, there are several Syrian merchants, furthering the interests of their native land, and extending their trading relations even to Scandinavia and North America. The prosperous condition of the Beyrút Christians is the natural result of their intelligent industry. Here are found none of those proletariate classes, who cause so much anxiety in the large European cities. Everybody is either a merchant, or else engaged in some settled industry, while still preserving the freshness of the simple patriarchal family life. The women are comely, and although without much book-learning, good mothers, thrifty housewives, and devotedly attached to their husbands. They associate little with the outer world, passing their days in happy seclusion in the midst of their families. Their reading is limited to their Arabic prayer-books, and the harmless *Beyrút Review*, while novel-reading and piano-strumming are accomplishments which are still rare, except, perhaps, where the superficial French culture has been introduced.

There is no lack of girls' schools, though instruction is here limited mainly to the study of English or French. The "Sisters of Charity," however, have an excellent training school, where woman's work is taught, and where native teachers are trained. The rival houses of the "Sisters of Nazareth" and of "Prussian Deaconesses" are also highly spoken of. Mrs. Nott, a rich Englishwoman, has recently contributed large sums to benevolent and religious purposes. The American missionaries are also doing good work, aiming especially at practical objects. The native Protestant community already numbers several hundred families in Beyrút, where the money flowing in from beyond the Atlantic has enabled them to build a handsome church, besides supporting several schools and a printing establishment.

Even in the Lebanon, Protestant views are making rapid progress, notwithstanding the existence of some good Roman Catholic institutions. Of these, the most noteworthy is the college of the Melchite Greeks, which is admirably conducted, and already numbers several hundred pupils. The Jesuit College at Ghazir is also efficiently managed, and this is also true of the Lazarist College at Antura. Both are exclusively French establishments, and as most of the young men of Beyrút have been educated at one or other of them, the French language has become very general amongst the upper classes. It has already almost entirely superseded Italian, which prevailed in the last generation.

In the year 1862 the district of the Lebanon was detached from the vilayet of Damascus, and formed into a separate pashalik, administered by a Christian governor under the control of the European legations. But the limits of this new government depart considerably from the natural limits of Mount Lebanon, having been laid down solely in accordance with the religious interests of the people. Hence districts where the majority were Muhammadans continue to form part of the Syrian province, while all the Christian communities were included in that of the Lebanon. But Tripoli, Beyrút, and Saida (Sidon), the three most important seaports, were also attached to Syria, so that the boundaries of the modern district of the Lebanon are extremely irregular.

It comprises an area of rather over 1000 square miles, with a population of about 300,000, all Christians, except 70,000 Druses and Mussulmans. Hence it may fairly be regarded as a Christian land, where Christianity has held its ground almost from the apostolic times.

The Lebanon Christians call themselves Maronites,

from the national saint, Maron, a famous recluse who flourished about the year 400. They are the direct descendants of the orthodox community as constituted in the seventh century, and although united with Rome since the time of the first Crusades, they still retain many local privileges and peculiarities, such as a married clergy, administration of the sacrament under both species, celebration of mass in the Syriac language, but otherwise according to the Latin rite, together with their own hagiology and national feasts. They are devotedly attached to their religion, and are in other respects a brave and energetic people. Their villages, and 200 monasteries, are perched like eyries on the spurs and slopes of the main range, and are often surrounded by cornfields waving over artificial terraces, so disposed as to prevent the rich loam from being washed away.

Unfriendly neighbours of the Maronites are the mysterious Druses, settled partly in the Acre district south of the Lebanon, partly in the remote Haurán uplands, on the verge of the desert. The origin and peculiar tenets of this half-pagan people have not yet been satisfactorily explained. Though apparently having some affinity in faith to the Mussulmans, they jealously preserve a sort of secret doctrine, said to have been handed down from the ancient Egyptians. In fact, however, they make no outward profession of any religion, although believing in a God. Physically they are a fine race, brave, with something of poetry and heroism, but also fierce, cruel, and treacherous.

Druses and Maronites lived for ages amicably together until bitter feuds sprang up between them during the present century. Sudden raids have been followed by sanguinary reprisals, and the restoration of order has frequently been attended with much bloodshed. Latterly the Druses seem to be gradually retiring altogether to the Haurán highlands, and in the Lebanon district they now number scarcely more than 40,000. The English missionaries claim to have made great progress amongst them.

There are some 50,000 Christian Greeks in the Lebanon. Some "Ishmaelites" also dwell here, descended from the murderous sect of "Assassins," who have given a familiar word to most European languages. Here also are some 15,000 Mussulmans on the skirts of the range, and about the same number of Meteollis or Shiah sectaries, who are generally regarded with suspicion by their neighbours.

North of the Lebanon we enter the domain of the mysterious Nusarieh race, which gives its name to the northern coast range, and forms the majority of the population along the whole of these uplands, and even beyond the Amanus mountains, right into Cilicia, as far as Adana and Tarsus. Here dwelt from the remotest times the Nazarini, of whom the ancients seem to have known as little as we do of their direct descendants, the Nusarieh. These highlanders live and die in their mountain homes, which they never willingly leave. Tillage and stock-breeding afford them a sufficient livelihood, but while conducting themselves as true followers of the Prophet in the presence of their Moslem neighbours, they maintain profound secrecy on the subject of their peculiar worship. Their speech is the Arabic dialect elsewhere current in the Syrian highlands. Throughout Syria, where they are called Fellahîn, and are said to number from 120,000 to 180,000, they have the reputation of being irreclaimable and desperate highwaymen.

The great bulk of the present population of Palestine, which scarcely exceeds 700,000 altogether, consists of Arabs, partly Bedouin nomads, partly Fellahîn, or settled agriculturists. They dwell mostly in wretched mud hovels, or amidst the ruins of old buildings. They all speak Arabic, and are mainly followers of Muhammad. A few Christian communities are found in Nazareth and elsewhere. But the Jews have almost completely vanished from the land of their forefathers. Except in Jerusalem, they are scarcely anywhere to be found within the limits of Palestine proper. In recent times a few men, inspired with religious enthusiasm, mostly from Würtemberg, have settled about Mount Carmel, in Jaffa, and a few other places. But they do not exceed 750 altogether, so that it is somewhat premature to speak, as some already do, of the German colonisation of the Holy Land.

8. *Topography: Damascus—Aleppo—Emessa—Beyrút—Nazareth—Jerusalem—Hebron—Jericho.*

In Anatolia and Mesopotamia most of the old cities have either disappeared or sunk to the position of obscure hamlets, whose sites have with difficulty been identified. In Syria, on the contrary, although Tyre, Tadmor, Baalbek, and some other famous places, have shared the same fate, many of the most venerable cities in the world, such as Damascus, Aleppo, Emessa, Beyrút, Jerusalem, not only continue to flourish but retain their ancient names in more or less modified forms. This, however, is true only of the region west of the El-Ghor and Bekáa depressions, beyond which hundreds of formerly prosperous towns have been swallowed up in the sands continually advancing westwards from the desert.

Here an almost solitary exception is Damascus, which claims to be the oldest city in the world, and which, owing to the favourable conditions of the soil and climate, still continues to maintain its political and commercial supremacy almost on the verge of the wilderness. It lies nestled amid gardens and orchards at an elevation of 2300 feet above the sea, in a district which owes its exuberant fertility to the Abana and Pharpar flowing eastwards from the Anti-Lebanon and Mount Hermon. Owing to its thoroughly Oriental aspect it is one of the most interesting cities in the East. The Arabs have corrupted its name to Esh-Sham, which term they have

DAMASCUS.

extended to the whole of Syria. But when they wish to speak more particularly of the capital, they lose themselves in raptures about "the breath of heaven," "the mole on the cheek of the earth," "the plumage of the peacock," "the necklace of beauty," and suchlike Oriental imagery. For them it is one of the four Edens, although the city proper enclosed within its crumbling walls and projecting towers is far from corresponding with the favourable impression produced by a more distant prospect. The irregular and narrow streets wind along between high dead walls, broken at long intervals by small grated windows, but nowhere relieved by any touches of art The monotonous piles of dull stone are varied only by a few ancient gateways, which alone make any attempt at architectural display.

Of great historic interest is the former Church of St. John, now the largest mosque in Islám. But more attractive are the numerous bazaars, in extent and richness surpassing most of those elsewhere met in Eastern cities. Amongst their motley throngs nearly all the peoples of the East are represented.

Nearly due north of Damascus are Homs (Emessa), Hamah (Epiphania), and Aleppo. Homs is still a considerable place on the right bank of the Orontes. Hamah, on the same river and a little farther north, is an almost exclusively Moslem town, in the neighbourhood of which is the interesting Ala district described farther back. Still farther north, and about midway between the Euphrates and the coast, is Aleppo, second only to Damascus in size and importance. Capital of a vilayet, it does a considerable local and transit trade, and is occupied with some long-established industries. An old aqueduct still supplies it with water from some perennial springs 8 miles off. Aleppo was wasted by a terrific earthquake in 1822, since which time it has never quite recovered its former prosperity.

All the chief seaports of Syria are still found on the coast of what was formerly the land of the Phœnicians, the most famous navigators of antiquity. Amongst them are Latakia (Laodicea), with a sheltered but shallow harbour; Tarabulus (Tripoli), at the foot of a spur of the Lebanon, nearly destroyed by the explosion of a powder-magazine in 1864; Beyrút (Berytus), 50 miles farther down, next to Smyrna the largest and most flourishing seaport in the Levant. It stands on a noble bay extending in crescent-shape between the spurs of the Lebanon and the sea, and boasts of some fine new quarters and splendid villas, interspersed with shady groves and gardens. Of the population about two-thirds are Syrian Christians.

All the southern ports—Sidon, Tyre, Acre, Cæsarea, Jaffa, and Ascalon—have gradually lost most of their trade since the stirring days of the Crusades, and are now little more than fishing villages with a small local traffic.

On the other hand, Alexandretta (Iskandrún), in the extreme north, has of late years acquired some importance as the outpost of Aleppo. Here is by far the finest harbour on the whole coast, and, notwithstanding its unhealthy climate, Alexandretta cannot fail to become a flourishing place should the projected railway line ever be executed which is to run from the coast at this point through Aleppo to the main trunk-line in the Euphrates valley.

In Galilee still the most important place is Nazareth (En-Nasirah), west of Mount Tabor, and 1100 feet above the sea. It has now a Christian population of about 7000. The chief place in Samaria is the busy little town of Nablus (Neapolis or Shechem), lying in a fertile and well-watered valley between Mounts Ebal and Gerizim, and on the route from Damascus to the coast. Here still survives a small community of about 200 Samaritans, who, like their forefathers, continue to worship on Holy Gerizim. Amongst them is jealously preserved the precious codex of the Pentateuch in the old Samaritan dialect and in the archaic Hebrew character. Samaria, which gave its name to the land, has dwindled to a hamlet now called Sebastieh, a little to the north-west of Nablus.

From the summit of Gerizim, looking southwards, the eye lights on a limestone plateau, rising 2600 feet above the Mediterranean and nearly 4000 above the Dead Sea, connected northwards with the great tableland of Judæa, and on the three other sides enclosed by rugged gorges. Here stands Jerusalem, to the Christian the most hallowed of all places. It is even by the worshippers of Allah regarded as El-Kuds, or "The Holy Place." Here are still the Holy Places, the Church and Shrine of the Holy Sepulchre, to which are ever turned the footsteps of thousands of pilgrims from the West.

JERUSALEM.

No writer has more vividly described the outward aspect of Jerusalem than Chateaubriand. "In the heart of a mountain range lies a desert basin, enclosed on all sides by yellow, rocky heights. These heights are open only towards the east, thus affording a prospect of the depression of the Dead Sea and the distant hills of Arabia. In the middle of this stony landscape, on an uneven and inclined plain, encircled by walls that once crumbled beneath the blows of the battering-ram, and are now propped by tottering towers, we behold some scattered heaps of ruins, ruins overgrown with a few solitary cypresses, aloes, and prickly pears, and overbuilt by Arab huts resembling whitewashed sepulchres: and such is the mournful picture now presented by Jerusalem. At the first sight of this forsaken spot, the heart is overcome by an overwhelming sense of despondency. But this feeling disappears as we gradually pass from desolation to desolation, and at last reach the boundless open space, which so far from oppressing, rather inspires us with a certain sense of cheerfulness and buoyancy. Unwonted sights everywhere reveal a land crowded with hallowed memories. The sultry sun, the fierce eagle, the modest hyssop, the stately cedar, the barren fig-tree—here are concentrated all the poetry and all the imagery of Holy Writ. In every name lurks a mystery, every cavern lifts a corner of the veil shrouding the future, every hill-top echoes with the song of the prophet. By these rushing waters God Himself has spoken to man, and their dried-up beds, the rocks rent asunder, the yawning graves, still bear witness to His voice. Still hushed seems the wilderness, awe-stricken, and as if afraid to break the silence; for it has heard the voice of the Everlasting."

The present generation has undertaken with thoughtful piety again to rescue the ancient sites of the Holy Land from the accumulated *débris* of ages, and to determine their identity with the actual spots traditionally bearing their name. Attention has naturally been centred in Jerusalem, and great results have already been achieved, especially by the English "Palestine Exploration Fund," which has been at work since 1875.

Of the 30,000 inhabitants of Jerusalem, 10,000 are Jews, and 5000 Christians of all sects. Here are made the crucifixes and rosaries of mother-of-pearl and olive wood, and eagerly purchased by the 6000 or 8000 pilgrims who annually visit the Holy Places. Solomon's tank and the old aqueduct have lately been restored, and the city is now supplied with water from these sources.

Six miles south of Jerusalem is Bethlehem, where the great Church of St. Mary marks the traditional site of the birthplace of the Saviour. Ten miles still farther south is Hebron, one of the oldest places in the world, and traditionally associated with the life and death of Abraham. The wretched village of Eriha (Riha), 18 miles north-east of Jerusalem, and near the north end of the Dead Sea, is supposed to occupy the site of the equally ancient town of Jericho. Beyond the Jordan there appear to be no inhabited places deserving the name of town. But this region, a survey of which has been begun by the Palestine Exploration Fund, is very little known, and rendered almost inaccessible by the lawless character of its Bedouin inhabitants. When M'Coan visited the Dead Sea a few years ago he was plundered by some Moab Arabs at the ford of the Jordan, near Jericho. But the marauders seldom extend their raids quite so far west.

9. *Highways of Communication.*

In the north a much-frequented caravan route runs from Alexandretta, the natural port of Aleppo, through that city eastwards to the Euphrates at Bír, here ramifying westwards to Diarbekr and Kurdistán, southwards to Bagdad. From Aleppo the great caravan and pilgrims' route to Medina and Mecca follows the Orontes valley by Hamah and Homs to Damascus, running thence through the Haurán southwards to the Arabian peninsula. Damascus itself is connected with the coast at Beyrút by a splendid specimen of French engineering, which is carried right over the Anti-Lebanon and Lebanon, and across the Bekáa and the plains of Phœnicia, for a total distance of 65 miles. This fine highway gives easy access to the magnificent ruins of Baalbek (Heliopolis), formerly the chief centre of the worship of the Sun God, whose temple is still justly regarded as one of the wonders of the world. Baalbek is also connected with Damascus by another road through the rocky valley of the Wady Yafu'ah (a tributary of the Leontes), near which it passes the village of Surghaya, 4500 feet above the sea, and the highest inhabited point of the Anti-Lebanon. The way lies thence across a stony upland plain to the village of Dumar, where it strikes the French main highway. Another well-known route runs from Damascus across the Upper Jordan valley and through Nablus south-westwards to the coast at Jaffa, where it converges on the main road from the coast to Jerusalem. But the highways are not kept in good repair, and most of the other routes across the country are mere caravan tracks or bridle paths.

NOTE.—*The Wady-el-Arabah* (see page 89).—This depression is regarded by Edward Hull as a line of main fault or fracture continuous with the Jordan Valley, which, however, never reached the Gulf of Akabah. Terraces of marl, silt, etc., occurring at 100 feet above the Mediterranean, show that the level of the Dead Sea itself was formerly 1400 feet higher than now, filling a depression 200 miles long north and south (*Mount Seir, Sinai, and Western Palestine*, 1885).

CHAPTER V.

ARABIA.

1. *Boundaries—Extent—Area—Coast-line—Islands.*

ALTHOUGH with no very clear limits towards the north, Arabia is on the whole one of the best-defined regions in Asia. In the north it falls, on the one hand, gradually towards the Mesopotamian plains, while on the other merging almost imperceptibly in the uplands of East Palestine and Syria. Here the so-called "Syrian Desert," reaching to about the 35th parallel, might with more propriety be regarded as the "Arabian Desert;" for in its physical and ethnical features it bears a much greater resemblance to the southern peninsula than to the surrounding regions of Syria and Mesopotamia. Like Arabia proper, its watercourses are mere "wadies;" its soil sandy, and in parts destitute of vegetation; its climate dry and almost torrid; and from time immemorial it has been exclusively occupied by nomad tribes of pure Arab stock.[1] Hence many geographers look upon it as merely a northern extension of the peninsula wedged in between the Euphrates and the Syrian highlands, and only in a conventional sense separated from Arabia proper. A convenient line, however, may be drawn from El-Arish on the Mediterranean to the Euphrates delta at the head of

[1] The Sebaa Bedouins, a branch of the great Anazeh family, reach northwards beyond the ruins of Tadmor, and are met even in the neighbourhood of Aleppo.

the Persian Gulf, leaving the vilayets of Damascus and Bagdad on the north, and including on the south all that has at all times and indisputably formed part of Arabia in the strictest sense. Elsewhere the peninsula is surrounded by water—the Persian Gulf and Gulf of Omán on the east, the Arabian Sea on the south, the Red Sea and Suez Canal on the west. Its great axis, running north-west and south-east, measures, as the bird flies, nearly 26 degrees of latitude, or about 1800 miles between the head of the Gulf of Suez (29° 58′ N., 32° 30′ E.) and the Ras-el-Had (22° 23′ N., 60° E.) The mean breadth between the Red Sea and Persian Gulf is about 600 miles, with a total area estimated at rather over one million square miles, and a population of probably not more than five millions.

The shores of Arabia, which stretch from Suez to the Euphrates delta for a total length of nearly 4000 miles, present on the whole a somewhat uniform aspect, and, except in the Persian Gulf, are diversified by few islands or inlets. In the Red Sea the coast is fringed by extensive coral reefs, forming here and there groups of sunken rocks and islets, which render the navigation very dangerous. Between the Straits of Bab-el-Mandeb and Omán, separating the peninsula from Africa and Persia, the coast is generally elevated and rocky, but low and flat thence to the head of the Persian Gulf. The whole coastline has been admirably surveyed by Moresby, Haines, Elwon, Saunders, Carless, Wellsted, Cruttenden, and other officers of the Anglo-Indian navy, mainly between the years 1819 and 1860.

Of the islands the chief are the small group marking the entrance of the Gulf of Akabah; Farsan, off the Tehama coast; Perim, in the Strait of Bab-el-Mandeb, where the English batteries completely command the approaches of the Red Sea; the Kuria-Muria (Kurian

Murian) group and Moseirah, in the Arabian Sea; lastly, in the Persian Gulf the Bahrein Archipelago, centre of an important pearl fishery. The large island of Socotra, although occupied by an Arab population, and politically attached to the peninsula, belongs geographically to Africa, and has accordingly been described in the volume of this series devoted to that continent.

2. *Relief of the Land: Mountains—Plateaux—Lowlands—Deserts—Volcanic Tracts.*

Arabia is with good reason regarded as one of the least inviting regions on the face of the globe. The large blank spaces which still meet the eye as it lights on a map of this peninsula bear silent witness to our scanty knowledge of the interior. The glowing and shifting sands of the great southern desert have scarcely ever been visited, and never yet traversed by any European traveller, and fully one-half of this enormous region still remains entirely unexplored. In its general physical aspect, its climatic conditions, fauna and flora, it so closely resembles the adjacent African mainland that it seems almost more like an eastern extension of this continent than an integral part of Asia.

There are volcanic islands in the Red Sea, and one of them, the Jebel-Tír, is still active. The rocks of Aden are also volcanic.

The bulk of the land consists of a quadrangular mass broadening southwards, and largely covered with arid plains, sandy in the south, gravelly or stony in the north, the whole constituting a vast plateau at a mean elevation of probably 3000 feet above the sea. The gravelly plain of El-Hamád in the extreme north falls to 2500; but the red sand desert of Nefúd between El-Hamád and the Jebel-Shammar rises to 3000 and 3200, while the land continues to rise thence southwards to 4000 and 5000 feet in the Wahhabi country. Blunt recently ascertained that from Meshed Ali near the Euphrates in Irak-Arabi (32° N.) there is a regular ascent of 10 feet in the mile to Haïl in the Shammar highlands (27° N.); and the whole peninsula may be said to culminate towards the extreme south-west corner, where the Yemen uplands attain an elevation of from 6000 to 7000 feet. Thus we see that the tableland is tilted somewhat uniformly towards the north-east and east, so that in a developed water system the drainage would mainly be to the Lower Euphrates and Persian Gulf.

As in the Sahara, the arid tracts are broken by hilly districts and even ranges, where the valleys are watered by short streams or rivulets, and occupied by settled populations residing in small towns and villages. Thus a large portion of the central plateau, comprising the so-called Nejd—that is, "High Land"—consists of fertile hilly tracts everywhere surrounded by uninhabitable wastes and intersected by several ridges running in various directions. The term Nejd is applied to several tracts of this character, hence a certain vagueness inseparable from its use. But the Nejd proper includes, according to Blunt, all the high-lying land enclosed by the Nefúds, or deserts proper. It thus comprises the three provinces of Jebel-Shammar in the north, Kasim in the centre, and Aared or the Wahhabi country in the south, and lies mainly between 24°-28° N. latitude. It is in no sense a political, but purely a geographical expression, by which may be understood the whole of the interior, bounded on the north by the red sand Nefúd, on the south by the great unexplored Dakhna, or sandy desert, eastwards by desert tracts separating it from the Turkish province of El-Hasa, westwards by the Turkish province of El-Hejas. The arable districts in Nejd, the Hejas, Yemen, and elsewhere, are so extensive as to raise the more or less productive lands to about two-thirds of the whole area, leaving not more than one-third of absolutely desert and uninhabitable wastes, lying chiefly in the south. These wastes are variously termed Dakhna (Dahna), Ahkáf, Nefúd, or Hamád, according to the greater or less depth or shifting nature of the sands, or the more or less compact character of the soil. The sands, which rest on basalt, limestone, but mainly granitic, beds, have, according to Palgrave, a mean depth of 400 feet, attaining in some places as much as 600 feet. They prevail in the vast unexplored region comprising most of the south, between Nejd and the Hadramaut coast range north and south, and between Yemen and Omán west and east. Here almost absolute sterility is the dominant feature, whereas in the northern Nefúd, between El-Hamád and the Jebel-Shammar, not only the hollows but all parts of the plain are well clothed with shrubs. "After a rainy winter I have little doubt that the whole of this Nefúd is covered with grass and flowers. Indeed the Nefúd explains to me the existence of horses and sheep in Nejd" (*W. S. Blunt*).

The most clearly developed and best-known mountain system is the extensive range skirting the Red Sea at a distance of one to three days' journey from the coast. In the Asir district, south of Mecca, this range attains an altitude of about 8500 feet, and between this point and the Strait of Bab-el-Mandeb it broadens out in the Yemen highlands, where every condition combines to render the south-west corner of the peninsula deserving of the name of Arabia Felix, applied to it by the ancients. These highlands are continued along the south coast by a series of disconnected ridges, which again rise in the extreme south-east to the Jebel-Akhdar, running at an elevation of 6000 to 7000 feet along the Gulf of Omán from the Ras Hadd to the Ras Mussendum. From this point to the head of the Persian Gulf the coast is generally low and flat.

In the interior the Nejd is crossed by several ridges, of which the largest and best known is the Jebel-Shammar, running nearly east and west under the 27th parallel at an altitude of about 6000 feet. Farther south the Jebel-Toweyk attains probably an equal elevation in Aared, on the northern skirts of the Great Desert.

Lowland plains occur chiefly in El-Hasa on the Persian Gulf, and along the shores of the Red Sea. Here the long narrow strip of the Tehama—that is, Low or Hot Land—stretches from Mecca to Mokha, between the coast and the Jebel-Hejas, or "Separating Range," as the term is commonly interpreted.

A conspicuous feature of the peninsula are the so-called *Harra*, or volcanic tracts, strewn with basalt and other igneous rocks. The northern harra south of the land of Bashan is described by Blunt as "a vast plain strewn with volcanic boulders—a black, gloomy region."

3. *Hydrography: Wadies Sirhán, Dawassir, and Er-Rumma—Coast Streams.*

Arabia is almost a riverless region, in which the *nahr*, or perennial stream, is mostly replaced by the *wady*, or intermittent and dried-up watercourse. These watercourses, generally dry for nine or ten months in the year, occur everywhere,—in the highlands, on the plateaux, in the lowlands, and even in the deserts, and especially in the northern Hamád. Here the great Wady Sirhán runs at an elevation of 1850 feet in a south-easterly direction from the Haurán highlands to the Jof district on the skirts of the Nefúd. It is fed by the Wady-er-Rajel in the extreme north-west, and for over 200 miles between Kaf and Jof the wells are plentiful along its whole course. Hence it is much frequented during the summer by marauding tribes, who claim the right of plundering all comers, and acknowledge no authority except that of the tribal chief. Less known is the Wady Dawassir, which receives the Nejran, Bisheh, and other streams on its left bank, and drains all the Asir and Southern Hejas highlands northwards to the Bahr-Salúmeh, the only known lake in the whole peninsula. The Aftán, another large wady, runs from the borders of Nejd and the southern desert eastwards to the Persian Gulf. But the most important watercourse in Arabia seems to be the unexplored Wady-er-Rumma, which flows between the Sirhán and the Dawassir from the Hejas coast range right across the peninsula in a north-easterly direction towards the Lower Euphrates, for a total length of nearly 800 miles. With a more abundant rainfall, this would augment the Shat-el-Arab, and give unity to the now disjointed water systems of south-west Asia. As it is, the Wady-er-Rumma, our knowledge of which is mainly due to Wetzstein's studies,[1] receives during the rains a vast quantity of water through countless affluents, some rising apparently in the far south.

Perennial coast-streams occur chiefly in Yemen, where their short courses have recently been accurately determined by Manzoni.

4. *Natural and Political Divisions : Peninsula of Sinai—West Coast (El-Hejas, Yemen)—South Coast (Belad-Aden, Hadramaut)—Nejd (Jebel-Shammar, Wahhabi Country)—East Coast (El-Hasa, Sultanate of Omán).*

If physically and ethnically one, Arabia is politically

[1] Wetzstein's views receive fresh confirmation from M. Huber, who in December 1880 penetrated to Kheibar.

a disjointed land. The bulk of the inhabitants being still in the tribal state, there can be no question of a common national sentiment as developed in the west. Hence nearly all the coast lands have fallen to the stranger, while even in the interior Nejd is distracted between the waning Wahhabi and rising Shammar rulers, the only two that here claim sovereign power.

By the ancients the whole peninsula was broadly divided into three great sections, *Arabia Petræa, Deserta,* and *Felix.* The first and last of these answer roughly to the modern divisions of the Peninsula of Sinai in the north-west, and Yemen in the south-west. But Arabia Petræa, which confounded the great central tableland with the surrounding wastes, highlands, and lowlands, must necessarily disappear as the collective name of a region which we now know to be composed of several sections differing widely in their physical features. Such are—in the centre the plateau of Nejd, the northern Nefúd, and the great Southern Desert; on the west coast El-Hejas; on the south and south-east coasts Hadramaut and Omán; on the east coast El-Hasa or Bahrein. There are no doubt many other geographical expressions of a more or less local character; but these may be taken as the great natural divisions of the land, and they have the convenience of also corresponding on the whole with its political distribution. Thus the coast lands of El-Hasa, Yemen, and Hejas answer to so many Turkish vilayets; Sinai is administered by Egypt; Hadramaut is controlled by England, firmly entrenched on the rock of Aden. Omán and Nejd are under more or less independent native rule; all the rest is a prey to the Bedouin or the sands.

A line drawn from the Dead Sea through the Wady-el-Arabah to the Gulf of Akaba will mark the natural limits of the Sinai Peninsula on the east. From its base on the Mediterranean this triangular section projects with its southern apex far into the Red Sea, thus developing east and west the Gulfs of Akaba and Suez. The triangle will be almost mathematically perfect, if we take the Suez Canal as its north-western limit. But here the conventional frontier between Egypt and Arabia is drawn from Suez through the sands north-eastwards to El-Arish,

SINAI.

on the Mediterranean. Hence the mouth of the little River Arish, which gives its name to the port, is the converging point of two continents, and of the three famous lands of Palestine, Arabia, and Egypt.

The Sinai Peninsula forms a rocky limestone plateau intersected by rugged gorges, and in the north comprising the extensive desert of Et-Tih, which ascends southwards to the alpine region of Sinai proper. This desert waste covers an area of some 10,000 square miles, where a sparse population of perhaps 4000 nomad Bedouins finds a difficulty in procuring sustenance from the arid soil.

Here the land derives its grandeur and peculiar charm from the very nakedness of its rocky heights. In some of the wadies the hillsides are scored by countless seams of the brightest hues, their fantastic designs producing an indescribable pictorial effect. What is seemingly the mere outline of a distant landscape reflects a charming and almost phantasmagoric vista, as if the bare rocks were clothed with woods or vineyards, or their summits capped with eternal snows.

It is remarkable that the scriptural name of Sinai given to the mountain where Moses communed with Jehovah and received the tables of the law from above, is now unknown in the land. When asked for Mount Sinai, the Bedouin will shake his head or point to the Jebel-Musa (Moses' Mount), one of the highest in the peninsula, where a shrine has been erected to the Jewish lawgiver. A Muhammadan mosque has also been erected there. But we do not know for certain that this is the Sinai of Holy Writ, which many have identified rather with the Jebel-Serbal (6734 feet), lying a two days' journey farther north, while Beke thinks it was the Barghir, or Jebel-en-Núr (Mountain of Light), a peak 5000 feet high in the range bounding the Arabah valley on the east. The view from the granite crest of the Jebel-Musa shows that it is eclipsed by several surrounding peaks, such as the Jebel-Katharine (8536 feet), the more southerly Um-Shaumer (8449 feet), and the Jebel-Gosh (8300 feet), none of which have yet been visited by modern explorers. In fact, this alpine region, whose geological formation corresponds with that of the European Alps, and which still bears traces of former glaciers, is still largely an unknown land.

The west coast of Arabia is comprised in the Turkish vilayets of Hejas and Yemen, which have no well-defined limits towards the interior. Theoretically El-Hejas stretches in the north half across the peninsula, where it is supposed to meet the eastern vilayet of El-Hasa, in the official maps now called Basra. But between these two provinces lies the powerful state of the Emir of the Shammar, which cannot properly be regarded as forming part of the Ottoman dominions. El-Hejas, however, extends beyond the coast range inland far enough to include the cities of Medina and Mecca and the whole of the El-Asir district bordering southwards on Yemen, which comprises the rest of the south-west coast down to the neighbourhood of Aden. The two vilayets have thus a total length of about 1300 miles, varying in breadth from 60 to 150 between the sea and the western limits of Nejd. El-Hejas consists mainly of the sandy, barren, and torrid plain of the Tehama, varying from 30 to 80 miles in width along the coast, and of the mountain range or highlands with a mean elevation of 3000 feet, separating it from Nejd. The Tehama, which term is by some writers applied to the whole seaboard from above Mecca to Mokha, but by others restricted to the southern section between Yemen and the coast, seems to have formerly formed part of the bed of the sea, from which it has been slowly upheaved. It abounds in marine fossils and saline deposits, and appears to be advancing according as the sea continues to recede. Although everywhere extremely hot and generally very unhealthy, it contains especially in the south many well-watered and fertile tracts, affording good pasturage and yielding heavy crops.

But the chief value of Hejas is rather of a political than an economical character, giving to the master of the "holy cities" a great prestige, and perhaps his best title to the Caliphate, or headship of Islám. Yemen is, on the contrary, valuable for its own sake,—a land of fertile and well-watered valleys, rich pastures, and perennial streams, and dotted over with numerous flourishing towns and villages. Fully one-fifth of the entire population is concentrated in this narrow corner of the peninsula, where settled and agricultural communities, elsewhere extremely rare, have existed from the dawn of history. This exceptional position is partly due to the greater mean elevation of the land, partly to its rich soil and happy configuration, calculated to receive from the Indian Ocean and retain in its sheltered valleys an abundance of moisture. Here the sovereignty of the Porte exists almost more in theory than in reality. Vigorous efforts have been made since 1868 to revive its old claim to absolute sovereignty, and at one time the Turkish and English forces had almost come into collision in the neighbourhood of Aden. Nevertheless the Imám of Sana is still the chief potentate in Yemen, which also acknowledges the sway of several other petty rulers, some tributary to the Turk, others allied by treaty with the British. "The native chiefs, locally called 'Sultans,' still exercise their old patriarchal sovereignty, and the writ of the Padishah runs little beyond the range of his cannon" (*M'Coan*).

The extreme south-west corner from the Strait of Bab-el-Mandeb to Cape Seilán, east of Aden, and reaching inland to the Jaffa range, comprises the so-called Belad-Aden, or Country of Aden, and besides Aden itself includes the Sultanate of Laháj, under British protection.

East of the Belad-Aden the little-known region of Hadramaut stretches between the great desert and the sea eastwards to Omán. The interior of this vast but sparsely-peopled tract was almost a perfect blank until some light was thrown upon it by the travels of A. von Wrede a few years ago. This explorer assures us that the term Hadramaut applied by geographers to the entire south coast is by the natives restricted to its inland or northern section. It is in any case a very old name, for Ptolemy places the Adramitæ in this very region between the Homeritæ of Yemen and the Omanitæ of Omán. The land here rises from the coast in a succession of terraces to the Jebel-Hamra (5284 feet), which is connected north-eastwards with the Jebel-Dahura, probably 8000 feet high. This is the highest of the terraces, and beyond it the land slopes gently northwards. Here Wrede descended by very difficult and dangerous tracks down to the Wady Doan, which flows through the land of the Yssa tribe (Belad-beni-Yssa) northwards, apparently to the verge of the desert. This district is bordered on the west by Belad-el-Hasán, on the east by Belad-Hamúm, all three being bounded on the north by Hadramaut proper. But how far this region extends northwards, and whether the sandy desert of El-Akkáf (Bahr-es-Saffi) really begins with the Wady Rakhiya, a branch of the Doan, are points on which Wrede throws no light.

The southern coast lands are on the whole level, and are succeeded by a hilly tract of moderate height, beyond which the upland plains or ranges begin to fall northwards to a depression between the highlands and the vast inland plain of El-Jauf (Gof).[1] A bold attempt to penetrate into the interior from the west coast was made by the French Jew Joseph Halévy in 1870.

Of all the main divisions of the peninsula the great central tableland of Nejd or Negd is certainly one of the most interesting. It has been fairly well explored by Sadleir, Wallin, Reinaud, Palgrave, Pelly, Guarmani, and most recently by Mr. and Lady Anne Blunt. During the early portion of the century the whole of this region belonged

[1] There are several Jaufs or Jofs in Arabia, because the word simply means "low land," in opposition to Nejd, or "high land."

to the powerful and fanatical Wahhabi State, whose capital, originally at Derayeh, is now at Riad. But of late years Muhammad Ibn Rashid, Emir of the Shammar nation in the extreme north, has not only asserted his independence, but is at present by far the most powerful potentate in Nejd. His territory is bordered southwards by the Kasím country, separating it from the Wahhabi State. Northwards his influence extends beyond the Nefúd right away to the oases of Kâf and Ittery in the Wady Sirhán in 38° E. long., 31° N. lat., east from the Dead Sea. The inhabitants of these oases acknowledge Ibn Rashid as their suzerain, "paying him a yearly tribute of £4 for each village" (*Blunt*). The people of the intervening district of Jauf on the northern verge of the Nefúd also acknowledge his authority, which reaches westwards to Teyma (27° 30′ N. lat., 37° E. long.), some 80 miles from the Red Sea. He further commands the new pilgrim road from Persia, which formerly passed southwards through Riad, but now runs through Haïl, capital of his dominions. This alone brings him in a revenue of £20,000, besides giving him enormous influence throughout the whole of the north from Mecca and Medina to the Lower Euphrates valley. Ibn Rashid's green and purple banner has thus become the symbol of authority in all the land enclosed by Hejas and Palestine on the west, the Syrian Desert on the north, Irak-Arabi and El-Hasa on the east. Yet he himself, although at present by far the most powerful personage in the peninsula, is content to pay a small annual tribute to the Sheríf of Mecca in recognition of the Sultan's suzerainty, such is still the potent influence of the acknowledged head of Islám.

"Although this Richard of the Nejd reached the throne over the murdered body of his young nephew Bender, and by the massacre of sixteen possible future pretenders, he governs his subjects wisely and firmly. His rule is described as mild and just, and the *mejliss* or public court of justice is still daily held in the palace-yard, where the Emir appears, just as Mr. Palgrave describes his predecessors, surrounded by officers of state, and a body-guard of 800 soldiers. . . . He is on terms of alliance with all the Bedouins south of the Nefúd, and every year brings him in fresh tributaries from among the former dependants of Ibn Saoud [the Wahhabi ruler]. Taxation is light, service in the army voluntary, and Ibn Rashid's government eminently popular. Nowhere in Asia can be found a more prosperous, contented, and peaceable community than in Jebel-Shammar" (*Blunt*).

And thus has statesmanship succeeded where fanaticism failed. For the once formidable but now almost extinct Wahhabi State had its rise in what was in its origin essentially a religious movement. It aimed at the reform of Islám, but it soon degenerated into a purely political system, upheld by terror and blind fanaticism. It was founded in the middle of the last century by Abd-el-Wahháb,[1] but owed its subsequent expansion to his successor Ibn Saúd, in whose family the office of Imám, or spiritual and temporal head, has since remained. By the beginning of the present century the Wahhabi empire had spread over most of the peninsula, and even aimed at creating a united Arabia by the expulsion of the Turks. But it never quite recovered the blow it received in 1819, when the Egyptian troops destroyed Derayeh, and dissipated the Wahhabite dream of universal empire. Their power is now virtually limited to the highland territory of Aared, bounded north-westwards by the independent district of Kasím, and encircled elsewhere mostly by the sands.

In their religious views the Wahhabi are the most rigorous of monotheists, setting their face against all undue veneration of the Prophet, saints, relics, or aught

[1] The *hh* represents the strong Arabic guttural ح = Hebrew ח.

else in the least savouring of idolatry. Their ceremonial is extremely simple, and they carry to a heroic degree the Eastern virtues of hospitality and almsgiving. Their political system is based substantially on the cultivation of the land, and thus was developed a powerful and industrious peasantry, said at one time to have numbered nearly 2,000,000, and capable of raising an army of 60,000 disciplined warriors. But in Central Arabia the seat of power has passed from Riad to Haïl; nor can the result be regarded otherwise than as satisfactory. The collapse of the Wahhabite movement, whose influence was at one time felt even in India, lessens the fear of the peace of the world being again threatened by a revival of Moslem fanaticism. The apprehension, however, of disturbance being produced by this cause, from time to time, is not extinguished.

The east coast of Arabia, which is washed by the Persian Gulf, projects almost to a sharp point at the Ras Mussendúm, where the Strait of Ormuz separates it from the Persian mainland, and connects the Persian Gulf with that of Omán. The northern section of this coast is officially included in the Turkish vilayet of Basra (El-Hasa), the southern and western frontiers of which are arbitrarily drawn according to the caprice of the Ottoman functionaries. But El-Hasa is naturally limited southwards by the projecting headland of Ras Rekkan, which encloses the group of the Bahrein Archipelago, claimed by Turkey, and valuable for its pearl fishery. Bahrein, as the strip of coast between Capes Rekkan and Mussendúm is sometimes called, forms a sort of neutral land between the vilayet of El-Hasa on the north and the independent native State of Omán, which comprises all the rest of the east coast and the whole of the south-east corner of the peninsula.

The Sultan of Omân, formerly more popularly known

as the Imâm of Maskat,[1] at one time ruled over an extensive territory on the East Coast of Africa. But this was assigned in 1856 to a brother of the reigning Sultan, and now constitutes the independent State of Zanzibar. He, however, still claims jurisdiction over Germansir, a strip of the opposite Persian coast, stretching from about the 28th parallel to the west frontier of Persian Mekrán, and including the port of Bundar Abbas (Gombrún) and the large island of Kishm.

On the Arabian mainland the north-east coast is rocky, but well supplied with good natural harbours, while the south coast west of Cape Hadd is flat, and sheltered only by the island of Moseirah. At a distance of 50 or 60 miles from the sea there runs a mountain range, the Jebel-Akhdar, parallel with the crescent-shaped east coast, beyond which the surface is dotted with a number of true oases, abounding in water, incredibly fertile, and covered with an exceedingly dense growth of vegetation.

The Sultan of Omân maintains a small navy to keep down piracy in these waters. He has also long enjoyed the benefit of a close alliance with England, which, while adding to his prestige amongst his own subjects, guarantees him from any overt acts on the part of Turkey or Persia. His States are fairly well governed, justice is efficiently administered, and peace secured within his borders, which verge everywhere inland with the desert.

5. *Climate: Rainless Zone.*

The prevailing climatic conditions are intense heat and dryness. The zone of maximum heat on the surface of the globe in July embraces the whole of the Persian Gulf, the greater part of the Red Sea, and of the inter-

[1] His proper title is *Sayid*, or "Sovereign." He never assumed the religious dignity of Imâm commonly attributed to him.

ARAB DRAW-WELL.

vening Arabian peninsula. This also comprises one of the rainless regions, where rain falls only at intervals of from one to three or four years. Even the periodical wet seasons, to which Yemen and some other favoured tracts are exposed, are occasionally interrupted by counter-atmospheric currents from Africa, and then whole provinces have to depend for two or three years at a time on their wells, tanks, and other reservoirs. It is the vicinity of the African Sahara that prevents Arabia from enjoying, as India does, the full benefit of the moist winds from the Indian Ocean. Hence in summer, when India is often deluged by tropical downpours, the south coast of Arabia swelters under the vertical rays of a fierce sun, and the parched-up land finds no shelter either in an overcast sky or a leafy vegetation. Thus has been developed in the course of ages the great Southern Desert, surpassing the Sahara itself in absolute aridity and barrenness. But as the land rises towards the Yemen highlands, the glass naturally falls, the nights become pleasantly cool, and the tanks here freeze in winter. At elevations of 6000 feet storms become frequent, and are at times accompanied by heavy showers. The heat probably reaches its maximum in the low-lying coast district of the Tehama on the Red Sea, and along the west coast of the Persian Gulf. From the bare rocky walls skirting both sides of these land-locked basins the sunbeams are reflected with redoubled strength on the glowing waters, thus producing an enormous evaporation, which converts the surrounding atmosphere into a vapour-bath. For Europeans a trip across the Persian Gulf is considered at these times extremely perilous, and the unhealthy climate of the Tehama has become proverbial. On the other hand, the high central plateau of Nejd enjoys a climate described by Palgrave as one of the most salubrious in the world. Here the pure air, dry atmosphere, and moderate temperature have proved highly favourable to the development of animal life, although the lack of moisture has prevented a corresponding vegetable growth. Altogether, the most favoured region in this respect is Yemen, where the glass, even in July, seldom rises above 90° F. At Sana, Niebuhr found that it did not exceed 85°, while in the neighbouring Tehama it stood at 98° F. in the shade. Here also snow falls occasionally, and it freezes during the three winter months, while at Loheia

(Tehama coast), the glass never falls much below 80° F. in January. Such is the astounding difference in temperature produced by the relief of the land in the same district.

Sand-storms prevail very generally, but are not dangerous to travellers, except, perhaps, in the great Southern Desert. On the other hand, the extent of the range of the simooms, or poisonous winds, seems to have been exaggerated.

6. *Flora and Fauna: The Horse and Camel.*

The most valuable plants are the date-palm, of which over 130 varieties are reckoned growing in all the oases, and supplying the chief staple of food; coffee, indigenous in Yemen, and largely exported from Mocha, whence the "Mocha coffee" of commerce; aromatic and medicinal plants chiefly along the south and west coast, producing frankincense, myrrh, gum-arabic, balsam, senna, which have supplied the markets of the world for ages. The vine is cultivated for its fruit; the peach, apricot, almond, fig, and other fruits of excellent quality, are produced in Yemen, and cotton is cultivated in Omán. Of the few forest trees the chief are the sycamore, the *nebek* or thorny lotus, the cassia, and the manna-yielding ash. Yemen, and some other parts, also yield maize, millet, wheat, barley, durra, lentils, tobacco, madder, indigo. Characteristic of Nejd is the *ghatha*, which grows to 12 or 15 feet high, and yields the purest charcoal in the world. It abounds in the northern half of the Nefúd, and is found as far north as Kaf in the Hamád.

Amongst the wild animals are the lion, panther, leopard, wolf, wild boar, jackal, gazelle, fox, monkey, wild cow, or white antelope (Beatrix antelope, genus *Ornyx ?*), ibex, webber (marmot ?), horned viper, cobra, bustard, buzzard, hawk. The locust abounds in Arabia, but is here rather preyed upon than the spoiler. "It is not generally known how excellent locusts are as food. . . . The red locust, which is, I believe, the female, is the best eating, and should be plain boiled. In taste it resembles green wheat, having a very delicate vegetable flavour. Horses thrive on them, and nearly every animal in the desert devours them. Our dogs caught and ate them greedily. A camel will occasionally munch them in with their pasture, and a hyæna I shot was found to be full of them. Locusts should be gathered in the morning while the dew is still on their wings" (*Blunt*).

The chief domestic animals are the ass, mule, fat-tailed sheep, and above all the camel and horse. Of the latter there are two classes: the kadîshi, of unknown pedigree, used for rough work; and the kokhlâni, or koheileh, whose genealogies have been recorded for over 2000 years, and which spring traditionally from Solomon's studs. They are mostly of small size, between 13 and 14 hands high, but symmetrical, hardy, and endowed with extraordinary staying power. The best breed, formerly in the Nejd, is now said to be found amongst the Anazeh and other Bedouin tribes of Mesopotamia. But opinions differ on the point, and while Blunt holds that "the tale of a distinct Nejd breed is entirely fabulous," Rawlinson still considers that the Anazeh is "of much inferior blood to a real Nejd horse." Nejd is supposed to abound in horses, but this would seem to be a mistake. Burckhardt long ago remarked that here they are comparatively rare, and that the Bedouins of the rich Mesopotamian plains own the largest stock. This view is now confirmed by Blunt, who asserts that "horses of any kind are exceedingly rare in Nejd." Here "the camel is the universal means of locomotion with the Bedouins. The townsmen go on foot."

But for the camel the desert would be absolutely

uninhabitable. Of this animal there are several species, or rather varieties, abounding especially in Nejd, hence termed *Omm-el-Bel,* or "Mother of Camels." The Nejd "Ship of the Desert" will pass four and even five days in the summer heats without a drop of water; but those most suited for riding are said to come from Omân.

7. *Inhabitants: Bedouin Life.*

Few Asiatic lands can boast of a more homogeneous population than Arabia. The whole peninsula belongs from prehistoric times to the great branch of the Semitic family, who have always called themselves Arabs, a term probably meaning nothing more than "people of the plains." Within this branch there are doubtless many divisions and subdivisions, which will be found elsewhere specified; but all are essentially one in origin, physique, speech, and religion. The only true distinction that can now be recognised is rather of a social than an ethnical character—that is to say, the distinction between the settled agricultural element residing in towns and villages and the nomad Bedouins of the wilderness. The former are met everywhere in more or less numerous communities, wherever the land is fit for cultivation—in El-Hejas (Mecca, Medina, Taïf); in Nejd (Haïl, Derayet, Riad); in Omán (Maskat); but especially in Yemen, where the settled political status preponderates over the tribal organisation of camp-life.

Some readers may possibly be surprised at the term "organisation" applied to the social condition of the free children of the desert; but the popular ideas regarding the habits, customs, and usages of the tented Arab are in many respects erroneous. He is usually represented as ceaselessly roaming with his tents and flocks from place to place, whereas there is perhaps no people less given to wandering, or more attached to their homes, than the true Bedouins.[1] Hence Arabic is almost the only language that has a perfect equivalent in the term *watan* (وطن)

ARAB SHEIKH.

to the English word *home.* They have their allotted winter and summer camping-grounds, seldom changing their settlements except when removing from one to the other with the seasons. While *en route* they never pitch

[1] This very word *Bedawi*, plural *Bedawiún*, means rather *pastor*, *stock-breeder*, than *nomad.*

their tents, sleeping in the open, wrapped in their flowing garments. Their encampments resemble those of the gipsies, only the occupants are perhaps somewhat wilder and more picturesque in appearance. Women in dark-brown cloaks grinding the corn with primitive hand-mills, or weaving cloth for the tents; children, goats, and dogs, all playing together in happy harmony; the men lounging about smoking, or drinking coffee, form on the whole a not unpleasant scene of homely life.

The Bedouins are often represented as highwaymen and robbers from their birth. Their ideas regarding the rights of property differ seriously from those prevalent in the West; but these very ideas of theirs are based on a keen sense of right, and grow out of the proud spirit which resents the intrusion of strangers or hostile tribes on their domain. It must be allowed that among them there are what may be called marauding tribes by profession; but even these have a certain traditional code of law and honour, strange as the word may sound in such an association, a code which all alike accept and implicitly obey. A curious illustration of this spirit is afforded by the circumstances attending the attack on Mr. Blunt's party in the Wady Sirhán in the year 1878. "Lady Anne and I," he writes, "happened to be separated from the rest of our party, and were sitting under a ghatha bush eating our midday meal of dates, when we suddenly heard the galloping of horse-hoofs in the sand. Looking up, we saw a dozen Bedouins bearing down on us with their lances. . . . Our thick cloaks saved us from the points of the lances, and my Bedouin head-rope saved my head; and when we had cried 'Dahil,' 'I yield,' and given up our mares, they left off knocking us about. It then turned out that our captors were a party of Roala, friends of our own and of Muhammad's, though they knew nothing of us personally; and after we had sworn to our identity, they brought us back our mares and everything that had been dropped in the scuffle."

Jonas Hanway also vindicates the Bedouins from the animadversions of some writers in the last century. "Their skill in horsemanship, and their capacity of bearing the heat of their burning plains, give them a superiority over their enemies. Hence every petty chief considers himself as a sovereign prince, and as such exacts customs from all passengers. When they plunder caravans travelling through their territories they consider it as reprisals on the Turks and Persians, who often make inroads into their country and carry away their corn and their flocks."[1]

Amongst themselves and towards all placed under their protection their sense of honour and trustworthiness are beyond suspicion. Owing to the fearful severity of the custom of blood-revenge, murder is of much rarer occurrence in the wilderness than in more civilised lands. The character of the country and their social habits develop a sort of clannish confederacy amongst the several tribes, as well as a certain common sympathy with all belonging to the Arab race. The Turi or the Maghrabi tribes have now a salutary dread of "the Consul."

In other respects the contrast between the social relations of the Bedouin and those of the "more civilised" inhabitants of the towns and villages is very much to the advantage of the former. Their simple diet and the pure untainted atmosphere which they breathe, render them healthy in mind and body. They are cheerful and even possessed of a fund of humour, and will often endure the greatest hardships without a murmur. Their demeanour is courteous and even refined.

As with most Eastern peoples, parents are treated

[1] *The Revolutions of Persia*, part v. pp. 221-2.

with the greatest respect by their children while under age. But as soon as the young Bedouin is old and strong enough to set up an independent establishment, he considers himself released from this duty, henceforth regarding himself in the light of an equal.

On the men naturally falls the care of supporting the tribe, the means of doing which are often scanty enough. Their chief source of wealth is derived from the camel. The escorting of travellers, pilgrims, and goods, is a profitable branch of industry, but restricted to the few tribes recognised as the duly authorised *ghufara*, or "protectors."

A limited trade is also carried on with Suez and Cairo, the Arabs supplying charcoal, millstones, ibex horns, gum Arabic, and the like, in exchange for corn and tobacco. A few inhabiting such fertile districts as the Feirán own a little land, on which they cultivate tobacco, bartering or selling it to the neighbouring tribes. Owners of sheep and goats turn the hair and wool of these animals to account, and use the milk in spring, but seldom kill them except in sacrifices. Another article of trade is the "munn," or manna, a glutinous saccharine substance exuding from the tamarisk tree for about two months, "while the apricot is in bloom."

8. *Topography: Mecca—Medina—Haïl—Maskat—Aden.*

In a land of which probably not more than one-tenth is arable, towns cannot be numerous. Nor are there more than two or three places in the peninsula with a fixed population of 50,000. The largest are Sana and Maskat, capitals of Yemen and Omán respectively, both of which may have 45,000 inhabitants. But by far the most important are the two "holy cities" of Hejas, Mecca and Medina, towards which the eyes of one hundred millions of Muhammadans are constantly turned, from the shores of Marocco to the distant islands of the Eastern Archipelago.

Mecca, the Rome of Islám, is an unwalled city situated in a narrow sandy valley enclosed by rocky eminences from 200 to 500 feet high, and about 65 miles from Jidda, its port on the Red Sea. The valley is scarcely 600 yards broad, narrowing southwards to about 300, where it is almost blocked by the Beit-Ullah (God's house), the great mosque enshrining the famous *Kaaba*. The whole building forms a rectangle 250 paces long by 200 broad, of which the north side is formed by four rows of pillars, the other three of three rows each, arched over, and so disposed that each group of four supports a little cupola, making altogether 152 of these structures. Along its entire length suspended from the arches are glass lamps, all of which are kept burning during the Ramadán, or fasting season. The oldest pillars are hewn out of the neighbouring rock; the others, consisting of marble, granite, and porphyry, are mostly offerings of the Faithful, and include some antiques from the old temples of Syria and Egypt.

Within the mosque is the *Kaaba*, or "Holy House," a small, massive building about 40 feet high. Tradition associates this unpretending and curious little structure with a multitude of marvels and legends, one more preposterous than another. On the north side is a doorway leading through steps inlaid with gold and silver to the inner sanctuary. In a corner lies the famous "black stone," supposed to have been given by God to Abraham, but now known to be a meteoric block descended, if not from heaven, at least from the interplanetary spaces. West of the Kaaba is the "golden channel," carrying off from the flat roof the rain-water, which is reputed to be endowed with miraculous properties.

Access to Mecca is rendered extremely difficult in

consequence of the ceremonies imposed on all wishing to visit the birthplace of the Prophet, and expressly designed to exclude unbelievers. Yet the feat has been accomplished during the present century by Burckhardt, Wallin, Palgrave, Burton, Keane (?), and perhaps by others, mostly disguised as pilgrims.

In Mecca resides the great Sheríf of Mecca and Medina, a far more important dignitary than the Turkish Vali or Governor of Hejas. As guardian of the holy shrine of Islám he receives a heavy annual stipend from the Porte, in return acknowledging the suzerainty and caliphate of the Sultan.

About seventy miles south-east of Mecca is the small but pleasant town of Taïf, to which the pashas condemned for the murder of Sultan Abdul-Aziz have been banished. It is one of the most interesting places in Arabia, surrounded by gardens and vineyards, from which Mecca has been supplied with fruits for ages.

Nearly under the same meridian as Mecca, and 240 miles farther north, lies the almost equally venerated city of Medina. Hither fled the Prophet when his obdurate fellow-citizens were deaf to his voice, and from this flight dates the Muhammadan era.[1] Here also is his tomb, a shrine second only in sanctity to the Kaaba itself. Medina lies at an elevation of 3000 feet above the sea, close to a range of hills running north and south between Hejas and Nejd. It is built of solid stone, but the streets are very narrow, and everywhere lined with lodging-houses for the convenience of pilgrims. The great mosque containing the Prophet's tomb is approached by the main street from the gate of Cairo. It is smaller than that of Mecca, and is supposed to have been really built by Muhammad himself. His coffin is encased in silver, and covered with a heavy marble slab.

[1] That is, the *Hijra*, or "Flight."

By far the most important place in the south-west is Sana, capital of Yemen, and seat of an Imám, who enjoys a widespread jurisdiction in this region. Sana, which is perhaps the finest and best-built city in the whole of Arabia, has been visited of late years by Wrede, Halévy, Manzoni, and several Englishmen. It lies in a fine, well-cultivated upland valley, over 4000 feet above sea-level, and about 110 miles north-east of Hodeida, its port on the Red Sea. Its walls, about six miles in circumference, are mounted with cannon, and they enclose two stone palaces of the Imám, besides a great number of highly-ornamental mosques, baths, and caravansarais.

In the interior the only noteworthy places are Haïl and Riad, capitals of Jebel-Shammar and Aared respectively.

Haïl lies 3500 feet above the sea, not to the south, as is usually stated, but to the east of the Jebel-Aja, a granite range 6000 feet high, which ends abruptly at this point. In this neighbourhood is the Emir's castle, a formidable stronghold occupying a position of immense natural strength in the Jebel-Aja. Blunt visited this place in 1878, but does not give its exact site, lest the information might be utilised by the Turks under possible future contingencies.

Riad, which has succeeded Dereyah as capital of the Wahhabi State, lies in the heart of the Aared country, enclosed north and south by the Jebel-Towoyk, and about 280 miles south-east of Haïl. It is a large place, with a population of probably 30,000; but nothing is known of its present state, as no European has visited it since the time of Palgrave.

On the east coast the only large place is Maskat, capital of Omán, which, although extremely hot and unhealthy, is the centre of nearly all the import and export trade in these waters. For this position it is indebted more to its well-sheltered and convenient harbour than to the fact that it is the seat of government and residence of the Sultan. It is built in a series of terraces rising one above the other on the side of the frowning precipices enclosing its picturesque bay. But though presenting a pleasant prospect from a distance, a nearer view reveals the usual features of large Oriental towns—narrow, dirty, and gloomy streets, tasteless buildings, and masses of dead walls, beyond which stretches a swampy suburb occupied by nomad Arabs and African slaves. The townspeople themselves are a motley mixture of Arabs, Persians, Syrians, Indians, and even Kurds and Afghans, who have either taken refuge here from oppression at home, or else have been attracted to the place by its great facilities for trade.

Politically by far the most important place in the southern section of the peninsula is Aden, occupied since 1838 by the English, who from this stronghold and the neighbouring island of Perim command the whole of the Red and Arabian Seas, and keep open the water highway to British India and the far East. But besides forming one of the most important links in the chain that girdles the eastern hemisphere from London to Hong-Kong, this Gibraltar of the Indian Ocean is also a free port, doing a considerable trade with the interior, and with a present population (including Perim) of 34,860. Yet it lies perched on a bare rock in an indescribably barren and desolate coast district, a hotbed of the most deadly diseases, altogether one of the most uninviting and unhealthy spots on the surface of the globe, and in summer sultry almost beyond endurance. The old town lies in the very crater of an extinct volcano, 1775 feet high, whose sides, which have partly fallen in, are crowned with formidable works bristling with cannon.

Aden lies well within the rainless zone, where no

rain falls at times for intervals of two or even three years. Hence for its water-supply it is dependent on wells, tanks, condensers, and the magnificent old reservoirs in the neighbourhood, which have been recently restored. Here are two good harbours, formed partly by the adjacent island of Sírah, and, thanks to its convenient and commanding position near the entrance of the Red Sea, Aden has become one of the chief coaling depots and calling

ADEN.

stations for steamers in the Indian waters. It is also an important political centre, barring the further advance of the Turk in this direction, and guaranteeing the independence and good government of Lahej and the other petty States along the south-west coast.

The settlement, which includes the rocky peninsula, 15 miles in circumference, and extends to the Khor Maksar Creek, two miles north of the defences, is politically subject to the Government of Bombay, and administered by a Resident with two assistants. Since the opening of the Suez Canal the shipping has steadily increased, and a vessel of war is usually stationed at the port, which is in charge of a Conservator and regulated by the Indian Ports Act.

9. *Highways of Communication.*

In Arabia there are scarcely any roads properly so called. But the peninsula is crossed everywhere except in the south by well-trodden caravan routes, whose direction is mainly determined by the greater or less abundance of wells and other reservoirs along their course. There is so little local trade and so much visiting of the holy places from all quarters of Islám, that these routes naturally converge on Mecca and Medina. The two main highways are what might be called the Sunnite and the Shiah *haj*,[1] the former from the north for the convenience of the orthodox Turk, the latter from the east for the heretical Persian.

The northern pilgrim road starts from Damascus and runs nearly due south through the Haurán highlands and the Roala, Sherarat and Harb Bedouin territories between El-Hejas and the Nefud to Medina and Mecca. The chief intermediate stations are Kalaat Belka, east of the Dead Sea; Maan, east of the Wady-el-Arabah, Tebuk; and Medain Salah, east of the Red Sea. The journey to Medina takes thirty days, and the pilgrim caravan is usually escorted by the governor of Damascus. But this route is not now so much frequented as formerly, the pilgrims from Anatolia and Syria preferring the less

[1] *Haj* means "pilgrimage," whence the "Haji," or pilgrim *par excellence*, who has visited the holy places, a personage who holds his head very high in the East.

fatiguing and more expeditious journey by steamer through the Suez Canal and Red Sea to Jidda, whence they reach Mecca by easy stages in three or four days. Since the opening of the canal the pilgrim traffic of Jidda by sea has increased to from 45,000 to 50,000 yearly. Many also still reach Jidda by land from the Barbary States, Sudan, and Egypt, by a route from Cairo across the Sinai Peninsula and down the coast of Hejas. But this line is also being gradually abandoned in favour of the sea voyage from Suez.

The eastern or rather north-eastern road from Persia runs from Bagdad through Kerbela and Meshed Ali, nearly due south, to the wells of Shaibeh (27° 10′ N., 44° E.), here turning west to Haïl and thence south-west to Medina. This line traverses the domain of the Montefik and Daffir Bedouins in the north, the Jebel-Shammar State, and the Harb Bedouin country in the west. In lat. 28°-29° and long. 44° 20′ it touches the famous reservoirs built by Zobeyde, wife of Harún-el-Rashíd, for the special use of the pilgrims. A caravan route from the Haurán through the Wady Sirhán, the Jauf Oasis, and the Nefúd, strikes this line at Haïl, and is continued thence south-eastwards through Bereydah and the ruins of Dereyah to Riad. A track from Riad through Yemamah reaches the Persian Gulf at El-Katif above the Bahrein islands. But no certain lines are known to run from this direction southwards. Riad, however, seems to be connected westwards through Taïf with Mecca, and this route, if it exists, nearly completes the main lines crossing the peninsula. Blunt denies the existence of a Roman road said to have formerly run from Melakh on the Syrian frontier across the Hamád to Basra on the Shat-el-Arab.

In the foregoing chapters—II. Asia Minor; III.

Euphrates and Tigris Basin; IV. Syria and Palestine—the headings 10, or Administration, and 11, or Statistics, have not been included; and these headings have yet to be given for the present chapter. They will now be given for all these chapters together. It is thought best to combine Arabia with Turkey in Asia for this purpose, because some part of Arabia belongs to, and still more is claimed by, Turkey, while other parts are independent.

10. *Administration: Turkish System in Asiatic Turkey generally—Social State—Taxation—Justice—Religion—The Ulema—Education.*

In theory the government of Asiatic Turkey is an absolute despotism, limited in practice by many social and religious checks. The Sultan's personal action is now largely controlled by that of the Grand Vizier and Divan (Prime Minister and Cabinet). But he still nominates not only all the members of the Divan but all the provincial governors and lieutenant-governors, whose tenure of office being precarious, the incentive to rapacity often becomes irresistible.

Since 1867 Asiatic Turkey is divided for administrative purposes into vilayets or provinces, sanjaks (" banners "), answering to the French arrondissements, kazas or districts, and nahiés or communes. The vilayets, mostly named from the chief town, are governed by valis, ranking as mushirs or pashas of the highest order; the sanjaks by caimacans or lieutenant-governors, ranking as mutessarifs or second-class pashas; the kazas by mudirs, elected in theory by the inhabitants, but in reality by the valis; the nahiés by muktars or mayors, ostensibly elected for a year, but really named by the mudir. There are a great number of other officials, of whom it may be affirmed that " not one owes his post to personal merit or qualification, but all to bribery or intrigue. The vali himself buys his appointment from some palace favourite or other patron at the Porte. . . . The same may be said of the cadis (magistrates), of the commandant of the police, and of the directors of the customs" (*M'Coan*).

Taxation is largely based on the old tithe system, and as the tithes themselves are farmed out, ample scope is given to extortion, the sum raised always far exceeding that imposed by the Treasury. Justice also, although the civil and criminal codes are based on sound principles of equity, is dispensed by servile ministers in such a manner as to become an additional instrument of oppression. In all the courts bribery is a recognised factor, and although the Turk is personally honest and upright, the Turkish official has under this system become the incarnation of servility and corruption. Even the Christian assessors associated with their Moslem rulers would seem to be deeply tainted by the prevailing laxity. M'Coan mentions the case of a Christian member of a civil court waiting on the advocate of some parties in a pending case, and arranging for a bribe of £100 to secure judgment in their favour. This judge is now "president of one of the Stambul courts, a rich and respected functionary."

The real grievance of the Christians is that their testimony carries but little weight, even when not absolutely rejected, in all the courts of the empire. "What we require," said one of them to Captain Burnaby in Smyrna, "is similar treatment for all sects, and that the word of a Christian when given in a court of law should be looked upon as evidence and in the same light as a Muhammadan's statement. If the Caimacans and Cadis were only compelled to do us justice in this respect, we should not have much cause to grumble."

Religion—The Ulema.

The Sultan is primarily not so much a temporal sovereign as the accepted Caliph, or spiritual head of Islám. Hence the organic laws of the empire are all based on the Koran, to which the last appeal must be made in all emergencies.

The cardinal doctrine of the Muhammadan religion is pure theism, formulated in the words, "There is but one God;" and besides the Prophet, it accepts the divine mission both of Moses and Christ. "The Son of Mary" especially is acknowledged as the Word proceeding from God, as the Messiah of the Jews, Mediator with God in heaven, and the appointed Judge of all. A final judgment, an after state, a heaven and a hell, good and bad spirits, and guardian angels, are amongst the tenets of this religion. Most of its rites, such as punctilious and ceremonious prayer, ablutions, circumcision, pilgrimage, abstinence from alcoholic drinks, are either positively good or at the least harmless; while some parts of its morality, inculcating the virtues of almsgiving, truth, sobriety, mercy to the brute creation, are to be commended.

"Islám," or the Muhammadan faith (literally "submission to God"), differs, in Asiatic Turkey, from most other religions in the absence of a true priesthood. For the Ulema[1]—that is, the "wise" or "learned"—were originally nothing more than a body of interpreters, instituted to study and expound the text of the Koran. But as the Koran contains the secular as well as the religious code, this body could not fail gradually to usurp a preponderating influence in the councils of the State. This

[1] From the Arabic root عِلْم *ilm*, knowledge, science, comes the adjective *álim*, learned, wise, of which the plural is *ulema*.

influence it still enjoys and exerts in a spirit hostile not only to Christianity but to all true progress not in accordance with the "letter of the law." At present the head of this college is the Sheikh-ul-Islám, or "Head of the Faith"—that is, next after the Caliph, but in purely spiritual matters enjoying a power almost paramount even to that of the Sultan himself.

Education is still in an extremely backward state, and must continue so until emancipated from the control of the Ulema, whose interest it is to restrict its range to the reading and expounding of the Koran. Attempts at reform were made so far back as 1845, when the principle of secularisation was adopted and a new university founded in Constantinople. Primary instruction was soon afterwards made compulsory, but through the influence of the Ulema it was restricted to reading, writing, elementary arithmetic, and the study of the Koran. Even in the *mekteb* or secondary schools, and to a large extent in the *medresseh* or colleges, the teachers are all members of the Ulema, with the inevitable result that education still resolves itself into a training calculated more to fill the mosques and uphold the old system than to produce enlightened and liberal-minded citizens. So much, however, has been secured that the bulk of the people, even in Asiatic Turkey, can now at least read and write.

The above description is applicable to Asiatic Turkey. But Arabia, which is included in this chapter, gave birth to a religion that has extended to several other countries besides the Turkish dominions. The Muhammadans are divided into two sects, the Sunnis and the Shiahs. The Sunnis are usually regarded as the orthodox party. They acknowledge the succession of the four Caliphs who inherited the spiritual and temporal supremacy bequeathed by the Prophet Muhammad himself. Their name indicates those who follow the true tradition. The Shiahs are usually regarded as sectaries, as their name implies. They are considered as heretics by the Sunnis, who formed the dominant party for many generations. In this age, however, they contribute an influential minority. Originally they followed Hasan and Hosen, the grandsons of the Prophet by his daughter Fatima and her husband Ali. The grandsons took up arms against the Caliphs, successors of the Prophet, and were slain in battle. Their memory is revered as that of martyrs. The two religious centres of the sect are Mashhad in the north-east corner of Persia, and Kerbela on the border of Arabia and Mesopotamia. In all the world there is no place more heartily venerated by millions of people than Kerbela. In the main the Shiah country is Persia; but Bokhara, Constantinople, Bagdad, Cairo, Delhi, Kábul, are Sunni. The Mogul emperors of India were Sunni, though, as will be seen hereafter, there are many Shiah sectaries in India. The Sunnis and the Shiahs in India have their respective watch-cries. The Sunnis say "Dam-i-chahâr yâr;" or, Hail to the four disciples of the Prophet (that is, the Caliphs). The Shiahs say "Dam-i-panj-tan;" or, Hail to the five relations of the Prophet; meaning that the descendants have a prior claim over those who were the disciples only. The Sunnis mean that the disciples were nominated as lawful successors, and that allegiance is therefore due to them.

In Asiatic Turkey the Muhammadan practice at least is understood to be becoming more and more tolerant.

Outside Asiatic Turkey, however, the Muhammadan faith maintains its hold upon the hearts and minds of the influential classes among its adherents. It has priestly classes bearing the names of Mufti, Molavi, Mullah. They are hearty and sincere zealots. Their religious sentiments, originally pure and lofty, often degenerate into bigotry and fanaticism. From time to time, as for instance the Wahhabi movement in Central Arabia, efforts are made to reinvigorate the austere strictness of the Prophet and the Caliphs, his immediate successors. But veneration for the Prophet, his Koran, and his Tradition, never causes the people to forget the attributes of God (Allah), which ever have been, and still are, defined and formulated with extraordinary accuracy and fidelity. The merits of such tenets still infuse potent life into the religion. Though the name of "the most merciful" is constantly invoked, yet something the reverse of mercy and charity, as understood by Christians, is really presented. Almsgiving is indeed proclaimed to be a duty in the loftiest terms. But kindness is really reserved for those within the pale. For all outside the pale, fierce intolerance and an almost sanguinary animosity is declared. These are charged with "unbelief," and the term Kâfir, or unbeliever, is still regarded as a severe inculpation. For all that, in countries such as British India, where Muhammadans are brought into contact with Europeans, the common humanity asserts itself, and there many good, faithful, and friendly Muhammadans are to be found.

The Muhammadan nations are retrograding, and the retrogression is in part attributable to their religion. The following sentences are taken from the Rede Lecture delivered at Cambridge in 1881 by Sir William Muir, one of the first living authorities:—

"Some, indeed, dream of an Islám in the future, rationalised and regenerate. All this has been tried already, and has miserably failed. The Coran has so encrusted the religion in a hard unyielding casement of ordinances and social laws, that if the shell be broken the life is gone. A rationalistic Islám would be Islám no longer. The contrast between our own faith and Islám is most remarkable. . . . There are in our Scriptures living germs of truth which consist with civil and religious liberty, and will expand with advancing civilisation. In Islám it is just the reverse. The Coran has no such teaching as with us has abolished polygamy, slavery, and arbitrary divorce, and has elevated woman to her proper place. As a Reformer, Mahomet did advance his people to a certain point, but as a Prophet he left them fixed immovably at that point for all time to come. . . . The tree is of artificial planting; instead of containing within itself the germ of growth and adaptation to the various requirements of time and clime and circumstance, expanding with the genial sunshine and rain from heaven, it remains the same forced and stunted thing as when first planted some twelve centuries ago."

11. *Statistics of Asiatic Turkey and Arabia.*

AREAS AND POPULATIONS.

	Area in sq. miles.	Pop.[1]
Anatolia	220,000	10,859,124
Turkish Armenia	30,000	674,608
Turkish Kurdistán	50,000	1,570,587
Mesopotamia	220,000	5,934,333
Syria	108,000	2,309,837
Palestine	12,000	700,000
Turkish Arabia	300,000	1,614,857
Independent Arabia	500,000	3,400,000
	1,440,000	27,055,346

APPROXIMATE CLASSIFICATION BY RACES AND RELIGIONS.

		Pop.
Moslem[2] 23,105,000	Turks: Anatolia, Armenia, Syria, etc.	11,500,000
	Arabs: Arabia, Mesopotamia, Syria	7,600,000
	Syrians: Syria	1,750,000
	Kurds: Kurdistán, Armenia, Anatolia	1,500,000
	Circassians and Abkhasians: Anatolia	400,000
	Yuruk Turkomans: Anatolia, Syria	200,000
	Lazis: Lazistán, Anatolia	200,000
	Meteollis: Syria	15,000

[1] Based on a careful estimate prepared in 1878 by Mr. Redhouse for the British Government, but never published. Mr. Redhouse, one of the first authorities on the subject, shows that the population of Asiatic Turkey, usually given at about 17,000,000, has been greatly underrated. There can be little doubt that it amounts to at least 23,000,000, which, with that of Independent Arabia, gives about 27,000,000 altogether, as above.

[2] All Sunnis, except the Meteollis, who are of the Shiah sect, but with peculiar rites.

		Pop.
Christian[1] 3,610,000	Greeks: Anatolia, Syria	2,000,000
	Syrians: Syria	350,000
	Armenians: Armenia, Anatolia	760,000
	Maronites: Syria	250,000
	Nestorians: Mesopotamia, Kurdistán	250,000
Sundries 260,000	Druses: Syria, Haurán	90,000
	Jews: Anatolia, Syria, Mesopotamia, Arabia	80,000
	Nusairieh: Syria, Anatolia	50,000
	Kizil-Bashis: Anatolia	30,000
	Yezides: Anatolia, Mesopotamia	10,000
	Ishmaelites: Syria	2,000
		27,550,000

Vilayets[2] and Separate Administrations.

		No. of Sanjaks.	Pop.
Asia Minor	Scutari	1	450,000
	Ismid	1	228,288
	Khodavendikiar	4	1,383,948
	Biga	1	193,500
	Aidin	4	1,611,366
	Kastamuni	4	1,258,953
	Angora	4	860,397
	Konia	5	1,260,003
	Janik	1	389,787
	Sivas	3	1,383,762
	Trebizond	3	856,141
	Adana	4	469,368
Ægean	Archipelago	5	460,611
	Samos (Principality)	...	[3] 45,000
Armenia	Erzerum	5	674,608
Kurdistán	Aziz	2	337,779
	Van	2	1,015,269
	Diarbekr	4	220,539
Mesopotamia	Bagdad	6	4,755,618
Irak & E. Arabia	Basra and El-Hasa	...	1,276,572
Syria	Aleppo	4	1,298,232
	Damascus	7	991,605
	Lebanon (Christian Protectorate)	1	330,000
Palestine	Jerusalem	1	390,000
West Arabia	Hejas	2	720,000
	Yemen	4	798,000
		78	24,165,624
Deduct, ceded 1878, to Russia, 415,178 (Batúm, etc.)			
,, ,, to England, 186,100 (Cyprus)			606,278
,, ,, to Persia 5,000 (Kotur)			
	Total		23,559,346
	Independent Arabia		3,400,000
Grand Total, Asiatic Turkey and Arabia			27,059,346

[1] Of these about 800,000 are "Uniates"—that is, in union with the Church of Rome.

[2] Both the number and names of the vilayets and sanjaks have been subject to considerable changes from time to time. The arrangement here adopted is substantially that of Mr. Redhouse.

[3] Before the earthquake of 1881. Present population about 40,000.

TOWNS WITH UPWARDS OF 4000 INHABITANTS.[1]

	Pop.		Pop.
Smyrna	150,000	Tripoli	25,000
Damascus	130,000	Rhodes	20,000
Aleppo	90,000	Gaza	20,000
Scutari	80,000	Homs	20,000
Beyrût	75,000	Medina	20,000
Brusa	70,000	Riad	15,000
Bagdad	60,000	Mardin	15,000
Kaisarieh	60,000	Ismid	15,000
Maskat	60,000	Bitlis	12,000
Erzerum	55,000	Saida	12,000
Trebizond	50,000	Nablus	10,000
Sana	50,000	Latakia	10,000
Mosul	50,000	Tarsus	10,000
Mecca	45,000	Jaffa	10,000
Diarbekr	45,000	Haïl	10,000
Urfa	40,000	Hillah	10,000
Manissa	40,000	Kurnah	10,000
Tokat	40,000	Rowandiz	9,000
Angora	38,000	Taïf	8,000
Edessa	35,000	Mokha	7,000
Aidin	35,000	Nazareth	7,000
Jidda	30,000	Yanbo	6,000
Aden	30,000	Basra	6,000
Aintab	30,000	Arabkir	6,000
Kiutayah	30,000	Bethlehem	5,000
Jerusalem	28,000	Hebron	5,000
Sivas	25,000	Acre	5,000
Amassia	25,000	Bayazid	5,000
Konia	25,000	Loheia	5,000
Hodeida	25,000		

[1] These figures are mostly approximative.

NOTE.—*The Harra* (see page 114).—These volcanic tracts appear to occupy a wider extent of North Arabia than is generally supposed. In some districts C. M. Doughty found lava beds in the lateral valleys 600 feet deep, with numerous extinct volcanoes and volcanelli. On the route from Kheiber to Hail he crossed a craggy lava field rising to a height of 6000 feet, and forming the divide between the waters running east to the Wady-er-Rumma and west to the Wady-el-Humth. Signs of volcanic action are still seen in the Harra, smoke issuing from the crevices of the Kheiber, and steam from the summit of the Jabel Ethnan ("Travels in North-West Arabia," in *Proceedings of the R. Geographical Society*, 1884).

CHAPTER VI.

PERSIA.

1. *Boundaries—Extent—Area—Iranian Plateau—Coast-line—Islands.*

EAST of the Persian Gulf and of the Mesopotamian basin, which may be regarded as its northern extension, the land rises abruptly to a vast upland region, occupying the whole space between the Tigris and Indus valleys. From its earliest known inhabitants, the Iranian branch of the Aryan race, this region has received the name of the Iranian plateau. In relation to the general highland system of the eastern hemisphere, it must be regarded as forming the connecting link between the great central and western tablelands. For it is united through the Paropamisus and Hindu-Kush eastwards with the Great Pamir, the focus of the Asiatic system, and through the Armenian highlands westwards with the Anatolian tableland, whence the uplands are continued across the Ægean to the Balkan ranges and the Alps, the focus of the European system.

This vast tableland, which has a total area of about one million square miles, presents the form of a trapeze, enclosed on the south by the Arabian Sea, on the north by the Aralo-Caspian depression, eastwards by the Indus valley, westwards by the Persian Gulf and Mesopotamian basin. It is encircled on all sides by distinct mountain ranges, which descend everywhere abruptly to the surrounding waters and depressions, except in the north-west, where they merge in the still more elevated Kurdistán and Armenian highlands. Through these the plateau is supposed to be connected with the Caucasus range traversing the Ponto-Caspian isthmus. But here there is a deep intervening depression through which the Kúr (Cyrus) flows east to the Caspian, while farther west the valley of the Rion (Phasis), draining to the Euxine, forms a less marked line of separation between the two systems.

The Iranian plateau thus forms a clear geographical unit. But ethnically and politically it is a divided land. Although the original home of Aryan peoples, it has for ages been the battlefield of "Iran" and "Turan"—that is, of the rival Caucasic and Mongolo-Tatar races. This struggle, combined with the spread of Islám in the seventh century, has brought about a final rupture of the old Persian Empire, which formerly gave political unity to the land. The eastern section of the plateau is thus at present occupied by the independent States of Afghanistán and Kelát (Baluchistán), the western by all that now remains of the ancient Persian monarchy, which at one time stretched from the Bosphorus to the Indus. And even here the sceptre of the "king of kings" has passed from the old native Persian dynasties to a house of the intruding Turanian race. The usurper Nadir Shah was khan of the Afshár Túrki tribe, and the present ruling family belongs to the rival Qájár Túrki clan.[1]

Within its present limits, as laid down by various treaties with Russia and Turkey, and by the Sístán and Afghan Boundary Commissions of 1870-2, Persia is bounded on the north—1st, by the Russian territory of

[1] Hence the title of the present Shah—

ناصر الدین شاه قاجار

Násr ud-dín Sháh Qájár.

Transcaucasia, the frontier line here following the River Aras (Araxes) for the greater part of its course to the plain of Mogan and the Lenkoran district on the Caspian, which are included in Transcaucasia; 2d, by the south coast of the Caspian; 3d, by the new Russian Transcaspian territory, formerly the Tekke Turkoman country. Here the frontier has not yet been determined by the Russo-Persian Boundary Commission of 1881; but it will probably run from the south-east end of the Caspian, along the Atrak River and Kopet-dagh, through Askabad to Sarakhs on the Tajand River. Westwards, Persia borders on Asiatic Turkey, the limits here following a line already laid down at p. 31. On the south-west and south, the Persian Gulf and Arabian Sea form the natural limits, which again become arbitrary, and in some places quite uncertain, on the east towards Baluchistán and Afghanistán. The line has been drawn in the south from Gwatar on the coast to the Maskid River, between which and Lake Sístán it is somewhat vague. Farther north it runs nearly under the 61st meridian to Ghurian, beyond which it follows the Hari-rúd to Sarakhs. It will thus be seen that in the south the frontier line should be drawn much farther to the east than is the case in most English maps, so as to include a large slice of Makrán (South Baluchistán) and most of Sístán proper, which has always been claimed and is now held by the Shah's government.

With this rectification of the east frontier considerably more than one-half of the Iranian plateau belongs to Persia. The Wazir of Karmán has even lately received the title of Sardár of Baluchistán, and attached to his government are the two large districts of Bampúr and West Makrán, which are practically Persian territory. Including these outlying tracts, the monarchy forms an irregular triangular mass with a base running from below

Mount Ararat for about 1000 miles south-east to the Gulf of Omán, and with nearly equal sides of 700 miles north and east from Ararat to Sarakhs, and thence to the south coast at Gwatar in 61° 30′ E. long. Its contour has been compared to that of a cat on a footstool, and as Persia is specially famous for its cats, the fitness of the resemblance cannot be denied. The total area is about 610,000 square miles, with a population variously estimated at from five to ten millions, or from eight to sixteen to the square mile.

Notwithstanding its extensive oceanic coast-line of over 900 miles from Fao to Gwatar, Persia is almost destitute of islands. In the Arabian Sea scarcely a reef exists, and in the Persian Gulf, besides a few rocks and the small but important island of Karak, nothing but the Kishm group of islands claimed by the Sultan of Omán. Off the Caspian coast also there is a total absence of islands, and even here the little rock of Ashurada in the south-east corner has been ceded to Russia. The importance of Ashurada as a Russian station is considerable.

2. *Relief of the land: Highlands—Plains—Deserts—The Kavírs.*

Since the surveys of Khanikoff, Lovett, St. John, and others, between the years 1858 and 1876, our notions regarding the extent, direction, and elevation of the Persian mountain systems have been fundamentally modified. Yet the old ideas still hold their ground in popular treatises, which continue to represent the country as mainly a vast sandy plain fringed on the north and west by continuous escarpments. The truth is that the land is almost everywhere traversed by lofty ranges to such an extent that the strictly highland would seem to prevail greatly over the plateau formation. These ranges, which, with few exceptions, run with surprising regularity in the direction from north-west to south-east,[1] are far higher than was supposed, and so perfect is the parallelism that it actually influences the direction of the atmospheric currents in all the central and western provinces. The disposition of the ranges, especially in the interior, is still far from being perfectly understood; but we now know that some of the ridges run for over 100 miles at mean altitudes of from 8000 to 10,000, rising in some places to 16,000 or 17,000 feet. The most extensive and loftiest seems to be the Kúh-Dinár,[2] traversing the western province of Fars, in the normal direction, at an elevation of perhaps 17,000 or 18,000 feet. Although still unexplored it is perfectly visible from the Persian Gulf at a distance of 130 miles over intervening coast ranges known to be 10,000 feet high. Yet this is about the height given on many old maps to a doubtful Mount Daena, assumed to be the culminating point of the Kúh-Dinár. So also the volcanic Damávand, highest peak of the Elburz chain fringing the south coast of the Caspian, usually marked 14,700 feet on the maps, has been fixed by the Russian Caspian Survey at 18,600, and Mount Savalán, between Tabríz and the Caspian, has been raised by the same authority from 11,000 to 14,000 feet. The *Kuru-Kúh* range, running south-eastwards to Yazd, maintains for a long way a height of 10,000 feet, and is

[1] It is curious to note that the same direction was followed by the repeated shocks of earthquake felt in September 1881 in the Khoi district, Azairbiján. The name Azairbiján (corresponding with the ancient Media) means "place of fire," from the old Persian word آزر (Azur), "fire," like Terra del Fuego.

[2] *Kúh* or *Kóh* (کوه) is the Persian term for mountain, as in *Kóh-i-Núr*, "Mountain of Light." Like the Arabic *Jebel* and Túrki *Dagh*, it is used also for a continuous chain.

continued towards the volcanic Kúh-i-Basmán (10,000 to 12,000 feet), by the snowy Kúh-Banán, and other lofty ridges, culminating with the Kúh-Hazár (14,550 feet). South-east of this point the Kúh-i-Naushádа volcano in Sarhád rises to 12,000 or 15,000 feet.

In the Bampúr or south-eastern corner of Persia the normal north-westerly direction is broken by the coast ranges, which run either south-west or west and east, parallel with the sea. The only other important exception to the general parallelism occurs in the north, where the eastern section of the Elburz sweeps round the Caspian in a north-easterly direction from Mount Damávand to the valley of the River Gurgan.

In the north-west the separate ranges merge in the general highland systems of Luristán, Kurdistán, and Armenia, where several snowy peaks fall little short of 15,000 feet. In the north-east the Khorasán frontier is usually supposed to be separated from the Turkestán depression by a continuous range running between Afghanistán and the Caspian, and connecting the Hindu-Kush through the Paropamisus and Ghor mountains with the Elburz range. But here also the main direction is south-east and north-west from the Hari-rúd valley to the Great and Little Balkans near Krasnovodsk on the Caspian. Thus the Kuren-dagh,[1] the Kopet-dagh, and the other unsurveyed sections of the north Khorasán highlands, run, not in a continuous line, but rather at an obtuse angle with the north Afghan system, and the break of continuity is well marked by the valley of the Hari-rúd and Tajand, giving easy access from Turkestán to Herat. It follows that the northern scarp of Khorasán forms an eastern continuation, not of the

[1] Although from 8000 to 11,000 feet high, the Kuren-dagh was scarcely known till its rediscovery by V. Baker in 1873 (see *Clouds in the East*, p. 289).

Elburz, but of the Caucasus, a fact which has only recently been determined beyond doubt.

The Persian or western section of the Iranian plateau has thus a mean altitude of probably not less than 5000 feet, with a general tendency to rise towards all the seaboards and the western and northern frontiers. Between the coast ranges and the sea there are scarcely any lowland or alluvial plains except those of Khuzistán at the head of the Persian Gulf, and a few strips north of Bushahr and east of Bandar-Abbas. On the Caspian also the only alluvial tract of importance is the delta of the Safíd-rúd, noted for its great fertility. But in the interior, besides the rich plain of the Urmia basin in the extreme north-west, extensive level tracts everywhere occur between the parallel ridges. Those of Isfahán and Shiraz in the west maintain an altitude of about 5000 feet, rising south-eastwards to perhaps 6000 feet. But eastwards and north-eastwards the land falls continuously to the two great depressions of Sístán and Khorasán, probably not more than 1300 or 1400 feet above sea-level. Here the plains become more extensive, but also more arid, the grassy valleys gradually yielding to sandy and saline swamps and wastes. Eastwards a perpetual struggle for the mastery seems to be going on between the arable tracts and the shifting sands, which have already absorbed even some towns and villages, such as Rhages south-east of Tehrán, and which are now threatening to swallow up Yazd in the very heart of the country.

Still farther east the sands themselves yield to those dreary saline marshy tracts so characteristic of East Persia, and which are termed *Kavír* in the north and *Kafeh* in the south. Here all the water from the surrounding streams and from the slight rainfall is collected in the depressions, where it forms a saline efflorescence with a thin whitish crust, beneath which the moisture is retained for a considerable time. Thus are produced those dangerous and impassable slimy quagmires, which in winter are covered with brine, in summer by a thick incrustation of salt.

By far the most extensive of these saline wastes is the Dasht-i-Kavír, or Great Salt Desert of Khorasán, which, with its southern continuation the Desert of Lút in Karmán, occupies probably the greater part of East Persia. The northern desert, which is much more salt than the southern, and apparently separated from it by a distinct water-parting, is divided into two great and several minor sections, drained by the Shuráb, Karasu, and other streams, which unite to form the Great Kavír. There are some other large formations of a similar character, north of Kúm, west of Yazd, and south of Khaf, while "the ordinary kavírs are innumerable" (*St. John*). Their mean elevation above the sea seems to be little over 500 or 600 feet, and some authorities have even asserted that the Great Kavír is actually at a lower level than the Caspian.

3. *Hydrography: Inland and Seaward Areas of Drainage—The Atrak and Tajand Rivers.*

In any case it cannot be doubted that the greater part of the interior has a distinct inland drainage like that of so many other Asiatic tablelands. For while the average elevation of the plateau is about 5000 feet, it rises to 8000 towards the Tigris valley and all the surrounding seas. In fact, the true basin-like character not only of Persia but of the whole Iranian plateau is fully established by a comparison of the inland and outer areas of drainage. Of this plateau about 230,000 square miles drain to the Indian Ocean, and 250,000 to the Aralo-Caspian depression, leaving no less than 550,000 to the inland drainage. Of this area over 200,000 belong to the Helmand or Sístán basin (160,000 in Afghanistán and Baluchistán, 40,000 in East Persia), and the total inland drainage of Persia proper has been estimated at somewhat less than two-thirds of its whole area, as thus:—

		Square Miles.
To the Indian Ocean		130,000
	Aral and Caspian	100,000
Inland.	Lake Sístán	40,000
	Lake Urmia	20,000
	Kavírs and other depressions	320,000
	Total Area	610,000

The rivers draining the western and south-western uplands to the Indian Ocean diminish in size from north-west to south-east. Thus by far the largest are the Karkhah, Karún, and Jaráhi, flowing from the Kurdistán and Luristán highlands through Khuzistán and Arabistán to the Shat-el-Arab at the head of the Persian Gulf. The Karún, which with the Diz forms a stream navigable to the first range of hills, formerly flowed direct to the sea, but now sends most of its waters through an artificial channel to the Shat-el-Arab at Mohamrah, close to the mouth of the Jaráhi. Farther south the Tab has helped to form the Arabistán delta, one of the most extensive and fertile alluvial plains in Persia. But from this point to the Indus not a single navigable stream reaches the coast. Noteworthy is the Mináb, which, though scarcely marked on the maps, drains all the extensive plains across the hills to the north-east of Bandar-Abbas.

Of those flowing to the Caspian by far the largest is the Kizil-Uzún (Safíd-rúd, or "white river"), which drains an area of 25,000 square miles east and south of Lake Urmia. The opposite or south-western coast of the Caspian is reached by the Gurgan and Atrak, the latter of which possesses great political importance, as marking the possible future Russo-Persian frontier line in this direction. In the absence, perhaps, of actual information, its course has been variously laid down, apparently according to the political proclivities of the cartographers. But it is now known to be identical with the Germeh-rúd (Germe Khaneh), which rises near Kabushán (Kushán) on the southern slope of the Kuren-dagh, 6000 feet above the sea, about 58° 50′ E. long., and 37° 30′ N. lat. It flows thence mainly north-west through Shirván and Bújnúrd along the southern base of the Kuren-dagh and through the Goklan Turkoman country to the Caspian at Hasan Kuli Bay. Although over 300 miles long, the Atrak is scarcely more than 30 feet broad at its mouth, except during the spring floods, when it overflows its banks to a width of 7000 or 8000 feet. At other times it is nearly exhausted by irrigation canals and evaporation before reaching the Caspian. Near Kabushán (Kushán) also rises the Kashaf-rúd, which, however, flows south-east past Mashhad to the left bank of the Hari-rúd, their junction forming the Tajand. This river, which offers the most accessible approach from Turkestán to Herat, does not end in the sands near Sarakhs, as is generally supposed, but expands into a swamp in the Attok country about 58° E. long. With a sufficiency of moisture it would doubtless reach the Caspian either directly or through the Usboi, or old bed of the Oxus, between the Little Balkan and Kopet-dagh.

The inland drainage, notwithstanding its vast extent, receives no rivers of any size, and most of them become brackish before losing themselves in the lakes or the desert. The largest are the Aji-chai and Jaghatu, flowing

MESOPOTAMIA, SYRIA AND PALESTINE, &c.

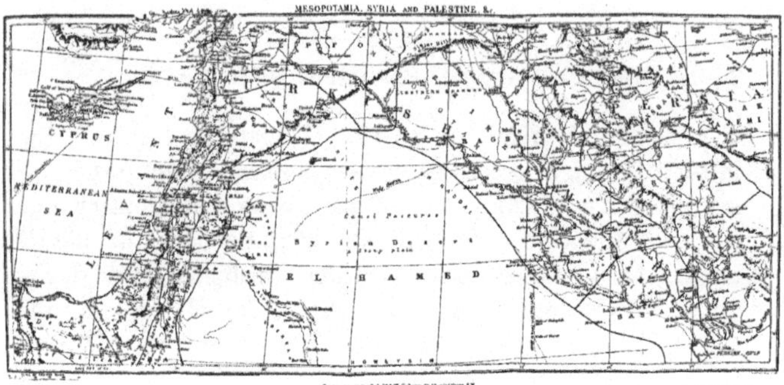

London: Edward Stanford, 26 & 27, Cockspur St. Charing Cross, S.W.

to the salt lake Urmia; the already-mentioned Kara-su (Hamadán-rúd) and Shuráb disappearing in the Great Kavír, the Zainda-rúd watering the Isfahán district, and running dry in the unexplored salt marsh of Gavkhána; the Kúr (Bendamír), chief feeder of the salt lake Bakhtegan, whose true name is Nírís; lastly, the Mashkid, which filters rather than flows along the Baluchistán frontier northwards to the Hamun-i-Mashkel in the Karan desert, which is separated by a range of hills from Lake Sístán.

4. *Natural and Political Divisions: Irak—Sístán—Khorasán—The Caspian Provinces—Urmia.*

From the foregoing account of its physical constitution, it is evident that the western section of the Iranian plateau presents two well-marked natural divisions only—the central and lowland plains, and the highlands, by which they are enclosed on every side except towards the Helmand basin. But while the plains are mainly comprised in the two great provinces of Khorasán and Karmân, the uplands with the narrow intervening coast strips are, for administrative purposes, subdivided into nine other governments, as shown in the statistical tables at the end of this chapter.

These "Memleket," as they are called, are grouped round Irak-Ajemi, which forms the political centre of the State. Here are situated both Isfahán and Tehrán, the old and new capital. Irak slopes from the Kurdistán highlands eastwards down to the Khorasán wastes, and rises northwards to the Elburz range, separating it from the Caspian. Southwards it reaches to the Kúh-Dinár range, thus including in its general administration the subordinate divisions of Ardalán, Luristán, and Kashán. Here are some rich grassy plains and fertile valleys, which when well watered yield excellent crops of cereals and fruits. But in the east most of the land is waste, and already invaded by the sands continually advancing westwards.

North and south of Irak-Ajemi lie the provinces of Azairbiján, Luristán, and Khuzistán, the latter including the rich alluvial plain of Arabistán at the head of the Persian Gulf. Along the Persian Gulf and Arabian Sea stretch the extensive governments of Farsistán, Karmán, and that portion of Makrán assigned by the Baluchistán Boundary Commission to Persia. These coast regions consist of lofty highlands rising in terraces rapidly inland, and with their main axes running north-west and south-east everywhere except in Makrán, where they run partly south-west and north-east, partly west and east parallel with the coast.

North of Makrán, and almost in the very centre of the Iranian plateau, lies the deep depression of Sístán, now partly included in Persian territory, but geographically belonging mainly to the Afghan drainage system. It is an extensive level and low-lying tract on the borders of Afghanistán, Persia, and Baluchistán, partly filled by the Hamún (Sístán) Lake or swamp, which receives the Helmand, Farah, and other large rivers from the east, but only a few insignificant streams from Persia. The basin does not, however, form a single expanse of water, but is divided into the three depressions fed by the Farah, Helmand, and Zirreh.[1] The so-called plain of Hamún is generally dry, and the presence of a large lake at this spot, as marked on most maps, can be explained only by supposing that in spring a few pools or tarns are formed in the channels about the river mouths, and occasionally united by floods in a continuous sheet of water some 70 miles long by 25 broad, but

[1] The Zirreh, formerly supposed to drain south to the Arabian Sea, flows, on the contrary, north to the Hamún-i-Sístán basin.

scarcely ever more than three or four feet deep. Its margin is covered with a dense growth of reeds, tenanted by numerous herds of wild hogs. The water is fresh about the river mouths, but elsewhere brackish, while its bed seems to be gradually rising, owing to the mass of detritus and mud brought down by its influents. Although fish are scarce, it is much frequented by geese, duck, and other water-fowl.

Sir Frederic Goldsmid of the English Boundary Commission distinguishes two districts in this region—Sístán proper and Outer Sístán. The former, with an area of perhaps 980 square miles, has a settled population of 35,000, besides 10,000 nomads,—one-third Persians, Baluchis, and Afghans, the rest "Sístánis." The country is generally flat, with a sandy alluvial soil, growing shrubs, but no trees. There is no lack of irrigation by means of rills and rivulets, and the land is fertile, yielding melons in profusion, besides the staple products, wheat and barley, and excellent pasturage.

Outer Sístán comprises the country stretching along the right bank of the Helmand some 120 miles farther up, and properly forming part of Afghanistán.

The whole of the north-east is comprised in the vast province of Khorasán, which was carefully explored by a Russian expedition conducted by Khanikoff in 1858, and since then visited by Baker, Napier, MacGregor, and several other travellers. Its eastern section, contrary to the general impression, has been found to be very hilly, and the southern portion even bears the name of Kuhistán,[1]

[1] From *Kúh*, mountain, and the ending *stán* or *istán*, so universal in Persian geographical nomenclature. This ending has the general meaning of *country*, as in Farsistán, Afghanistán, Turkestán, etc. It comes from the Aryan root *tan*, as in the Latin *tendo*, with the primary idea of extension, whence a large open space, a plain, and land in general. In this sense it has travelled with the spread of the Aryan race eastwards to *Hindu-stán*, westwards to *Aqui-tania*, and *Bri-tania*.

or "Highlands." But between the ranges there extend broad tracts of waste lands eastwards to Afghanistán, westwards to Irak-Ajemi.

In this region traces are still everywhere met of the recent famine as well as of the abject fear hitherto inspired by the neighbouring Turkoman hordes, whose predatory raids the feeble Persian Government was powerless to resist. But this source of trouble has been lessened since the Russian occupation of the Attak in 1881. But so great was the distress from hunger that it got the better of the intense fear with which the people regarded their hereditary foe. Bellew tells us that the inhabitants of Mashhad crowded out of the gates of the city in the hope of being seized and carried into captivity by the Turkoman marauders, prefering a crust of bread in exile and slavery to a lingering death by starvation at home. The Turkomans spared none but the Arabs, paid no respect to sex or age, and all unable to pay a sufficient ransom were carried off and publicly sold in the slave-market of Khiva before its suppression by the Russians in 1874. But since the Perso-Turkoman has become a Perso-Russian border-land, something has been, and will yet further be, done towards the suppression of slavery and of predatory raids within Turkestán. And the establishment of order there must sooner or later lessen, if not stop, the practice whereby the Turkomans carry off Persians into slavery.

The open country visited by Bellew was found to be dotted over with a peculiar kind of tower, formed by a round mud wall 14 feet high enclosing an empty space open to the sky, and with a low entrance accessible only on all fours. The moment the Turkoman horsemen were detected, the people took refuge with their flocks in these buildings, which offered them a safe if temporary refuge.

The provinces of Ghilán and Mazandarán comprise the wooded northern slopes of the Elburz, besides a more or less extensive strip of flat alluvial coast-land between that range and the Caspian. This tract, often swampy and exposed to deadly fevers, and producing chiefly rice, cotton, silk, and some sugar-cane, is mostly covered with dense forests, like the neighbouring mountains themselves. Herein it presents a striking contrast to the bare, desert or arid regions to the south of the Elburz—that is to say, to the rest of Persia, which has been caustically described as a land divided into two sections—a salt waste and a saltless waste.

The extreme north-west between the Caspian and Turkey—that is, the "cat's head" in the general contour—is comprised in the province of Azairbiján. It is partly cut off from the Caspian by the Russian district of Lenkoran, and the Russo-Persian frontier is here traced by the River Aras almost from the foot of Mount Ararat nearly to its junction with the Kúr. In this upland region, where Mount Savalán attains an altitude of 14,000 feet, the great feature is the remarkable closed basin of Lake Urmia, alike interesting in a geological, ethnical, and economic sense. In this comparatively narrow tract several streams rise almost in close proximity, which nevertheless flow in four opposite directions—the Safíd-rúd east to the Caspian, the Kara Rud north to the Aras, the Aji-chai west to Lake Urmia, the Zab south-west to the Tigris. Here also, after a lapse of thousands of years, the surrounding antagonistic ethnical elements have hitherto failed to establish an equilibrium—Turkoman, Kurd, Nestorian, Armenian, and Persian still struggling for the supremacy, and apparently unconscious that the shadow of the northern colossus has already fallen on the land. The recent ravages of the Kurdish marauders in this district have already been referred to.

The lake, which although 4750 feet above the sea is a completely closed basin some 80 miles long by 20 broad, is extremely salt and very shallow. The average depth scarcely exceeds 6 feet, and is nowhere more than 24. It lies in a district of almost unrivalled fertility, covered with vineyards, orchards, gardens, rice grounds, and thickly studded with towns and villages. Urmia, the largest of these, whence the lake and district take their names, is the centre of an American mission, which has for many years worked earnestly and successfully in the cause of true progress and enlightenment.

5. *Climate: Rainfall—Prevailing Winds.*

The climate of Persia is on the whole continental, great dryness being combined with excessive heat, and in many of the uplands with extreme cold. On the northern ranges snow falls as early as November, and it sometimes freezes in Tehrán as late as the middle of March. Between these ranges and the Caspian the heat is almost tropical, with an abundance of rain, resulting in the rich and varied vegetation of Ghilán and Mazandarán. The sultry and unhealthy climate of the Persian Gulf seaboard has already been noted (p. 127). That of Sístán in the extreme east is also very unhealthy, and subject to great extremes of temperature.

North-west and south-east winds prevail throughout the year with great uniformity, their direction being largely determined by the Black and Arabian Seas at these two quarters, and by the remarkable parallelism of the intervening mountain ranges. The atmosphere of the central plateau being rarefied by the great heat of the sun, the cooler currents from the Euxine and Indian Oceans set in to fill up the vacuum; and as the former are the colder of the two, the north-west naturally prevail over the south-west winds. On the south-west coast these two currents often meet, so that a gale from the north-west is often raging at Bushahr in the Persian Gulf when Bandar-Abbas on the Strait of Hormuz is exposed to the fury of a south-easter. As most of the moisture is also brought from the latter quarter, it follows that the prevailing winds are dry, especially as the rain-clouds from the Black Sea and Caspian are mostly arrested by the Armenian and Elburz highlands, while much of the moisture from the Indian Ocean itself is precipitated on the southern and western coast ranges. Hence, excluding the Caspian basin and a few other more favoured tracts, the average annual rainfall on the Persian plateau is probably less than ten inches, and in the eastern Kavír region and Sístán, not more than half that quantity. "Were it not that the lofty hills store the moisture in the shape of snow, nine-tenths of Persia would be the arid desert that half of it now is. As it is, cultivation over the greater part of the country is possible only by artificial irrigation, either by canals or by the system of wells connected by underground channels called *Kanát* or *Karíz*, and peculiar to the Iranian plateau" (*St. John*).

6. *Flora and Fauna : The Camel.*

The results of this deficient rainfall are seen, not only in the undeveloped water system, but also in the vegetation, which is characterised by the absence of trees and even large shrubs almost everywhere except on the outer slopes of the coast ranges. The date-palm flourishes along the sandy shores of the Persian Gulf, but the oaks and other trees of the Bakhtiari and other inner ranges are mostly stunted, and true forests are found only on the northern slopes of the Elburz. Here large tracts are covered with dense plantations of magnificent timber, especially cedars, elms, oaks, the walnut, beech, and the valuable box tree. Wheat and barley are here cultivated to a height of several thousand feet, while the lowlands yield cotton, sugar, silk, grapes, figs, cherries, peaches, plums, and other fruits, in great profusion. Indigo, rice, tobacco, and madder are also cultivated in this region, as well as in the Urmia basin, and on the Isfahán and Shiraz plains, which are almost the only other really fertile tracts in the whole kingdom. Pasture lands are much more extensive, occupying most of the elevated longitudinal valleys and slopes of the parallel ranges in the west and north-east. Hence the Kurdish, Lúr, Farsistán, and North Khorasán highlands have been held for ages by nomad pastoral tribes both of Iranian and Turanian stock. The eastern low-lying plains of Khorasán and Karmán are almost destitute of vegetation, producing little beyond sands and salt. In Sístán, tamarisks and dwarf mimosas are a prevailing feature.

Farsistán is still haunted by the lion, while the tiger, leopard, chitah (used for hunting), hyæna, wolf, lynx, jackal, and some smaller beasts of prey, infest the northern provinces skirting the Caspian seaboard. The *Capra ægagrus*, the supposed ancestor of the domestic goat, is spread over the whole Iranian plateau, and the bustard (*Otis houbara*), here indigenous, is hawked or followed with the gun. Here also the pheasant is indigenous, and the woodlands are enlivened by the song of blackbird, thrush, and bulbul or Eastern nightingale. Fish abound in the Persian Gulf and Caspian, and the sturgeon fisheries of the rivers flowing to the latter sea are very productive. But fresh-water fish are rare, and Urmia and the other lakes are almost destitute of animal life.

With the exception of the Mazandarán black cattle and the familiar "Persian Cat," the domestic animals are mostly of inferior stock. The goats, however, of Karmán yield a hair equal almost to that of Kashmir, while the fat-tailed sheep supplies the chief staple of animal food. Its wool, also, is of good quality, and either woven into fabrics of various sorts or else left on the skins, which are then cut into garments much affected especially by the nomad Iliat tribes. The chief beast of burden is the mule—a strong, hardy, and sure-footed animal, well adapted to the rude tracks in the highland districts. The camel is also employed for the caravan trade across the sandy plains, and there is a useful breed of small horses, crossed with Arab blood and noted for their speed and shapely forms. The hair of the camel forms the woof and cotton the warp of the camel's hair cloth for which Persia is famous. It is woven very closely so as to be quite waterproof. The camel is not only a very unmanageable beast, but also extremely timid and scared by the least unwonted sound or sight. The jambaz, or riding camel, is, however, an exception, and this breed is also noted for its speed and endurance.

7. *Inhabitants: The Modern Persian.*

In its ethnical constituents Persia presents a most striking contrast to the Arabian peninsula. Here all the inhabitants belong to one primeval stock, with scarcely any intrusion of foreign elements. But in Persia not only are the two fundamental Asiatic types fully represented, but several distinct branches of each divide the land into a number of ethnical groups presenting almost as great a variety of races as is found in any other part of the continent. This will be at once evident from the subjoined table of the inhabitants of Persia classed according to their several racial affinities:—[1]

[1] The justification of these and subsequent groupings will be found in the Ethnological Appendix to this volume.

CAUCASIC TYPE.	ARYANS.	*Iranic Branch.*	Tajiks (Persians).
			Kurds.
			Lúrs.
			Laks or Leks.
			Baluchis.
		Indic Branch.	Gipsies.
			Jats.
		Haïk Branch.	Armenians.
	SEMITES.		Arabs.
			Jews.
			Chaldeans ("Nestorians").
MONGOLO-TATAR TYPE.	MONGOL BRANCH.		Aymaks.
			Hazárahs.
	TÚRKI BRANCH.		Turkomans.
			Kizil-Bashis.

PERSIAN LADY AT HOME.

Nevertheless the bulk of the people still belong to the old indigenous Iranian stock. These western Iranians, or Persians proper, are everywhere throughout Central Asia known exclusively as Tajiks, and in West Irania as Tats, possibly a contracted form of the same word. In the north and east Turan has largely encroached on Iran; but elsewhere the old race has held its ground, hemmed in between the Arabs on the west, the Armenians on the north, and the Turkoman tribes on the north-east and east.

PERSIAN LADY IN WALKING DRESS.

In religion the Persians belong almost exclusively to the Shiah sect, and often harbour feelings of rancour towards their western neighbours the Sunnite Turks. The Persians, often called Qájár,[1] from the tribal name of the reigning dynasty, and usually recognised by the *kulah*, or characteristic black lambskin head-dress, are very extravagant in their dress, the *jube*, or outer garment, often costing from £40 to £50. On the other hand, they seem to be economical in respect to under linen. In fact, when judged by the Persian standard, the Turk himself appears to be a model of cleanliness.

The domestic garb of the women is unattractive, the smock reaching only to the hips, from which hangs a short and very wide skirt.

The Persians, especially of pure blood, have readiness of wit and persuasiveness of manner. More nervous in action, more animated in conversation, and of quicker apprehension, yet in their moral and mental temperament they stand on a lower level than the descendants of the Tatars and "White Hordes." The Turk is a man of few words and serious speech, the Persian is at once a fluent rhetorician and a skilful sophist. He possesses more taste and a greater natural sense of beauty than the Osmanli; in these respects often betraying a strong likeness both to the Greek and the Jew. And though he is apparently more fanatical than the Sunnite Muhammadan, yet European ideas ought to find more acceptance with him than under the sway of the Crescent. The splendour and the power of the state seem with the Turk to be bound up with his religion, for "the glory of Islám is the glory of the Osmanli." But not so with the Persian, whose forefathers were Persians before the appearance of Islám, and whose nationality had already acquired a recognised political status ages before the days of the Prophet. The Ottoman, again, is a stock-breeder, a husbandman, and a soldier; the

[1] Pronounced in some districts "like our word cudgel" (*St. John*).

Persian, above all, a trader and an artist. And that the natives of Irania descend from an ancient and long-civilised race is agreeably evident to the stranger in the politeness, courteous and even refined demeanour of the people, whether they belong to the urban or rural classes.

But from the old despotic systems they have inherited the taint of cruelty. The savage sentences imposed, especially for murder, theft, and political offences, are carried out in a cold-blooded manner, which implies that some feelings at least have long been deadened to all sentiments of humanity. A frightful story is told of a slave in Shiraz, twelve years old, who had accidentally shot his master's son, and who was sentenced by the governor to be crucified. Here also Bower witnessed the execution of eleven robbers in one morning. The criminals being arranged all in a row and smoking a *kalián*, the executioner walked up, slipped their heads under his left arm, and cut their throats one after the other in the most matter-of-fact way.

8. *Topography: Tehrán—Kúm—Isfahán—Shiraz—Persepolis—Mashhad—Kelát—Tabríz—Seaports.*

From the outward conditions of soil and climate it naturally follows that nearly all the settled population and large towns are found concentrated in the western provinces, where the land contracts between the Caspian and Mesopotamian basin. The desert region east of the 53d meridian, comprising about two-thirds of the kingdom, contains scarcely any noteworthy places except Mashhad, Tabbas, Yazd, and Karmán, forming so many stepping stones across the saline and sandy wastes from north to south. But west of that line are situated not only the mediæval and modern capitals, but also the ruins of the ancient Persepolis, besides Shiraz, Rasht, Kasvin,

Tabríz, Bushahr, Shustar, Karmanshah, Hamadán, Kashán, and several other towns, which either still are or have been important centres of trade and culture during the long annals of the Persian Empire.

GATEWAY, TEHRÁN.

Tehrán, the present capital, lies far to the north, almost at the foot of the Elburz mountains, where they culminate in the majestic Damávand. From the summit of this quiescent volcano the city is perfectly visible.

lying in the midst of an arid steppe, apparently one of the most unlikely spots to form the political centre of a large monarchy. Although it has been the capital since the year 1788, Tehrán has scarcely a respectable building to show except the quadrangular palace of the Shah, absorbing nearly one-fourth of the enclosed space, and the mansions occupied by some of the nobles and the European legations. The streets are mostly narrow, crooked, and badly paved, and lined with mean houses, whose uninviting exterior corresponds with their miserable internal appearance and fittings. The bazaars, however, contain a good show of the various artistic objects for which Persia has at all times been famous. Thanks to its political importance, Tehrán has considerably increased in size of late years. The old walls, four miles in circumference, have everywhere been encroached upon, and the new quarters have now been enclosed by an outer wall and ditch enclosing a space much larger than the whole of the old town. In summer, when the heat is almost intolerable, the Court, embassies, and wealthy citizens retire to Gulahak and other pleasant retreats on the neighbouring hills. The road to these places passes the Kasr-i-Qájár, or "Palace of the Qájár," which, though now seldom occupied by the Shah, stands on an imposing site in the midst of beautiful grounds, that have been compared by Oriental fancy with those of Versailles.

About 85 miles on the road from Tehrán to Isfahán lies the town of Kúm, which is held next in sanctity to Mashhad "the Holy." Here is the famous shrine of Masúma Fatima, the sister of the Imám Riza, the gilded dome of which has been completed by the present Shah, and which also contains the remains of ten kings and 444 "saints." It is usual to visit this shrine before proceeding to Mashhad or Kerbela, and Kúm has become a favourite spot for the interment of the Faithful, whose bodies are brought hither from great distances. But the town itself is mostly in ruins, of its 20,000 houses not more than 4000 being at present occupied. "Its streets and bazaars are deserted, and dangerous from the innumerable holes and pitfalls with which they abound; and its general condition provides an impressive commentary on the state of absolute stagnation, which seems to be one of the chief characteristics of the Muhammadan religion" (*Major E. Smith*).

On the western border of Irak-Ajemi lies Hamadán, the ancient Ecbatana, where the Jews still show the tombs of Esther and Mordecai. Farther south, the apex of an isosceles triangle, whose base connects Hamadán with Tehrán, marks the site of Isfahán, the mediæval capital of the kingdom, and the centre of Muhammadan culture in Persia. Isfahán, which was at one time probably larger than any of the old or more recent capitals, lies in a pleasant, well-cultivated plain, almost midway between the Caspian and Persian Gulf, and between Karmán and the Turkish frontier south-east and north-west, about 300 miles from all these points, consequently in the most central habitable part of the State. Notwithstanding the many calamities it has suffered and the loss of prestige following the removal of the seat of government northwards, it is still a large place and the centre of many flourishing industries. It is still adorned by several magnificent edifices, dating mostly from its former periods of prosperity, conspicuous amongst which are the large royal palace, the Chhar Bagh, the royal mosque (Masjid-i-Shah), said to be the most sumptuous in the whole Muhammadan world, and the great bazaar of Shah-Abbas. Under Shah-Abbas (1587-1628), who made it his capital, Isfahán reached its greatest splendour, and had at that time no less than 1800 caravansarais, 270 public baths, over 100 large mosques, and a population estimated at 750,000.[1] Even still it ranks with the foremost cities of the East, and, according to the local saying, but for Lahore it would be equal to half the world. It suffered severely during the famine of 1871, but has since then sufficiently revived to give the general impression that it must have fitly represented the regal state and grandeur of modern Persia.

Shiraz, capital of Farsistán, occupies one of the most favoured sites in Persia, at an elevation of 4500 feet above the sea, about 220 miles south of Isfahán, and 120 east of Bushahr, its port on the Persian Gulf. Nestling amid rose gardens, vineyards, and cypress groves, Shiraz, although like Isfahán a mere shadow of its former greatness, still retains a certain importance, due largely to its excellent wine, in flavour like the royal Hungarian Tokai, and to its rosewater and attar of roses industries. Its delightful situation has been the everlasting theme of the Persian poets, and the first sight of its soft dark-green vegetation above which towers the lofty dome of the Shah-Cherak mosque, is naturally calculated to produce some enthusiasm after the traveller's eye has lighted for weeks together on nothing but arid sandy wastes. The abundance of water here produces a flora of tropical luxuriance, and to the charms of a magnificent and varied vegetation are added those of a limpid blue sky and a perennially mild atmosphere. Unfortunately a soft climate, a fertile soil, and an easy life, have had an enervating effect on the inhabitants.

South and east of Shiraz are the two salt lakes Mahalu and Bakhtegan (Niris), and 25 miles to the north-east lie the extensive ruins of Persepolis. Conspicuous amongst them is the so-called palace of Darius,

[1] Even when captured by the Afghans under Mir Mahmúd in 1722, "it was esteemed the largest and most magnificent city in Asia, with 600,000 inhabitants" (Jonas Hanway, iii. 122).

said to have been destroyed by Alexander, and occupying a terrace 1430 by 800 feet, approached by steps cut in the rock. Vast portals and sphinxes, with many still standing pillars and walls covered with sculptures and cuneiform inscriptions, still attest the former magnificence of the royal palace of the Achæmenides.

In the north-east the only noteworthy place is "Mashhad-i-mukaddas," or "Mashhad the Holy," capital of Khorasán, and the religious and trading centre of East Persia. Next to Mecca and Kerbela, this is the most hallowed spot in the Moslem world, for here reposes under a gorgeous gilded dome their most revered saint, the Imám Riza. His shrine, to which no "infidel" is allowed access, is yearly visited by over 100,000 votaries from all parts. Although slumbering in his sumptuous tomb for centuries, Riza is still treated as if he were actually living. "His shrine is enormously rich, possessing land and property in all parts of Persia, and attached to it is a large establishment of officials and servants" (*Major E. Smith*). This traveller adds: "Holy as Mashhad is said to be, we were struck with the great amount of drunkenness prevalent there amongst the followers of the Prophet."

About 50 miles north of Mashhad, and 60 west of Sarakhs, in 37° N. lat. and 60° E. long., lies the extraordinary natural fortress of Kelát, about 3400 feet above the sea, and close to the new Russo-Persian frontier. Very little was known of this marvellous place until it was recently visited by Colonel MacGregor and Valentine Baker, the latter of whom calls it "one of the wonders of the world," describing it as a gigantic stronghold formed by Nature herself, with very little aid from man. "The walls are mountains of from 800 to 1200 feet high, and with a sheer perpendicular scarp between 300 and 600 feet. It is an irregular oblong about 21 miles long by

5 to 7 broad. There are only five entrances, through narrow natural scarps, and these are fortified. The ground enclosed within is very rich, and it might be a perfect garden, and self-supplying. A stream runs right through the place, in at the southern entrance and out at the northern. There are also several springs within the fortress, and an ample supply of good water could thus be obtained for the cultivation of the whole interior. But everything about it now betokens utter ruin and neglect" (*Clouds in the East*, p. 201). Owing to this neglect, the fortress, where a battalion of troops with cavalry and some guns are maintained, has become so unhealthy that the garrison is often decimated by typhus, and constantly deserting its post.

Near the north-west frontier lies Tabríz (Taurus), the largest city and principal commercial emporium of the kingdom. It stands at the base of the high and rocky Mount Sahend, about 5000 feet above the sea, and on the Aji-chai, 36 miles above its entrance into Lake Urmia. Tabríz, which contains no remarkable buildings except the citadel, originally a mosque, over 600 years old, at one time possessed a large number of khans, splendid mosques, public baths, and a population of 550,000, now reduced to one-fifth of that number. The neighbourhood is extremely fertile, producing large quantities of magnificent grapes and other fruits.

Of the seaports, the most noteworthy are Rasht and Barfrush on the Caspian; Bushahr and Bandar-Abbas on the Persian Gulf. Rasht, capital of Ghilán, stands at the head of the shallow lagoon or backwater of Enzeli, where all the shipping stops. It is a thoroughly Persian town, with dirty, close streets, and very unhealthy, as is most of this low-lying, swampy coast. Its importance is due chiefly to its large export silk trade, all the silk of the province being shipped here.

Barfrush lies at the mouth of a large sluggish stream 300 feet broad, here crossed by a solid brick bridge. It is surrounded by dense forests, is noted for its numerous schools and colleges, and does a considerable trade in silk and cotton. The population, said at one time to have amounted to 200,000, has now fallen to less than one-fourth of that number.

Bushahr, the chief port on the Persian Gulf, lies 150 miles from the Euphrates delta. It is the great emporium of the British and Indian trade with the southern provinces; but although presenting a pleasant appearance from a distance, a nearer approach reveals the usual uninviting features of Persian towns. From this point to the Indus the only port of the least importance is Bandar-Abbas, a small place facing the island of Kishm in Hormuz Strait. It is inhabited chiefly by Arabs, who carry on a considerable coasting trade in fish, salt, and fruits; but the heat is so intense that the place is almost deserted in summer. The Sultan of Omán has long claimed jurisdiction over Bandar-Abbas and the neighbouring strip of coast and adjacent island of Kishm.

9. *Highways of Communication.*

In Persia there are one or two good roads of short length—as, for instance, that which runs from Tehrán for a few miles to the villas and villages on the southern slope of the Elburz. But all the rest are mere caravan tracks or bridle-paths, whose character depends more on the nature of the land than on the hand of man. The wretched state of these routes is the universal theme of travellers, who are more surprised to find any attempts at repairs than disappointed at the universal neglect. "The absence of roads is the curse of the country. The whole traffic is carried on by mules on the mountains, and camels on the plains, no wheeled carriages existing" (Baker's *Clouds in the East*).

The main highways, such as they are, run in all directions, and even across most of the kavírs between all the large towns and the Russo-Turkish frontiers. Towards Afghanistán and Baluchistán there seem to be scarcely any recognised tracks, and those that formerly existed have been mostly closed and lost through political jealousies. A Persian army could no doubt again find its way from Mashhad to Herat; but for much of the way the route for baggage and artillery would have to be rebuilt. The English Boundary Commission, coming up from the coast to Sístán, was guided in many places more by compass and the stars than by any perceptible paths, and it would probably be impossible to get from Yazd or Karmán direct to the Helmand basin. Here the best tracks, starting from the Hamún swamp, seem to run through Birjand and Kaian, or more to the west through Tún northwards to Mashhad, south-westwards across the Sarhád country to Bam, where it strikes the path running from Bampúr north-westwards to Karmán, and so on through Yazd, Agda, and Nain to Isfahán, and thence through Kashán and Kúm northwards to Tehrán. Here it would meet the northern route continued from Mashhad through Sabzawar, Shahrúd, Damghan, and Samnán, thus completing the circuit of East Persia. A pilgrims' route from Mashhad to Yazd and Isfahán follows the watershed between the northern and southern deserts, the chief stations being Tún, Tabas, and Gustán, with a branch at Tabas, passing direct either through Ardakan or Nain to Isfahán. The two capitals are connected by trade routes, with Bushahr and Bandar-Abbas on the Persian

Gulf, with Rasht and Enzeli on the Caspian, and through Tabríz and Erzerum, with Trebizond on the Euxine.

After the present Shah's return from Europe in 1875 an extensive railway scheme was projected, which began and ended with a small line of a few miles, opened in 1876 at Rasht. But a tolerably complete telegraphic system has been developed under the direction of Sir F. Goldsmid, by which Persia is brought into direct communication with the rest of the world. The lines are laid down in duplicate, one ashore and one submarine, from Karachi, Indus delta, along the coast to Jashak, whence both are submarine to Bushahr. Here they bifurcate, one branch running through Shiraz, Isfahán, Tehrán, and Tabríz, to the Russian system at Tiflis, the other crossing the Gulf to Fao, and thence running through Bagdad, Diarbekr, and Constantinople, to the various Western systems.

10. *Administration: Social State—Army—Education.*

The Government of Persia has ever been an absolute despotism in the strictest sense, the head of which bears the title of Shah-in-Shah, or "King of Kings." A revenue is raised of about £2,000,000, a sum probably equal to £8,000,000 or £10,000,000 in England. But it proves often insufficient to meet the requirements of the State. The country suffers from defects in the administration, the administration from faults in its subjects, the subjects from disadvantages of soil and climate,—a vicious circle, from which there seems no escape.

Among the physical disadvantages, the drifting of the sands is prominent. "The sands are in many places visibly gaining on the arable land, and even on the walled towns themselves. It is, in fact, in the process of changing from a series of rocky ridges to one of undulating sandy wastes. . . . You see the sand blowing against the wall, gradually getting higher and higher till it blows over, and then forms a mound in the field beyond, which gradually increases its height till all trace of wall and field is lost, and you have before you a sand-heap. I can quite believe now the story of towns being buried, having myself seen the thing on a small scale" (*Col. MacGregor*).

To these physical causes of decay must be added the foreign wars and internecine feuds, by which the monarchy was wasted throughout the whole of the last, and for many years during the present century. On the death of Nadir Shah in 1747 it was distracted by a series of fierce dynastic struggles between the rival Afshár and Qájár Túrki houses, attended by excesses of every kind, which caused Jonas Hanway almost to despair of its future. "These intestine broils," he exclaims, "have extinguished the glory of Persia. What the fate of that wretched country will be Heaven only knows. But this is evident, that the splendour of their monarchy, all their monuments of art and labour, with all the industry of past ages, are swallowed up by the ravages of war. What numbers of their towns, their cities, their fruitful plains and delicious mountains, are become a dreary waste, and the habitation of wolves!" (iv. 301).

The nominal strength of the army is 100,000 men, of whom perhaps not more than one-third are ever under arms at a time. The rest form a sort of reserve, which, though mostly unarmed and engaged in husbandry, are liable to be called out at any moment. Their arms consist of old English or French muskets, supplemented by a few thousands of home make, and perhaps a hundred available guns of small calibre, with a few Uchatius rifled cannon introduced in 1881. The officers are mostly ignorant and untrained, while the men, with their shabby and tattered uniforms, look more like half-starved mendicants or highwaymen than guardians of the peace. They are drilled after the English fashion, but in a very lax way, and are seldom regularly paid. But their physique, being drawn almost exclusively from the hardy Turkoman, Kurd, and Lúri tribes of Azairbiján, Kurdistán, and the Bakhtiari highlands, is magnificent. "It is the concurrent testimony of all who have been connected with the Persian army, that no people in the world present better rough material for soldiers than the Persians" (*Times* Correspondent, Sept. 8, 1881).

Public instruction, which had hitherto been mainly confined to religious teaching, is at present being thoroughly revised and improved. The nucleus of a university was formed in 1881 in Isfahán, where colleges are in course of erection for the teaching of the Oriental and European languages, besides various branches of art and science, mostly under European supervision.

11. *Statistics—Areas and Populations.*

VARIOUS ESTIMATES.[1]

	Area in sq. miles.	Population.
Polak	468,000	5,000,000
Kolb	595,000	5,000,000
Wagner	637,000	6,000,000
Ritter	645,000	5,500,000
Almanac de Gotha	660,000	7,000,000
St. John	610,000	10,000,000
Blackie	396,000	8,000,000
Dickson (1885)	660,000	7,653,000

[1] The prevailing ignorance regarding even the main features of Persia is well illustrated by these widely-diverging estimates. The area (660,000 sq. miles) as given by Dickson, and the population (7,653,000) based on Schindler's careful estimate, seem to come nearest the truth, and are accordingly here provisionally adopted.

APPROXIMATE CLASSIFICATION BY RACES AND RELIGIONS.

Moslem, mostly Shiahs, 6,770,000	IRANIANS.	
	Tajiks and Tats (Persians), all the Towns and Agricultural Districts	5,000,000
	Kurds proper, Persian Kurdistán	300,000
	Mikri Kurds, Azairbiján	50,000
	Shadilu and other Kurds, N. Khorasán ranges	50,000
	Lúri proper, Luristán	370,000
	Bakhtiari Lúri, Pish-i-Kúh	250,000
	Laks or Leks,[1] Fars, Irak, Mazandarán	100,000
	Makráni Baluchis, Sístáni Baluchis } Makrán, Sístán, Karmán	100,000
	TATARS.	
	Túrki Iliats, Irak, Khorasán, etc.	500,000
	Turkomans (Gokláns, etc.), Mazandarán, Astrabád	50,000
	MONGOLS.	
	Taemuri Aymaks,[2] South and East Khorasán	250,000
	Hazarahs,[2] towards Herat frontier	50,000
	SEMITES.	
	Arabs,[3] Arabistán, Fars, Laristán, etc.	350,000
Christians, 175,000	Armenians, Isfahán, Tehrán, Urmia	150,000
	Chaldeans ("Nestorians"), Urmia	25,000
Sundries, 53,000	Jews, the large towns	16,000
	Kizil-Bashis, Khorasán, Karmán	12,000
	Ghebrs,[4] chiefly Yazd	7,000
	Gipsies and Jats, Karmán, Irak, etc.	?20,000
	Total	7,650,000

APPROXIMATE AREAS AND POPULATIONS OF THE MEMLEKET (PROVINCES).

		Area in sq. miles.	Population.
North	Astrabád	10,000	150,000
	Mazandarán	8,000	350,000
	Ghilán	6,000	400,000
	Azairbiján	35,000	1,400,000
West	Irak-Ajemi	115,000	1,300,000
	Ardelán	6,000	150,000
	Khuzistán	30,000	600,000
	Luristán	30,000	300,000
	Farsistán	60,000	1,300,000
South	Laristán	20,000	90,000
	Karmán, with Kohistán, Makrán, and Sístán	150,000	750,000
East	Khorasán	140,000	860,000
	Total	610,000	7,650,000

[1] Many of the Laks, known as "Nasari" and "Ali-Iláhi," reject the Prophet, hence are not regarded as true "Believers."

[2] All Sunnis, although the Hazárahs of Afghanistán are Shiahs.

[3] Many of these Arabs have become Shiahs, and are in other respects also assimilated to the Persians.

[4] Descendants of the old Persian fire-worshippers. Their numbers have been greatly over-estimated. Blackie gives 40,000; but in the town and district of Yazd, where they are chiefly concentrated, Major E. Smith found they had dwindled to 3800 in 1871. They are easily recognised by a uniform turban of a drab or dust colour.

Towns with upwards of 10,000 Inhabitants.

	Pop.		Pop.
Tabríz	164,630	Kashán	30,000
Tehrán	120,000	Dizful	25,000
Isfahán	65,000	Khoi	25,000
Mashhad	60,000	Zenjan	24,000
Barfrush	50,000	Shuster	20,000
Karmán	41,000	Burujird	20,000
Yazd	40,000	Kúm	20,000
Urmia	40,000	Bandar-Abbas	20,000
Rasht	40,000	Sari	15,000
Kazvín	40,000	Mohammera	15,000
Hamadán	30,000	Nishapúr	11,000
Karmanshah	30,000	Bushahr	10,500
Shiraz	30,000	Astrabád	10,000

Revenue.

Income.	Expenditure.	Debt.
£2,250,000	£2,000,000	*Nil.*

Trade (1883).

Imports.[1]	Exports.[2]
£3,853,000	£3,264,000

Postal Service (1885).

Post Offices	73	Telegraph Offices	82
Letters forwarded	1,680,000	,, Lines, miles	3,824
Receipts	£13,611	,, Wires, ,,	6,124
		Despatches	600,000

Distances in English Miles.

Tehrán to Kúm	87	Rasht to Tehrán	180
Kúm to Isfahán	158	Tabríz to Tehrán	360
Isfahán to Yazd	191	Mashhad to Tehrán	558
Yazd to Karmán	219	Mashhad to Sístán	582
Karmán to Bam	136	Sístán to Karmán	360
Bam to Bampúr	242	Sístán (Nasirabad) to Bam	246
Bampúr to Gwadar	220	Bam to Bandar-Abbas	248
Bushahr to Shiraz	120	Shiraz to Isfahán	220

1 Chief imports—Cottons and other woven goods, hardware, glass, paper, metals, tea, sugar.

2 Chief exports—Silk, tobacco, skins, carpets, rugs, opium, gums, wool. Trade mainly with England, India, and Russia.

SECTION B.

SOUTHERN ASIA: BRITISH POLITICAL SYSTEM.

CHAPTER VII.

AFGHANISTÂN AND BALUCHISTÂN (KÂBUL AND KELÂT).

1. *Boundaries—Extent—Area.*

ALTHOUGH we now enter a new political world, the countries forming the subject of this chapter still belong geographically to the previous section. They were even for many centuries comprised within the Persian monarchy, from which they are now separated by little more than conventional frontiers. Nevertheless the valley of the Hari-rúd, the Sístán depression, and the change of direction in the mountain system of West Makrán, offer a sufficiently defined physical parting line between the western and eastern divisions of the Iranian plateau. The eastern section, stretching thence to the Indus valley and bounded on the south by the Arabian Sea, northwards by the Hindu-Kush and its little-known western extensions to the Hari-rúd, forms a quadrangular mass about 600 miles long north and south, and 550 broad east and west, with an area of some 400,000 square miles.

Of this area about 170,000 square miles are comprised in the southern division forming the Khanate of Kelát, and 230,000 in the northern division forming the Amirate of Kábul, States more commonly known as Baluchistán and Afghanistán respectively. But since 1873 a large tract, about 70,000 square miles in extent, lying beyond the natural limits of the Iranian plateau, has been recognised as politically belonging to the Amir of Kábul. This tract, known as Afghan Turkestán, lies mainly between the northern scarp of the plateau and the Upper Oxus, the boundary here following the left bank of the river from its source on the Pamir to Khojah Saleh. The northern frontier line runs thence across the Dasht-i-Chul desert a little north of Maruchak on the Murgh-ab River to Zulfikar on the Persian frontier Afghanistán has thus a total area of 300,000 square miles with a population variously estimated to from 5,000,000 to 6,000,000. The boundary between Kábul and Kelát is scarcely anywhere clearly determined, but may be said to follow the 30th parallel from Persia to within 30 miles of Quetta, whence it runs north-east to an undetermined point on the frontier of British India. The boundary of both States towards British India is more carefully laid down, and generally follows the course of the Indus at a mean distance of 50 miles from its right bank along the foot of the hills from above Peshawar to Cape Monze a little west of Karachi. Within these limits the two States have probably a total joint area of 470,000 square miles, and a population of about 6,500,000.

2. *Relief of the Land: Highlands—Hindu-Kush—Paropamisus—Sufed-kóh—Suliman Mountains—Hala and Coast Ranges—Desert.*

The eastern section of the Iranian plateau rises from the central Hamún depression towards the highlands, by which it is enclosed on the south, east, and north, and which in the north-east gather to a head in the Hindu-

Kush, connecting the whole system and the tableland itself with the Pamir and great central Asiatic plateau.

The recent surveys of the Afghan highlands, covering an area of nearly 30,000 square miles, have shown that while the southern ranges are more elevated, the Hindu-Kush, at least in its western section, is a far less formidable barrier between India and Central Asia than had been supposed. "Throughout the whole length of it visible from the Kábul plain, it is by no means an imposing range. No part of it is snow-covered, except for a few months in winter; there are no grand peaks, no magnificent altitudes. Previous estimates of its general altitude must be reduced by from 1000 to 2000 feet at least. . . . It is crossed by mountain paths at intervals along its whole length, from the Irak Pass leading to Bamian to the Khawak Pass, east of which the Hindu-Kush rises into a really formidable mountain chain, increasing gradually eastwards till we arrive at peaks of truly Himalayan proportions. The Tirieh Mir, at the Nuksan Pass, is fixed now at 25,000 feet, and others have been seen not far west which cannot differ by many thousand feet. Still, so far as the Kóh-i-Dáman or the plains of Kábul are concerned, the line of the Hindu-Kush is hardly a defensible, and is certainly a most undesirable, military frontier" (*Capt. T. H. Holdich*).

At its north-east end the Hindu-Kush is crossed by the Baroghil Pass, leading from India, Chitral, and Kashmir, to the Upper Oxus valley, Kashgar, and Yarkand. To the south-west of the Tirieh Mir stretches the still unexplored Kafiristán section of the system, where, however, at least one pass, the Apaluk mentioned by Major Raverty, leads to the Oxus basin.

From the south-west corner of the Pamir the Hindu-Kush runs mainly south-west to about the 68th meridian, whence it is continued for over 100 miles westwards by the Kóh-i-Baba. This range, apparently the Paropamisus of the ancients, still remains one of the least-known highland regions on the globe. Its three western ramifications—the Tirband-i-Turkestán, Sufed-kóh, and Siah-kóh—have never yet been explored. But they seem to run nearly parallel through the Hazárajat and Zamindáwar to Herat and the Hari-rúd valley, whence they are continued by the Dáman-i-kóh system north-westwards through the Little and Great Balkans to Krasnovodsk on the Caspian. Between the Kóh-i-Baba and Herat they throw off numerous spurs running almost uniformly north-east and south-west, and forming longitudinal valleys, which drain through the Helmand and other rivers to the Hamún depression.

The orography of the extreme north-east is still imperfectly known, but its leading features can be fairly traced. From the angle formed by the converging Hindu-Kush and Muz-dagh ranges spring a number of lofty spurs separating the head-streams of the Gilgit River. One of these, with many peaks over 20,000 feet high, forms the water-parting between the Chitral and Gilgit basins, and is crossed by the Darkot and Moshabar Passes. Just south of the 36th parallel a remarkable transverse range runs from the Indus at Bunji nearly to Chitral, throwing off a succession of spurs between the Kandia, Swat, Panjkora and Chitral (Kunar) river valleys. Here the peaks diminish from nearly 20,000 feet to between 4000 and 7000 as we proceed southwards to the Kábul River. This transverse range, supposed by Major Tanner to be the Hindu Roj of the Afghans, is an important feature in the physical geography of the Hindu-Kush, as it separates the comparatively rainless tracts of Gilgit, Hunza, and Yasin, from the well-watered southern valleys of Panjkora, Bashkar, and Swat.

From the junction of the Hindu-Kush and Kóh-i-

Baba an important spur, running eastwards between the Helmand and Ghorband basins, sweeps round the head-waters of the Arghand-ab to the north of Ghazni, and thence trending north-east follows the 34th parallel as the Sufed-kóh ("White Mountains") between the Kábul and Kuram river basins, eastwards to the plains of Peshawar. From this range, which culminates with Mount Sikaram[1] (15,620 feet), the whole system of the Suliman Mountains projects southwards between the Iranian plateau and the valley of the Indus. The main range of this complicate system runs from near the Shutar-gardan Pass (10,900 feet) southwards to the great Kund Peak, where it branches off into a number of minor spurs, ultimately merging in the east Baluchistán highlands, which continue to skirt the Indus valley to the coast. Besides the main chain forming the watershed between the Helmand and Indus, it is now ascertained that a continuous system of parallel ranges runs from the gorge of the Gomul to about the 30th parallel.

South of the Gomul Pass run two main ranges nearly 12,000 feet high, which include several remarkably parallel ranges, increasing in number southwards, till no less than twelve distinct ridges are observed where the Nari River pierces the whole system. Many other streams or torrents rising on the eastern slopes of the Inner Sulimans, when swollen with the rains or melting snows, penetrate across the intervening ridges down to the Indus. These *darahs*, or river gorges, afford easy access at many points from India to the Afghan uplands, so that the whole frontier from Peshawar to Jacobabad is now found to be traversed by a large number of "excellent natural roads and passes" (*Holdich*).

[1] At right angles with this peak runs the Peiwar range, a well-wooded spur crossed by the Peiwar Pass, the scene of General Roberts's signal victory over the Afghans on December 2, 1878.

Between the Gomul and Kuram (Kurmah) valleys lie the Waziri highlands, and the Sufed-kóh skirting the Kuram River on the north maintains a uniform level of 12,500, culminating eastwards in a double-peaked mountain 14,680 feet high. North of the Sufed-kóh project three important spurs, one east of the Logar River traversed by the ill-omened Khurd Kábul defile, another (the Karkacha ridge) washed by the Tezin and Surkh-ab Rivers, and a third springing from the intersection of the meridian of 70° 45′ with the main range, and dividing the Khaibar from the Bazar valley.

Towards Baluchistán the most prominent and important range is the Khoja Amrán, running nearly north and south between the Pishin valley and the Kandahar country, and forming in this direction the present political frontier of Afghanistán. It culminates with the Khoja Amrán Peak, and is crossed in the north by the Psha Pass, in the centre by the Khojak (8000 feet), in the south by the less known but easier Gwaja, through which it is proposed ultimately to carry the Indo-Afghan railway, at present arrested at Quettah within 110 miles of Kandahar.

North of the Shál district the Khoja Amrán ramifies northwards into the Toba and Surkh-ab ridges, the latter enclosing the Pishin valley on the north-east and sloping gently towards Shal (Quetta). Eastwards the hills fall more abruptly, and here the chief approaches from India are through the famous Bolan and Mula river gorges.

The Suliman system, which culminates with the Takht-i-Suliman (11,298 feet), and which has many other peaks, such as the Takatu between Pishin and Quetta, Chapar and Kalipat farther east, and several others ranging from 10,000 to 12,000 feet, has a mean width of about 150 miles between the Indus and the desert. The whole distance from Sukkur on the Indus to Kandahar through the Bolan and Khojak Passes is 410 miles, of which 140 are comprised in the alluvial riverain tract and the Kachi desert as far as Sibi, which is still only 700 feet above sea-level. Beyond the Khojak Pass, which is 90 miles from Kandahar, the land again falls rapidly towards the central desert, so that the true highlands between the southern end of the Bolan Pass and the Khoja Amrán range [1] between the Pishin valley and the Kandahar district are about 180 miles wide.

The southern section between Baluchistán and the Lower Indus has no general native name, but is variously known to Europeans as the Brahui or Hala range.[2] This highland region, which is politically included in the territory of the Khan of Kelát, is approached from the Indus valley through short steep watercourses to a height of 1200 feet. The main ridge, running north and south, throws off various branches east and west. Eleven such offshoots occur between Kelát (6700 feet) and Khozdar (3800) at the foot of the Mula Pass, forming in a tract scarcely 100 miles long as many as thirteen upland plains at various elevations. From Khozdar the route surveyed by Bellew in 1872 descends towards the coast through the Purali valley, and towards India through the dangerous Mula Pass.

The Baluchistán southern highlands run mainly east and west parallel with the coast from the Indus delta to the Persian frontier, where they change abruptly to the south-west. The intervening valleys ascend successively

[1] The Khoja Amrán has no general native name, and the term Khoja (properly Khwaja) is merely the name of a peak in the Gwaja Pass at its southern end. Khojak also is rather the name of the river, the bed of which forms the pass, than the pass itself.

[2] The *Hala* seems to be properly only a small ridge running from Kelát southwards to the Baghwana River, lat. 28° to 29° N. lat., and 66° 30′ E. long.

inland to a height of 2500 feet, and are often of great length. One of them runs from the Khelát hills uninterruptedly westwards nearly to Bampúr in Persia, and 70 miles south of it is another stretching for 250 miles westwards to Kasr-Kand also within the Persian frontier, where all these valleys are closed in by the intricate highland system of West Makrán. "No difficulty exists for wheeled traffic from one end to the other of these two valleys" (*Major Lovett*). Farther inland a third parallel range, the Wushuti or Mue Mountains, stretches along the border-land of the two states at a distance of about 280 miles from the coast.

Most of the inner space enclosed between the northern, eastern, and southern highlands consists of an extensive sandy plateau, at a mean elevation of perhaps 3000 feet above the sea, and sinking everywhere towards the central Hamún depression. Except along the river banks, this region may be regarded as waste; and south of the Helmand, where there seem to be no more rivers, the desert formation is complete. It begins at the foot of the Khoja Amrán range, and stretches thence almost uninterruptedly along the Afghan and Baluch border-lands eastwards to Sístán and Persia. No European has yet ventured across this almost impassable wilderness, which still remains nearly a blank on our maps. Seen from the neighbourhood of Kandahar, it presents the appearance of endless undulating sand-hills rolling up from the far south.

Similar desert tracts are found within the uplands themselves—as, for instance, the Kachi desert below Sibi, 90 miles long, and now traversed by a railway, and the Dasht-i-Be-daulat ("Desolate Plain," or, more exactly, "the plain without wealth") in the very heart of the highlands above the Bolan Pass and south of Quetta, 200 square miles in extent.

3. *Hydrography: Inland and Seaward Drainage—The Hari-rúd—Helmand and Kábul Basins.*

The East Iranian drainage system is threefold—two inland to the Hamún Aralo-Caspian and some smaller depressions, one to the Indian Ocean either directly or through the Indus. Afghanistán belongs to all three, but mainly to the Hamún basin, while Baluchistán drains almost exclusively seawards.

Afghan Turkestán is comprised entirely within the Aralo-Caspian basin, all its rivers flowing from the northern slopes of the Hindu-Kush and Paropamisus to or towards the Oxus and Aral or Caspian. Here we again meet with the same undeveloped or partially dried-up water system which was found prevailing in Arabia and Persia, and which forms such a striking feature of the great central Asiatic tableland. In the east the Kokcha and Kundúz still reach the Upper Oxus, but as we proceed westwards we find that all the rivers flowing north fail to reach either the main stream or either of the great inland seas. Thus the Dehas-rúd (Balkh), rising in the Kóh-i-Baba, gets no farther than Mazar-i-Sheríf, where it takes the name of Band-i-Barbari, and runs dry in the Siyagird district after a course of over 180 miles; the Sir-i-púl is lost in the sands beyond Shabirkhan; the Murgh-ab, after irrigating the Merv oasis, disappears in the Karakum desert, and the same fate overtakes the Hari-rúd (Tajand) after skirting the Dáman-i-kóh on its way to the Caspian.

The Hari-rúd, or river of Herat, has its source in a deep valley 9500 feet above the sea at a point where the Kóh-i-Baba ramifies into the Siah-kóh and Sufed-kóh. It flows thence rapidly through an unexplored region down to the town of Obeh, where its waters are largely diverted into irrigating rills. Its course lies thence westwards to Herat and Ghorian, where it turns abruptly northwards along the Persian frontier to its junction with the Keshef-rúd above Sarakhs. The united streams now take the name of the Tajand, whose course has been described at p. 160.

The southern slopes of the Hindu-Kush all drain to the Indus through the Kábul River, which also receives on its right bank several streams from the Sufed-kóh. Thus the north-eastern portion of Afghanistán is comprised in the Indus basin, to which also belongs the eastern slope of the great watershed of the Suliman, as well as all the intervening outer parallel ridges. But nearly all the land west of this parting line, and south of the northern scarp of the plateau, an area of about 200,000 square miles altogether, drains to the great Hamún depression. Of this vast basin the chief stream is the Helmand, which flows from the west side of the Pughman range through a deep channel in the Hazárajat south-westwards to within 40 miles of Girishk, where it enters the plains which merge southwards with the Baluchistán desert. Here it is largely utilised for irrigation purposes, and at Girishk is crossed by the great caravan route from Kandahar to Herat. It then sweeps southwards through the fertile Garmsel country, beyond which it turns north-west to the Hamún or Sístán swamp. The Helmand, which has a course of about 700 miles, is never without an abundant supply of water, but in winter after the floods it comes down with great rapidity, sometimes overflowing its banks in consequence of the neglected state of the old embankments. Its chief tributaries are the Arghand-ab, Tarnak, and Dori, whose united stream joins it from Kandahar a few miles below Girishk. West of the Helmand the Kash-rúd, Farrah-rúd, and Harút, all flow from the Ghor highlands in nearly parallel beds southwards to the Hamún swamp.

To the same system belongs the lagoon Abistáda, the

only other body of water in East Irania deserving the name of lake. It lies over 7000 feet above the sea some 60 miles south-west of Ghazni, and is fed by the river of that name. It is about 17 miles by 15 in extent, and, although hitherto supposed to be a closed basin, there is little doubt that during the floods it overflows to the Arghasan, a tributary of the Arghand-ab. Its water is brackish and very shallow, nowhere exceeding 5 or 6 feet in depth.

The crest of the water-parting between the Helmand and Kábul basins is marked by the Sher-i-Dahan ("Lion's Mouth") Pass, crossed by the road going south to Ghazni. Rising at a height of about 8400 feet above the sea, the Kábul flows mainly east by Kábul, Jelalabad, to the Indus at Attok. During its rapid course of about 250 miles it receives from the Hindu-Kush the Swat, Kunar (Chitral), Alingar, Tagao, Panjshir, and Ghorband; from the Sufed-kóh the Logar, Surkh-ab, Bara, and Tira. Of the northern affluents the most important is the Kunar, which flows from the Baroghil Pass through the Chitral valley for nearly 300 miles down to the main stream, a few miles below Jelalabad.

South of the Kábul River are the important Gomul and Kuram basins, the former of which covers an area of perhaps 13,000 square miles between the western and eastern Suliman ranges, along which the great trade route from Central Asia to India passed for centuries. The Kuram, which rises on the eastern slopes of the great water-parting between the Indus and Helmand basins, is joined on its course to the former river by numerous affluents from the Sufed-kóh on the north, and from the hilly country of the Mangal tribes on the west and south.

The Lower Indus receives no important stream from Baluchistán, which seems to be almost as riverless a country as Arabia itself. To its inland drainage belongs the Lora, which rises with several head-streams on the east slope of the Khoja Amrán, and after watering the Pishin valley, escapes through the Tang gorge in the Tang range south-westwards to the Hamún Lora morass in 29° 30′ N. lat., 65° E. long. Its lower course, like those of Bale and other streams flowing in the same direction, still remains to be traced. On the Makrán coast the only noteworthy river is the Dasht or Bhingwur, which is supposed to rise far inland, and to make its way through all the intervening ranges and valleys to the sea at Gwattar Bay in 61° 40′ E. long. But here scarcely any perennial streams seem to exist, and few of them flow through regular or well-defined beds.

4. *Natural and Political Divisions: Wakhan—Badakhshan—Afghan Turkestán, Afghanistán Proper—Kafiristán—Chitral—Swat—Baluchistán—Kelát—Makrán.*

Afghanistán as at present constituted comprises three physically distinct regions—the northern slopes of the escarpment forming geographically a part of Western Turkestán, the basin of the Helmand embracing most of the central plateau, and the eastern highlands mostly included in the Indus basin. But to these natural regions the political divisions correspond in part only. Since the foundation of the modern Afghan State by Ahmad Shah in 1747, comparatively little progress has been made towards moulding it into one political system. So heterogeneous are its ethnical components, so inaccessible many of the highland tracts, and so persistently upheld is the old tribal organisation of the dominant Afghan race itself, that in many places the Amir's authority is merely nominal, in others openly defied, in some never yet recognised. Protracted internecine feuds between the rival branches of the royal Durani (including the Barakzye) tribes, combined with several disastrous foreign wars, first with the Sikhs of the Panjáb and then with the British Government of India, have added to the confusion to such an extent that disintegration rather than consolidation of empire has lately seemed imminent. As it is, the Amir has been fain to sacrifice external independence, and to accept a somewhat indefinite position of subordinate relationship to the Suzerain of India.

In the north, Afghan Turkestán, comprising that portion of the land included in the Aralo-Caspian basin, possesses a certain ethnical as well as physical unity, for here the bulk of the people belong to the Usbeg branch of the Túrki stock. It is administratively divided into a number of provinces corresponding with the old Usbeg khanates, all of which have completely lost their autonomy.

In the extreme north-east is the alpine territory of Wakhan, which consists of two upland valleys traversed by the Sarhad (Panja), the chief southern head-stream of the Oxus. On either side the valleys are hemmed in by lofty mountains, those to the south forming the northern section of the Hindu-Kush, here crossed by some difficult passes, the easiest of which is the Baroghil (12,000 feet) leading to Chitral and Gilgit. The chief resources of the people are derived from their flocks, mainly sheep and the Tibetan yak. The land is too elevated and sterile for tillage, but yields a pasturage like that of the Pamir, possessed of peculiar fattening qualities. In this alpine region the lowest hamlet is 8000, and Sarhad, the highest, no less than 11,000 feet above the sea, or higher than the loftiest peaks of the Pyrenees. Yet a little pulse and barley are grown in a few sheltered glens. When Lieutenant John Wood, discoverer of the source of the Oxus, visited Wakhan in 1838, he estimated the population at about 1000; but Forsyth, thirty-five years later on, raised the number to 3000, which agrees with the Russian estimate. The mir or chief, who, like so many others in this region, claims descent from Alexander the Great, resides in Kila-Panja ("Five Forts") on the Oxus, and close to the Pamir. In Wood's time he was almost independent; but since then has become tributary through Badakhshan indirectly to Kábul.

Badakhshan, adjoining Wakhan on the west, comprises the valley of the Kokcha and the little-known tracts enclosed on one side in the great northern bend of the Oxus, and stretching on the other to Kafiristán. Besides the Kokcha, it is watered by the Wardoj, and both streams unite a few miles above Faizabad, the capital, before joining the Oxus. In the upper parts the crops are often nipped by summer frosts. But lower down the more favoured sites yield wheat, barley, mulberries, walnuts, pistachios, and pulse. The country is exposed to earthquakes, one of which in 1832 was very destructive to life, and was felt as far south as Lahore. Badakhshan is noted for its mineral wealth, salt, sulphur, iron, and especially the ruby[1] and lapis lazuli,[1] prominently mentioned by Marco Polo. The lapis lazuli mines, which lie close under the crest of the Hindu-Kush, have been fully described by Wood.

West of Faizabad the road diverging to the right through Rustak crosses the Oxus to Kulab, Karateghin, and other Trans-Oxian districts. The gold-washings in a small stream between Rustak and the Oxus yielded a revenue of about £100 in 1874. The main road beyond this point still runs westwards over the Lattaband Pass,

[1] These rubies, which are of a delicate rose colour, were formerly known as *balais* or *balash*, a corrupt form of Badakhshan, which Marco Polo calls Balacian. The lapis lazuli takes its name from the district of Lajúrd or Lazúrd, whence both the words *lazuli* and *azure*.

through Talikhan down to Kundúz. Here the descent from the Badakhshan highlands to the marshy plains of Turkestán, here scarcely 500 feet above sea-level, is attended by a marked change of climate, that of Kundúz being excessively hot and unhealthy especially in summer.

Kundúz is watered by the river of like name, which rises with several head-streams in the Kóh-i-Baba. Below the town of Kundúz it joins the Oxus below Hazrat-Imám. Here are extensive undulating plains yielding good pasturage, and tenanted by nomad Usbegs and Hazaras. From Talikhan and Kundúz there diverge to the left routes which lead over the Sir-alang and Khawak Passes to Kábul. But here the chief highway is the historical route which passes through Bamian and Heibak, joining the Badakhshan road at Khulm or Tashkurgan. This route was traversed for some distance by the British troops with horse artillery in 1839. Near Bamian are two gigantic idols, one of which is said to be 100 feet high, cut in bold relief in the face of the cliff skirting the road. They stand in deep niches, and are clothed in flowing drapery. These idols and caves are generally supposed to be of Buddhist origin, but all memory of the time and hands by which they were executed has long perished. Here also are the stupendous ruins of Ghulgulch destroyed by Chingiz Khan, besides many other remains, which have been fully described by Masson.

Adjoining Kundúz is the smaller but not less populous khanate of Khulm. It occupies a vital position in the heart of the ancient Baktriana, the converging point of all the natural highways from India, Persia, and Central Asia. Here are the ruins of Baktra and of its successor Balkh, now supplanted by the modern towns of Siyagird, Mazar-i-Sherif, and Khulm. The country has been largely encroached on by the desert, and the Khulm River, flowing from the Kara-kóh hills, now no longer reaches the Oxus. In the plains the river of Balkh, here called the "Band-i-Barbari," or "Dyke of the Barbarians," is soon absorbed in irrigation works in the gardens interspersed amidst the vast ruins and flourishing towns of this historic land.

West of Balkh are the four petty Usbeg States of Akcha, Shabirkhan, Sar-i-pul with Andkhui, and Maimana, lying mostly between the outer spurs of the Paropamisus and the sands by which they are now cut off from the Oxus. This tract is very fertile and well watered by the streams from the mountains, but it is also proverbially unhealthy. Nevertheless here are the populous towns of Shabirkhan and Andkhui, lying on the very skirt of the Turkoman domain.

Of these khanates Andkhui alone has retained a certain measure of independence. All the rest are absolutely controlled, and even administered, by Kábul, though the old geographical and political divisions are still preserved. The village of Gurzivan and the Darzat valley in the hills south of Sar-i-pul have also lost their autonomy, though still retaining the empty titles of khanates.

The Usbeg inhabitants of these districts are not called upon to render military service; but, according to the authority of Grodekov, which, however, is not above suspicion, they are so heavily taxed that they are impatiently awaiting the arrival of the emancipating Russians.

On the southern slopes of the Hindu-Kush, bordering eastwards on Kashmir, south-eastwards on the Panjáb, are the territories of Kafiristán, Gilgit, Chitral, and Swat, conventionally supposed to belong to Afghanistán, but *de facto* not only independent but still to a great extent unexplored regions.

The Kafiristán highlands lie north of Lamghan, and

occupy the watershed between the Khawak Pass and Chitral. The upper slopes are snow-clad, but lower down they are covered with dense forests of magnificent conifers and other trees. Tillage is but scantily practised, owing to the absence of level ground; but the orchards yield fruits in great abundance. The inhabitants, who have recently been visited by Major Tennant and W. W. M'Nair, are independent of Kábul, and claim to have enjoyed this freedom for centuries. They have reminiscences of the Græco-Baktrian empire, and the chiefs, like those of the surrounding districts, trace their descent from Alexander the Great. They are strictly honourable in their dealings, extremely hospitable, and cordial haters of their Muhammadan neighbours.

A southern branch of the Siah-Posh Kafirs, or "Black-Clad Infidels" as they are called by the surrounding Muhammadans, are the Safis and Chagnans, whose domain reaches down to the Kábul River. Masson describes them as a straightforward manly race: but very little is known of them or their country.

East of the Panjkora valley lies the Swat country, unexplored till 1878, when one of the Indian native surveyors mapped it out from the source of the Swat River in the great transverse range between Bunj on the Indus and the Chitral valley to its junction with the Panjkora. Swat formed at one time a powerful state under a venerable chief of great repute for sanctity, called the Akhund, who exercised considerable influence over the unruly tribes of the district.

The Chitral country comprises the upper course of the Kunar River, which is here called the Chitral, or Kashkar. It produces orpiment, or yellow arsenic, in great quantities, and the natives manufacture coarse carpets, chogas or cloaks, daggers, and sword hilts. They are on good terms with their Kafir neighbours, who acknowledge a sort of allegiance to Aman-ul-Mulk, "King of Chitral."

East of Upper Chitral lie Yassim and Gilgit, the latter of which during the last few years has been the site of a British residency under Major Biddulph. This officer was here stationed on the Kashmir frontier with a view to control the tribes occupying a district of some strategic and political importance. The River Gilgit drains eastwards to the Indus near the Nanga Parbat peak, which marks the north-western extremity of the central Himalayan chain. This region, where the Dard and Afghan races meet about half-way up the valley, has at all times proved most inaccessible to external influences.

In Afghanistán proper the political divisions are often far less distinctly defined than in its outlying Turkestán possessions. Some regions in the Hindu-Kush, such as Kafiristán and Swat, as well as nearly the whole of the northern highlands between Bamian and Herat, besides many tracts in the Suliman Mountains, have never acknowledged the Amir's authority, and must be regarded as *de facto* independent territory. Elsewhere, as in the districts of Herat and Kandahar, and even in Kábulistán itself, the tribal organisation still largely prevails, so that the limits of the provinces are scarcely anywhere carefully laid down, and it becomes impossible to speak of provincial administration in the ordinary sense of the term. Hence, instead of taking the various provinces separately, it will be more convenient to deal with them in connection with the chief towns round which they are grouped.

In Baluchistán, although more order has recently been introduced, a similar state of things still largely prevails. The Khan of Kelát, who may be said to have frankly accepted the suzerainty of the Kaisar-i-Hind (Empress of India), is nominal ruler of the whole land.

But his authority has often been confined to the Kelát district itself, and is still challenged by many of the tribal chiefs, especially towards the Persian frontier. The natural divisions of the country, the eastern and southern highlands merging inland with the desert, are grouped in seven recognised provinces: Saráwan and Katch-Gandava, including the Mari and Bugti country on the north-east; Kelát, between these two; Jhalawan on the east: Lus on the south-east: Makrán, comprising the southern coast region; Kohistán, or the "highlands" of the west.

Most of the land is still practically unknown. The north-eastern section lying between the Indus and the Pishin valley, along the Afghan border, and thence southwards to Kelát, has been thoroughly surveyed, and a military station has even been established by the British at Quetta, above the Bolan Pass, and overlooking the Pishin valley. The south coast has also been carefully surveyed by the Admiralty, and somewhat farther inland by the English Telegraph Staff; while the country has been crossed, chiefly from east to west, by Grant, Pottinger, Ferrier, Goldsmid, Bellew, Lovett, and a few other explorers during the present century. Still, most of the interior has never been visited, and the sandy plains stretching beyond the hills towards the Hamún depression remain a blank on our maps. Elsewhere the highland formation everywhere predominates, although in the south the parallel ridges are separated by long and almost level valleys reaching from the Persian frontier to the eastern uplands. This southern region, from the sea to the desert, is usually spoken of collectively as Makrán; but the term should properly be restricted to the strip of land between sea and the first parallel ranges. Here the geographers of Alexander placed the Ichthyophagi, or "Fish-eaters," apparently a mere translation of the local name.

The country is almost entirely occupied by pastoral tribes under semi-independent sirdars and chiefs. Hence the so-called provinces are not administrative divisions in the ordinary sense, and should be more properly called territories. Besides those above mentioned there are several others current amongst the natives as applicable to particular cantons, especially in Makrán and Kohistán. Here there are several semi-independent chiefs, of whom the most powerful was, till recently, the Khan of Kej, in central Makrán. But the native ruler was, some ten years ago, replaced by a direct nominee of the Khan of Kelát, and although the change was at first followed by disturbances, it has had the effect of somewhat consolidating the Khan's authority, and thus barring the further progress of Persia in this direction.

The Khan or Mir of Kelát, who belongs, not to the Baluch, but to the Brahui stock, concluded a treaty in 1877 with England, in virtue of which he has become a feudatory of the Empress of India. The right had already been secured of occupying at pleasure the mountain passes between Kelát and Afghanistán. But the new treaty places the whole country at the disposal of the British Government for all military and strategical purposes. In return the Khan has acquired a certain prestige amongst the tribal chiefs and sirdars, who no longer seriously question his supremacy, and his subjects have begun to enjoy the blessings of peace.

5. *Climate: Sand-storms.*

In Afghanistán the prevailing climatic conditions are dryness combined with great extremes of temperature. Snow lies on the ground for three months in the Kábul and Ghazni districts, and many of the peaks from the Hindu-Kush to Kelát rise above the snow-line. But so much depends on elevation that Jelalabad, 2000 feet above the sea, is scarcely colder than India, while the winters are almost as severe as those of Russia on the neighbouring Kohistán uplands. The summer heats, on the other hand, are everywhere intense, more so, in some places, even than in Bengal. At Kábul (6500 feet) the glass rises to 90° and 100°, and in Kandahar even higher. Yet the country is on the whole decidedly salubrious, in this respect presenting a marked contrast to the fever-stricken lowland districts of Afghan Turkestán.

In Baluchistán also intense heats are followed by almost equally intense colds, the snow lying for two months on the ground even in the Shal and other valleys. The Kej district and some other parts of Makrán are said to be the hottest places in the whole of Asia. Even in March Major Lovett registered "125° F. in the shade in the neighbourhood of Kej." On the other hand, Pottinger found it so cold in February at Kelát that water poured on the ground froze instantaneously. Owing to its proximity to the ocean, Baluchistán receives on the whole more moisture than Afghanistán. The dry season lasts from March till September, but rain or snow falls intermittently throughout the winter, and often heavily in February and March. Unfortunately most of it is precipitated on the outer ranges, leaving little for the deserts of the interior, where the sultry heats are intensified by fierce sand-storms.

6. *Flora and Fauna: The Karez Irrigation System—The Uromastix Lizard.*

Bare, treeless mountains, sandy and absolutely unproductive plains, fertile valleys and riverain tracts, producing enormous quantities of magnificent fruits and vegetables, besides cereals of various kinds, are the prevailing features almost everywhere from the Upper Oxus to the Indian Ocean, and from Persia to the Indus valley. In the north, however, the southern slopes are often clothed with forests of walnut, birch, oak, and conifers, the latter growing to a height of 10,000 feet, In Afghanistán the asafœtida covers extensive tracts, and here the most productive districts are those of Herat, Kandahar, the Lower Helmand, the valleys of the Kábul and Logar Rivers, and the Kóh-i-Dáman. Wheat, maize, and rice are the staples of food; the vine and many other fruits are indigenous; cotton, sugar, and tobacco thrive in the well-watered low-lying tracts, and the melon and many other vegetables arrive at astonishing perfection. The apples, the grapes, the pomegranates of Afghanistán are celebrated throughout India.

In Baluchistán wheat, barley, rice, cotton, pulse, madder, indigo, and tobacco, are cultivated; the date-palm prevails in Makrán; magnificent fruits and vegetables are grown in the valleys. Asafœtida abounds, and of forest trees the chief are the plantain, walnut, sycamore, wild fig and olive, mulberry, tamarisk, and mimosa.

The "karez," or underground system of irrigation peculiar to the Iranian plateau, is well suited to this region, and extensively practised. "The soil being naturally open and porous, composed of water-worn stones embedded in a sandy soil, which, however, having a large admixture of lime, hardens at a short distance below the surface into an impermeable conglomerate, it is easy to understand how flowing water may in many places be found 20 or 30 feet from the surface, while on the surface itself for miles round there is nothing but an arid plain. The water thus found is led gradually towards the surface through the karez. A series of wells are dug at intervals of 15 to 25 yards, and connected below by an underground passage, through which the water runs till at last it reaches the surface and is utilised for irrigating the fields" (*Capt. R. Beavan*).

In East Irania wild animals are comparatively scarce. Lions and leopards of a small type haunt the upper

WILD ASSES.

valleys of the Hindu-Kush, where are also met the wolf, and two species of bear. The so-called Angora cat is indigenous in Kábulistán, and in the plains the dromedary or one-humped camel is the chief beast of burden. Here the horse is far inferior to the Turkoman breed. The ass exists in the wild and domestic state, but sheep and goats form everywhere the chief resources of the pastoral tribes.

In Baluchistán the lion, tiger, leopard, and wolf are occasionally met, the jackal, wild dog, fox, wild goat, and ass more frequently. There is a distinct species of gazelle (*Gazella fuscifrons*), and both species of camel occur, the Baktrian or two-humped on the uplands, the dromedary on the plains, where it is highly valued for its speed by the marauding tribes. Serviceable horses are bred in the north and west, but those of Makrán are small, weak, and spiritless. In Baluchistan is found the curious Uromastix lizard, one of the most remarkable animals in the world. It looks at a distance somewhat like a rabbit in appearance and size, but is really a sort of diminutive saurian, called by the Persians búz-miji, or goat-sucker, from its peculiar habit of bleating like a kid to attract the goats, whose teats it then sucks. The Makrán coast abounds in fish, where it still forms the staple food of its ichthyophagous inhabitants.

7. *Inhabitants: The Afghans—The Brahuis, Baluchis, and Lúri.*

East Irania presents a greater complexity of races even than Persia itself. For to nearly all the elements contained in the west must here be added at least three others—the Galcha of the Hindu-Kush, the Hindu of the large towns, and the Brahui of Kelát; this last being distinct in speech, not only from all the others, but from all other known linguistic groups. The subjoined table comprises all the races in the region classed according to their most probable ethnical affinities:—

ARYANS	*Galcha Branch.*	Wakhis	Hindu-Kush (northern slopes).
		Badakhshis	
		Chitralis	Hindu-Kush (southern slopes).
		Swatis	
		Siah-Posh Kafirs	
		Safis	
		Chagnans	
		Kohistanis	Hills north of Kábul.
	Iranic Branch.	Afghans	Kábul; Suliman Mountains, Kandahar, Helmand basin; Herat.
		Tajiks	Herat; most towns and settled districts.
		Baluchis	Baluchistán lowlands; Makrán.
		Sístánis	Lower Helmand; Hamún.
		Kurds	Baluch Kohistán.
	Indic Branch.	Hindkis	Most large towns.
		Lassis	Prov. Las, So. Baluchistán.
		Jats	Makrán chiefly.
		Lúris	
MONGOLO-TATARS.	*Mongol Branch.*	Hazarahs	N. highlands between Bamian and Herat.
		Aimaks	
	Túrki Branch.	Usbegs	Afghan Turkestán.
		Turkomans	Herat, Maimana, and Andkhui.
		Kizil-Bashis	Kábul chiefly.
	?	Brahuis	Mainly East Baluchistán highlands.

Of these various peoples four only possess a decided political or social preponderance in their respective areas —the Usbegs in Afghan Turkestán, the Afghans in Afghanistán, the Brahuis and Baluchis in Baluchistán. The Usbegs, here represented by the Kateghán family, differ in no material respect from their kinsmen of the adjoining khanates of Bokhara and Khiva, and will therefore be more conveniently dealt with in the chapter devoted to that region.

The Afghans, commonly known in India as Pathans, are all Sunnis in religion, but are socially still in the tribal state, a fact that is not sufficiently taken into account in estimating the political situation of the country. There is an Afghan *race*, one in physical type, speech, religion, and culture; but there is, strictly speaking, no Afghan *nation* possessing a distinct sense of its unity as a whole, with common political sentiments and aspirations.

Such common sentiments are scarcely felt beyond the several great sections into which the race continues to be divided. The Duranis, the Ghiljis, the Waziris, the Afridis, the Mangals, Momands, Jusafzais, and others, form so many States, as it were, within the State, each with its own separate interests, and each capable of combining rapidly for some common tribal object, but all in-

AFGHAN CHIEF.

capable of combining together and acting in concert for a common national object. When Ayub Khan of Herat moved in 1881 against Abdur-Rahman of Kábul, the people of the intervening Kandahar district refused to pay revenue, not through any love of the Amir, but through indifference to the claims of the rivals for supreme authority. For both Abdur-Rahman and Ayub are chiefs rather of the Durani tribe than of an Afghan nation. And the Duranis themselves are regarded by other almost equally powerful sections merely as usurpers of the sovereignty, their usurpation dating only from the death of Nadir Shah in 1747, when their chief, Ahmad Khan, took advantage of the disorders in Persia to raise the royal standard in Kandahar. Ahmad

AFGHAN GENTLEMAN.

endeavoured to give a national importance to his tribe, not only by changing its name from Abdali to Durani,[1] but also by associating with it some other sections, such as the Jusafzais, Momands, Afridis, Shinwaris, Orakzais,

[1] Derived not, as is often stated, from the supposed custom of wearing a pearl (*durr*) in their right ears, but from the title of Durr-i-Dúrán ("Pearl of the Age"), adopted by Ahmad when he assumed the royal power.

and Turkolanis, under the common designation of Bar-Duranis. But the attempt failed, these sections still retaining their tribal integrity, and refusing to be fused into a common Afghan nationality.

In the Durani tribe there are several sections, among which are included the two royal branches—the Suddozais and the Barakzais. It was to the Suddozais that Shah Sûja belonged, who was placed on the throne by the British in 1839, after the first Afghan war. It is to the Barakzais that the equally well-known Dost Muhammad and his successors on the throne of Kábul belong.

The sections themselves are divided into a multiplicity of minor branches, septs, and clans,[1] offering still further obstacles to a general amalgamation of the whole race. And the race itself is everywhere opposed to other races speaking different languages, such as the Tajiks, Hindkis, Usbegs, Siah-Posh Kafirs, Hazaras, and Aimaks, which, although numerically inferior, possess greater national cohesion, and which in some cases have been able to maintain their independence.

But for these untoward circumstances the Afghan race, by its warlike spirit and remarkable physical vitality, might seem destined to subdue the surrounding peoples. But their national resources continue to be frittered away in internecine broils and struggles for the local independence of individual chiefs and tribes.

Yet it would be a mistake to suppose that the Afghans are absolutely incapable, under proper conditions, of turning from turbulent to peaceful ways. Although surrounded by hostile and marauding tribes, the Povindahs of the Suliman inner ranges have for ages occupied themselves with tillage, stock-breeding, and trade. These itinerant and warlike dealers, who claim descent from a

[1] Usually termed *zais* or *khels*, as in Barakzai, Abdur-Rahmánzai, Ali Khel, Utman Khel, etc.

goatherd of Ghor in the days of the Ghaznevid Mahmud, follow their industrious pursuits in the face of extraordinary difficulties. In the summer they pitch their tents on the plains near Kalát-i-Ghilzai and Ghazni, where they pay £60 to the Amir's government for grazing rights, and where the women and children remain under a sufficient guard, while the men are away trading at Samarkand, Bokhara, Herat, or Kábul. In the autumn they repair to the Indian plains through the Gomul route, fighting their hereditary foes, the Waziris, on the way, and encamping on the Derajat plains. From this point the men again disperse towards Multan, Lahore, Benares, retailing their raw silk, druggets, clothes, saddlery, horses, saffron, dried fruits, and other wares. In April the Povindahs reassemble for the return journey, and ascend the pass towards Kandahar and Ghazni.

Many other promising elements of future progress exist in the land, such as the Kakar and Tajik agriculturists, the Hindki traders, met with in every large town, and even the despised Kizil-Bashis of Kábul.

In Baluchistán the ruling race are not the Baluchis, but the Brahuis, who are moreover both the aboriginal and the most numerous element. Hence the term Baluchistán, unknown in the country itself, is altogether inappropriate, though it may now be too late to substitute the expression Brahuistán, as some geographers have proposed. The Brahuis, whose racial and linguistic affinities still remain an unsolved problem, are predominant in all the eastern highlands; the reigning Khan and most of the chiefs and nobles are of Brahui stock, and this race still continues to control the destinies of the land. The Baluchis still dwell mainly in the lowlands, and form the rural population both in the direction of India and Persia. Both races are Muhammadans, the Brahuis like the Afghans being Sunnis, the Baluchis like their Persian kinsmen Shiahs.[1] There can be little doubt that the latter penetrated eastwards originally from Karmán, and they are still predominant in the adjoining districts of Makrán and Sístán. Bellew describes them all in two words—needy and hungry. They are true nomads, migrating like many Afghan tribes with their families and flocks from the uplands to the lowlands. But some few are settled in villages.

Distinct both from the Brahuis and Baluchis are the Lúri, a sort of gipsies of Indian origin scattered in single families all over the country. They are generally met with as strolling minstrels, potters, tinkers, rope-makers, weavers of mats, and pedlars. They own no lands, never cultivate the soil, and are regarded as outcasts by the rest of the people. Each troop has a "king," and Pottinger noticed their "marked affinity to the gipsies of Europe."

8. *Topography: Khulm—Mazar-i-Sheríf—Balkh—Herat—Kandahar—Kábul.*

In Afghan Turkestán the chief places are—1. Khulm, at the junction of the Bamian and Badakhshan routes, where the Khulm River emerges from the mountains. It is three miles in circumference, and its houses are built of clay or adobe. The inhabitants are chiefly Tajiks, Kábulis and Hindkis, trading in live-stock, cottons, leather ware, fruits, and melons. Four miles south of Tashkurgan, as this place is now called, is Old Khulm, formerly noted for its excellent fruits. 2. Mazar-i-Sheríf, 50 miles west of Tashkurgan, capital of Afghan Turkestán and residence of the Governor-General. It is surrounded by fields and orchards, and in 1878, when Grodekov passed through it, the population amounted to 25,000

[1] Some writers class the Baluchis as Sunnis.

Usbegs, Afghans, and Tajiks. Two miles off is the fort of Takhta-pul, with gun factories, cannon foundries, and manufactories of swords, knives, and felt helmets. But Mazar is chiefly noted for its mosque, held in great veneration for a tomb supposed to be that of Ali; and for the shrine of Hazrat Shah, a famous Moslem "saint." 3. Baktra and Balkh, both now ruins on the Dehas or Balkh River, a few miles west of Mazar. Baktra, capital of the Græco-Baktrian monarchy, was one of the oldest cities in Central Asia, and its successor Balkh still bears the title of "Mother of Cities." It was the chief town in Afghan Turkestán till 1872, when a terrible outbreak of cholera caused the seat of government to be removed to Mazar, and in 1878 Balkh was an insignificant village, whose former greatness was attested only by numerous canals and miles of ruins. Here are buried the travellers Moorcroft and Guthrie. 4. Andkhui, on the verge of the desert due west of Balkh, a large but proverbially unhealthy place, of which the Persians say that with its salt water, its scorching sands, venomous flies and scorpions, "it is a real hell on earth." 5. Maimana, on a plain near the foot of the Kóh-i-Baba, noted for its excellent horses and textiles woven of wool and camel's hair. Previous to 1874 Maimana was a very large place, with a population estimated at 60,000. But in that year it was besieged and nearly destroyed by the Afghans, who massacred 18,000 of its inhabitants. Since then it has somewhat revived, and must always enjoy a certain importance from its position at the junction of the routes from Herat and Kábul.

In Afghanistán the three cities of Herat, Kandahar, and Kábul stand out conspicuously as at once the chief centres of power and population, as well as the most important strategical points in the country. They occupy the three angles of a triangle, whose base crosses the northern scarp of the plateau, and whose apex lies nearly in the centre of the State. Thus Herat and Kábul at the west and east ends of the base respectively are separated by intervening impassable highlands occupied by the hostile and independent Mongolo-Tatar Hazaras and Aimaks. Hence the route from one to the other is deflected southwards to the apex, where Kandahar thus occupies the key of the whole position. North of the scarp is the Turkoman country, now entirely absorbed in the recently organised Russian Trans-Caspian territory. From this direction the plateau can be approached in the east only by the difficult "Gate of Bamian," in the west by the easy Tajand and Murgh-ab valleys. Here, therefore, the importance of Herat becomes obvious. And this circumstance itself enhances the importance of Kandahar, which bars the direct and only route from Herat to India, and which lies on the flank of the not impossible route through Bamian and Kábul to India. It is satisfactory to know that under these circumstances the railway is already completed from the Indus to Quettah above the Bolan Pass, and that the ground thence to the Gwaja Pass on the way to Kandahar has been surveyed (see diagram), more or less completely.

The city of Herat lies in the well-watered valley of the Hari-rúd, or Upper Tajand, which is extremely fertile, and capable of furnishing supplies for an army of occupation of 150,000 men. This, coupled with its lofty ramparts and fortifications, and its central position as the converging point of routes from the Caspian, Mashhad, Merv, Bokhara, and India through Kandahar, has invested it with a strategic importance which has earned for it the title of the "key of India." In Pottinger's time it was the great emporium of trade in Central Asia, and though its many vicissitudes of fortune and innumerable sieges have caused its population to fluctuate excessively, it still contained 50,000 inhabitants in 1878, a motley gathering of Afghans, Indians, Tatars, Turkomans, Jews, Tajiks, and others. Carpets of soft texture and brilliant colour are here produced, and the district is noted for its excellent horses, wheat, water, and

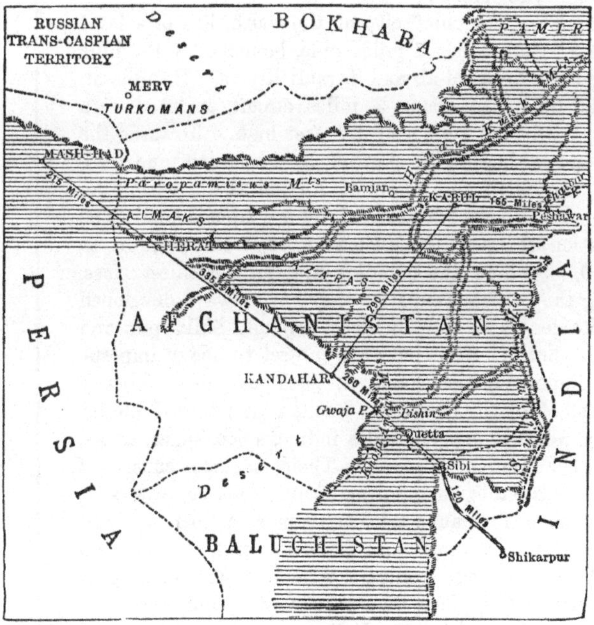

grapes, of which there are no less than seventeen varieties.

The road from Herat to Kandahar lies through the fertile but little cultivated Zamindawar country, peopled by the Duranis as far as the fort of Girishk, near the fords of the Helmand. Although its fortifications are slight, the strong position of Girishk on the main route and in the vicinity of supplies has at all times invested it with strategic importance. About mid-way between it and Kandahar is Khushk-i-Nakhud, mournfully signalised in 1880 by the defeat of General Burrows at the hands of Ayub Khan.

Kandahar, the chief city of the south, lies in a level cultivated plain about 7 miles wide, bounded by low hills between the Arghand-ab and Tarnak Rivers. It forms an irregular oblong of about 3 miles circuit, enclosed by a substantial baked-mud wall 27 feet high. Towards the north end is the citadel, shut off by a massive mud wall, and to the west the tomb of Ahmad Shah, within an octagonal structure surmounted by a golden dome. The population is variously estimated at from 50,000 to 80,000, of whom the Hindkis are the wealthiest class. During the British occupation they have always developed a profitable trade with Bombay through Shikarpur and Karachi, but at other times are subject to heavy imposts. They import British produce, such as silks, calicoes, muslins, chintzes, merinoes, woollen and broad cloths, cutlery, needles, paper, besides indigo, spice, sugar, drugs, and other Indian produce. Their exports consist of madder, asafœtida, wool, dried fruits, tobacco, raw silk, besides such Persian goods as carpets, copper utensils, arms, turquoises, gold and silver braid, horses and "yabus," or baggage ponies. Whenever the railway is completed to this place Kandahar must become the great emporium of British and Indian produce for Central Asia. Its chief manufactures are silks, felts, and rosaries of a soft crystallised silicate of magnesia. The melons, grapes, and other fruits of the district are abundant, and of excellent quality.

The old citadel, of which many ruins are still standing, is situated a few miles outside the walls of the present city. This citadel has been the scene of remarkable sieges and defences.

The strategical value of Kandahar is increased by the fact that it is the first place where an army advancing from Herat towards the Indus would naturally rest to recruit its strength. It also gives access to the Ghazni and Kábul road through the Tarnak valley.

KANDAHAR.

Its proximity to the desert on the south renders one at least of its flanks safe from being turned. As it is further accessible from Persia and India west and east, it has incessantly changed hands during the period of its history—Persians, Usbegs, Afghans, and in recent times the English, having more than once occupied and relinquished it.

On the great military and trade route between Kandahar and Kábul the chief stations are Kalát-i-Ghilzai and Ghazni, the former of which is a strong fort standing on a commanding plateau on the right bank of the Tarnak. The fortified town of Ghazni, about as far south-west of Kábul as Kalát-i-Ghilzai is north-east of Kandahar, lies on the left bank of the Ghazni River near a spur of the Gilkoh range, and 7730 feet above the sea. It is surrounded by a lofty stone and brick wall on the top of a mound, and contains a citadel erected on an abrupt knoll at its north end. Besides Afghans it is inhabited by Hazaras and a few Hindki traders, dealing chiefly in corn, fruits, madder, sheep's wool, and camel's hair cloth from the Hazara country. Ghazni is memorable for its brilliant capture by Sir John Keane in the first Afghan war. Three miles to the north-east are the ruins of the old city, destroyed in the 12th century by the Prince of Ghor, who, however, spared the tomb of the renowned Mahmud of Ghazni. The entrance to this mausoleum, which is still preserved with careful veneration, was formerly closed by the famous sandal-wood gates brought by Mahmud from Somnath in Guzerat, but sent back to India in 1842.

Crossing the watershed north of Ghazni, between the Helmand and Indus basins, we enter the territory of Kábulistán, which takes its name from the ancient city of Kábul, the present capital of the State. Kábul lies seven miles above the confluence of the Logar and Kábul rivers, at the western extremity of a spacious plain in an angle formed by two converging ridges. It is now an open town, though formerly encircled by brick and mud walls. There are no noteworthy public buildings, and its interest, apart from its being the seat of government for nearly a hundred years, arises chiefly from its position at the junction of routes from Turkestán, Herat, Ghazni, the Kuram valley, and the Panjáb by way of Jelalabad. This happy situation has made it an emporium of Central Asiatic trade, notwithstanding the difficulty of the passes connecting it with the Oxus valley. To the south-east stands the Bala Hissar, or citadel, on a commanding knoll at the extremity of the spur overlooking the city Kábul imports from India calicoes, indigo, spices, drugs, and all kinds of British goods; from Russia broadcloth, silks, velvets, gold and silver lace, paper, and hardware, mostly by the long and toilsome route through Bokhara and Bamian. The province yields wheat, barley, and fruits, in considerable quantity and of good quality.

Of the alternative routes between Kábul and Jelalabad, the northern and more frequented leads over the Lattaband Pass, while the southern follows the narrow Khurd Kábul defile, where about 3000 men, women, and children perished on the occasion of Elphinstone's ill-fated retreat in January 1842. East of the Karkacha hills lies the equally ill-omened Jagdalak Pass, where the massacre of the retreating troops was continued, a few officers and men alone escaping to Gandamak. At this place, where the last treaty with the British was signed in 1880, a rapid descent leads down to the well-watered plain of Jelalabad, about midway between Kábul and Peshawar. This town forms an irregular quadrilateral surrounded by walls at the junction of roads from India, Kábul, and over the Hindu-Kush from Yarkand. It is noted for the brave and successful defence by Sir Robert Sale in 1841-2 against overwhelming numbers of Afghans. Hither it was that Dr. Bryden, sole survivor of the Kábul disaster, found his way in January 1842.

Between Jelalabad and Peshawar are the towns of Lalpúra and the fort of Ali-Musjid. The latter was stormed by the British at the beginning of the war with the late Amir Shír Ali.

Kelát, the capital of Baluchistán, is almost the only town in the country. It is a small fortified place in the centre of the province of Kelát, 6000 feet above the sea, badly built, and presenting an appearance of extreme squalor and decay. Its 12,000 inhabitants include representatives of nearly all the surrounding races—Brahuis, Baluchis, Afghans, Tajiks (here called Dehwars), Jats, and Hindus. Here resides the Khan, surrounded by a bodyguard of troopers, described by Bellew as tatterdemalions.

A far more important place strategically is Quetta (Shal), the capital of a district near the head of the Bolan Pass and close to the Pishin valley, from which it is separated by Mount Takatu, 10,504 feet high. By treaty with the Khan, Quetta has become an advanced British military station at a vital point on the southernmost route from India to Afghanistán, and about midway between Shikarpúr and Kandahar. Its occupation secures the Pishin valley, holds all the unruly Marri, Bugti, and other border tribes in check, keeps open the roads of the Khojak and Gwaja passes over the Khoja Amrán range, and thus facilitates a rapid advance on Kandahar. The valley of Quetta lying 5500 feet above the sea, and enclosed by mountain ranges which rise from 5000 to 6000 feet higher, is an extremely romantic spot, surrounded by rocky mountains.

In Makrán, Kej and the other so-called "towns" are mere clusters of hamlets, or insignificant fishing villages on the coast.

9. *Highways of Communication: Passes.*

One of the chief results of the recent hostilities in Afghanistán was the revelation that not two or three, as had been supposed, but at least a score of practicable routes give access from the plains of India to the Iranian plateau. From above Peshawar nearly to Karachi the intervening highlands are almost everywhere pierced by rivers and mountain torrents flowing down to the Indus, many of which run through gorges and ravines affording good passes to the interior. "What we have learnt chiefly in connection with them is this—that most excellent roads are easily constructed along even the worst of them" (*Captain Holdich*).

Still more surprising was it to find that the Hindu-Kush itself is crossed throughout its whole length by mountain tracks more or less practicable during the summer months from the Khawak westwards to the Irak Pass leading to Bamian. The Paghman range also, parallel and equal to it in height, is crossed by "durras" or paths leading from nearly all the large villages north of Kábul over the intervening Ghorband valley and Hindu-Kush down to Afghan Turkestán.

At the western extremity of the Paropamisus the Tajand and Murgh-ab valleys also afford ready access from what is now Russian Turkmenia to the Herat district. Two comparatively easy routes lead southwards to this city, one from Sarakhs through Zulfikar and over the Borkhut hills (900 feet), the other from Merv through Penjdeh and up the Kushk valley.

From Peshawar, north-western terminus of the Indian railway system, the great historic route to Kábul enters the Afridi hills near Ali-Musjid, thence following the Khaibar River over the Khurd Khaibar Pass (3370 feet) north-westwards and south of the Kábul River to Jelalabad. Here, crossing the Nangnahar plains, it ascends through the narrow Jagdalak defile to the Karkacha hills and dangerous Khurd Kábul Pass, with an alternative northern route over the Lataband Pass and the hills near Butkhak down to Kábul.

Farther down, the scarcely less important Kuram route to the capital runs by Thal and the Kuram River to near Fort Kuram, north-westwards, over the Paiwar range and Pass to Ali-Khel at junction of Rivers Karaia and Hazardarakht. Thence it follows the latter river over the Surkai Kotal between the Kuram and Kábul basins to the Shutargardan Pass (10,800 feet), and so on by Dobandi, Khushi, and the Logar valley, north to Kábul.

Between these two the ancient Gomul route ascends from the Derajat plains over the Kotal-i-Sarwandi water-parting to the Gomul or Gwalari Pass and thence to Ghazni.

The southernmost route to Afghanistán follows the new line of railway, now completed, from Sakkar on the Indus across the plains and Kachi desert to Quettah, near Pishin, at the head of the Bolan Pass. There are also alternative routes through the Bolan to Quettah, and through the Nari River valley to the Lora River and Pishin valley, and thence over the Khoja Amrán range, by the Khojak and Gwaja Passes, and across the Dori and other streams to Kandahar. The latter, though the longer, is the easier of the two, and will probably be followed by the railway through Quettah and over the Gwaja Pass to Kandahar.

The usual routes from Herat to Kábul are the northern, round by Maimana and Bamian, and the southern round by Kandahar and Ghazni. But the direct route across the Aimak and Hazara highlands up the Hari-rúd and east of Obeh is also occasionally used by the natives, and has been frequently traversed in eight days on horseback. The southern crosses the Zamindawar Durani domain by Farah, Girishk, and Khushk-i-Nakhud, to Kandahar. Here it follows the Tarnak valley to Kalát-i-Ghilzai, Ghazni, and over the Sher-i-

Dahar Pass between the Helmand and Kábul basins, and down the Shiniz River valley to its junction with the Logar, where it bifurcates through the Wardak and Logar valleys to Kábul.

Beyond those from the Indus through the Bolan and Mula Passes to Quetta and Kelát, no regular routes are yet laid down in Baluchistán ; but the longitudinal valleys running east and west parallel with the coast are often traversed, and give easy access from the eastern highlands to Persia.

10. *Afghan and Baluchi Administration.*

During the last twenty-five years, since the death of the Amir Dost Muhammad (Barakzai), Afghanistán has been so torn by internal and dynastic feuds that there is not much of civil administration can be said to exist beyond the collection of the revenue. Life and property are protected by the strong right arm ; there is no other protection. In so far as the late Amir Shir Ali had consolidated his administration, it is understood to have been unpopular. The tendency of Afghan society is towards minute tribal organisation, or to what might be termed a loose democratic federation—and all this militates against anything like a centralised administration.

In Baluchistán, the Khan is suzerain over a number of feudal chiefs, retaining a part of the territory under himself direct. The relations between him and them were so bad as to threaten the disruption of the State. But during recent years these arrangements have become much improved, and the satisfactory condition of internal affairs was evident during the Afghan war of 1878-80.

11. *Statistics.*

AREAS AND POPULATIONS OF EAST IRANIA AND AFGHAN TURKESTÁN.

	Countries.	Area in sq. miles.	Population.
Aralo-Caspian Basin.	Wakhan	3,000	3,000
	Badakhshan	8,500	158,000
	Kundúz	11,000	400,000
	Balkh	15,000	64,000
	Andkhui	6,000	60,000
	Shabirkhan	2,500	270,000 (Shabirkhan, Akcha, Sar-i-pul, Maimana)
	Akcha	3,300	
	Sar-i-pul	2,000	
	Maimana	15,000	
	Gurzivan	2,500	5,000 (Gurzivan, Darzab)
	Darzab	1,500	
	Total Afghan Turkestán	70,000	950,000 [1]
Hindu-Kush (Southern Slopes).	Kafiristán	7,000	5,000,000 (Kafiristán, Chitral, Swat, Afghanistán Proper)
	Chitral, Swat	6,000	
	Afghanistán Proper	217,000	
	Baluchistán	170,000	500,000
	Grand total	400,000	6,450,000

APPROXIMATE CLASSIFICATION BY RACES AND RELIGIONS.

Fire-Worshippers	Wakhis	Galcha Stock (Wakhis to Badakhshis)	3,000
Pagans	Chitralis		250,000 (Chitralis, Swatis, Siah-Posh Kafirs)
	Swatis		
	Siah-Posh Kafirs		
Muhammadans.	Safis and Chagnans		100,000 (Safis and Chagnans, Kohistanis)
	Kohistanis		
	Badakhshis		160,000
	Afghans	Iranic Stock (Afghans to Kurds)	3,000,000
	Tajiks (Dehwars)		800,000
	Baluchis		200,000
	Sistanis		50,000 (Sistanis, Kurds)
	Kurds		
	Hindkis	Indic Stock (Hindkis to Jats)	500,000
Pagans	Lassis		50,000 (Lassis, Lúris, Jats)
	Lúris		
	Jats		
Muhammadans.	Hazarahs	Mongol Stock (Hazarahs, Aimaks)	500,000
	Aimaks		
	Usbegs	Tatar Stock (Usbegs, Turkomans, Kizil-Bashis)	400,000
	Turkomans		50,000
	Kizil-Bashis		150,000
	Brahuis	Mongoloids ?	350,000

Pagans and Fire-worshippers	203,000
Muhammadans, mostly Sunnis	6,260,000
Total	6,463,000

[1] Grodekov's estimate.

CHIEF TOWNS

	Pop.		Pop.
Kábul	75,000	Khulm	10,000
Kandahar . . .	60,000	Ghazni	8,000
Herat	50,000	Kelát	6,000
Mazar-i-Sheríf . .	25,000	Sar-i-pul . . .	3,000
Andkhui . . .	15,000	Kundúz	3,000
Shibirgan . . .	12,000	Faizabad . . .	2,500
Maimana . . .	?	Jelalabad . . .	2,000

DISTANCES.

	Miles.		Miles.
Kábul to Herat . . .	600	Kábul to Peshawar . .	165
Herat to Mashhad . .	200	Sukkur to Sibi, by rail . .	140
Kandahar to Sukkur . .	410	Sibi by projected railway to Kandahar	300
Kundúz to Balkh . . .	105	Kandahar to Khoja Amrán .	90
Kandahar to Herat . .	335	Kohat to Ghazni	264
Kandahar to Kábul . .	290	Dera Ismail Khan to Ghazni, *via* Gomul . . .	250
Balkh to Bokhara . .	260	Dera Ismail Khan to Kandahar, *via* Sakhi Sarwar .	424
Balkh to Andkhui . .	100		
Kábul to Ghazni . . .	60		
Kábul to Jelalabad . .	75		

CHAPTER VIII.

THE INDIAN EMPIRE.

1. *Boundaries—Extent—Area.*

WITH a few comparatively unimportant exceptions, British India forms a vast geographical and political system, which, if it cannot everywhere yet boast of strictly "scientific frontiers," enjoys none the less at many points the advantage of the grandest natural boundaries of any region on the globe. For it consists mainly of a vast peninsular mass shut off from the Asiatic mainland by the lofty Brahui and Suliman ranges towards the north-west, and on the north by the still loftier Hindu-Kush and Himalaya, while it is elsewhere washed by the Arabian Sea and Bay of Bengal. In outline it presents the form of a somewhat irregular equilateral triangle with its base rooted in the Himalayas, whence it tapers across 28 degrees of latitude southwards to its apex in the Indian Ocean. Of this triangle the three sides fall about 100 miles short of 2000 miles each, the distance between the extreme frontiers of the Panjáb and Assam west and east, and from these points to Cape Comorin, at the apex, being as nearly as possible 1900 miles. The coast-line, although broken on the east side only by the small Chilka lagoon near Jaganath, and on the west by the more important inlets of the Katch and Cambay Gulf, has a total length of nearly 9000 miles. The land frontier is conterminous for no less than 5600 miles with the surrounding States of Baluchistán, Afghanistán, China, Burma (or Ava, as contradistinguished from British Burma), and Siam. Within these borders there is an area estimated at 1,490,000 square miles, with a population, according to the census of 1881, of 252,000,000, or about one-sixth of mankind.

2. *Relief of the Land: The Himalayas—Plateau of the Deccan—The Mountains of the Vindhya, the Satpura, the Ghats, the Nilgiris, the Aravalli.*

A good idea of the general relief of the land will be had by supposing it to subside about 500 or 600 feet below its present level. Such a slight subsidence, altogether imperceptible in the northern highlands, would have the effect of flooding all the plains at their base and converting the rest of the triangular mass into an island, shorn of a narrow strip along the east coast, but elsewhere almost intact. In other words, the Himalayas in the north would continue to present much the same outlines that they now do. The southern region of the Deccan also, forming an elevated plateau 2000 to 3000 feet above the sea, fringed on the north by the Vindhya range, and on the west by the Western Ghats, would be materially affected only on the east side, where a strip of low-lying and partly alluvial coast-lands intervenes between the low and interrupted range of the Eastern Ghats. But the space occupied by the Indus and Ganges valleys, known emphatically as the "Plains of India," and lying mainly between the Himalayas and the northern scarp of the Deccan, would disappear altogether, their place being occupied by a broad strait or channel connecting the Arabian Sea with the Bay of Bengal.

That such was the actual condition of things has

until recently been the generally accepted conclusion of geologists, who hold that the Indus and Ganges valleys are mainly the alluvia brought down by those great rivers and their numerous tributaries from the Himalayas and Vindhyas. And although this view is now combated by Mr. W. T. Blandford, its mere expression serves to give us a clear conception of the physical geography of India. For we thus see that this region consists of four highland systems—the Himalayas, Vindhyas, West and East Ghats; one vast plain, that of the Indus and Ganges valleys; and one vast plateau, the Deccan.

From the Great Pamir, focus of the continental highland systems, the Himalayas seem to break away south-eastwards in three main parallel lines—the Karakorum and Kailas or Gangri ranges, enclosing between them the valley of the Shayok, and the Himalayas proper, enclosing with the Gangri the Upper Indus valley. The Karakorum or northernmost range is known as the Tsung-ling, or Muz-dagh ("Ice Mountains"), to the natives, who reserve the term Karakorum to the pass of that name. Beginning at the knot of Púsht-Khar in 74° 30′ E. long., it forms an eastern continuation of the Hindu-Kush, sweeping round the northern frontier of Kashmir, and stretching thence in a south-easterly direction to the neighbourhood of the sources of the Indus in Tibet. Of its eastern continuation beyond the Chang-Chenmo Pass nothing definite is known, and it is still uncertain whether it forms a connection with the Kailas range about the sources of the Indus and San-po, or merges gradually with the Tibetan plateau. The highest elevations occur in the section between the Karakorum Pass and the Gilgit valley, where the Dapsang, near the pass (28,000 feet), and the crest marked K^2 on the Indian Survey maps (28,278), are, next to Mount Everest, the highest peaks on the globe. The northern extremity is broken by long transverse

THE HIMALAYAS.

To face page 232.

valleys, while the southern presents much more abrupt escarpments towards the Indus valley.

The general direction of the Muz-dagh from north-west to south-east is maintained at a mean elevation of 18,000 to 19,000 feet for some distance beyond the Karakorum Pass, after which it trends southwards, and again rises to imposing heights along the southern edge of the Tibetan plateau. The snow-line seems to rise on the north side to 18,000, on the south to 18,600 feet, and the Karakorum Pass leading from the Shayok valley to Yarkand is no less than 18,200 feet above sea-level. The Karakorum, rather than the more northern Kuen-lun, forms the true water-parting between the inland Asiatic and southern drainage. All the streams flowing from its southern slopes make their way through the Indus to the Indian Ocean, while those rising on its north side belong to the closed basins of Tibet or Eastern Turkestán.

Noteworthy in this alpine region are the numerous glaciers, the largest of which is the Baltoro, 33 miles long, and flanked on either side by two giant peaks over 27,000 feet high. Yet, vast as they are, these glaciers are mere remnants of the enormous ice and snow fields, which formerly covered the whole region of the Western Himalayas. These highlands are also exposed to sudden floodings, avalanches, and landslips, often causing wide-spread ruin in the upland valleys.

The Himalayas proper—that is, the "Abode of Snow,"[1] as they have been named by the Aryan inhabitants of the plains—constitute, if not the largest, by far the most elevated highland system on the globe. With a breadth varying from 180 to 220 miles, they stretch in a continuous curve of about 1500 miles along the Indo-Tibetan frontier between 72° to 96° E. long. from the western

[1] From the primitive Aryan root *hi, hu*, preserved in the Greek χεῖμα = Latin *hiem-s* = *winter, storm.*

limits of Kashmir to the eastern extremity of Assam. The main direction for nearly two-thirds of the distance to Mount Everest (29,002 feet), culminating point of the globe, is north-west and south-east, and thence nearly due east to the Indo-Chinese frontier. Throughout this vast distance a mean elevation is maintained of from 17,000 to 19,000 feet, while as many as forty peaks are known to exceed 24,000 feet—that is, a height greater than the loftiest summits of the Andes, or probably any other range beyond the Asiatic continent. The Himalayas, which do not form a single chain, but a number of more or less parallel ridges, with spurs often projecting in various directions, may be regarded as forming the southern scarp of the great Central Asiatic tableland, towards which they slope gently, while falling abruptly down to the Indian lowlands. Far inland lie the inmost ridges, which from the coast cannot be distinguished from the more advanced chains and transverse sections, often projecting far into the plains, above which they rise in a succession of steep rocky barriers to the Tibetan tableland. The southern foot of the main ridge is skirted by the marshy "Tarai," forming a watery hollow trough of great depth, extremely favourable to the growth of a luxuriant and even rank vegetation, but also perpetually shrouded in noxious exhalations rising from the dank ground. The Tarai, which traverses the British and Nepal frontier for nearly 500 miles east and west, lies at a lower depth than the plains from which it is separated by the outer and lowest ridges of the system.

A prominent feature of the Himalayas consists of the narrow gorge-like valleys of the advanced spurs, entirely destitute of waterfalls, and seldom presenting favourable sites for human abodes. But a few of the more gently sloping valleys, at elevations of from 6000 to 7000 feet, have been chosen for the summer retreats and sanitaria of the English officials, and even these are occasionally subject to sudden and destructive landslips.[1]

The Himalayas may be divided into a western, a central, and an eastern section The first begins at Mount Nanga-Parbat (26,629 feet), where the Indus suddenly trends southwards between Kashmir and Gilgit. Although there are here no well-defined ridges, there are several longitudinal valleys between which the Indus and other rivers flow for hundreds of miles before they can find an outlet southwards. Here also several peaks, besides the Nanga-Parbat, rise above 23,000 feet, the Nanda-Devi attaining an elevation of 25,661 feet.

The central section, forming the so-called Nepal highlands, and stretching from the source of the Indus to the Tísta, a tributary of the Brahmaputra, is intersected by numerous transverse valleys running north and south, and contains the highest summits on the globe. The most conspicuous peaks are the Dhawalagiri (26,826 feet) in the west, Gauri-sankar or Mount Everest (29,002) in the east, and Kanchinjinga (28,156) north of Darjiling on the Sikkim frontier. In the extreme north tower the glittering summits of the main chain, forming, as it were, the topmost foamy crests of these billow-like formations, which, after sinking twice to a depth of 10,000 feet, again suddenly fall to little over 1000 feet above the level of the plains. But before reaching the lowlands there is another abrupt rise to from 3000 and 4000 feet, formed by a long sandstone ridge rolling away towards the so-called "Bhaver," a dry wooded tract, which in its turn sinks through a succession of long undulations down to the Tarai.

The eastern section of the Himalayas, running west

[1] In the year 1880 the station of Naini Tal was partly destroyed by one of these landslips, which partly filled in a lake at the foot of the hills.

and east through Sikkim, Bhutan, and North Assam, while maintaining a mean elevation of 16,000 feet, presents no peaks comparable to the giants of the central and western sections. The highest known summit is Chumalarhi (23,650 feet). But much of this region still remains unexplored, and the eastern uplands, where the San-po suddenly disappears in a profound abyss, have never yet been visited by European or native surveyors.

South of the northern plains rises the triangular plateau of the Deccan, which has a mean elevation of from 2000 to 3000 feet, with a general incline eastwards to the Bay of Bengal. The northern scarp of this extensive tableland is formed by the Amarkantale Vindhya range, whose secondary sandstone formations are continued north-eastwards beyond Panna and Rewah nearly to the Ganges below Benares. Here is the water-parting between the streams flowing north to the Ganges basin and west to the Arabian Sea. The left bank of the Sone, which joins the Ganges above Patna, is skirted by the Khaimur range, separated by a broken plateau from the Panna ridge, which traverses Bundelkhand, and is noted for the deep gorges and isolated crags on its north-western slopes.

The steep southern slopes of the Vindhyas present the aspect of a weather-beaten coast-line, as if the valley of the Nerbadda now flowing at their base had once formed a deep inlet of the sea. This valley is separated southwards from that of the Tapti River by the parallel Satpura range, which runs east—from the classic Amarkantak, the source of the Nerbadda — westwards for nearly 600 miles at a mean elevation of 3000 feet, and culminates with the Pachmarhi hills (4500 feet), rising abruptly from the Nerbadda valley at Dhupgarh, east of Betul.

This culminating point of Central India is one of the

most hallowed regions in the Hindu world. Here is the renowned shrine of Siva, the *Mahadeo*, or "Great God," a term sometimes applied to the whole range. The road from Jilpa, the last village on the plains, lies through a romantic region that has been vividly described by J. Forsyth. After crossing the jungle it surmounts the scarp of the Pachmarhi plateau, which presents the aspect of a beautiful English landscape; and here, through breaks in the dense woodlands, a first glimpse is had of three isolated peaks all aglow in the fiery sunset, and standing out from the purple clouds banked up in the background. East of the plateau the rocky heights descend from an altitude of 2000 feet down to the vast level forest of Sal, while the scarps of the plateau are furrowed with mysterious abysses, one of which, the sacred and almost inaccessible Jambo-Dwip, forms an awe-inspiring natural marvel on the path of the pious pilgrim. These woodlands are the home of the bison and "sanbar," prince of red deer.

East of Asirgarh the Bombay-Allahabad Railway, and the main highway to Central India, cross this chain at a depression 1240 feet above the sea. But west of this point the system is continued to the Western Ghats by a highland tract 40 to 50 miles broad, with a mean height of 2000 feet, and several peaks from 3000 to 4000 feet.

The Western Ghats begin immediately south of the Kandeish valley, which separates them from the Satpura Mountains. From this point the Ghats—that is "Passes"—run close to the coast along the western edge of the Deccan southwards to the Nilgiri hills, where they meet the Eastern Ghats. The prevailing formation is trap, and indurated lava in the northern and central parts, culminating with the Mahabaleshwar Peak (4800 feet), and succeeded by sandstones and granites in the southern part. Like most coast ranges, the Ghats slope gently inland towards the central tableland, but fall abruptly down to the narrow strip of lowlands separating them from the sea. Here they are scored by the beds of deep watercourses, which in the rainy season are flooded by foaming torrents rushing over precipices and romantic waterfalls down to the coast. From the Tapti valley to the Nilgiris the Ghats maintain a mean elevation of about 4000 feet at a uniform distance of 30 to 40 miles from the sea.

The Nilgiris, or "Blue Hills," which culminate with the Dodabetta (8760 feet), form the converging point of the Western and Eastern Ghats, by which the plateau of the Deccan is enclosed. They cover an area of 700 square miles, and are noted especially for their genial and healthy climate, rendering them a favourite resort of Europeans enfeebled by the enervating heats of the plains.

The Eastern differ from the Western Ghats chiefly in three respects. They are much less elevated, with a mean height of scarcely more than 1500 feet; they do not form a continuous chain, being broken up into distinct sections by the valleys of the Godavari, Kistna, and other streams flowing to the Bay of Bengal; lastly, they run at a much greater distance from the coast, the intervening lowlands averaging from 50 to 80 miles. They stretch from the Mahanaddy River valley near Kattak for about 500 miles south-eastwards to the nucleus of the Nilgiris, beyond which they fall abruptly southwards to the so-called "Gap," a narrow, deep, transverse fissure, scarcely 400 feet above sea-level. North of the Godavari the Eastern Ghats attain an elevation of over 5000 feet.

South of the Nilgiris the Palni hills to the west of Madura are crowned by peaks 6500 and 7100 feet. These hills, like the Western Ghats, are extremely salubrious, and form a sanitarium for Europeans.

Beyond the above-mentioned "Gap," the extremity

of the peninsula is occupied by the independent system of the Cardamum Mountains from about the 8th parallel to Cape Comorin. In these highlands, which culminate in the lofty Anamalli hills (9700 feet), are found the highest elevations south of the Himalayas. They seem to be connected with the mountain system of Ceylon by "Adam's Bridge," a chain of rocky islets stretching between the Gulf of Manaar and Palk Strait from the mainland to the northern extremity of the island.

There remains to be noticed the somewhat isolated Aravalli range, running north-east and south-west across the Rajputana country, which they separate into two natural divisions—desert plains in the north-west, fertile and well-watered rolling lands in the south-east. At their southern extremity is the somewhat detached Mount Abu (5653 feet), highest point of the system, which has a mean elevation of about 2000 feet. Between Meywar and Marwar, where they rise to 4330 feet near the village of Jargo, the hills are crossed by the Dasuri Pass, which is alone practicable for wheeled traffic. The isolated character of the Aravalli range would be made evident by the already suggested subsidence of 600 feet, when they would appear as a long narrow rocky island about midway between the Baluch and Vindhya hills at the western entrance of the strait connecting the Arabian Sea and Bay of Bengal.

3. *Hydrography: The Indus, Ganges, Brahmaputra, Godavari, Kistna, Nerbadda, and Tapti Rivers.*

In its water system, as in many other respects, India presents a most striking contrast to the Iranian tableland. While this arid upland region is characterised chiefly by an inland drainage, and by a deficiency of large rivers, the Indian peninsula has absolutely no inland drainage at all, and possesses, in proportion to its size, a greater number of streams, all flowing seawards, than perhaps any other country in Asia. In the north nearly all these streams are collected into three vast systems, flowing either through the Indus to the Arabian Sea, or through the Ganges to the Bay of Bengal, or through the Brahmaputra and its affluent the Megna to the same bay. Even the Brahmaputra forms no exception to this general disposition, for its numerous channels are mingled with those of the Ganges delta before reaching the coast. But in the southern plateau of the Deccan there are almost as many river mouths as there are rivers, most of the large streams here forming separate systems, and finding their way in independent channels to the sea. This is true not only of the Mahanadi, Godavari, Kistna, Pennar, Kavari, and others, draining eastwards to the Bay of Bengal, but also of the Nerbadda, the Tapti, and the innumerable little mountain torrents rushing from the Western Ghats to the Arabian Sea. Thus it happens that, whereas the coast north of the Vindhya hills is broken only by the Indus and Ganges-Brahmaputra deltas, the southern seaboard is scored by at least fifty watercourses from the mouth of the Nerbadda to that of the Mahanadi. At the same time the volume of water sent seawards through the two great northern deltas is vastly greater than that of all the southern estuaries combined.

The Indus, like nearly all the great Asiatic rivers, has its farthest sources, not on the seaward slopes of the outer range, but behind the Himalayan escarpment of the Tibetan tableland itself. It rises on the north side of the Kailas range in 31° 20′, 82° E., near the sources of the Satlaj and San-po, and within 60 miles of the Karnali, farthest head-stream of the Ganges. The Indus flows first north-west through Ladak between the Kailas and main Himalayan range nearly to Gilgit, in 36° N., 75° E. Here it trends sharply southwards, maintaining this direction for the rest of its course through the Panjáb and Sind to its delta in the Arabian Sea between Katch and Karachi. In its upper course it receives no important tributary except the Shayok joining its right bank from the Karakorum range. But on emerging from the Himalayas it collects all the southern drainage of the Hindu-Kush through the Kábul River, which joins its right bank at Attock, almost on the frontier of British India. Lower down it receives the waters of the Suliman uplands mainly through the Kuram and Gomul Rivers. But the chief accession to its volume is from the united waters of the Jhelum, Chenab, Ravi, and Satlaj, all flowing from the western Himalayas and through the Panjnad joining the left bank of the main stream at Mithun-Kot in the Derajat, towards the Sind frontier. These four great tributaries, with the Indus itself, give their name to the Panjáb—that is, the "Five Waters"—beyond which province the united stream receives no further affluents.

It is remarkable that throughout its entire course of 1800 miles the Indus flows by no important towns, the only places of any consequence on its banks being Sakkar on its right bank, with the opposite town of Rori on the left bank, and Hyderabad near the head of its delta. Multan, Lahore, Amritsar, Wazirabad, and all the other large cities in its basin, which has a total area of 373,000 square miles, lie not on the main stream, but on or near the Chenab or other great tributaries. This fact may seem remarkable, inasmuch as the Indus was the first great stream occupied by the Aryans during their migrations from the north-west into the peninsula. The real cause is the shifting character of the banks below Kalabagh. During the rainy season the Indus is subject to sudden inundations which spread for miles along both banks, often causing great devastation, and preventing the foundation even of villages in its immediate vicinity.

From its source to the sea the Indus has a total fall of about 18,000 feet—that is, 8000 to Leh in Ladak, 9000 between that place and Attock, 1000 thence to the coast, a distance of nearly 950 miles. Hence the current in the upper reaches is extremely rapid, and even below Attock it runs at the rate of 6 miles an hour, mostly between high cliffs as far as Kalabagh. Here it enters the plains, suddenly widening to an average breadth of from a half to over one mile, with a mean velocity of rather less than 3 miles an hour. At low-water the tides are felt for nearly 80 miles from the mouth, and the Indus, like most of its great tributaries, is navigable to the foot of the hills for light craft. The delta is very extensive, reaching inland to Hyderabad, and from Karachi to the Rann of Katch, or about 130 miles both ways. The mean annual discharge through the mouths or through irrigating canals is estimated at over 150 billion tons, being about 41,000 cubic feet per second in December, and fully ten times that quantity during the August floods.

Although taking their name from the Indus, the Hindus still regard the Ganges as pre-eminently the great river of India. And in this they are so far justified that, although of shorter length than the Indus, it has a larger area of drainage, comprised entirely within the limits of the peninsula. For the Ganges differs in this respect from the other great Asiatic streams, that it rises, not behind the scarp of the plateau, but on the seaward face of the higher Himalayan range. Its two chief head-streams, the Bhagirati and Alaknanda, flow from an immense mass of snow 14,000 feet above the sea in the native Garwhal district, 31° N., 79° E. After a southerly course of about 80 miles the two streams unite a little above Hardwar, 30° N., where they burst through the outer barrier of the Himalayas, thenceforth flowing in a south-easterly direction through the rich alluvial plain of Northern India to the head of the Bay of Bengal.

SOURCE OF THE GANGES.

In the same district of Garwhal rises its great tributary, the Jamna, which pursues a nearly parallel course south of it to their junction at Allahabad. The Jamna carries to the common artery the drainage of Rajputana, Sindhia, and Bandelkand, collected by the Chambal,

Betwa, and Ken, all of which join its right bank below Agra. Below the junction the united stream still continues to receive several large affluents, of which the chief are the Són from the south, the Gumti, Gogra, Gandak, and Kúsi, from the Himalayas. At Hardwar the Ganges has a discharge of 7000 cubic feet per second, in the cold season, when the water is at its lowest, which at Benares has increased to 19,000, with a breadth in the rainy season of 3000 feet and a rise of 43.

For about 500 miles from its mouth it maintains a nearly uniform depth of about 30 feet, and a width of over one mile, while the fall from Hardwar to the sea scarcely exceeds 1000 feet. Hence the Ganges would afford one of the finest water highways to be found in any country but for the troublesome and even sometimes dangerous navigation of its shallow tortuous channel and numerous mouths. Of these the southernmost and most frequented is the Hugli, which gives access to large vessels for 100 miles as far as Calcutta. Beyond this point large boats ascend for upwards of 1000 miles along the main stream, and for perhaps five times that distance along its numerous tributaries, northwards to the Himalayas, southwards to the Vindhyas. "The navigation of the Brahmaputra and its affluents, of the Lower Ganges and its many branches, is quite magnificent, and offers probably one of the finest spectacles of its kind to be seen in the world. Not only every trader and landholder keeps many vessels, but every cultivator or peasant has his boats, and almost every labourer his canoe; thus the craft may be reckoned by hundreds of thousands. At several points on the great rivers the vessels congregate for several months consecutively, and form floating cities and marts, where many thousands temporarily dwell, where much barter takes place, where monetary transactions are arranged, and banking business is done" (*Sir R. Temple*).[1]

The united Ganges-Brahmaputra delta is of vast extent, probably the largest in the world, and of most complicate character, with the annual inundations constantly shifting its channels, and continually advancing towards the sea, which it discolours for a distance of 60 miles with over 235,000,000 cubic yards of matter yearly brought down to the coast. The delta extends for over eighty miles along the Bay of Bengal, and reaches for 200 miles inland, discharging through its innumerable channels 100,000 cubic feet per second during the dry and 500,000 during the wet season from April to August. At this time there is a rise of 32 feet above its ordinary level, which is sufficient to flood the whole country for 100 miles about the junction of the Ganges and Brahmaputra, leaving nothing visible except the tree-tops and the villages built on mounds raised above the highest level of the floods. Besides these inundations the delta is exposed to cyclones and to the phenomenon known as the "Bore," when a tidal wave five to ten feet high rushes up the Hugli with a roar at the rate of 18 miles an hour, often causing a rise of several feet as far up as Calcutta, or even 20 miles above it.

The upper course of the Brahmaputra had long been one of the most interesting geographical problems awaiting solution. But the recent explorations of the natives employed by the Indian Survey Office have at last practically confirmed the generally accepted view that the San-po of Tibet, the Dihong of Assam, and the Brahmaputra, form a continuous water highway, which has been traced throughout its whole course with the exception of a small gap, where the San-po plunges into a ravine and traverses a still unexplored region of the Himalayas.

[1] *India in* 1880, p. 319.

Although belonging properly to Tibetan geography it will be convenient here to deal with the San-po,[1] as forming undoubtedly the true upper course of the Brahmaputra. Its source has not been visited, but from information obtained by the Pundit's journey in 1865, it may be fixed with tolerable certainty in 82° E., 30° 35′ N., at a height of nearly 16,000 feet above sea-level, a little east of Lake Mansaraur, source of the Satlaj. This lake with the Rakus-tal partly fills the depression between Mounts Gurla and Kailas sources of the Ganges and Indus. Between the lake and the San-po the water-parting is very low; yet it suffices to send the Satlaj on a journey of 1000 miles to join the Indus on its way to the Arabian Sea, and the San-po for 1800 miles in the opposite direction to the Ganges delta.

Flowing first eastwards along the northern base of the inner Himalayan range, the San-po receives several tributaries on both banks, and near Shigatze it trends north-east with a huge bend, the apex of which lies above the intersection of the 94th meridian with the 30th parallel. It then turns south-east, passes through the above-described gorge in the Himalayas, and apparently reappears about 100 miles lower down as the Dihong of Assam. The unexplored gap is occupied by fierce and lawless tribes, whose hostile spirit, combined with the rugged character of the land, has hitherto defeated every effort to penetrate into this region, and clear up the mystery by actual observation.

Near the Buddhist monastery of Tadum (13,000 feet above the sea), where the Mariam-la route enters its valley, the San-po is already navigable for light craft; but lower down the navigation is obstructed at several

[1] The word, which in Tibetan means "holy water," occurs in a great variety of forms, such as Tsangbo, Tsambo, Tsanpu, Dzangbo, Sampo, Sambo, Sampu, Sanpu, Sanpo, etc.

points by shoals and rapids. It is also crossed by ferries and by light suspension bridges at many places where the stream is narrowed by projecting bluffs. At Chetang, a little below the junction of the Kichu from Lassa and 600 miles from its source, it is as large as the Rhine, and in the dry season 1400 feet wide, with a discharge of about 30,000 cubic feet per second, which during the summer rains is probably increased to over 700,000. Yet it is still uncertain what becomes of this vast body of water. For even accepting without reserve the theory, that it is the true upper course, not of the Irawady or Salwen, but of the Brahmaputra, a mystery still hangs over the true connecting link by which the junction is effected in Assam. The balance of opinion, however, inclines in favour of the Dihong or western branch, although the claims of the Dubong, Subansiri, or Lohit, are still advocated by some well-known names. Hence it will be wise to suspend opinion in the hope of some decisive information being obtained. The one point so far established by the Pundit Krishna, known as A. K., is that the San-po can have no possible connection with the Irawady, from which it is separated by a lofty range of mountains crossed by this explorer on his return to India from his travels in Tibet during the years 1878-82.[1]

But even independently of the San-po, the Brahmaputra proper represents a vast river system, filling the whole of the Assam valley, where it collects the waters of the Eastern Himalayas from the north, and those of the Naga, Khassia, and Garo hills, from the south and east. Here its chief affluents are the Dubong, Subansiri, Lohit, or Brahmakunda, the latter of which was long regarded as its true upper source. The main stream flows through the centre of Assam, nearly due west, to the 90th meridian, where it turns sharply to the south to the

[1] *Proceedings of the R. Geographical Society*, February 1885.

already described Ganges delta. Including the San-po, it has a total course of 1800 miles, an estimated area of drainage of rather over 360,000 square miles, and a mean discharge of 146,000 cubic feet per second in the dry season. But this discharge is vastly increased during the summer months, when the incessant rains of this watery region convert the river into a great inland sea, flooding the whole of the Assam lowlands, and cutting off all communication except by boats and causeways raised 10 or 12 feet above the level of the roads.

In the Deccan six considerable rivers find their way in independent channels to the coast—the Nerbadda and Tapti westwards to the Arabian Sea; the Mahanaddy, Godavari, Kistna (Krishna), and Kavari, eastwards to the Bay of Bengal. Of these by far the largest are the Godavari and Kistna, which are 900 and 800 miles respectively, and jointly drain the greater part of the region between the 14th and 22d parallels, representing a total area of over 206,000 square miles. They both have their farthest head-streams on the slopes of the Western Ghats, whence they follow nearly parallel winding courses through the Bombay Presidency and the Nizam's Dominions. As they approach the Eastern Ghats, the main streams gradually converge, and after traversing the narrowest part of the Madras Presidency, they enter the sea through two large deltas, which, during the floods, overflow into the intervening Lake Colair. This lake, or lagoon, which is the largest in India, is 47 miles long by 14 broad, and is entirely formed by the overflow of the Godavari and Kistna, whose lower courses between the Ghats and the coast are dammed by enormous dykes and connected by an extensive system of canalisation available both for irrigation and navigation purposes.

A somewhat similar parallelism is observed by the Nerbadda and Tapti, whose valleys are separated by the

Satpura range. The two streams rise from the Satpura range in the very heart of the peninsula, and gradually converge towards their respective estuaries near Baróch and at Surat in the Gulf of Cambay. The Nerbadda presents scenes of enchanting beauty, especially in its upper reaches about Jabalpur. Here it winds for a short distance with a narrow transparent stream of greenish-blue waters, between two glittering walls of snow-white marble, with here and there a vein of dark-green or black basaltic rock considerably heightening the effect of the marble. Near its mouth a fine prospect is also commanded from the noble railway bridge which crosses the estuary at Baróch. Although much obstructed by rapids, it is navigable by boats for 250 miles to the falls of Dari. It has a length of 800 miles, or about double that of the Tapti, and the united drainage of these two rivers is somewhat over 63,000 square miles.

4. *Main Natural and Political Divisions.*

To the three main physical divisions of the peninsula correspond, on the whole, its three great political administrations. Thus the northern highlands and lowland plains are mostly comprised in the Presidency of Bengal with its dependent feudatory States, while the southern plateau of the Deccan is divided between the Presidencies of Bombay and Madras, the former embracing the Western Ghats and Malabar coast-lands, the latter including the Eastern Ghats and Coromandel coast-lands. Through the universal acceptance of British rule the whole region has also in recent times acquired a general political unity, answering to such general physical unity as is derived from its peninsular form and tropical climate. But beneath this broad uniformity we are everywhere confronted with a dualism, betrayed especially in the social, religious, and political worlds. Thus we find a society based on caste intermingled with communities which ignore all class distinctions—Brahmanism invaded even in its most hallowed precincts by Islám; territory administered directly by the paramount power everywhere in contact with tributary and even with semi-independent States.

In the northern highlands, besides the independent States of Bhutan and Nepal, we have the semi-independent feudatory States of Sikkim, Garwhal (Tirhi), and Kashmir, collectively occupying most of the southern slopes of the inner Himalayan ranges. These States are attached to the Presidency of Bengal, an expression the original meaning of which has become considerably modified. Bengal is still understood to comprise all the British territory not included in either of the other two Presidencies of Bombay and Madras. But as a matter of fact the Bengal Presidency is itself now subdivided into nine distinct administrations (including the three Lieutenant-Governorships), which, with those of the two southern Presidencies and attached Native States, will be found fully tabulated at the end of this chapter.

Kashmir and Jammu.

In the extreme north-west the basin of the Upper Indus, probably the grandest alpine region in the world, is almost entirely comprised in the territory of the feudatory prince Golab Singh, Maharajah of Jammu and Kashmir. This State, which by the treaty of Amritsar, March 16, 1846, accepted the paramount sovereignty of England, embraces within its borders a great variety of climatic, physical, and ethnical conditions, stretching as it does from the hot plains of the Panjáb for 280 miles to the eternal snows and glaciers of the Western Himalayan and frontier Karakorum ranges, and from the Hindu-Kush for 400 miles east to Tibet. It is essentially a highland region, almost everywhere mountainous, but having one splendid valley (Kashmir), broad, long, and populous. Moreover, there are many broad upland valleys, extremely fertile, well sheltered by the towering Himalayan crests from the northern blasts, and watered by copious streams all draining to the Indus or to its tributaries.

Physically speaking, the whole country may be divided into three zones, rising in successive terraces from the Panjáb lowlands to the Karakorum range. The lowest and southernmost of these zones comprises the more advanced hilly districts with a mean elevation of 2250 feet above sea-level. This is succeeded by the central zone between the Himalayas proper and the Kailas range, from 7000 to 9000 feet high, beyond which follows the truly alpine region of Baltistan, or "Little Tibet," between the Kailas and Karakorum, with a mean elevation of 11,000 feet, and culminating with the nameless peak K^2, 28,278 feet, next to Everest the highest point on the globe. Here also are the Baltoro and many other glaciers, which, vast as they are, seem to be but the poor remains of the prodigious icefields which must have formerly covered the whole region of the Himalayas. The melting of the snows in the fierce summer sun, combined with the precipitous slopes and the silent action of underground waters, exposes all these upland valleys to sudden floodings, avalanches, and landslips, often causing widespread ruin.

In the central zone lies the lovely vale of Kashmir, 4500 square miles in extent, hemmed in on all sides by snow-clad peaks and watered by the Jhelum, which in its placid winding course flows through the Wular and several other beautiful lakes. Thus pent up, and with an elevation of over 5000 feet above the sea, this romantic valley presents somewhat the appearance of a vast cirque with a narrow southern outlet, through which the Jhelum escapes towards the Indus. Kashmir has ever been the theme of Eastern song, an earthly Eden, where prevails a perennial spring, and of which the Mogul emperor, Shah Jahan, was wont to say that he would prefer to sacrifice all his vast Indian dominions rather than be deprived of this delightful retreat. Here the picturesque elements are the snowy peaks, the romantic gorges, the numerous lakes, streams, and waterfalls, the magnificent woodlands, and rich flowery meads, —a combination of natural beauties scarcely to be found elsewhere concentrated in an equal area.

Of the numerous passes leading into the Kashmir valley, and practicable for pack animals, the chief are the Banihal (9700 feet) from the south, the Punch (8500) from the west, and the Pir Panjal (11,500) from Gujarat in the Panjáb. The present Maharajah seems no doubt busy with reforms of all sorts, and holds open weekly courts accessible to the humblest of his subjects. But the land tax and other parts of the fiscal administration are not thought to be successful on the whole, and the valley has suffered from fearful visitations of famine and pestilence. The famous fabrics and other indigenous products of the valley are maintained in all or nearly all their beauty, but the trade as a whole is hardly increasing. The shawl-weavers receive very low wages, and are prohibited either from leaving the country or changing their occupation. The finest goats' hair used in the manufacture of these shawls comes, not from the country itself, but from Turfan in Yarkand.

About one-third of the whole population, or 500,000 souls, are concentrated in the vale of Kashmir, which might easily support a much larger number. Female infanticide probably exists in the sub-Himalayan tracts, and till recently "suttee," or widow-burning, was still practised in Jammu, where it was more fanatically enforced than elsewhere in India.

Nepal.

South-east of Kashmir follow the Native States of Garwhal (Tihri), Nepal, Sikkim, and Bhutan, occupying most of the southern slopes of the Himalayas nearly to the great southern bend of the San-po, where it disappears in the North Assam highlands. The central section is almost exclusively comprised in the nominally independent kingdom of Nepal bordering east on Sikkim and Darjiling, west and south on British territory. Nepal thus consists of a comparatively narrow strip 550 miles long and 160 broad, limited northwards by the crests of the inner Himalayan range (here culminating with Gaurisankar, the highest peak on the globe), and falling in a series of five continuously diminishing terraces and deep intervening troughs down to the Indian plains. One of these troughs is the already described Tarai, which sinks even to a lower level than the open plains, and forms the chief physical feature along its southern border. In some places it is overgrown with a low jungle, very sparsely inhabited, while it consists elsewhere of uninhabitable wastes covered with a coarse growth of grass. Some parts of the Tarai frontier, however, are very fertile. The heart of the country comprises a delightfully well-watered and productive caldron-shaped valley, in which is situated Katmandu, capital of the State.

Nepal, which is despotically ruled by a hereditary minister of the warlike Ghurka tribe, under a titular Maharaja, abounds in mineral wealth, including copper, iron, lead, arsenic, and sulphur. The principal valley of Katmandu is well known; it is a sort of second Kashmir.

Beyond that, the interior is very little known, and such is the jealousy of the Government that no Englishman is allowed to pass beyond the Katmandu valley under any pretext. Hence no surveys have here been yet carried out, and the British Government is still without exact information regarding the relations between Nepal and China. A mission bearing presents of a prescribed value is said to proceed every five years from Katmandu to Pekin, although Nepal, while independent in internal administration, is in the position of a feudatory State to India.

Sikkim.

A somewhat similar position is held by the adjoining petty State of Sikkim, stretching for 52 miles between Nepal and Bhutan, and for 66 between Tibet and India. Physically, Sikkim forms an eastern continuation of Nepal, from which, however, it is separated by a lofty transverse ridge of the Himalayas, 11,000 to 12,000 feet high, and culminating northwards with Mount Kubru (24,015 feet). The crest of this ridge is so sharply defined that it may be traversed for 40 or 50 miles at a uniform level almost without a break. But politically, Sikkim owes allegiance both to Tibet and India. The Rajah, who resides at Tamlang (Tumlong), on the Indian side of the Chola range, in winter, and in summer in the Chumbi valley on the Tibetan side, accepts allowances from both countries, about £200 from Lhassa, and £1200 from Calcutta, the latter grant being made on the condition of his affording every facility for the trade between the two countries. He is, however, absolutely a feudatory of the British Empire. The Chumbi valley, traversed by the Am-mo-chu, a head-stream of the Brahmaputra, is a south-eastern corner of Tibet, wedged in between Sikkim and Bhutan, near a lovely lacustrine district in east

Sikkim. The lakelets, which lie at elevations of 10,000 to 16,000 feet at both sides of the border range, are mostly tarns or closed basins, evidently due to glacier action.

Bhutan.

Still less known than Nepal is the State of Bhutan, lying mainly between Tibet and Assam, east, north, and south, and stretching from Sikkim for 400 miles eastwards to the unexplored region separating the San-po from the Dihong. The surface is intersected by two parallel ranges intervening between the inner Himalayas and Assam, the first enclosing a bleak and almost uninhabitable tableland; the second skirting the "Duars," a fertile tract ceded in 1866 to the British in return for a yearly subvention.

In the north the country is extremely wild, but elsewhere Bhutan affords some of the grandest and most romantic scenery in the world.

In the more sheltered districts Bhutan produces millet, wheat, and rice in abundance. But it is mostly uninhabited, with a population of scarcely more than 300,000 (?) altogether. It is nominally ruled by the Dharm Raja, a sort of incarnate Buddha. But the real head of the State is the Deb Raja, elected every three years by the Penlows or chiefs from their own body. Commercial relations are confined mainly to Tibet and Assam, to which countries musk, madder, coarse cloth, and horses, are exported in exchange for cottons, woollens, tea, gold, silver, and embroidered work. The capital, Punakha or Dosen, occupies a position of great natural strength on the Bugni River, 96 miles north-east of Darjiling. But a better-known place is Tasichozong (Tassisudon), on the Gudada River, centre of the peculiar form of Lamaism prevalent in the country.

India Proper—The Panjáb.

From the foregoing account of the highland northern States, it appears that the Himalayas still belong to some extent politically, as they mostly do physically and ethnically, rather to Central Asia than to India proper. They form, in fact, the outer and southernmost barrier of the great Central Asiatic tableland, and although Indian influences and elements of race are here everywhere more or less perceptible, the bulk of the Himalayan aborigines belong not to the Aryan, but to the Tibetan Mongoloid stock. Hence the Indian peninsula may be said properly to begin with the plains, which stretch along the base of the great northern barrier from the eastern scarp of the Iranian plateau to the Bay of Bengal.

These plains are wholly comprised in the Indus and Ganges basins, and lie nearly altogether in the Presidency of Bengal. In the extreme west, however, the lower portion of the Indus valley, embracing the province of Sind, is exceptionally included in the Bombay Presidency. But higher up the whole of the Indus valley, as far as the Hindu-Kush and Kashmir uplands, is included in the Panjáb Province, one of the separate administrations of the Bengal Presidency. In the north this province is very hilly, comprising several ranges separating the upper courses of the large rivers between the Jamna and the Indus. In the extreme north-west the hills beyond the Indus form a sort of connecting link, broken by the Kábul River valley, between the Himalayas and the Suliman range, which skirts the province along its western border for nearly 400 miles. Farther south the open plains fall very gradually from an elevation of about 1600 to 200 feet above the sea at the confluence of the Indus, with its united tributaries. Here the riverain tracts are generally extremely productive, whereas the so-called "doabs," or "two-waters"—that is, the spaces enclosed between the great rivers, which give their name to the province—often consist of mere wildernesses of scrub and jungle, but generally afford extensive pasture for cattle, camels, sheep, and goats. The country between the Indus and Jhelum is generally rugged, cultivated in parts only, and thinly inhabited, intersected by the Salt range, with a mean elevation of 3000 feet, and culminating with the Sakesar peak, 5010 feet. On these heights the salt frequently crops out, and there are several large salt deposits. West of the Indus the range is continued under the name of the Kalabagh hills as far as the Suliman highlands, and the long narrow alluvial strip between these highlands and the river is known as the Upper and Lower Derajat.

Besides Kashmir there are over thirty Native States attached to the Panjáb. Among the most important of these is Bahawalpur, occupying a strip about 17,000 square miles in extent between Rajputana on the south-east, and the Satlaj, Panjnad, and Indus on the north-west. Practically the most important are the so-called Phulkhian States (Patiala, Jhind, and Nabha), with a joint area of 7500 square miles in the hilly and plain tracts between Delhi and Lahore. North of these are the "Simla Hill States," traversed by the Upper Satlaj, and reaching eastwards to the Tibetan frontier. None of these are of any size except Basâhar (Rampur), which has an area of over 3000 square miles.

Sind.

The southern province of Sind, although, as stated, included administratively in the Bombay Presidency, cannot be separated physically from the Panjáb. They both merge eastwards in the Rajputana wastes, and the western parts of the Sind lowlands still consist of waterless steppes yielding little beyond a scanty pasture for herds of buffaloes, asses, and camels. Even in the more productive Indus delta barren and swampy tracts are everywhere intermingled with the cultivated fields.

With but few exceptions, the cultivation in this province depends on a large series of canals drawn from the Indus, which river is to Sind what the Nile is to Egypt, and has caused the province to be called "the lesser Egypt." These channels are filled when the river rises in the summer and are dry when it subsides in the winter.

The whole of this low-lying delta region is for a long way up stream exposed to the inundations of the Indus, which reach east to the "Thar" or great Rajputana desert, and west to the foot of the Baluchistán hills. East of the delta the Gulf of Katch penetrates far inland, skirting the north side of the Gujarat or Kattywar peninsula, and gradually merging in the so-called "Ranns" of Katch. These remarkable formations, consisting of two portions, the Great and Little Rann, with a total area of 9000 square miles, become sandy, saline swamps, and inland lagoons or arms of the sea, according to the season of the year. When flooded they connect the Gulfs of Katch and Cambay, thus converting the Native States of Gujarat into an island. Northwards they are confined by the Allahband—that is, the dam or mound of Allah, and both dam and lake owe their existence to an earthquake which occurred in the year 1819.

Rajputana.

In most coloured maps of the peninsula a large space towards the north-west will be noticed marked off from British territory proper, as belonging to Native States.

In the very centre of this region is the small British enclave of Ajmir-Merawa, about 2700 square miles in extent, which is administered by a British officer. It consists partly of an elevated plateau, partly of a picturesque hilly district at the northern extremity of the Aravalli range, and is politically distinct from the surrounding plains; for with the exception of this enclave the whole of the region in question is divided amongst a large number of feudatory Native States attached to the Bengal Presidency. These States are disposed in two distinct geographical and political groups, under the control of the Rajputana and Central Indian Agencies respectively.

The Rajputana Agency has a total area of no less than 129,000 square miles, and is bounded north-west by the Panjáb and Sind, north-east by the North-West Provinces, south-west by Sind and Gujarat, south-east by the Central Indian Agency. Rajputana is divided by the Aravalli range into two unequal parts, of which the north-western or larger consists to a great extent of sandy, arid, and unproductive wastes, with some arable and even fertile tracts towards the north and north-east. Here is the *Thar*, or great sandy desert of Northern India, intersected everywhere by long parallel dunes 50 to 100 feet high, with few streams or wells, and a scant vegetation of tufty grass and scrub.

Considerably more elevated and fertile is the south-eastern division of Rajputana, which is diversified by wooded rocky hills, and watered by the Chambal, Banas, and some other large rivers flowing north to the Ganges basin. The country between the Chambal and Patar consists of a rich black loam, highly productive and well cultivated. But even in this division most of the surface is stony, rugged, under jungle, and unfertile, except close to the river banks.

Of the twenty Rajputana States the largest are Udeypore (Meywar), Jeypore, Jodhpore (Marwar), and Jeysulmir. But all except Shahpura and Lawa belong to the first rank in the empire, being under treaty with the Imperial Government. Of the Bhíl tracts between Sirohi and Dungarpur some are directly administered by British Commissioners, while others are either tributary to Udeypore or under the control of Rajput princes.

The Marwar and Bikanir tracts are essentially pastoral, abounding in cattle, sheep, and a superior breed of camels. But elsewhere Rajputana is mainly agricultural, yielding grain, cotton, and opium in considerable quantities.

Central India Political Agency.

The Central India Agency comprises all the region lying between Rajputana and the British Central provinces, and stretches from Gujarat eastwards to Chota-Nagpore and the North-West Provinces. It is divided by British territory into two sections—native Bundelkhand and Baghekhand lying to the east, and the Central India portion to the west of the Jhansi and Lalatpur districts of the North-West Provinces. The eastern section forms a part of the Deccan plateau, here intersected by the Khaimur and Panna ranges, and watered by the Ken, Betwa, and Són, flowing to the Jamna and Ganges.

The western section consists of a broken upland tract stretching from the Vindhyas northwards to the Jamna, mostly fertile, and well watered by the Chambal, Parbatti, Sind, and Betwa. Wheat, rice, cotton, sugar, and especially opium, are the staple products, and iron, copper, and coal, besides diamonds, exist in many places. The diamantiferous district, yielding several thousand pounds' worth yearly, lies some 14 miles north-east of

Panna, and the stones here found are of four different tints, all, however, inferior to those of the Kistna valley.

The Central India Agency has a total area of 86,000 square miles, and comprises nearly sixty feudatory States, of which the largest and most important are Gwalior (Sindhia's), Indore (Holkar's), Bhopal, and Rewah. Most of the numerous petty States have feudal relations with one or other of the larger ones, while their autonomy is guaranteed by the paramount power.

North-West Provinces—Behar—Bengal.

The Ganges valley forms a well-defined natural region, occupying the whole space between the Himalayas and the Deccan, and comprising the separate governments of the North-West Provinces, Oudh and Bengal, including Behar, in the Bengal Presidency. These vast alluvial plains, watered by the Ganges, Jamna, and their numerous tributaries, form the very heart of British India, in which nearly one-half of its entire population is concentrated. The fertility of the soil and the thrifty habits of the people have here produced the same results as in the rich alluvial valleys of the Yang-tze and Hoang-ho. In both of these regions the density of the population is estimated by several hundreds to the square mile, amounting in the North-West Provinces to 380, in Oudh to 470, and in Bengal to 484. These proportions are considerably more than treble those of France and other European States, which are regarded as fairly well peopled, and the census returns for 1881 showed a total of no less than 114,000,000 for the three above-mentioned provinces alone. Yet many parts are covered with dense jungle, tenanted only by wild animals, while others are either barren wastes, uninhabitable swampy tracts, or rugged uplands, necessarily but very thinly peopled. Hence the density in the more populous alluvial and cultivated districts is much higher than might be supposed even from these astonishing figures. Such teeming multitudes, which the official returns show to be steadily increasing, especially in Bengal (including Behar), could not possibly be supported even in these exuberant lands except on the most frugal diet, and as a matter of fact rice, the cheapest of all grains, forms the staple, in many cases almost the exclusive, article of food for the great majority throughout the Ganges basin.

In this basin are comprised, besides the lowland plains, extensive highland tracts, consisting in the north chiefly of the outer Himalayan ridges, and in the south of the more advanced spurs and offshoots of the Vindhyas, which near Benares and other points approach to within a few miles of the Ganges. About Darjiling in British Sikkim the hills rise to elevations of from 6000 to 8000 feet, commanding a view of Kanchanjanga (28,156 feet), lying 45 miles due north on the frontier of Tibet and independent Sikkim. These Darjiling hills are extremely interesting, not only for their magnificent scenery and glorious vegetation, but also as forming the "divide" between the Ganges and Brahmaputra basins and between the Hindu and Buddhist religious worlds.

It would be difficult to conceive a greater contrast than that which exists between these cool or cold mountains and the watery plains of the Lower Ganges, whence we pass by a natural transition to the still more watery region of the Lower Brahmaputra valley. All this low-lying tract, as far east as the Garo hills and British Burma, is included in the Bengal Provinces, which thus consist altogether of four different sections—Behar, enclosed west and north by the North-West Provinces and Nepal; Chota-Nagpore, stretching from Behar southwards to the Central

Provinces ; Orissa, lying between Chota-Nagpore and the coast ; and Bengal proper, comprising the united Ganges-Brahmaputra delta, and stretching north to Sikkim, west to Behar, east to Assam, the independent Lushai hill tribes, Burma, and Arracan, but overlapping or enclosing the independent Hill Tipperah country on the south-west frontier of Assam.

Below the Himalayas Behar is mainly alluvial and level, but rises westwards to the Rajmahal hills in the Santhal Parganas. Chota-Nagpore, on the contrary, is an upland rugged country, embracing the eastern spurs of the Vindhyas, with elevations of from 2000 to 4400 feet ; while Orissa consists of a flat diluvial tract between these hills and the coast, and an extensive hilly district in the interior occupied by petty tributary States. These hills also form a continuation of the Vindhyan system, which here culminates in the Parasnath (4480 feet), close to the East Indian Railway, about midway between Benares and Calcutta. Far higher are the elevations on the east frontier of Lower Bengal, where the Tipperah highlands, forming a continuation of the Lushai and Manipur ranges, attain an altitude of from 11,000 to 12,000 feet. West of these hills stretches the great delta with its "thousand mouths," its intricate network of countless channels and backwaters, its highly-cultivated and densely-peopled inland Backergange tract, its almost impenetrable coast region of the Sundarbans, covered with dense jungle, and still a prey to wild beasts, terrific cyclones, and deadly exhalations. Here land and water still struggle for the mastery, while unbridled nature laughs at the feeble efforts of man to tame the jarring elements. The work of the day is swept away by the raging night storm, and the patient labour of years often suddenly disappears in a chaos of widespread ruin. Nevertheless, the Saugor Light, firmly established at the entrance of the Hugli, shines like a beacon of future promise, that one day even this wild region will be brought under the control of man.

Assam.

Beyond the Brahmaputra section of the delta lies the province of Assam, occupying the north-eastern extremity of the empire between Bhutan and Tibet on the north, Burma on the east and south-east, the Manipur Lushai and Manipur hill States on the south. The administrative province embraces the Brahmaputra and Surma (Barak) river valleys, with the intervening Naga, Jaintia, Khasi, and Garo hill tracts.

But Assam proper is confined to the Brahmaputra valley, an extensive alluvial plain 450 miles long by about 50 broad, everywhere enclosed by lofty ranges, except towards the west, where the Brahmaputra escapes towards the Ganges. This plain, however, is diversified by innumerable rivers, boundless woodlands, extensive prairies, and even by isolated ridges at some points approaching close to the Brahmaputra. The number of watercourses is probably greater than in any other country of equal extent, no less than sixty considerable streams having been enumerated in this narrow tract, all connected together by a labyrinth of channels and branches. Thanks to this superabundance of water, Assam is one of the most fertile regions in India. But owing to the dense forests and lofty enclosing ranges impeding the free circulation of the air, the moist climate is oppressively hot, and all the more unhealthy because the rainy season continues here longer than in any other part of the empire except British Burma. It lasts from March to November, when the low-lying riverain tracts are often completely flooded. The slow evaporation of these liquid masses charges the atmosphere with dank vapours, generating ague, dysentery, and other malarious disorders. Even during the cold season dense fogs usually prevail in the plains from midnight to noon, enveloping everything in an impenetrable misty veil. Assam, however, is beyond the reach of the hot winds, which in May and June convert many parts of the peninsula into a glowing furnace. It abounds in coal, iron of excellent quality, sulphur, salt, and petroleum, and after the rainy season the natives search the streams for the gold-dust brought down with the alluvia from the hills.

British Burma.

From the Chittagong district forming the south-eastern limit of Bengal proper there stretches a territory for about 1000 miles between the Bay of Bengal on the west, and on the east Upper Burma and Siam southwards to about the 10th parallel. This region forms the province of British Burma, and consists entirely of Burmese territory at various dates during this century ceded to the English. It comprises three divisions, Arakan, Pegu, and Tenasserim, which completely shut off Upper Burma from the sea, and about the 12th parallel nearly reach across to the Gulf of Siam, between Siam proper and its lower province at the neck of the peninsula of Malacca. In Malacca itself the Straits Settlements, terminating at Singapore at its southern extremity, continue the British domain almost uninterruptedly round the Bay of Bengal to the Eastern Archipelago.

Arakan, the northernmost division of British Burma, presents from the coast a fine appearance. The mountains forming a southern continuation of the Lushai hills, and clothed to their summits with a rich forest vegetation, rise in a succession of parallel ridges from the plains to a height of from 5000 to 6000 feet. The plains themselves are of small extent, being mostly either limited by offshoots of the lower coast ranges, or else hemmed in by wooded tracts, which on the coast consist exclusively of mangrove trees. The lowlands are intersected by countless streams from the hills, while the spring-tides flood extensive low-lying districts, forming a labyrinth of channels and backwaters. These watercourses take the place of highways, serving as a means of rapid intercourse between the towns and villages. Mud volcanoes occur both along the coast and on the neighbouring islands, and coal, iron, and petroleum are found in many places, while salt of a fine quality is obtained by evaporation in the numerous tidal estuaries.

Pegu comprises the region of the Lower Irawady and Sittang Rivers, which here form a common wide-branching delta, in its main features resembling that of the Brahmaputra-Ganges. The land is mostly low, sandy, or muddy, and in the wet season exposed to destructive floods. But it is well suited for the cultivation of rice, which is here produced in superabundance. Its trade and industries are also furthered by the railway, 163 miles long, between the capital, Rangoon, near the coast, and Prome, higher up the Irawady, towards Upper Burma. This is the only line that has yet been permanently opened in any part of the Asiatic mainland east of Cis-Gangetic India.

The mountain system throughout the whole of this coast region is of a very simple character, consisting of regular and parallel ridges running uniformly north and south, and forming water-partings between all the large rivers, which thus find their way independently to the Bay of Bengal.

For a portion of its lower course the Salwin forms the

border line between Siam and Tenasserim, the southernmost of the three great divisions of British Burma. This division, whose southern extremity approaches the insular region of Malaysia, is itself fringed along its entire length by a vast number of islands forming in the north the Moscos, in the south the much larger Mergui Archipelago. A few only of these little-known islands are inhabited, chiefly by Burmese and Karens from the opposite mainland. They are all hilly, with peaks 2000 to 3000 feet high, and are often densely wooded with the caoutchouc and other valuable trees. They are said to abound in minerals, and are tenanted by the tiger, rhinoceros, deer, and a great variety of reptiles.

The eastern frontier of Tenasserim is formed by a mountain range 5000 feet high, which again acts as a water-parting between the Tenasserim and the Siamese river systems. On the British side the chief river is the Tenasserim, named from the capital, and flowing between the hills and the coast, mainly south, for over 230 miles, of which about 100 are navigable.

The Central Provinces.

The Deccan, in the largest sense of the term geographically, is distributed politically between the Central Provinces, forming a portion of the Bengal Presidency, and the two southern Presidencies of Madras and Bombay, together with the Native States, of which by far the most important are the Nizam's Dominions and Mysore.

The Central Provinces form a British enclave almost everywhere cut off from British territory proper by intervening feudatory States. But the broad political boundaries are the Chota-Nagpore States of Bengal on the north, the tributary Native States (of which some belong to the Orissa province and others to the Madras Presidency) on the east, the Nizam's territory on the south, the Central Indian Agency on the west and north-west. They form an irregular square (almost a triangle) about 600 miles long east and west, by 500 north and south, with a total area, including the Berars, of over 130,000 square miles. They constitute the northern portion of the Deccan plateau, here divided into two sections by the Satpura range, with a mean elevation of 1500 to 2000 feet, and a general eastward tilt. But the surface is everywhere diversified with hilly plains and river valleys, and on the south it is enclosed by the upland Bastar tract reaching from the coast to the Godavari, and stretching thence, under different names, westwards to the Khandeish plateau.

This extensive region is traversed by the Nerbadda, Mahanadi, Wainganga, and Wardha, flowing generally in deep beds, and navigable for long distances during the rainy season. Large tracts are still covered with dense virgin forests, but others are well adapted for the cultivation of cereals, and cotton of the best quality is grown, especially along the right bank of the Wardha.

The Nizam's Dominions.

In the Nizam's Dominions are included East and West Berar, forming the Hyderabad Assigned Districts. These were, under the treaties of 1853 and 1861, assigned by the Nizam to the British Government, which, on its part, undertook to maintain a body of troops, to be styled the Hyderabad Contingent; but the sovereignty is still retained by the Nizam.

The Nizam's Dominions still form by far the largest and most important of all the Native States in the empire. They comprise the very heart of the Deccan, lying mainly between the two great rivers Godavari and Kistna north and south, and between the Presidencies of Bombay and Madras west and east, and stretching about 475 miles both ways, with an area of close on 100,000 square miles. It consists of an elevated plateau, sloping from 2500 down to about 1000 feet towards the Eastern Ghats which skirt its south-eastern frontier. Much of the surface is still waste and covered with low brushwood, but the soil is naturally fertile, and where irrigated produces heavy crops of cotton, cereals, oleaginous plants, and even dates.

In the Kistna valley within the Nizam's territory are the famous Partial and Kollur diamond-fields, where the "Great Mogul," the "Koh-i-nûr," the "Pitt" or "Regent," the "Orloff," and many other magnificent gems were found. The rough stones yielded by these mines were formerly cut and polished in the town of Golconda, about 100 miles farther north, and from this circumstance the diamonds were popularly supposed to be produced at or near Golconda, which is not a diamantiferous district.

The Madras Presidency.

This Presidency comprises roughly the whole of the Eastern and a considerable section of the Western Ghats, the coast-lands stretching thence to the Bay of Bengal, and all the southern portion of the peninsula from the Nizam's territory to Cape Comorin. It is thus bounded by the sea on two sides, and landwards by Orissa, the Central Provinces, the Nizam's Dominions, and the Presidency of Bombay. Besides Mysore, Kúrg, Travancore, and some other smaller Native States, the chief historic divisions are the Northern Sircars and the Carnatic on the Coromandel coast, South Kanara and Malabar on the west coast, the Balaghat or uplands near the junction of the Eastern and Western Ghats, the Nilgiri highlands connecting the Eastern and Western Ghats south of Mysore. The Presidency stretches across 12 degrees of latitude (8° to 20° N.) on the east side, and across 6 (8° to 14° N.) on the west side, with an extreme length of 1000 miles, breadth 380 miles at the parallel of the capital, and total area 148,000 square miles. It is traversed on both sides by the Ghats, and is generally mountainous towards the south, where the Nilgiris, Palni, Shevaroy, and other hills, occupy most of the apex.

The Malabar or West Coast is a narrow but very fertile and highly-cultivated tract intervening between the shore and the Ghat Mountains. Several shallow inlets, called "backwaters," run sometimes for 150 or 280 miles parallel with the coast. For 170 miles from Cape Comorin the east coast is also low, but rocky and fringed with reefs. Navigation is here further obstructed by the so-called "Adam's Bridge," a sandbank stretching from the mainland to the northern extremity of Ceylon, with two open channels only. But even these are too shallow for large vessels, so that all the deep-sea navigation is diverted round the island from the Gulf of Manar and Palk Strait. The Coromandel coast, running from Point Calymere nearly due north to the Kistna delta, retains the character of a low sandy seaboard, the beach shoaling very gently, and preventing large vessels from approaching the land. Beyond the Kistna delta the Golconda coast, as it used to be called, trends north-eastwards for nearly 300 miles to Vizagapatam, and although the Ghats here approach somewhat nearer to the sea the intervening strip is deltaic, being so low and flat that it is sometimes subject to inundation. But on the northern coast, stretching for 230 miles from Vizagapatam to the Chilka Lake or Lagoon on the frontier of the Bengal Presidency, the hills approach nearer to the shore, which is unbroken by any inlets or large river mouths.

Mysore and Kúrg.

A large portion of the interior is occupied by the Native State of Mysore and the province of Kúrg, which are enclosed by the Madras Presidency on all sides except towards the north-west, where Mysore impinges on the Bombay Presidency. It consists of an extensive table-land 290 miles by 230, with an area of over 27,000 square miles, filling the angle where the Western and Eastern Ghats merge in the Nilgiris. The surface is very undulating, and diversified in many places by the remarkable rocky formations known as *Drugs,* huge piles rising either isolatedly or in clusters from 1000 to 1500 feet above the plateau. Many of these have perennial springs on their summits, which have often been converted into almost impregnable strongholds.

The plateau culminates at Bangalore, which lies due east of Madras, over 3000 feet above sea-level. But the highest peaks are found more to the west, where the Kuduremukha, 6215 feet, forms a striking landmark visible from the sea. Here also two peaks in the Baba-Budan Mountains rise to elevations of 6214 and 6317 feet, and on the former is the tomb of Baba-Budan, a Muhammadan saint, from whom the range takes its name.

The chief rivers of Mysore are the Kavari, Penner, Paler, and Pennair. None of the streams are here navigable, and many are utilised to form artificial reservoirs, of which there are no less than 38,000 in the State. Some of these are of considerable size, the Sulikere tank, which is the largest, having a circumference of 40 miles.

The Native State of Mysore forms no part of the Madras Presidency, but is separately administered on the model of the Panjáb by a Chief Commissioner, directly responsible to the Supreme Government.

A similar arrangement has been adopted for the adjoining territory of Kúrg, a hilly tract 2000 square miles in extent, occupying the crests and eastern slopes of the Western Ghats between 12° and 13° N. latitude. The term Kúrg, generally written Coorg, is a corrupt form of *Kudagu* or *Kodumale*, meaning "steep mountains," which is a sufficiently accurate description of the land. It consists of a series of steep ridges and deep gorges, densely wooded in the east on the Mysore frontier, more open towards the west. Nearly the whole surface is drained by the head-streams of the Kavari, which here rises in the Brahmagiri range at Tale Kavari, a place of great repute among the Hindus. Here are some temples yearly visited by thousands of pilgrims, who regard the Kavari as a holier river than even the Ganges itself. In their course to the main stream all the mountain torrents form romantic waterfalls, conspicuous among which is that of the Jessy near Merkara, the capital of the territory.

Respecting sanctity, however, opinions differ; some Hindus claim it for one river, others for another. But the Ganges still maintains its supremacy among the majority of Hindus.

The cardamom plant is indigenous in Kúrg, where it is extensively cultivated at elevations of from 2500 to 5000 feet above the sea.

Travancore.

The two Native States of Cochin and Travancore, occupying the south-western extremity of the peninsula, are under the direct control of the Madras Government. They have a joint area of 8000 square miles, of which about five-sixths are comprised in Travancore, which embraces the western slopes of the hills and the low-lying coast-lands from Cochin to Cape Comorin. The lowlands are very fertile, and watered by numerous small streams flowing from the hills, which have here a mean elevation of 4000 to 5000 feet, culminating with the Augustier Peak 7000 feet high. Here the sovereignty as well as the inheritance of property passes in the female line, a custom probably due to the practice of polyandria, formerly universal along the Malabar coast, and still surviving among many of the low-caste hill-tribes in this region.

Presidency of Bombay.

Including the already described northern province of Sind, the Bombay Presidency, occupying the north-western section of the peninsula, stretches from the Panjáb and Baluchistán for 1100 miles southwards to Mysore, with an average breadth of about 200 miles between the Arabian Sea (or Indian Ocean) and Central India. But towards the interior the frontier line is extremely irregular, being determined in this direction by the limits of the Panjáb, Rajputana, the Central Indian Agency, Berar, the Nizam's Dominions, the Madras Presidency, and Mysore. It has a total area of nearly 200,000 square miles, and, exclusive of Sind, comprises three distinct natural divisions—Gujarat, the western portion of the Deccan, including Khandesh, and the Konkans.

Gujarat.

Politically, Gujarat comprises mainly the feudatory States of Kathiawar and Katch, and Baroda, besides a portion of British territory proper about the mouth of the Nerbadda—some of the richest lands of the empire—and the Gulf of Cambay. Physically, it includes the Katch and Kathiawar peninsulas, which consist mostly of rich and highly-cultivated alluvial plains, varied by a few low ridges and isolated eminences. Towards Central India it is skirted by a chain of hills running from Mount Abu at the southern extremity of the Aravallis southwards to the western extremity of the Vindhyas.

The Deccan and Konkan.

The Deccan, including the plains of Khandeish, stretches thence over the vast upland between the Eastern and the Western Ghats, southwards to Mysore and Madras, including the Nizam's Dominions, and a part of the Bombay Presidency. Geographically, too, it includes the Balaghat districts of the Madras Presidency. This region is watered in the north by the Tapti, in the south by the head-streams of the Godavari and Kistna, which drain the whole of the plateau eastwards to the Bay of Bengal. The hilly district of the *Dangs* in the north and the Kanara district in the south are mostly covered with dense forests, and the Western Ghat mountains are mostly wooded; but elsewhere the country is open, generally well cultivated, and very fertile along the banks of the rivers.

The Konkan comprises the narrow strip of coast-lands extending from Bombay between the Western Ghats and the sea southwards to the Portuguese territory of Goa. These coast-lands are everywhere intersected by creeks and short rapid streams or torrents, flowing from the Ghats in separate channels to the sea, and in some places form tolerably sheltered harbours. Hence this rockbound coast is mostly of difficult access, and along the whole seaboard of the Presidency the ports of Karachi, Bombay, and Karwar, alone afford a complete refuge to shipping during the prevalence of the south-west monsoon. The lesser harbours are, however, being improved, and at several seasons are already useful.

French and Portuguese Possessions.

The only remaining political divisions on the mainland are the fragments of territory still remaining to France and Portugal. Here the French possessions consist of five isolated portions, with a joint area of 178 square miles, all subordinate to a Governor, residing at Pondicherry. The several settlements are: in Bengal—Chandernagore, on the right bank of the Hughli, 17 miles north of Calcutta; in Madras—Pondicherry, considerably larger than all the rest put together, on the Coromandel coast, 12° N. lat.; Karakal, next in size, in the Kavari delta on the same coast; Yanaon, or Yanan, on the old Golconda coast at the northern extremity of the Godavari delta; lastly, Mahé, on the Malabar coast, nearly under the same parallel as Pondicherry.

The Portuguese settlements consist of Goa, Daman, and Diu, all on the west coast, and within the Bombay Presidency, with a total area of 1096 square miles. Goa comprises a small territory, 64 miles long north and south, by 20 broad, on the Malabar coast near the southern limits of the Presidency. It is a fertile, well-watered, and cultivated tract, divided into the two districts of Salsette and Bardes, with a small sheltered harbour, five miles from the now deserted town of Old Goa. The new town of Panjim, or Villa Nova de Goa, lies at the entrance of the harbour, which is defended by several forts. Daman is situated on the Gujarat coast, at the entrance of the Gulf of Cambay, over against Diu, which is a small island on the Kathiawar coast, 170 miles northwest of Bombay.

Ceylon.

With the exception of Diu, all the islands in the Indian waters are either British territory or subject to the

Supreme Government. They consist of one large island, Ceylon, at the apex of the peninsula; the four groups of the Andamans, Nicobars, Mergui, and Moscos, in the Bay of Bengal; and the two groups of the Laccadives and Maldives in the Indian Ocean.

Ceylon, "Pearl of the Eastern Seas," is almost connected with the mainland by Adam's Bridge, a chain of low coral reefs and sandbanks, 62 miles long, running between the Gulf of Manar and Palk Strait. But from the northernmost extremity to Point Calimir on the Coromandel coast the distance is only about 40 miles. The island has the form of a pear, tapering northwards with a total length of 270 miles, an extreme width of 146, and an area of over 24,000 square miles. The surface rises gradually from the northern plains to the central highlands, which consist of a series of ridges and intervening upland valleys, culminating with the Pedrotallagalla Peak, 8260 feet, which overlooks the elevated plateau of Nuwara Eliya, itself 6000 feet above the sea. The other chief summits are Tolapella (7720 feet), Kirrigalpota (7810), and Adam's Peak (7420), an isolated mass on the south-western edge of the central highlands, long supposed to have been the highest point in the island.

The central highlands form a complete water-parting, whence a large number of rivers flow in every direction seawards, thus rendering Ceylon one of the best-watered countries in the world. The largest of these streams are the Mahavila-Ganga, running from the Nuwara Eliya plateau northwards to the east coast near Trincomali; the Kalani-Ganga, Kala-Ganga, and Maha Oya, draining to the west coast.

The soil is extremely fertile, even in the upland districts, and is almost everywhere clothed with a rich and varied vegetation. The chief sources of wealth are the coco-nut, cinnamon, and coffee plantations, besides tobacco in the northern lowlands. The cinnamon groves are restricted chiefly to the south-western districts about Colombo, and the coffee plantations to the upland valleys and mountain slopes, which yield about 50,000 tons of berries for exportation chiefly to the European market. In the forests, satin-wood, ebony, calamander, and other valuable trees, arrive at great perfection. Ceylon also abounds in minerals, such as plumbago, iron, manganese, nitre, alum, and salt, besides a great variety of precious stones—rubies, sapphires, amethysts, garnets, and the cat's-eye.

Ceylon is a Crown colony, entirely separated from India, and administered by a Governor appointed by the Queen, an Executive Council of five, and a Legislative Council of fifteen. It is divided into six provinces under Government agents, with a supreme civil and criminal court in the capital, Colombo. This city is connected with the old highland capital, Kandy, by a railway, which has recently been extended to the coffee plantations of the central province. On the south-west coast is the important harbour of Point de Galle, a port of call for all the large lines of steamers plying in the Eastern waters.

Maldives and Laccadives.

Nearly 500 miles due west of Ceylon is the group of the Maldives—that is, Maledíva, or "Thousand Islands." It forms a chain of coral islets, comprising 17 atols, each enclosing deep lagoons fringed with reefs, and are richly clothed with coco-nut palms. They also yield millet, fruits, and edible roots. The group is governed by a hereditary Sultan, who resides in the island of Male (Mól), and pays a yearly tribute to the Ceylon Government.

Some 200 miles north of the Maldives are the Laccadives, also of coral formation, comprising 20 atols, besides numerous islets and reefs, mostly barren, or producing nothing but coco-nuts. They form five separate groups, which, with Minicoy, midway between the Laccadives and Maldives, are now attached to the South Kanara district of the Madras Presidency on the opposite Malabar coast.

The inhabitants of the Maldives are Muhammadans of Malay stock; those of the Laccadives, a half-caste Indo-Arab race, also Muhammadans. All alike are extremely inoffensive, hospitable, and friendly to Europeans. The Maldivians especially are noted for their kind treatment of shipwrecked sailors, seldom accepting any pecuniary return for the care bestowed on them.

The Andamans and Nicobars.

The Andamans, with the little Cocos group at their northern extremity, and the Nicobars farther south, form the scattered links of a chain suggesting a former connection of Pegu with Sumatra. The Andamans consist of four large and several smaller volcanic islands, some 200 miles west of the Tenasserim coast, with a total length of 200 miles, and an area of 2700 square miles. They are surrounded by dangerous coral reefs, generally mountainous, culminating with Saddle Peak, 2400 feet, in North Andaman, and mostly clothed down to the water's edge with a dense tropical vegetation. In South Andaman is the well-sheltered harbour of Port Blair, chosen as a penal settlement for all India in 1868, when this archipelago was annexed and placed under a "Chief Commissioner of the Andaman and Nicobar Islands," responsible to the head Government. At the foot of Mount Harriet (1200 feet), in North Andaman, Lord

Mayo, Governor-General of India, was assassinated while on a tour of inspection, by one of the convicts in 1872. East of Middle Andaman, the largest of the group, is Barren Island, a remarkable active volcano, 7 miles in circumference and 1700 feet high.

The Nicobars, lying nearer to Sumatra than to the Andamans, form two groups, separated by Sombrero Channel—Great and Little Nicobar in the south; Nancowry, Kachal, Camorta, Car Nicobar, and a few others, in the north. Nancowry, about the centre of the archipelago, is 225 miles from Port Blair and 550 from Rangoon. The hills in the south are generally covered with forests to their summits, and in the north with grass. The whole group is under the Commissioner resident in Port Blair, while the Mergui Archipelago, as already stated, is attached to British Burma.

5. *Climate of India.*

The general features of the climate of India are mainly determined by four conditions — latitude, the northern highlands, the elevation of the Deccan plateau, the neighbourhood of the western desert, and proximity to the Indian Ocean. The latitude produces tropical heats, tempered on the southern plateau by the general elevation of the land, intensified on the northern plains by the Himalayas, which refract the vertical summer solar rays, while in winter intercepting the cold atmospheric currents from the bleak central Asiatic tablelands. The great desert intervening between the upper basin of the Ganges and the lower basin of the Indus helps to cause the hot blasts to blow over the North-West Provinces. The Indian Ocean, surrounding the peninsula on two sides, supplies a superabundance of moisture during the prevalence of the southern monsoons. None of the

Ghats or southern highlands are sufficiently elevated to arrest any large portion of the rain-bearing clouds, which at this time roll up continuously from the seething surface of the surrounding seas, sweeping over the Deccan plateau, penetrating far northwards through the head of the Bay of Bengal, and precipitating all their remaining humidity on the southern slopes of the Himalayas. From these conditions it results that, while great heats prevail everywhere, the provinces south of the Satpura range are, on the whole, cooler than the Indus and Ganges basins, and that an unusual quantity of moisture is pretty evenly distributed throughout the peninsula. At certain points the amount of this moisture surpasses the records taken on any other part of the earth's surface, varying on the Malabar coast from thirty to forty feet, and in the caldron-like Assam valley exceeding fifty.

But in such a vast region, stretching across thirty degrees of latitude, in Ceylon approaching to within six degrees of the equator, in Kashmir impinging upon the Pamir, in Nepal rising to the highest summits on the globe, there is necessarily much diversity amidst this general uniformity.

6. *Flora and Fauna.*

Although less than half of the peninsula lies within the tropics, the average temperature of the land is everywhere so high that the organic world of the torrid zone naturally predominates greatly over that of the temperate. Owing to their low elevation the eastern part of the great northern plains are in this respect quite as tropical as the southern plateau of the Deccan. In the north-western part of these great plains and in the upper part of the Deccan, wheat, barley, millets, pulses, European vegetables, and other plants characteristic of the temperate zone are cultivated successfully. The development of the wheat culture within the last few years has been remarkable. But the great staples of food and commerce are rice, jute, indigo, oilseeds, poppy, betel, all distinctly tropical growths. It is noteworthy that, with a few signal exceptions, the vegetable products of India are on the whole inferior in quality to those of other countries. Thus the cotton, tobacco, sugar, and rice here grown are all surpassed by those of America, while the maize, wheat, wine, fruits, and vegetables cannot be compared with those of Europe, and the betel-nut, cinnamon, spices, and dates are excelled by the corresponding products in the Eastern Archipelago and other parts of Asia. The most notable exceptions are the Malabar coco-nuts, the Bengal indigo, jute, and opium, the coffee and tea of Ceylon, the Nilgiris, Western Ghats, Assam, and Himalayas, all of which are unsurpassed, in some cases unapproached, in flavour. The indigenous uncultivated plants also, such as the cedars, pines, teak, ebony, india-rubber, rhododendrons of the Himalayas are fully equal, if not superior, to those of other regions.

Amongst the useful plants whose cultivation has been more recently developed, coffee, tea, cinchona, and the Australian *Eucalyptus globulosa* take a conspicuous part. The eucalyptus has already been naturalised in the Nilgiris, and according to the official report for 1881 there are now over 4,500,000 cinchona trees in Southern India, yielding a sufficient supply of bark for the medical depots of all the Presidencies, with a surplus of 3000 lbs. for sale to the public.[1] The cultivation of coffee has become one of the staple industries of Ceylon and Southern India. The plantations now extend almost continuously along the

[1] In connection with this industry it may be mentioned that Dr. King, head of the Government cinchona factory in Darjiling, succeeded in 1881 in producing sulphate of quinine from cinchona bark.

slopes and crests of the Western Ghats from North Mysore to Cape Comorin, and occupy in the Nilgiris alone no less than 12,000 acres, yielding an annual crop valued at £200,000.

Here in the south about 7000 acres are devoted to tea. But the culture has acquired its greatest development especially in Assam and along the lower slopes of the Himalayas. Assam now yields over 20,000,000 lbs. yearly, and nearly half as much again is produced in the Darjiling and other districts of British Sikkim, where the plantations cover about 38,000 acres. The Indian teas are on the whole superior to those of China imported into England, where they are now largely used for flavouring the Chinese varieties. Hence the future of this rapidly-increasing industry seems to be ensured. "In thirty years it has risen from nothing up to 34,000,000 lbs. annually, valued at £3,000,000. Many of the earlier European planters struggled towards the goal of this great success through a maze of difficulties, errors, and disappointments. They thus lost much capital, which has been replaced by capital generally yielding good returns. Their more fortunate successors form a regular profession of specially-trained and qualified men. The tea-gardens are now scientifically managed, improved processes are adopted in the factories for the preparation of the leaf, and steam machinery is beginning to be used."[1]

Notwithstanding the reckless destruction of timber that has been going on for ages, large well-wooded tracts are still found, especially on the slopes of the hills in every part of the country. Measures have of late years been taken for the preservation and increase of the forests, which cover an area of about 70,000 square miles altogether. The chief species are the conifers (cedar, pine, fir), the oak, elm, maple, plane, ash, ebony,

[1] *India in* 1880, by Sir Richard Temple.

TIGER-HUNTING—INDIA.

To face page 283.

teak, banyan, sandal-wood, mango, bamboo, sâl, and palms, including the date, palmyra, and betel-nut, and other varieties.

These forests, with the jungle of the plains, are still tenanted by vast numbers of wild animals, birds, and especially reptiles. So destructive are many of these, that about 20,000 human beings and 50,000 head of cattle are yearly destroyed by wild beasts and venomous snakes. Man suffers mostly from the cobra and other reptiles, while the herds are ravaged chiefly by the tiger, panther, and other large beasts of prey. India is probably the indigenous home of the tiger, which is found in every part of the country, and which, in the Royal Bengal species, attains his highest development. He preys chiefly on deer, flocks, and herds, but will sometimes turn upon man, and once he has tasted human flesh prefers it to any other. The "man-eater," as he is then called, is one of the greatest scourges of the villages lying on the skirt of the jungle. At present special measures are taken by the authorities to secure the speedy destruction of these animals.

Scarcely less formidable is the gray panther, or rather leopard, which also occasionally becomes a man-eater. But the cheeta, a somewhat smaller tawny-coloured species, is kept by native princes and trained for hunting. He is conveyed blindfolded in a cart to within a short distance of a herd of deer, when the hood is suddenly removed. In a few wonderful bounds he has seized the quarry, or, missing it, abandons the pursuit, having spent all his energy on a single effort.

Other large wild animals are the bear and wild boar, very generally the rhinoceros, chiefly in the woods at the foot of the Eastern Himalayas, the bison, gayal (*Bibos frontalis*) in the Ghats and North Assam hills; the elephant still met in large herds in Nepal, the hilly districts of Eastern Bengal, the Nilgiris, and some other parts; two species of the alligator, the harmless "sharp-nosed," and dangerous "snub-nosed," frequenting not only most of the large rivers, but many of the numerous tanks scattered over the country. Deer and antelopes abound in immense variety, while the ibex, ovis ammon, and fine-fleece-bearing goat and sheep are numerous, especially in the Western Himalayas.

As a rule, the domestic animals, like the cultivated plants, are inferior to those of most other countries. The sheep, oxen, camels, and especially the horses, are generally of indifferent stock, although some hardy breeds of ponies occur in the Himalayas, and the camel of Bikanir in Rajputana is noted for its great size, strength, and swiftness. Large herds of oxen, camels, sheep, and goats occur chiefly in the Panjáb and Rajputana, but India is on the whole more an agricultural than a grazing land. Hence, although there are a vast number of wild tribes in the more inaccessible hilly districts, there are, strictly speaking, no pastoral nomads, except, perhaps, the Ladakhi Champas and a few others of Mongoloid stock, who according to the seasons migrate between the southern and Tibetan slopes of the inner Himalayas in search of a scanty pasturage for their flocks.

7. *Inhabitants: Hindus, Dravidians, Kolarians, and Tibeto-Burmese.*

It is not to be supposed that the inhabitants of India belong to one homogeneous type. There is scarcely a country in the world containing a greater diversity of tribes and races in every stage of civilisation, from the cultured European and philosophic Hindu down to the most degraded savages. A certain outward physical uniformity, noticeable especially in the prevailing brown, olive-brown, and dark-brown complexions, has no doubt been brought about during the course of ages by the climatic conditions. It is also true that the great bulk of the population is ultimately reducible to two distinct stocks—the Hindu,[1] chiefly in the northern plains; and the Dravidian, in the Deccan. But besides these at least two others are also largely represented—the Kola-

A HINDU OF WESTERN INDIA.

rian, chiefly in the Vindhyan and Satpura ranges between the Aryans and Dravidians, north and south; and

[1] To avoid confusion it is necessary carefully to note the twofold meaning which this term Hindu has acquired. In its original ethnical sense it means, as here, the Aryan as opposed to the non-Aryan peoples of India. Hence the word *Hindustan*, or "Country of the Hindus," is properly restricted by native writers to the Indus and Ganges basins, the true home of the Indian Aryans. But in a religious sense Hindu is synonymous with the Brahmanical cult, and is opposed, not to the non-Aryans, but to the Muhammadan and other forms of belief prevalent in the peninsula.

the Mongoloid, inhabitants of the Himalayas, the Assam highlands, and British Burma.

Whether the absolute aborigines or not, the Kolarians are at all events the first arrivals in the peninsula, where they have scarcely anywhere risen above the lowest grades of human culture. Next came the Dravidians, some of whom, if true Dravidians, still remain at the same low level as the lowest Kolarians, while the great majority became in course of time susceptible to the civilising influences of the Hindus, who were the last arrivals from the north-west. The land was now full except on the remote northern and north-eastern frontiers, which were gradually occupied by Mongoloid Tibeto-Burman tribes from Central and South-Eastern Asia.

The subdivisions of the Kolarians and Tibeto-Burmans are chiefly of a tribal—that is, social—character, while those of most Dravidians and all the Hindus are based essentially on linguistic considerations. The Kolarians and Tibeto-Burmans themselves speak a great variety of different dialects, but their classification depends even more on the tribal organisation than on the diversity of those dialects. This is also true of many low-caste Dravidian tribes, especially in the Nilgiris and Malabar highlands. But the vast majority of the Dravidians and all the Hindus are grouped in different branches bearing much the same relation to each other that, for instance, the great branches of the Latin family bear to each other in Southern Europe. All have long been fused together in one common ethnical, social, and religious system, while still separated one from the other mainly by their different languages, all derived in Europe from the common Latin stock, in India either from a common Sanskritic, or from a common but now extinct Dravidian mother-tongue. These points should be borne in mind in estimating the value of the subjoined general grouping of all the Indian races. It is also to be noted that in a comprehensive classification of the human family the Hindus and Tibeto-Burmans would appear as mere branches of the Caucasic and Mongol stocks respectively, whereas both the Dravidians and Kolarians would stand quite apart, their possible affinities to the other great families of mankind being still undetermined (see Appendix):—

I. Hindus.

(Aryan Stock.)

Kashmiri	Rambani; Bhadarwahi; Padari; Doda; Kishtwari	Kashmir; Jummu	1,500,000
Panjábi	Sikhs; Jats; Changars	Panjáb	17,000,000
Sindi		Sind	2,000,000
Gujarati and Kachi		Gujarat	7,000,000
Marathi and Konkani		Bombay; Central Provinces; Berar	15,000,000
Hindi	Urdu; Marwari; Gwalior, etc.	N.W. Provinces; Rajputana; Upper Bengal	100,000,000
Bengali		Lower Bengal	40,000,000
Uriya		Orissa; Ganjam; Chota-Nagpore	5,500,000
Assamese		Assam Lowlands	1,500,000
Nepali (Parbhatia)		Nepal	2,000,000

II. Dravidians.

Tamil		Karnatic, Travancore; Mysore, N. Ceylon	20,000,000
Telugu		Sirkars, Nizam's; Berar, Mysore	16,000,000
Kanarese		S. Kanara, Mysore, Kurg	9,500,000
Malayalim		Malabar Coast	4,000,000
Tulu		Kanara, Malabar Hills	300,000
Kodagu		Kurg	150,000
Orâon	Tirki; Ekhar; Barar; Minjar	Chota-Nagpore	265,000

Rajmahâl		Rajmahâl Hills, N. of Chota-Nagpore	42,000
Khondi	Betiah; Beniah; Maliah	Ganjam; Orissa	270,000
Gondi	Dher; Gottur; Koi; Badiya; Madi; Wardha, etc.	Chota-Nagpore; Central Provinces	1,650,000
Tuda		Nilgiris	750
Kota		Nilgiris	1,100
Sinhalese ?		S. and W. Ceylon	1,700,000
Veddhas ?		Travancore and E. Ceylon	? 3,000

III. Kolarians.

Santhal	Saran; Murmu; Marli; Kisku	Baghalpur	1,250,000
Munda	Bumij; Ho; Larka	South of the Santhals	850,000
Kharia		Singbhum district; Chota-Nagpore	? 1,000,000
Mal-Paharia		Orissa, N. of Kattak	
Juang			
Gadaba		Bustar Hills, left bank lower Godavari	
Korwa		Barwah	
Kurku		About source of the Nerbadda	
Mehto		Chota-Nagpore	
Savara		N. Sirkars	
Bhîl	Kala; Ujvala; Mina	Vindhyas; Gujarat; Malva, Bundi	900,000

IV. Tibeto-Burmans.

(*Mongol Stock.*)

Ladakhi		Ladakh	100,000
Gaddi			
Champa			
Bunan			
Khamba			
Balti		Baltistán	
Garwhali	Rongbo; Kohli; Kakka	Garwhal	? 50,000
Kanawari			

Magar; Sarpa; Gurung; Pahri; Sunwar; Kachari; Chepang; Kusunda; Newar; Bhramu; Kiranti		Nepal	200,000
Lepcha; Yayu; Murmi; Limbu		Sikkim	?80,000
Lhopa; Dimal; Towang; Mechi; Char Duar; Thebengea		Bhutan	?750,000
Mikhir; Miri; Lutukotia; Angka; Mishmi; Dafla; Deori; Abor; Mijhu		North Assam Highlands	?50,000
Kachâri	Hojai; Garo; Mech; Koch; Rabha; Chutia; Tipperah	Goalpara and Garo districts, S.W. Assam	230,000
Kuki	New and old Kuki; Manipuri; Looe; Lushai; Sokte; Shindu; Kabui; Cheru; Maring	S. Assam Highlands, between the Khasi and Naga Hills	? 50,000
Mikir		Nowgong district, Central Assam	62,000
Khasi		Khasia Hills, S. Assam	140,000
Nagas. Central.	Hatigorria; Sema; Lhota; Banpar; Rengma; Primi, etc.	Naga Hills, S. and S.E. Assam	?200,000
Nagas. Eastern.	Tablung; Sangloi; Tengsa; Banfera; Mutonia; Mohongia; Namsang		
Nagas. Western.	Angami; Liyang; Arung; Mao; Muram; Luhupa; Maring		250,000
Rakhaingtha ("Mugs")		Arakan Plains	500,000
Karen	Sgan; Pwo; Bghai; Tari; Mopgha; Tungthus	S. Arakan Hills, Tenasserim	?250,000

Khayeng	Mru, Kyau, Sak	Hills N. of the Karens	?50,000
Kumi	Awa Kumi, Aphya Kumi	Kolading River, N. Arakan	?20,000
Talaing (Mon)		Pegu	500,000
Burmese		Tenasserim, Pegu, Arakan	250,000

V. SUNDRIES.

Shans	Ahom, Khampti	Assam	?100,000
Malays		Maldives, Nicobars ?	250,000
Negritos		Andaman Islands	10,000
Indo-Arabs		Laccadives	
"Moormen" (Arabs)		Malabar, Ceylon, etc.	
Baluchis	Mari, Bugti, etc.	Sind and Derajat	300,000
Afghans	Afridis, Orakzais, Yusafzais, etc.	Derajat and Peshawar	
Swatis		Peshawar	
Persians		Sind	50,000
Parsis		Gujarat, Surat, Bombay	
Eurasians		India generally	536,000
Europeans			

Caste.

The religious and social system of the Hindus is everywhere in India based on the institution of caste, which was originally introduced to uphold the political supremacy of the fair Aryan intruders over the dark aborigines. But before its introduction a considerable intermixture had already taken place, except perhaps amongst the very highest classes of the Aryan conquerors. The indigenous elements being by far the most numerous, the Aryans were thus threatened with ultimate absorption, and in fact had in many places become largely assimilated to the natives. They could be saved from extinction only by checking further alliances. Marriage with the dark races was accordingly forbidden,[1] and a definite rank assigned to each shade of colour, which had already been developed, while the prohibition itself was referred to divine prescription. Hence caste originally meant colour, and had therefore an ethnical value. But once established, the institution gradually acquired an indefinite development, and the four original castes of *Brahmans* (priestly order),

HINDU TYPES—TRAVELLING COACH.

Kshatryas (warriors), *Vaïshyas* (citizens, traders, agriculturists), and *Sudras* (the menial classes), have in course of ages expanded into minute subdivisions almost past counting. The process seems to be even still going on, and the last census returns give 2500 main divisions, and in Madras alone nearly 4000 minor distinctions.

[1] The interdiction seems to date no further back than the laws of Manu, 1000 or 1200 years before the Christian era.

These, however, probably include the *Pariahs,* or outcastes, a term which originally simply meant "hillmen," and which thus throws considerable light on the institution. It shows that while the three highest orders were reserved for the ruling Aryans, the Sudra mainly comprised the aborigines who had been reduced to a state of thraldom or Helotism; whereas the Pariah embraced the independent highlanders who were excluded from all the social privileges of the Hindu system. Refusing to submit to the conquering race, and successfully maintaining their independence in the inaccessible mountain fastnesses of the Vindhyas, the Satpuras, and the Ghats, both Eastern and Western, where so many of them are still found, they were declared to be outlaws; and the term pariah, or highlander, thus came to be synonymous with outcaste. Hence the outcastes must, to some extent at least, be regarded as the last remnants of the aboriginal elements, and the surviving representatives of a pre-Aryan or prehistoric culture.

Although still flourishing, the institution of caste has been somewhat though not largely affected, first by the settlement and spread of Muhammadans in the land, and then by the establishment of British rule. Hinduism, as a religious system, has always met with the utmost possible toleration both from the Moslem and Christian governments. Hence the Brahmanical or Sacerdotal caste has survived all the political changes by which the land has been convulsed during the past twelve hundred years; but the Kshatrya, or military caste, naturally lost its vitality under Muhammadan princes, and under the present political system, except in the feudatory Hindu States of Rajputana and the Deccan.[1] On the other hand,

[1] Some of the Kshatryas, such as the Khatri of the Panjáb, have even turned to trade. Some of the Khatris hold the same social rank in the north that the Baniyas (Banians) do in the Central Provinces and Southern Presidencies, while others still possess an important status civilly and politically.

sacerdotalism and secular tradition have been strong enough to maintain the class distinctions of the Vaïshya, and especially of the Sudra order, which last, with the Pariah, comprises most of the Dravidians and Kolarians of the Deccan. Its main subdivisions at present are: — 1. Husbandmen; 2. Graziers; 3. Artisans; 4. Writers: 5. Weavers; 6. Field labourers; 7. Potters; 8. Mixed, or broken, mainly sects who have discarded caste and attend to the service of the temples; 9. Fishers and hunters; 10. Barbers; 11. Washermen; 12. Low castes of various degrees merging in the outcastes.

Redemption from this social yoke will ultimately be found in the spread of education, in such internal upheavals as are foreshadowed by the Brahmo Samâj and other monotheistic movements, in the silent influences of the higher European culture, quickened by the development of the railway system and other levelling institutions.

The Brahmo sect is described by Professor Monier Williams in a paper read before the Royal Asiatic Society, in November 1880, regarding Hindu religious reformers.

Many of the castes still preserve clear indications of the physical distinctions on which they were originally based. This is especially true of the Brahmans, who are everywhere in the peninsula conspicuous for their intelligence, retaining much of the common Aryan inheritance, and displaying the noble cast of countenance which is characteristic of that race.

Religious Sects.

The religious system of the Hindus retains little of the primitive belief of the Aryan race, a few Vedic hymns and formulas recited without being understood by the priests being nearly all that survives of the old cult. In modern Brahmanism there are many sects, some of which have sunk to the lowest depths of the grossest superstition. Such are the Aghoris (Aghor-Pants), many of whom belong to the Brahman order. But the two most widespread sects are the worshippers of Siva and Vishnu, who typify the opposite poles of religious thought, the Vaishnava appealing to the deity he worships as the author of all good, while the followers of Siva seek in man alone and his efforts the attainment of supreme happiness. But apart from this fundamental difference, the two sects often meet on common ground. By Hindus generally Brahm is regarded as the creator, Vishnu the preserver, and Siva the destroyer.

All the civilised Dravidian races—Telugus, Tamils, Kanarese, Malayalims—have long conformed to the Hindu religious system. But nature-worship of a very crude type still prevails among the wild tribes—Tudas, Kolas, Kudagus, Gonds, Khonds—as well as among most of the Kolarians. All these rude hillmen still retain their primitive usages, practise sorcery, and believe in evil spirits. English and German missionaries, however, have been for some time at work amongst them, and have already succeeded in forming a large number of Christian communities, especially in the Santhal and other Kolarian districts.

Within the last twenty years, some judicial trials have disclosed practices of the worst social tendency among a sect in Western India called the Maharajas.

The Kashmirians and Nepalese.

On the other hand, many of the Hindus, especially in the north, have accepted Islam. The Kashmirians, among the finest of the Hindu races, became Muhammadans some centuries ago, and are mostly Sunnis. They are described as almost European in appearance, and in Kashmir we miss the slender frames, prominent cheek bones, and other unpleasant features so prevalent in other parts of India. The men are of a square, herculean build, well proportioned, and with a frank expression, while the women are fresh-looking and often decidedly beautiful, with an almost Jewish cast of countenance. Those of the better classes are scarcely darker than the average natives of Italy.

The Kashmirians are a shrewd, witty, and cheerful people, but superstitious and somewhat sensuous. They

A NATIVE LADY.

are skilful artisans and traders, but over-shrewd perhaps in bargains; and although crime in the ordinary sense is almost unknown in the country, Wilson, in his *Abode of Snow,* draws a far from flattering picture of their present social state.

Still there are some Kashmiri Brahmans remarkable for their intellectual ability.

The non-Aryan or Tibetan inhabitants of Kashmir

are found chiefly in Ladakh and Baltistan. Frederick Drew, who has carefully studied this region, attributes to the prevalence of polyandry the sparse population of these upland tracts. Here the partly Hinduised Gaddi tribe produces a startling effect by probably the most astonishing of all head-gears, while the Khampas are noted chiefly for their remarkable lung-power. Living in alpine valleys towards the Tibetan frontier, they find respiration difficult at any lower elevation than 11,000 feet above sea-level. Their favourite camping-grounds are the shores of an extensive salt lake in a secluded valley over 13,000 feet above the sea, where they live in tents and practise polyandry. But the kindred Balti race, having adopted the Muhammadan faith, have become polygamists, and are consequently now so numerous that many are compelled yearly to migrate southwards.

In Nepal there is a mixture of races, languages, and religions. The ruling people are the warlike Ghurkas of mixed Tibetan stock, but assimilated in speech and religion to the Hindus. The rest of the population are partly a mixed Indo-Tibetan race, like the Buddhistic Newaris, partly Bhutiyas—that is, pure Tibetans—who are mostly nomad shepherds, speaking ten or twelve distinct languages.

The Panjábi Hindus and Sikhs.

In the Central Panjáb the chief ethnical element consists of the Jats, a tall, hardy, and robust race, with genuine Caucasic features. These Jats of the Indus valley have never adopted the institution of caste in its integrity, and are regarded by the rest of the Hindus with a feeling which is embodied in the expression "Baheka"[1] or "aliens." The

[1] Baheka corresponds exactly to the term "Overn," applied in a like sense by the natives of the Isle of Wight to intruders from the English mainland.

Hindus themselves are of a superstitious type, addicted to many peculiar observances. At the birth of a son the priest is summoned to cast his horoscope (yaman-putri), and after forty days to give him a name. During the five first years his hair remains uncut, after which he is usually taken to Ivalamuki, where flames are often seen bursting from the ground, and here his hair is cut with much ceremony by a Brahman. Before his twelfth year his head is shaven, and he is instructed by the family priest in the "sandhya" and "gayatri," or sacred texts from the Vedas, and then receives the sacred thread. He is now considered to have reached his majority, and has to observe the six duties incumbent on all Hindus. He wears a solitary tuft of hair on the crown of his head and assumes the "dhoti," or loin-cloth, with the holy marks in red or white on his forehead.

On attaining his fourteenth year his parents cast about to find him a suitable wife of equal rank with the family. The father of the girl sends the family barber with six dates and a rupee to the boy's house in token of his willingness to accept the alliance. The inmates welcome the messenger by smearing the entrance with oil, after which the friends meet, the barber throws the dates and rupee into the bridegroom's lap and makes the sacred marks on his forehead.

The usages of the Sikhs differ greatly from those of the orthodox Hindus. They never employ the services of Brahmans, nor do they pay any attention to the Vedas or other sacred Hindu writings, replacing them by a so-called "granth" or "book" of their own, which contains their religious code. They marry somewhat later in life than the Hindus, and are a far more vigorous race. The wedding is conducted by the "granthi," who simply reads some appropriate text from the granth. The Sikhs, who never cut the hair or beard, wear close-fitting trousers and a high turban, invariably containing a bit of steel which must never be laid aside.

Amongst the Muhammadans, Hindus and Sikhs there are a great number of men called by the name of fakir, and many other names, who lead a life of pious indolence and contemplation at the expense of the poorer classes. These ascetics wander about over long distances

A FAKIR.

or pass their days under the trees, amid the tombs, or at the burning-places, or else herd together like monks in a monastery, under a "mahant" or abbot. Most of them prefer begging under a religious cloak to honest work. In times of political danger or excitement they are mischievous in carrying news, false or exaggerated, from place to place.

The Assamese.

Assam presents even a greater variety of races, religions, and languages, than Nepal. The bulk of the

CREMATION SCENE—INDIA.

inhabitants consists of Hindus, Muhammadans, immigrants from Bengal, and numerous Tibeto-Burman tribes on the highlands enclosing the Brahmaputra basin on three sides. The Muhammadans generally understand

Hindustani, which serves as the common medium of intercourse throughout most of the peninsula; and since the Government schools have been opened the educated classes have become familiar with English. But the language of the great majority is the Assamese, a Prakrit dialect closely allied to Bengali. Assam takes its name from the Ahoms,[1] the former rulers of the country, who were originally of Shan (Siamese) stock, but who have become nearly everywhere assimilated in speech and religion to the Hindus. They are a very fine, strong-built race, of rather fair complexion, extremely intelligent, and capable of a high degree of culture. The Ahom dynasty was overthrown in 1810 by the Burmese, who were in their turn ejected by the English in 1827. Since then the Ahoms have become some of the most loyal subjects of the Queen.

The surrounding hills are still peopled by numerous semi-independent wild tribes, such as the Garo, Khasi, Naga, Mishmi, Abor, Kuki, and others, mostly, if not altogether, of Tibeto-Burman stock, whose habits and customs are still but little known. Much valuable information, however, has lately been supplied regarding the Nagas by G. H. Damant and some of the officers engaged in suppressing the unruly Angami tribes in 1879-80.

The Talaings and Karens.

In British Burma the leading races are the Burmese, who are found everywhere in the open country, the Rakhaingtha, popularly known as "Mugs" in the Arakan plains, the Talaings or Mons of the Irawady delta, and the Karens of the coast ranges in Pegu and Tenasserim. Bengali immigrants and Muhammadan Hindus are numerous in

[1] *Ahom* is the same word as *Assam* or *Assom*, *h* interchanging with *s* in Burmese phonetics.

Arakan, where, however, the indigenous Mugs still constitute more than half of the population. They have a strong family likeness to the Burmese, but are of smaller stature and darker complexion. They speak a monosyllabic language accompanied with great emphasis and much gesticulation. Closely akin to them are the Kayans (Khayengs), a rude but inoffensive hill tribe, who live mostly on game killed with poisoned arrows, and resemble the Chinese in their partiality for dog's flesh.

The Talaings or Mons, if not the aborigines, are at least the earliest known immigrants into Pegu, where they form an isolated linguistic group, now restricted to the east and south of the Irawady delta in Martaban and North Tenasserim. Wearing the same dress, they differ little from the Burmese in appearance, but are generally of lighter complexion, with more delicate features and a slight growth of beard. But the two races live so intermingled, and alliances are becoming so frequent between them, that the time is perhaps not distant when the Talaings will have become absorbed in the dominant Burmese race. The two languages differ fundamentally, and affinities have been sought for the Talaing as far east as Cambodia and westwards amongst the Kolarians of Central India.[1]

The Tenasserim highlands are occupied exclusively by the aboriginal Karen, who still continue to live in the greatest seclusion. Having been formerly subjected to much harsh treatment and oppression by the Burmese conquerors of the land, they now avoid, as far as possible, all intercourse with them. They, however, occasionally visit the towns in the lowlands for the purpose of procuring by barter the indispensable articles of domestic use. Settling in small communities of twelve or fourteen

[1] Captain C. J. F. S. Forbes' *Comparative Grammar of the Languages of Further India.* London, 1881.

families near some stream in the higher woodlands, they clear the ground with fire, and cultivate rice, bananas, betel-nut, sweet potatoes, and other vegetables, on the reclaimed space. These products, with some poultry and game, suffice to supply all their daily wants. They are of a less robust build, with less prominent cheek-bones, less oblique eyes, and a fairer complexion, than their neighbours, thus approaching in some respects to the south European type. The high colour of the cheeks, often suffused in the young women by a slight blush, is very striking in a region inhabited mostly by yellow or olive-brown races. Possessing no writing system for their rude uncultivated speech, and being destitute of all instruction, they lack all the higher religious conceptions. In the natural phenomena surrounding them, recognising agencies inexplicable to their untutored minds, they attribute them to the *nats*, or good and evil spirits.

The Gonds and Bhíls.

Few races present matter of greater interest to the student of human culture than the uncivilised Dravidian and Kolarian tribes of Central India. Many of the Gonds, whose domain in the highlands north of the Deccan is from them called Gondwana, were formerly employed in the coal-pits of the Nerbadda valley and its tributaries. From their infancy they are accustomed to look on every rock, every river, gorge, and cavern, as the abode of a special spirit, who may be propitiated and rendered harmless by some simple rite.

Amongst the Bhíls of the Vindhyas there are many superstitions showing a striking analogy to those of the West. When a Bhíl goes out to fight or rob, if the byru bird is on his right hand he will prosper; if on his left, nothing will induce him to go. The belief is very strong in witchcraft, and in the powers of the Burwa, or witch-finder, who is consulted in all important cases. Should any person die without apparent cause, the friends inquire of the burwa, who selects the ugliest old woman in the village, and oracularly attributes the death to her spells. She is thereupon seized and tried, much in the same way as in Europe two centuries ago (*Col. Kincaid*).

The Nilgiri Hill Tribes.

Many of the dark aborigines of the Nilgiris and other parts of the south, although classed with the Dravidians, seem to bear a much greater resemblance to the Kolarians. Such are the *Kallar*, or "Robbers," on the Tanjore frontier, who "by no means disown their profession or consider it discreditable. Indeed the caste ranks high among the Sudras, and they have a king, the Tondiman Raja, who has always been a faithful ally of the British. The present well-educated and enlightened Raja receives a salute of twelve guns when visiting Madras. Unscrupulous as thieves, they are men of their word, and to this day are employed by the English residents of Trichinopoly to watch their houses — a trust they faithfully fulfil, and keep off all other thieves. Their skill in tracking equalled that of any savages. Their ordeals and marriage customs agree generally with those of the Bhíls, and like them they live in continual dread of witchcraft, being often driven to cruel deeds in revenge for supposed injuries. They are now fast becoming peaceable cultivators" (*M. J. Walhouse*).

In the Nilgiris dwell the Tudas, Kotas, and one or two other remarkable tribes, Dravidian in speech, but otherwise quite distinct both from the Dravidians of the plains and in some respects even from each other. The practice of polyandry would seem to point at the Bhutyas of Tibet as their remote ancestry, but for the fact that this custom is not confined to that race. Some of them, although not leading a nomad life, present in many particulars a great resemblance to the European gypsies. They betray no trace of a religion beyond what may be implied in a firm belief in witchcraft. They are all very peaceful and inoffensive, occupied either with agriculture or stock-breeding, but these hill tribes seem to be dying out. The Kotas number little over a thousand, while the Tudas have been reduced to about 700 or 800.

The Parsis.

Amongst the minor heterogeneous ethnical elements of the peninsula, one of the most interesting are the Parsis of Bombay, the direct descendants of the Ghebrs, or Old Persian fire-worshippers, who fled to India when the Muhammadan invasion burst with all its fury over the Iranian tableland in the seventh century. Since then they have kept entirely aloof from the surrounding peoples, preserving their race and religion alike intact from all extraneous influences. They are remarkable for their general intelligence, business habits, and commercial ability; and they sympathise with the English far more than with any of the native races. In proportion to their numbers, they are the wealthiest and most influential section of society in Bombay, as well as the most loyal and devoted subjects of the Queen in India.

The Indian Muhammadans.

In this respect they present a striking contrast to the Muhammadans, who must always be regarded as a dangerous element in the peninsula. The Indian Muhammadans, who are chiefly Sunnis, with an influential Shiah minority, are concentrated chiefly in Bengal, the North-West Provinces, and the Panjáb, and number altogether nearly 45,000,000, so that the Empress of India rules over far more Mussulman subjects than any other sovereign in the East. The Lieutenant-Governor of Bengal alone "has in his jurisdiction as many millions of Moslems as the Sultan of Turkey, and thrice as many as the Shah of Persia. The Indian Muhammadans are met with on all the coasts, and are emphatically the sailors of the Indian seas. In the interior they are urban rather than rural, employed in some branches of commerce, in retail dealing, in skilled and refined industries, in the army, in public and private service, but seldom connected with agriculture, save in the capacity of landlords. In Sind, however, the agricultural population is Muhammadan, both landlords and cultivators. In eastern and northern Bengal, in the region comprising the Brahmaputra basin, and in the united delta of the Ganges and Brahmaputra, the tenants and cultivators are also Muhammadan, while the landlords are Hindu, with the exception of some prominent and meritorious gentlemen of the Muhammadan faith."

"Elsewhere in India the Muhammadans, being scattered, do much to leaven the mass of native opinion. Besides the discontent engendered among them by historic memories, there is one special circumstance affecting their contentment. Under native rule they enjoyed a large portion, perhaps the lion's share, of the state patronage, and at the outset of British rule were found in the front everywhere. But nowadays they are beaten by Hindus in the open competition of mind with mind. It is to this that the Muhammadans themselves attribute the fact that they are falling in wealth and status while the Hindus are rising."

"The temper and disposition, politically, of the Muhammadans form one of the many sources of anxiety in India. Some years ago the religious revival commenced by the Wahhabis in Arabia, the breeze of fanaticism which ruffled the surface of the Muhammadan world, and other causes difficult to define, excited the Indian Muhammadans considerably. Plots were discovered and state trials instituted; some grave and melancholy events occurred which need not here be recounted. Within the most recent years, however, the Indian Muhammadans have become comparatively well affected. Be the reasons what they may, the symptoms of disaffection among them have of late abated" (*India in* 1880, by Sir Richard Temple).

An offshoot of the Muhammadan community are the Khojahs, the real descendants of the famous assassins of the Middle Ages.

The English and Eurasians.

The dominant English race are still almost aliens in the land. They have nowhere formed any agricultural settlements or permanent trading communities, nor is it likely that any serious attempt will ever be made by them to colonise even the more healthy and temperate upland districts in the Himalayas or highlands of the Deccan. Numerous sanitaria have almost everywhere been established in the more favourable sites in these districts. But such places are merely visited periodically by the officials and military, who escape during the summer season from the almost intolerable heats of the plains. It may be questioned whether three generations of Englishmen are anywhere to be found in Simla, Darjiling, Mussurie, Utacamand, or any of the many other health-resorts dotted over the uplands of the peninsula.

Nor has any advance been made towards a fusion of

the ruling and subject races. The Anglo-Saxon holds his head even higher than the haughtiest Rajput chiefs claiming descent from the gods and demigods of Hindu mythology. In former times alliances and other connections used to be formed between Europeans and Native females, but the result has not been such as to encourage a general spread of the practice. The offspring of European fathers and native mothers, called East Indians or Eurasians,[1] hold much the same position in relation to English society that the quadroons or octoroons do to the white classes in the United States. They do not exhibit any marked idiosyncrasy of race. Although both parents may belong to the Aryan stock, and although the English fathers are often distinguished by their physical qualities, and their Indian mothers by personal attractions, the Eurasians themselves do not generally display a striking appearance. They possess many intellectual endowments; but though quick of apprehension, they seldom acquire solid knowledge so well as Europeans, nor have they equal perseverance. From their mothers they seem to inherit gentleness and amiability. Among them individuals are found eminent in character and ability.

It fares still worse with the pure-blood European children who are constitutionally unable to struggle against the enervating effects of the climate, especially in the Ganges valley. Till their sixth year they retain the high complexion of the race and seem healthy enough, but on entering their teens they begin to lose their fresh colour, their features grow pale and wan, the weariness of premature decay, or some unaccountable secret blight, steals over them. Without any decided outward symptoms of disease, they droop and pine away like hothouse plants deprived of light and air. This light and air must be sought in the home of their forefathers before they

[1] That is, *European-Asians.*

attain their sixth year, for nothing but a speedy removal to the fickle but invigorating climate of England will now save them from an early grave or from physical deterioration. Of the Anglo-Indian children thus brought up in Europe, many of the young men return to India before their twentieth year in order to make a career for themselves in the civil and military services, or else to fill positions secured for them in commercial houses or other employments.

8. *Topography: Srinagar—Lahore—Delhi—Karachi—Agra—Cawnpore—Lucknow—Allahabad—Benares—Patna—Murshedabad—Dacca—Calcutta—Jaganath—Rangun—Moulmain—Prome—Nagpur—Jabalpur—Bhopal—Indore—Gwalior—Jyepur—Udeypur—Madras—Bellary—Trichinopoly—Madura—Tanjore—Calicut—Mysore—Seringapatam—Bangalore—Hyderabad—Secanderabad—Bombay—Ahmadabad—Baroda—Surat—Poona—Sholapur—Bijapur—Satara—Ahmadnagar.*

Although consisting mostly of agricultural and rural elements, the population of India is so enormous that enough still remains to overflow into many cities of the first magnitude. The number of large towns is also increased by the many administrative divisions and native feudatory States, each with its special capital, the centre of the government of the civil and criminal courts, and of other independent local interests. Thus it happens that besides nearly half a million rural villages there are no less than forty cities with populations of 50,000 and upwards. Many of these are mere aggregates of houses built of dried earth, with roofs of tile or thatch. But others, not a few, are of vast antiquity, the changeless capitals of shifting empires, reflecting in their monuments the varied tastes of many successive cultures, abounding in antiquarian and art treasures of every sort.

Srinagar.

In the upland regions of the Himalayas one of the most interesting places is Srinagar, the capital of Kashmir, which stands on the banks of the Jhelum in the midst of some of the loveliest scenery in the world. Like Venice, Srinagar is a city wherein the streets consist of numerous canals, or rather branches of the river, which traverse the place and connect it with the neighbouring Lake Dal. The canals are flooded by means of sluices from the lake when the Jhelum is low. But when the river rises above the level of the lake, the sluices are closed by the pressure of the back flow. There are shady poplar avenues in the neighbourhood. The lake is enlivened by the presence of water-fowl with brilliant plumage, while above its lotus-fringed banks majestic trees stand out against the azure sky. On the Dal itself, which is 5 miles long by 2 broad, the fantastic floating gardens recall the chinapas or swimming islands of Lake Tezcuco, near Mexico. Amongst the varied vegetable growths that here delight the eye, conspicuous is the thorny water-nut (*Traba bispinosa*), yielding a delicious flour and bread. About 60,000 tons of this substance are yearly produced at the larger Lake Wular in the same district. The Jhelum is spanned by several picturesque wooden bridges. Near the capital is the Takht-i-Suleman hill (Solomon's throne), from the top of which is seen the panorama of Kashmir, the finest landscape in the Indian Empire.

The only other large town in the Himalayan States

is Katmandu, the present capital of Nepal, including the old capital of Pâtan close by, lying in a productive and well-watered valley in the heart of the country. The

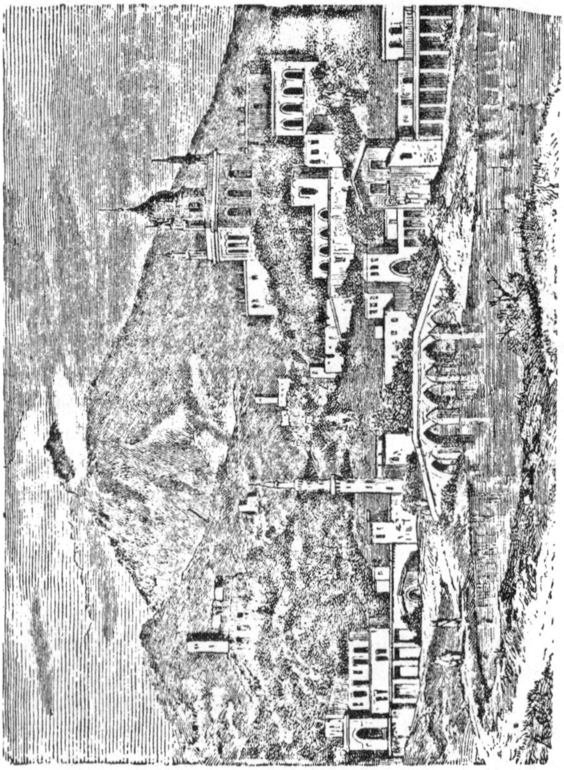

A HIMALAYAN TOWN.

two capitals together make a very interesting locality, with good streets, pleasant houses, many temples of unique style and beauty, in its appearance betraying a certain mixture of Indian and Chinese elements.

Lahore—Delhi.

In the Panjáb, the most considerable places are

RANJIT SING'S TOMB, LAHORE.

Peshawar on the Afghan frontier, the present terminus of the Indian railway system towards northern Afghanistán; Lahore, capital of the province, on the Ravi, nearly due south of Srinagar; Amritsar, the old religious capital of the Sikhs, a few miles farther east; Multan, near the united Chenab and Jhelum, a few miles above the junction of the Satlaj; Delhi on the border of the North-West Provinces in the Jamna valley. Lahore, which is a great railway centre, has in its neighbourhood many ruins of former brilliant epochs, and is still adorned with some fine palaces, mosques, mausoleums, and bazaars.

But none of these places can compare in interest with Delhi, which was for centuries the proud capital of the Mogul Empire, and the centre of the Moslem world in India. The present city occupies a circuit of little over 7 miles, in the midst of vast ruins, covering an area of 20 square miles. Yet its former greatness is still attested by several magnificent buildings, conspicuous amongst which are the Jama-Masjid, the largest and finest mosque in India, and the palace of the emperor Shah Jehan. The canal, 120 miles long, conveying water from the Jamna where it enters the plains, has been restored by the English. But Delhi never really recovered from the blow inflicted on it in 1739 by Nadir Shah, who carried off vast treasures in gold and precious stones, estimated at from eighty to over a hundred millions sterling. Amongst the prizes of conquest was the famous Koh-i-nûr diamond, the most highly esteemed heirloom in the family of the Mogul dynasty. After a series of almost fabulous accidents, this gem ultimately became an appanage of the Queen of England, who, as Empress of India, inherits all the possessions of that dynasty.

The Jamna is here spanned by a railway viaduct in front of the old palace of the Moguls.

Karachi.

Hyderabad at the head, and Karachi at the western

extremity, of the Indus delta, are the chief places in the province of Sind. Karachi, which lies close to the Baluch frontier, is the terminus on the Arabian Sea of the Indus Valley State Railway. Defensive works have here been undertaken, and much has lately been done to improve the harbour, which is somewhat obstructed by a bar, and affords room only for a limited number of large vessels.

Agra—Cawnpore—Lucknow—Allahabad—Benares.

Few regions in the world present such an array of splendid cities as those which line the banks of the main streams along the Ganges-Jamna valley for a distance of considerably over 800 miles. Between Delhi, capital of the old empire, now arbitrarily included in the Panjáb province, and Calcutta, capital of the new Imperial India at the opposite extremity of this vast river basin, there follow in majestic procession such memorable places as Agra, Cawnpore, Lucknow, Allahabad, Benares, Mirzapur, Patna, Murshedabad, and Dacca.

At Agra, which, like Delhi, stands not on the Ganges but on its great tributary the Jamna, artistic interest must ever be centred in the Moti-Masjid and Taj-Mahal, two buildings of surpassing loveliness, in which Muhammadan architecture reached its acme under the Mogul emperor Shah Jehan. The Moti-Masjid, or "Pearl Mosque," stands within the enclosure of the old imperial palace, and though inferior in size, is perhaps superior in design and harmony of proportions to its rival, the Jama-Masjid of Delhi. It is built entirely of white marble, and, with its glorious cupolas, arcades, and lovely surroundings, presents a picture of enchanting beauty, surpassed only, if surpassed, by the peerless Taj-Mahal.

This mausoleum, raised by Shah Jehan at a cost of three millions sterling over the grave of his beloved empress Mumtaz-i-Mahal, combines within itself more varied elements of beauty than almost any other building in the world. Site, size, general design, symmetry of parts, exquisite finish of details, choice materials, play of colour, and all the delightful surroundings, afford a vision of supreme loveliness, which, seen especially when bathed in the liquid atmosphere of a clear moonlight night, leaves an undying impression on the memory of the spectator.

The Jamna is here spanned by a railway viaduct right opposite the Mogul fortress and city.

Cawnpore and Lucknow are names inseparably associated with the most thrilling events of the Indian Mutiny. The most sacred sight in India for men of English blood still must be the monument raised over the well at Cawnpore to the memory of the slaughtered innocents, whose piteous fate inspired their avenging fellow-countrymen with the heroism displayed in the defence and relief of Lucknow. Both places present in other respects many points of interest, although the ambitious palatial structures of Lucknow plainly mark a period of decadence in the Muhammadan architecture in India.

Standing at the confluence of the Ganges and Jamna, and nearly mid-way between Bombay and Calcutta on the Great Indian Peninsular Railway, Allahabad occupies perhaps the most central point of the empire. The Jamna is here spanned by a very fine railway viaduct, commanding a view of the tongue of land at the confluence of the two rivers, which is held by a strong fortress containing an arsenal. The native city of Allahabad is not handsome, and has no buildings of note except the Muhammadan tombs in the Khusrû gardens. But the European quarter has of late years become very fine, with its railway station, its military barracks, and its civil structures.

The first great city on the united Ganges and Jamna below Allahabad is Benares, which holds the same position in the Brahmanical that Delhi does in the Moslem world. It is crowded with palaces and Hindu temples, and, although none of these are of great size, the numerous towers, cones, spires, minarets and porticoes, and flights of steps, present an almost unrivalled river-frontage, nearly three miles in extent. The river view of Benares is one of the most characteristic in the empire.

But the interior of this city is far from inviting, with its close, dirty, and irregular streets, rickety houses, nauseous smells, repulsive mendicants, and stifling atmosphere. The great number of palaces is due to the fact that the Hindu chiefs and princes in every part of the empire endeavour to secure a residence in this sacred city, which during the festivals is crowded by pilgrims from all quarters. The innumerable little temples are compared by Bishop Heber to so many shrines "stuck in the angles of the streets and under the shadow of the lofty houses. Their forms however, are not ungraceful, and many of them are covered over with beautiful and elaborate carvings of flowers, animals, and palm-branches, equalling in minuteness and richness the best specimens of Gothic or Grecian architecture." This description, though written many years ago, is applicable to this day.

Benares has always been a chief centre of Hindu learning, and the Sanskrit College founded here in 1792 is still the principal seat of native instruction in India.

Muhammadanism is also largely represented in Benares, where there are said to be as many as 300 mosques, including a structure with lofty minarets, erected by Aurengzeb on the site of a demolished Hindu temple.

Patna—Murshedabad—Dacca.

Lower down the river is the great trading city of Patna, capital of Berar, where the produce of the poppy is collected in order to be prepared as opium, to be sent to Calcutta for exportation to China. The city proper, within the crumbling old fortifications, occupies a comparatively small space; but the handsome suburbs, with their numerous mosques, temples, streets, and gardens, stretch nearly 10 miles along the river-bank. Patna is much more a Moslem than a Hindu town, and its Mussulman inhabitants have the reputation of being amongst the most fanatical in India. It is also a large industrial centre, and many of its linens, lacquered and other wares, find a ready sale at the great annual fair held at Hajipur on the opposite side of the river.

Near Patna is the railway viaduct over the Sone river, an affluent of the Ganges, one of the longest to be found in any country.

Between Patna and Calcutta the most important town is Murshedabad, on the Bagarathy near the head of the delta. It is a very large place, extending some 8 or 9 miles along both sides of the river. But though it has a large trade, and is in some respects flourishing, those parts which depended on the former court and camp of the Muhammadans present an appearance of decay. The Nawab's palace, however, is a fine structure, built in the European style.

Dacca has always been the centre of the Muhammadan world of Eastern Bengal. It has a flourishing trade, though some of its fine and delicate manufactures have decayed. Its climate is unfavourable. It presents a handsome frontage towards the river.

Calcutta.

On the Húgli, westernmost and largest branch of the Ganges delta, and about 100 miles from its mouth, stands Calcutta, the modern capital of the Indian Empire. It is divided into a European and native city, jointly covering an area of some 15 square miles. The European quarter, which is inhabited not only by the English, has a Western aspect, being laid out with fine spacious thoroughfares, which in the Chowringhee or aristocratic quarter are lined with many fine public buildings and large private residences. It is the frontage of Chowringhee which has caused Calcutta to be called "The City of Palaces." The houses are built in an architectural style peculiar to Bengal and suitable to the climate. This style is handsome as well as commodious, and may be regarded as an instance of originality on the part of the English. The native city also, in which the native population is collected, has broad straight streets, well laid out, and in that respect differs from the aspect of an ordinary Eastern town. It is interspersed with fine public buildings and some native houses built in the English style; otherwise the native houses are poorly built, quite inferior to those of the other capitals in India, the climate of Bengal being unfavourable to native architecture.

But all alike have easy access to the pleasant Eden gardens, which with their tropical vegetation and refreshing ornamental waters form a charming foreground to the surrounding government buildings. Here the winding waters, the varied foliage, the amphitheatre of handsome edifices, the forest of masts from the shipping in the near distance, the guns of Fort William overlooking the animated scene, produce a very pleasing impression.

Calcutta is fairly well supplied with water pumped from the Húgli into filtering beds, whence it is conveyed through pipes for a distance of 14 miles to the city. Above the harbour the river is crossed by a pontoon bridge, which is one of the best works of its kind existing in any country, and gives easy access to the large and rapidly-increasing suburb of Howra on the opposite bank. Below this bridge the ships are moored together with strong chain cables along the quays and jetties, an arrangement adopted as a precaution against the tremendous cyclones to which the delta is exposed. The intricate navigation of the Húgli, with its treacherous sands and constantly-shifting shoals, is conducted by a pilot service admirably organised by Government, and composed exclusively of Europeans with their headquarters in Calcutta. Hence this great capital may be regarded as tolerably safe from the attacks of hostile fleets, which would be wrecked were they to venture into the river without competent pilotage.

Jaganath.

Jaganath (Juggernauth), the most celebrated shrine in India, lies on the Orissa coast not far from the Madras frontier, and about 50 miles south of Kattak. Twelve great festivals, attended by over a million pilgrims, are here annually held in honour of Vishnu. Here is to be seen the huge car which, according to tradition, was supposed to be dragged over the bodies of devotee-victims. The great temple, which was finished at enormous cost in the twelfth century, stands at the head of the main thoroughfare.

Near here, on the sea-shore, stands the grand Hindu ruin known as "The Black Pagoda."

Rangun—Moulmain—Prome.

In British Burma the only large places are Rangun and Moulmain, the two great seaports of Pegu almost facing each other across the Gulf of Martaban. The Shway Dagohn pagoda at Rangun is one of the most remarkable structures of the kind in the Buddhist world. It stands on a wooded eminence, above which its gilded "htee" or umbrella shoots up to a height of 300 feet. From a distance it seems to flash in the sunlight above the dark foliage like a fiery meteor. The hills about Moulmain are also crowned with Buddhist pagodas, whence an extensive and varied prospect is commanded of the city, and the plains watered by three converging streams and enclosed eastwards by the distant Siamese frontier range. There is some literary activity in Rangun, which is gradually becoming the centre of intellectual life for the Buddhist world in Indo-China. Here are four vernacular presses, which have already issued a good many theological, literary, and scientific works, including dramas chiefly adapted from the Sanskrit, Buddhist tracts, often of a very polemical character between the rival Mahagandi and Sulagandi sects, an encyclopædia of Burmese knowledge (the Kawi Lekhana Dípaní), many translations from English works, and some periodical literature. Thus while British Burma is attracting all the material wealth and enterprise from the misgoverned kingdom of Burma, Rangun is in the same way draining it of all its intellectual vigour.

In the middle valley of the Irawady stands the old Burmese city of Prome, with gilded pagodas and wooden pinnacles on the summits of a cluster of hills near the river's bank.

Nagpur—Jabalpur—Bhopal—Indore—Gwalior—Jyepur—Udeypur.

In the Central Provinces the only places with populations exceeding 50,000 are Nagpur and Jabalpur. Both are connected with Bombay by north-eastern extensions of the Great Indian Peninsular railway system, one branch of which has its present terminus at Nagpur. Not far from the northern frontier of these Provinces are the important towns of Indore and Bhopal, in the political system which is termed the Central Indian Agency. Near Bhopal are the Buddhist remains known as the Bhilsa Topes. Indore is the capital of Holkar's possessions, with one of the finest British Residencies in the peninsula.

Due north of Bhopal and about 70 miles south of Agra stands the famous fortress of Gwalior, one of the largest and strongest in the empire. It occupies the level summit of a steep rocky hill 350 feet high, rising abruptly from the surrounding plain, and completely commanding the city of Gwalior, capital of Sindhia's dominions, which lies at its base. Perennial springs, reservoirs, and cultivated grounds, are enclosed within the walls of the stronghold, which is accessible only by steps hewn in the perpendicular side of the rock on which it stands. Yet this apparently impregnable fastness was twice stormed by the British—in 1780 by Major Bruce with a handful of native troops, and again in 1858 by Sir H. Rose, when held by a strong body of mutinous sepoys.

Jyepur is the principal town in Rajputana. It is quite modern and well laid out. In respect to arrangement of streets, it is superior to any native city in the empire. It is the seat of much wealth and commerce.

Udeypur is the very focus of heroic and chivalric traditions. Its palace-crowned hills, its tombs, its lakes and islets, make it the most picturesque city in the empire.

Madras—Bellary.

In the Madras Presidency there are few cities of large size, but many of great historic and antiquarian interest. None have a population of more than 60,000 except the capital and Trichinopoly. There are many disadvantages in the site occupied by the city of Madras on the open, surf-beaten shores of the Coromandel coast, exposed for months together to the full fury of the north-eastern monsoons. Nor is the climate much more favourable, being intensely hot in summer and not entirely free from the malaria so prevalent along the eastern seaboard. Yet in spite of these adverse outward conditions, Madras has under British rule expanded into a flourishing city of nearly half a million inhabitants, with many stately public buildings, literary and scientific institutions, educational and charitable foundations. Something has even been done to improve the harbour, or rather to create one by the construction of a large pier of great strength and size, which is capable of further extension. But with nothing but an open and shelving roadstead Madras can never become a great seaport, and must depend for its future expansion almost entirely on the system of railways by which it is already connected across the peninsula with Bombay, northwards with the Nizam's Dominions and Central Provinces, southwards with Pondicherry, Mysore, Trichinopoly, Madura, and Tuticorin. Madras is protected by Fort St. George, one of the earliest strongholds of the East India Company, and at present one of the arsenals of the empire. In its vicinity, on the sea-shore, are the rock-cut temples of Mahabalihuram, celebrated by Southey's poetry.

Near the main line from Madras to Bombay are the native rock-fortress and the European cantonment of Bellary. Near Bellary, again, are the wonderful extensive ruins of the Hindu city of Bijayanagar.

Trichinopoly—Madura—Tanjore.

Trichinopoly, the next largest place in the Presidency, lies in the fertile Kavari valley, a few miles west of Tanjore, and close to the famous temple of Srirangam. These Hindu buildings, which are amongst the most remarkable of their kind in India, occupy the western extremity of a large island in the Kavari, where the chief pagoda stands in the centre of seven separate square enclosures, with a total circuit of nearly four miles. It is a vast structure, surmounted by a gilded dome, beneath which is the statue of the presiding deity, one of whose glittering eyes, abstracted in the last century by a French deserter, proved to be a diamond of almost matchless purity. This gem, known as the Orloff diamond, now figures as the chief ornament in the imperial sceptre of Russia. Trichinopoly is commanded by a strong fort, perched on a steep granite peak, 500 feet high. It is noted for its peculiar style of gold work. In the Protestant Church of St. John repose the remains of Bishop Heber, interred here in the year 1826.

South of Trichinopoly, and connected with it by rail, lies the ancient city of Madura, with its truly magnificent temples, and other monuments of Hindu art. The palace, built by Tirumal Naik, a former ruler, the finest structure of its kind in India, never fails to excite the astonishment of visitors, who stand amidst its vast arcades, courtyards, vestibules, reception chambers, and halls, with their vaulted roofs and arches. Tanjore, at the head of the Kavari delta, also possesses some famous Hindu monuments, including a sacred bull 20 feet high, hewn out of a single granite block. It stands in one of the palace courtyards, but even modern engineers still marvel how it was carved and transported to its present site.

Calicut—Mysore—Seringapatam—Bangalore.

Calicut, on the Malabar coast, lies in one of the most fertile districts of the peninsula, yielding pepper, ginger, cotton, cardamoms, and other tropical products in vast profusion. This was the first Indian seaport visited by Vasco de Gama in 1498, and from the peculiar cotton fabric here formerly manufactured. the *calicoes* of the modern European looms take their name.

A line drawn from Calicut north-eastwards will very nearly intersect Mysore, Seringapatam, and Bangalore, in every respect the three most interesting places in the State of Mysore, which has recently been again placed under the administration of the native Rajah. The city which gives its name to the State forms a pleasant aggregate of regular streets, avenues, gardens, and temples, the whole commanded by a strong fort, constructed from European designs. This stronghold, which is separated by an esplanade from the city, encloses within its precincts the Rajah's palace, besides the dwellings of many wealthy citizens and other private buildings. But the British Residency lies some 5 miles farther south, on the summit of Mysore hill, 1000 feet above sea-level.

Seringapatam, on the main head-stream of the Kavari, is chiefly noted for its fortress, which figured so prominently in Indian history during the closing decade of the last century. This formidable stronghold of Tippu Sultan occupies the west side of a large island in the river, and although considered quite impregnable, was finally stormed by the British in 1799. Its streets, houses, and fortifications remain, but it is now a city of the dead.

Bangalore, which lies almost exactly midway between Madras and Mangalore on the opposite coast, and nearly 200 miles from both points, is by far the largest city in the interior, south of the Kistna valley. Yet it is quite a modern place, having been founded by Hyder Ali about 1780 as a bulwark against the English. The fort has long been disused; but, thanks to its central position in the midst of an extremely fertile district, the town soon acquired a rapid expansion. Lying at an elevation of 3000 feet above the sea, on the Mysore plateau, it enjoys a delightful climate, and is consequently a favourite resort of Europeans. Here is a large British cantonment, with extensive barracks, library, public gardens, racecourse, and other attractions. From a combination of happy circumstances, Bangalore has thus become, in a few decades, the chief centre for the diffusion of Western ideas amongst the Dravidian inhabitants of the interior of Southern India.

Hyderabad—Secanderabad.

Haidarabad (Hyderabad), capital of the Nizam's Dominions, occupies a somewhat central position in a fine climate, the eastern terminus of the native state railway running thence to join the Madras main line. It is the largest city in the whole of the Deccan, with a present population of over 400,000, and with a picturesque situation. There is a handsome British Residency, one remarkable mosque, and one fine gateway. Hyderabad is much more a Moslem than a Hindu city, Pathans, Arabs, and Rohillas being here numerous.

The neighbouring town of Secanderabad may be regarded as a European quarter, this being the headquarters of the British subsidiary force in the Nizam's territory. Here are some of the largest and best-constructed cantonments in India, with extensive barracks, hospital, Protestant and Roman Catholic churches, Masonic Lodge, promenades, public libraries, racket courts, lawn-tennis grounds, and racecourse.

Near here are the old citadel and the mausolea of Golconda.

In the north-west corner of the Nizam's Dominions are the rock-cut temples and caves of Ellora (Hindu) and Ajanta (Buddhist), also the hill-fortress of Daolatabad.

Bombay.

Bombay, capital of the Presidency, is not only the most flourishing city in the Indian Empire, but possesses probably more elements of future greatness than any other city in Asia. It occupies the south-east end of the island of like name, which is 8 miles by 3, and which is connected by a mound with the larger island of Salsette. These, with Elephanta and two or three others, form a little group close to the Konkan coast, in 18° 53′ N., 72° 48′ E., jointly enclosing with the mainland one of the most commodious and expansive harbours in the world.[1] The space available for shipping is nearly 14 miles long and about 5 broad, with an average depth of 10 to 12 fathoms. This splendid natural position has been greatly improved by artificial works, including extensive quays, wharves, and several docks, the finest of which is the Prince's Dock, with an area of 30 acres, recently completed at a cost altogether of over a million sterling.

[1] The word Bombay—*i.e.* Bom Bahia—means in Portuguese "Good Harbour." Otherwise it is thought to be a corruption of Mumbai, a small island named after the goddess Mumba.

The city consists properly of two parts, a native and European quarter, the latter stretching along the shore of the bay, where a line of magnificent buildings presents an imposing view when seen from Malabar Hill, at the south-west point of the island. The native city has several long streets, which are the finest in the Indian Empire. It is well supplied with good water, brought through pipes from two large artificial lakes embosomed in the picturesque wooded hills forming the advanced spurs of the Western Ghats, which here approach to within 20 miles of the coast. Although the scheme of defences is still incomplete, Bombay is already defended by several formidable batteries, as well as by two iron-clad turret-ships permanently stationed at this port. In case of danger the whole of the shipping might also find absolute security in the inner waters behind the island of Elephanta. When we add that Bombay is the first important place reached by vessels from Europe and the Suez Canal, and that it is directly connected by several railway systems with every part of the peninsula, it will be seen that this great seaport lacks none of the elements calculated to secure it a foremost position amongst the cities of Southern Asia. Between it and Calcutta there is an honourable rivalry for the first position. Each city has advantages peculiar to itself, and it is hard to say which of the two will ultimately prevail.

Ahmadabad—Baroda—Surat.

The Presidency also contains several other large cities, the most important of which are Ahmadabad, Baroda, and Surat in the Northern, Poona and Sholapur in the Central, Dharwar and Belgaum in the Southern Division.

Ahmadabad, which is a very large place at the neck

of the Gujarat peninsula, equidistant from the Rann of Katch and the Gulf of Cambay, contains many beautiful specimens of Muhammadan architecture. Unfortunately some of these monuments, including the great mosque of Sultan Ahmad, were shattered or destroyed by the terrible earthquake which seriously injured the place in 1819. Still many fine structures remain to delight the student of architecture. The city is now a great centre of Oriental art, producing exquisite specimens of damascened metal work, gold and silver plate, mother-of-pearl objects, rich trappings and caparisons for the native princes.

The late Gaikwar of Baroda was deposed by the paramount power for maladministration. He was held by the Government of India to have been guilty of an attempt to poison the British Resident by a dose of diamond-dust. Baroda itself, which lies nearly midway between Ahmadabad and Surat, has prospered in the sunshine of Maratha royalty, and is a fine city, though not remarkable for architecture.

Surat occupies a convenient position near the mouth of the River Tapti, about 160 miles by rail due north of Bombay. It is the natural emporium of the rich Kandeish valley, and covers a large space some 8 miles in circumference on the left bank of the river, 20 miles from the Gulf of Cambay. In the early days of the East India Company it was the principal trading place on the west coast, but during the last century it has become quite secondary to Bombay.

Poona—Sholapur—Bijapur.

No place in the Central Division of the Bombay Presidency can compare in importance with Poona, which is delightfully situated about 80 miles south-east of Bombay on the Deccan plateau, some 2000 feet above the sea. With its large British cantonments, hospitals, libraries, churches, colleges, and missionary schools, this famous capital of the Peishwas, or heads of the great Maratha confederacy, has in our days become the chief centre for the spread of European culture among the brave but somewhat turbulent Maratha races of Western India. The palace of the Peishwas, built of teak-wood, a noble specimen of Maratha architecture, was burnt in 1879.

South-east of Poona and close to the Nizam's frontier lies the town of Sholapur, a former stronghold of the Marathas, with two distinct lines of fortifications.

A far more interesting place is Bijapur, which lies some 60 miles farther south on a small tributary of the Kistna, and was the capital of the Muhammadan kingdom, which comprised the western Deccan, before the establishment of the Mogul Empire. The extent and splendour of the ruins attest the former greatness of this "Palmyra of the Deccan," as it has been called. These ruins,—which are remarkable especially for their great solidity and simple grandeur, and yet a suitable degree of ornamentation,—consist of Muhammadan palaces, mosques, and other structures, many of the domes, spires, and minarets of which are still standing. Among these is a mausoleum, with a cupola, the admiration of architects and the largest yet constructed in the world.

Satara—Ahmadnagar.

Satara and Ahmadnagar are the only other places in the Presidency which call for special mention. They both lie on the Deccan tableland and on small head-streams of the Kistna, the former 70 miles south, the latter 80 miles north-east of Poona. Ahmadnagar has a few good streets and substantial buildings enclosed by a wall, beyond which are a strong stone fort of historic celebrity, a finely-built palace, and on the crest of a neighbouring hill the tomb of Salabat Jung. Satara—much associated with stirring passages of Maratha history—is clustered round the base of a rocky eminence rising 800 feet above the surrounding plain, and crowned by the ruins of an ancient citadel. In the neighbourhood are European cantonments, which enjoy a favourable and healthy climate.

9. *Highways of Communication: Canals—Roads—Railways.*

Under the British administration a system of internal communication has been rapidly developed, which in this respect places India nearly on a level with the most civilised regions of the globe. Apart from the natural channels of the great rivers and their affluents, affording over 10,000 miles of navigable water highways, the irrigation canals, which are constantly increasing, are often navigable by small craft for hundreds of miles. Many of the larger ones have been specially adapted to this purpose, and by a wise provision have thus been made to serve a twofold object. The canals near Calcutta and in Orissa, and those of the Madras Presidency, are largely utilised in this way. The irrigation system has already assumed magnificent proportions. The chief scenes of these operations are the country between the Jamna and the Ganges, several parts of the Panjáb, and the deltas of the Mahanady, the Godavari, the Kistna, and the Kavari on the east coast, and the delta of the Indus on the north-west coast. Of main channels, great and little, there are no less than 13,000 miles completed, besides countless distributing rills with a total length of nearly 9000 miles in the north alone. Thus have been brought under irrigation about 7,000,000 acres, mostly of extremely fertile land, at a total expenditure by the State of £21,000,000 sterling, yielding an average interest of 6 per cent.

Although occupied for ages by settled communities, which had attained a high degree of culture long before Britain had emerged from barbarism, India seems to have possessed scarcely any roads before the advent of the English. Neither the ancient Hindu dynasties nor their Moslem conquerors paid any attention to this primary condition of true civilisation. Many of the petty rulers were even directly opposed to the development of easy lines of communication, which would have the immediate effect of opening up the country to the attacks of hostile neighbours.

Now all this is changed, and although the system is still far from complete, over 20,000 miles of metalled or macadamised highways have been constructed, mostly within the last fifty years. Thus all the great cities have been brought into direct communication with each other, and the uttermost limits of the land have been made accessible to trade and to defensive or offensive warfare.

The great trunk lines are those running from Calcutta for 1000 miles to Delhi, and thence through Lahore to the frontier at Peshawar; from Bombay for 900 miles to the last-mentioned at Agra; from Bombay for 800 miles over the Western Ghats and across the Deccan to Madras; from Bombay through Gujarat; from Madras northwards to Bengal, southwards to Trichinopoly and Madura, westwards through Bangalore to the Malabar coast. Important sections of the system are also the routes running in the Himalayas from Amballa to Simla and beyond it towards Chini, and from the Bengal plains to Darjiling and thence to the Chola range on the Tibetan frontier; in the Deccan the roads connecting Mirzapur on the Ganges through Jabbalpur and over the Satpura range with Nagpur in the Central Provinces; the line running from Poona southwards to Mysore, and that ascending from Coimbatore to Utacamand in the Nilgiris.

Most of these highways are solidly constructed, and often present splendid specimens of engineering skill in their gradients, cuttings, causeways, and bridges. Like the old Roman roads, they are in many places carried right over the Ghats, Vindhyas, and other ranges, and through such difficult passes as the Thal and Bhore in the Western Ghats. The section between the Jhelum and Indus, in the extreme north-west, consists of an almost continuous series of cuttings and embankments for a distance of over 150 miles.

The Indian railway system, carried out mainly on the wise plans laid down by Lord Dalhousie some thirty years ago, has already assumed considerable proportions, and will in a short time form a total mileage of over 9000 miles. The base of the system is the great trunk line running from Calcutta for 1500 miles up the Ganges valley through Allahabad and Lahore, and across the Indus at Attock to its present terminus at Peshawar on the Afghan frontier. From Allahabad, on this base, the Great Indian Peninsula runs first over the Bandelkand hills, down the Nerbadda valley and through the Satpura range and Western Ghats for 700 miles to Bombay, and thence again over the Western Ghats, through Poona, and across the Deccan for 800 miles to Madras. Thus the three capitals are brought into direct communication with each other and with all the more central and populous parts of the empire.

Of the other lines, some are of great length and of much commercial and strategical importance. Of these perhaps the most vital is the Indus valley line connecting Lahore with the sea at Karachi, the nearest port to England, and with a projected branch of 400 miles from Sakkar to Kandahar already completed to Sibi near the foot of the Bolan Pass. This line has much political importance in reference to the completion of the Russian Trans-Caspian line in 1881 to Kizil-Arvat and Bami within measurable distance of Herat.

Another great section runs from Bombay along the west coast, across the Lower Tapti and Nerbadda valleys, across Gujarat and Rajputana to the northern trunk line at Agra with a junction to Delhi. Several minor branches ramify from these main lines northwards to the Himalayas at Kurseong for Darjiling, southwards to Gwalior, eastwards to Nagpur and Secanderabad close to Hyderabad, capital of the Nizam's Dominions. From Madras two independent lines radiate, one right across the Deccan through Vellore and Coimbatore to Beypore near Calicut on the Malabar coast, the other southwards through Tanjore, Trichinopoly, and Madura, to Tuticorin and Tinnevelli near the apex of the peninsula. On these lines there are branches to Bangalore, Pondicherry, and Negapatam for Karikal.

Lastly, in British Burma a line runs up the Irawady valley from Rangun to Prome, near the Burma frontier, and another is in progress, also from Rangun up the Sittang valley to Tungu, near the frontier of the Karen country.

Most of these lines, the materials for which had to be brought mostly from England, were built at an average cost of about £14,000 per mile, the total capital already expended amounting to over £123,000,000. They are constructed partly on a narrow, but chiefly on a broad gauge, the former mostly by the State, the latter by private companies, to whom a rate of 5 per cent interest is guaranteed by the Government of India. Railway travelling is growing in popularity, and the various lines already convey over 43 million passengers yearly, the vast majority of whom are Natives.

The telegraphic system, originally planned by Sir William O'Shaughnessy, may be regarded as complete, comprising a total length of over 19,000 miles, exclusive of two lines to England, and others to China, Japan, and Australia.

10. *Administration: The Native States—Social Progress—Education.*

After the mutiny of the Native army of Bengal in 1857 the administration of the country passed from the old East India Company to the Crown, and on 1st January 1877 India was constituted an empire, the Queen of England assuming the title of Kaisar-i-Hind, or Empress of India. The sovereign is represented on the spot by the Viceroy and Governor-General, whose headquarters are at Calcutta, but who ordinarily resides during the summer months at Simla in the Himalayas. The Governor-General and the Governors of Madras and Bombay are each aided by Executive Councils, which are like Cabinets on a small scale. There is one Legislative Council of the Governor-General for legislation regarding imperial matters. There are also three local Legislative Councils, sitting at Calcutta, Madras, and Bombay respectively. The members of all these Legislative Councils are appointed by Government and not elected. The Government of India—that is, the Governor-General in Council—is subordinate to Her Majesty's Government in England, represented by the Secretary of State for India in London, who is assisted by a Council.

Subordinate to the Governor-General are the Governors of the two Presidencies of Madras and Bombay, the Lieutenant-Governors of Bengal, the North-West Provinces, and the Panjáb, the Chief Commissioners of the Central Provinces, Assam, and British Burma, the Resident at Hyderabad, the Agents to the Governor-General in Rajputana, Central India, and other Residents and Political Agents of the first rank. There are four High Courts of Judicature at Calcutta, Madras, Bombay, and Allahabad respectively, and one Chief Court for the Panjáb. There are three armies, belonging to the Presidencies of Bengal, Madras, and Bombay respectively, each army having a Commander-in-chief; but the Commander-in-chief of the Bengal army has a supreme command over all the Royal troops in the empire, as contradistinguished from the Native. There are five Bishoprics, exclusive of the Missionary Bishoprics. The Bishop of Calcutta is also Metropolitan in India. At each of the three Presidency towns there is a bank connected with the Government.

This is an outline of the machinery by which England from a distance of 8000 miles administers the affairs of 253,000,000 people, including a large number of Native States, which recognise the supremacy of the paramount power. As, on the other hand, there are not more than a few thousand Europeans, exclusive of the military, and only a few hundred European civil officers in the whole empire, it will readily be imagined how arduous must be the task imposed on the Government of keeping order amongst such a mass of human beings, consisting of heterogeneous elements. Not perhaps unnaturally, the imperial race to which such an inheritance has fallen, feels at times more oppressed by a deep sense of its overwhelming responsibilities than elevated by the commanding position it thus takes amongst the nations of the world.

All the Native States (some three hundred in number, great and small) may be regarded as placed under the protection of the suzerain power, the only really independent elements being some of the wild and often troublesome hill tribes on the frontiers. Of these States there are three categories—the allied, the tributary, and the protected. The allied are provided by the British Government with a regular contingency of subsidiary troops, for which a fixed charge is made. These represent a total population of over 20,000,000. In the tributary States the Government maintains no regular troops, but undertakes to defend them from any possible attacks from without, receiving in return a regular tribute. Of such States there are about fifty, with some 12,000,000 inhabitants. The protected States, exempt from tribute, stand in the same relation to the supreme authority, and number upwards of ninety, with a joint population of perhaps 18,000,000.

All three have renounced the right of self-defence and of independent diplomatic representation abroad, England guaranteeing them from attack, and acting as mediator in all the differences arising among them. They also maintain troops numerous enough to preserve peace within their borders. The English Government, moreover, reserves to itself the right of interfering in the internal administration whenever the native rulers become the oppressors instead of the protectors of their subjects. In fact, however, the Native States are becoming well governed.

The chiefs, princes, and other representatives of these various Native States appear from time to time at the "durbar," or public audience of the Viceroy, for the purpose of paying homage to the Empress through her representative.

Under this administration, ensuring the blessings of

peace at home and presenting a firm front to any possible assaults from without, the country has made astonishing progress both materially and morally in recent times. A far more radical transformation has taken place than might be suspected at a cursory glance. The removal of the centre of authority from the old inland capitals to the seaboard,—the general disarmament of the people, and the establishment of lasting peace and security in the remotest corners of the empire,—the suppression of savage rites such as human sacrifices amongst some wild hill tribes, and of Suttee[1] amongst the Hindus,—the surveys, trigonometrical, topographical, and geological,—the enlightened legislation, and the establishment of a system of civil and criminal justice,—the releasing of trade from transit duties and other fetters,—the assessment of the land-tax for long terms of years, and the recognition of proprietary right in the land,—the construction of highways, railroads, and telegraphic lines, and the extension of artificial irrigation,—the introduction of education on English principles—are all unmistakable evidences of social progress.

Some of the old native manufactures are dying out in many places, partly through the competition of the English looms, and partly through the introduction of modern machinery, while many of these manufactures continue to flourish. On the other hand, nevertheless, thousands are employed in the jute, cotton, and sugar factories, in the coal-mines, and in the plantations of tea and coffee. Nor has the traditional skill of native craftsmen and the hereditary genius of native artists succumbed to Western influences. Thus in Orissa and Southern India the handloom still maintains its place, and the most delicate

[1] *Suttee*, or rather *Sati* (that is, "the pure one"), properly means the widow who immolates herself on the death of her husband, but is commonly applied in English to the act itself.

muslins in the world may still be procured from the Dacca weavers, although at very high prices.

Another result of the English rule is the increased sense of unity that has been developed amongst the various nationalities. The same tendency is shown in the cultivation of the native languages (both classical and living), which formerly received little encouragement from the various Persian, Hindi, or Marathi speaking conquerors, but which are now fostered in the national education. Three Universities have been established (much upon the model of the London University) at Calcutta, Madras, and Bombay respectively, to each of which several colleges, belonging to Government and to private bodies, are affiliated. Three Medical Colleges of the best possible kind, with several Medical Schools and two Colleges of civil engineering, also several technical and industrial schools, have been established. Much satisfactory progress has been made in the popular instruction, although much remains still to be done in this direction, and especially as regards female education. In the Panjáb the schools and attendance have greatly increased; yet 60 per cent of the children are said to be still unprovided with instruction of any sort. In Bengal also the elementary schools have been greatly multiplied, and here the wish to learn English is increasing among the middle classes. Still more satisfactory has been the progress in Madras, where the schools increased from 9274 to 10,533 in 1880, and the attendance from 237,838 to 268,379. Here the greatest development is in the primary schools and amongst the native Christians. A really sound beginning has been made with female education. As regards the highest education, 1094 out of 3309 candidates passed the university entrance examinations, as compared with 356 and 2597, the corresponding figures in 1879.

For British Burma an Educational Syndicate was established in 1881 for the purpose of controlling the public examinations, which under new regulations will be especially designed to encourage the study of law, medicine, engineering, and the technical arts.

The State expenditure is largely incurred in the shape of grants-in-aid. About half of the educational expenses in the interior of the country are defrayed by the State, and the other half by the people.

The progress of Christian missions of all Protestant denominations is considerable. The Native Christians are about 400,000 in number, and the children under Christian instruction, though not actually Christians, are about 200,000. These numbers have been increasing at the rate of 50 per cent in each decade during the generation ending in 1880. There are about 450 mission stations and 500 European missionaries, 3 missionary bishops, and 300 native ordained clergymen. Several colleges and training institutions belong to the missionary bodies. The Vernacular Education Society conducts extensive operations in the publication and colportage of books for Christian instruction. The total income of the Protestant missions has been computed at something between £300,000 and £400,000 annually.

The Roman Catholic Church has real vitality, and includes Europeans, East Indians, and natives. It has archbishops, bishops, vicars apostolic, and lady superiors. It has many missionary stations, besides colleges, schools, convents, and other religious establishments.

Further proofs of material and social progress will be revealed in the subjoined tabulated statements of population, trade, education, etc.

11. *Statistics of British India.*

AREAS AND POPULATIONS.

BRITISH TERRITORIES.

Province.	Area in sq. miles.	Population according to Census of 1881.	Population according to previous Census, 1868-1871.
Bengal	155,997	68,829,920	63,061,518
Assam	45,303	4,815,157	4,056,054
North-West Provinces	81,748	32,699,436	30,769,056
Oudh	24,213	11,407,625	11,219,675
Panjáb	107,010	18,786,107	17,611,498
Central Provinces	84,208	11,505,149	9,251,229
British Burma	87,220	3,707,646	2,747,148
Ajmir	2,711	453,075	426,268
Madras	140,333	30,839,181	31,597,872
Bombay } Sind }	126,445	{ 13,978,488 { 2,404,934	14,038,359 2,192,415
	893,176	199,426,718	186,971,092
Andaman and Nicobar Islands	3,285	26,000	25,945
	896,461	199,452,718	186,997,037

NATIVE STATES.

State.	Area in sq. miles.	Population in 1881.[1]	Population from 1868-1871.
Central India	89,098	9,200,884	8,360,571
Rajputana	130,989	11,005,512	10,192,871
Hyderabad (Nizam)	80,000	9,167,789	9,000,000
Berar	17,711	2,670,982	2,227,654
Baroda	4,399	2,154,469	2,000,000
Mysore	24,744	4,186,399	5,055,412
Kúrg	1,572	178,283	168,312
Manipur	7,584	126,000	126,000
Bombay (Lesser States)	66,498	6,941,631	6,784,482
Panjáb (Kashmir and other States)	114,742	5,300,000	5,370,096
Travancore	6,730	2,401,158	2,308,891
Cochin	1,361	600,278	601,114
Nepal	85,000	2,000,000	2,000,000
Bhutan	19,000	750,000	750,000
North-West Provinces (Lesser States)	5,125	745,675	636,543
Total	654,553	57,429,060	55,581,946

British India.	Area in square miles.	Present Population.
British Territories	896,461	199,452,718
Native States	654,553	57,429,060
Total	1,551,014	256,881,778

[1] The returns of the census of 1881 have not yet been received for some of the Native States, and for some outlying places. But, so far as is yet known, the result shows that there has been an increase of at least 12,788,565 souls in British India, or in round numbers 13 millions, within the ten years from 1870 to 1880, equal to an addition of 6 per cent to the population notwithstanding two very severe famines. As there have been no famines since 1880, it is inferrable that this ratio of increase must be going on.

British Political System.

SOUTHERN ASIA.

	Area in sq. miles.	Population.
British India	1,551,014	256,881,778
Ceylon and other islands	30,000	2,630,000
Burma (exclusive of British Burma)	190,000	5,000,000
Afghanistán and Baluchistán	400,000	6,450,000
Straits Settlements, Malacca	47,000	600,000
Aden and Hong-Kong	200	300,000
Grand Total	2,218,214	271,861,778
French Possessions in India	178	285,022
Portuguese " "	1,086	444,617

Towns with upwards of 30,000 Inhabitants.

Panjáb.

	Pop.
Delhi	160,000
Amritsar	142,000
Lahore	128,000
Peshawar	58,000
Multan	50,000
Jullandar	50,000
Lhudiana	39,000
Bhiwani	33,000
Sialkot	32,000

N. W. Provinces and Oudh.

Lucknow	284,000
Benares	175,000
Agra	150,000
Allahabad	150,000
Cawnpore	113,000
Bareilly	91,000
Mirut	81,000
Shahjahanpur	72,000
Mirzapur	67,000
Farukhabad	65,000
Mattra	64,000
Moradabad	62,000
Koil	57,000
Gorakhpur	51,000
Saharanpur	40,000
Ghazipur	38,000
Faizabad	37,000
Budaun	33,000

Bengal.

	Pop.
Calcutta	686,000
Purneah	160,000
Patna	158,000
Howrah	97,000
Dacca	69,000
Bhagalpur	69,000
Gya	66,000
Monghyr	59,000
Kattak	50,000
Durbhunga	47,000
Murshedabad	46,000
Chupra	46,000
Behar	44,000
Arrat	39,000
Mozufferpore	38,000
Húgli	34,000
Burdwan	32,000
Midnapore	31,000

Central Provinces.

Nagpur	84,000
Jabalpur	55,000
Kampti	48,000
Sangor	45,000

British Burma.

Rangun	91,000
Moulmain	54,000

Rajputana and Central India Agency.

Ujein	100,000
Bhartpur	?90,000
Gwalior	?90,000
Jyepur	60,000
Jodhpore	60,000
Bikanir	60,000
Ajmir	31,000

Berar and Nizam's.

	Pop.
Hyderabad	400,000
Secanderabad	34,000

Madras and Mysore.

Madras	397,000
Bangalore	142,000
Trichinopoly	76,000
Mysore	57,000
Tanjore	52,000
Madura	51,000
Bellary	51,000
Salem	50,000
Nagapatam	48,000
Calicut	47,000
Combaconum	44,000
Cuddalore	40,000
Vellore	38,000
Conjeveram	37,000
Masulipatam	36,000
Coimbatore	35,000
Vizagapatam	32,000
Cannanore	31,000

Bombay Presidency.

	Pop.
Bombay	644,000
Baroda	140,000
Poona	118,000
Ahmadabad	116,000
Surat	107,000
Karachi	56,000
Sholapur	53,000
Hyderabad	41,000
Shikarpur	38,000
Hubli	37,000
Bróch	37,000
Ahmadnagar	36,000
Cambay	33,000
Belgaum	32,000

Himalayan States.

Katmandu	? 100,000
Srinagar	51,000

Ceylon.

Colombo	97,000
Point de Galle	47,000

Population of India classed according to Races.[1]

Hindus	190,000,000
Dravidians	53,000,000
Kolarians	4,000,000
Tibeto-Burmans	4,000,000
Shans	500,000
Afghans and Baluchis	300,000
Malays	250,000
Parsis and Persians	50,000
Indo-Arabs	50,000
Negritos	10,000
Africans	3,600
Europeans, Australians, and Americans	336,000
	252,699,600

Population of India and Ceylon classed according to Religions.

Hindus	190,000,000
Muhammadans	50,000,000
Buddhists and Jains	5,500,000
Nature-worshippers	7,500,000
Sikhs	1,250,000
Christians	1,100,000
Parsis	60,000

[1] This classification is based mainly on language. For details see p. 287.

Chief Rivers of India.

	Length. Miles.		Length. Miles.
Brahmaputra and San-po	1800	Tapti	440
Indus	1800	Kavari	470
Ganges	1500	Pennar	355
Irawady	1000	Luni	320
Godavari	900	Sitang	230
Kistna	800	Brahmani	410
Salwen	750	Mahi	350
Nerbadda	800	Baitarani	345
Mahanaddi	520		

Canalisation (1880).

Capital expended	£20,500,000
	Miles.
Main Canals and branches in the three Presidencies	4900
Panjáb and Sind	2250
Tanjore or Kavari system	700
Distributing Canals in North India	8300
	16,150

		Acres.
Area of Irrigation	Madras and Bombay	1,900,000
	Panjáb and Sind	2,600,000
	N.W. Provinces	1,450,000
	Behar and Orissa	360,000
		6,310,000

		Miles.
Railways (1885)	East Indian Lines	1509
	Guaranteed Lines	4528
	Assisted Lines	869
	State Imperial Lines	3036
	State Provincial Lines	1509
	Native States Lines	664
	Total	12,005

Passengers (1884)	73,815,119	Gross receipts (1884)	£16,006,225
Under construction	650 miles	Capital expended	142,423,903

Budget (1884-85)	Receipts	£71,727,000
	Expenditure	70,340,000
	Debt	161,300,000

Chief Heads of Income.		Chief Heads of Outlay.	
Land Tax	£21,546,000	Army	£15,972,000
Opium	8,850,000	Productive Works	12,358,000
Salt	6,350,000	Interest on Debt	4,404,000
Stamps	3,578,000	Civil Departments	11,474,000
Excise	4,014,000	Non - Productive Works	6,571,000
Provincial Rates	2,794,000		
Customs	1,030,000	Charge of Collection	8,125,000
		Exchanges	3,253,000

ARMY—NORMAL STRENGTH.

British.	Bengal.	Madras.	Bombay.	Total.
Artillery	6,879	2,811	2,549	12,239
Cavalry	2,898	966	483	4,347
Infantry	29,420	8,271	8,271	45,962
Staff and Sundries	1,086	685	407	2,178
	40,283	12,733	11,710	64,726
Natives	63,933	34,293	26,645	124,871
Total British and Natives	104,216	47,026	38,355	189,597

OCCUPATIONS OF MALE ADULTS IN BRITISH INDIA.

Agriculture	Large landlords	750,000	
	Lesser proprietors	5,750,000	
	Peasant proprietors, sub-proprietors, cultivators	28,500,000	
			35,000,000
Labouring Classes			8,000,000
Industries			8,000,000
Domestic Pursuits			4,500,000
Trade			3,500,000
Professions, Employees			2,500,000
Independent			2,000,000
Mendicant			1,000,000
			64,500,000

EDUCATION (1884).

Colleges	95	Attendance	8,717
Secondary Schools	4,182	,,	347,950
Primary Schools	101,342	,,	2,507,000
Technical Schools	309	,,	11,400
Universities	3[2]	Matriculated	3,593
State Expenditure on Public Instruction			£2,165,000
Proportion of Population receiving Instruction			11 per 1000

LITERATURE.

Average yearly publications—	English	500
	Vernacular languages	3000
	Classical languages of India	730
	In more than one language	570
		4800

POSTAL AND TELEGRAPH SERVICES (1884).

Post Offices and Letter Boxes	14,305	Telegraph Stations	350
Postal Lines, miles	63,500	,, Lines, miles	23,341
Letters, etc., carried	203,340,000	Messages forwarded	1,837,000

[1] Since 1885, in reference to the north-west frontier, some additions to this force are being made.

[2] Calcutta, Madras, and Bombay. Not more than one-fourth of the matriculated students become graduates.

Progress of Foreign Trade and Shipping.

Year.	Imports.	Exports.
1839	£5,240,000	£11,774,000
1851	15,370,000	20,194,000
1860	40,622,000	28,889,000
1878	58,819,000	67,433,000
1885	67,026,000	85,001,000

Annual Trade (1885), £150,000,000.

Present yearly average of sea-borne trade £70,000,000 exports, £80,000,000 imports, of which 40 per cent is with Great Britain.

Chief Exports (1885).

Opium.	Dyes.	Cotton.	Jute and wool.	Rice.	Wheat.	Tea.
£10,883,000	£4,069,000	£15,524,000	£5,374,000	£7,192,000	£6,310,000	£4,045,000

Shipping.

Year.	Vessels entered.	Tonnage.	Vessels cleared.	Tonnage.
1878	167,002	8,062,000	152,622	7,670,000
1884[1]	110,343	7,293,000	107,963	7,153,500

The foreign trade, of which 88 per cent is British, employs at present 12,500 vessels of 5,500,000 tons burden, of which 2000 are steamers with 2,250,000 tonnage. Of the steamers two-thirds use the Suez Canal.

Miscellaneous.

Villages (British India), 494,000; houses, 37,000,000.
Municipal population, 1,500,000.
Towns with 10 to 50,000 inhabitants, 1360.
Towns with upwards of 50,000 inhabitants, 46.
Mean density of the population, 212 per square mile.
Land under cultivation, 300,000,000 acres.
Waste or unproductive, 290,000,000 acres.
Under crops of all sorts, 188,000,000 acres.
Under food crops, 166,000,000 acres.
Yield of food crops, 52,000,000 tons; value, £332,000,000.
Yearly output of coals, 1,000,000 tons.
Spinning and weaving mills, 53; with 1,500,000 spindles.
Police, 158,000.
Rural police (village watchmen), 442,000.
Yearly criminal charges, 970,000.
Yearly Convictions, 550,000.
Prisoners and convicts, 118,500.[2]
Yearly Civil lawsuits, 1,500,000.
Charitable dispensaries and hospitals, 1150—indoor patients, 270,000; outdoor relief, 6,500,000.
Lunatic asylums, 22; inmates, 3500.
Annually destroyed by wild beasts, 20,000 persons; 50,000 cattle.
Emigration from 1869 to 1879, 173,420 persons.
Circulation of Government paper currency, average 13 millions sterling.

---

[1] Including coast traffic.

[2] Of these 5500 only are females.

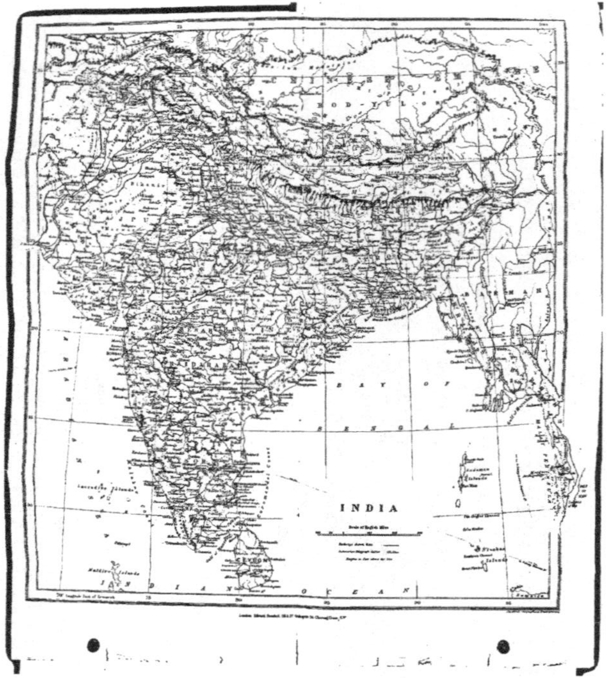
INDIA
BAY OF
BENGAL
INDIAN
OCEAN
CEYLON

SECTION C.

NORTHERN ASIA: RUSSIAN POLITICAL SYSTEM.

CHAPTER IX.

CAUCASIA.

1. *Boundaries—Extent—Area.*

NORTHERN ASIA forms one vast political system, comprising nearly one-third of the whole continent, and, with a few trifling exceptions, directly administered by Russia. It embraces three distinct geographical regions—Caucasia, Turkestán, and Siberia—which will here be treated under three separate chapters.

Caucasia consists, broadly speaking, of the Ponto-Caspian isthmus—that is, of the narrow neck of land separating the Euxine (Pontus) from the Caspian Sea, and connecting the south-east corner of Europe with South-Western Asia. From its peculiar geographical position and intermingled ethnical elements, this region has been regarded as a sort of neutral or debatable border-land between the two continents. But one marked physical feature seems to be decisive in favour of its claim to be included within the limits of Asia. This is the deep depression of the Manich steppe river, which may be geologically looked upon as a survival of the broad strait formerly connecting the two seas. When flooded in spring by the swollen waters of the Kalaus from the northern slopes of the Caucasus, the eastern branch of this stream still finds its way to the Kuma delta on the Caspian, while the western branch reaches the left bank of the Lower Don at the head of the Sea of Azov.

The Manich depression thus clearly indicates the former direction of the Ponto-Caspian Strait, which in a not very remote geological epoch flowed in a broad channel between the two continents. And as this Ponto-Caspian Strait lay entirely to the north of the Caucasus, it follows that this great highland region belongs physically not to Europe but to Asia. Hence the Manich depression may now be taken as at once the parting-line between the two continents and the northern boundary of Caucasia. Its western and eastern limits are marked by the Euxine and Caspian respectively. But the southern frontier line between the Turkish and Persian States is somewhat irregular, and even arbitrary. The new boundary towards Turkey has already been determined at p. 31, while the frontier towards Persia follows the windings of the River Aras from Mount Ararat to within a short distance of its junction with the Kura. Here the line is deflected south and east to the Caspian, thus including, in Russian territory, the hilly coast district of Lenkoran, which forms a part of the Iranian plateau.

At its narrowest point, between Poti on the Euxine and Derbent on the Caspian, the Ponto-Caspian isthmus is about 350 miles broad west and east. But from the Strait of Kerch to the mouths of the Kuma the distance is nearly 500 miles in a straight line, and from the low water-parting of the eastern and western Manich to Mount Ararat, on the Perso-Turkish frontier, a similar line will measure 420 miles north and south. But even these dimensions fall considerably short of the actual length of the Great Caucasus, whose axis stretches in an oblique line for 720 miles across the isthmus, from the Taman peninsula, between the Black and Azov Seas, to the Apsheron peninsula in the Caspian. By this central range the whole region is divided into two unequal parts —Cis-Caucasia and Trans-Caucasia, with a joint area of 186,000 square miles.

2. *Relief of the Land: The Great and Little Caucasus—Armenian Plateau—Ararat and Ala-göz.*

The Caucasian region has been during recent years visited by several English travellers—Moore, Grove, Bryce, and especially by Freshfield, who ascended apparently for the first time the summit of Elburz, and has presented to English readers a charming narrative of his proceedings.

The Caucasus presents in its general outlines one of the best-defined mountain systems in the world. Approached from the northern steppes, it everywhere offers the appearance of an unbroken rocky barrier, rising rapidly from the plains, and surmounted all along the line by a series of magnificent snowy peaks. Southwards, also, it falls everywhere abruptly towards the valleys of the Rion and Kura, which form a nearly continuous trough or depression running from sea to sea between Poti and the Kura delta. This southern depression answers somewhat to that of the Manich on the north, the whole of the intervening highlands constituting the Caucasus proper, or the Great Caucasus. They take the latter name in contradistinction to the Little or Anti-Caucasus, which consists of the spurs rising in confused masses beyond the Rion-Kura depression. The connection between the two systems is effected by the Suram or Mesk range, which forms the Rion and Kura water-parting east of Kutais

Except at this point the Great Caucasus is thus completely isolated from the southern Lazistán and Armenian highlands. The direction of its main axis, which is continued by a submarine ridge across the Caspian to the Balkan hills, about Krasnovodsk, also shows that the Great Caucasus forms the real north-western continuation of the north Iranian escarpment. This escarpment, which is itself a western continuation of the Hindu-Kush and Paropamisus, broken only by the Tajand (Hari-rúd) valley, may now be regarded as stretching under the name of the Kuren-dagh along the northern frontier of Khorasan to within a short distance of the south-eastern shores of the Caspian. Here it ramifies into two branches, one sweeping round the south coast of the Caspian, as the Elburz range, and merging north-west in the Armenian highlands, while the other effects through the Little and Great Balkans and the already-mentioned submarine ridge a junction with the south-eastern extremity of the Great Caucasus in the Apsheron peninsula. We thus see that the Great Caucasus forms the direct continuation of the Central Asiatic systems north-westwards to the Taman peninsula, and beyond it to the South Crimean highlands. The continuity of the whole system is clearly shown by the underground fires, naphtha and oil wells, mud volcanoes, and other still active igneous agencies, occurring at intervals all along the line.

The Great Caucasus bears in many respects a striking resemblance to the Pyrenees. Both run between two marine basins, both are marked by the Sierra formation in their higher crests, and both are divided into two sections of unequal length. But in the Caucasus the break formed by the tremendous fissure of the Dariel Gorge lies almost exactly midway between the two seas. Through this pass runs the great military highway from Vladikavkaz to Tiflis, the respective capitals of Cis- and Trans-Caucasia. It was also owing to this remarkable geological fault that the Russians were enabled to divide the Caucasus, so to say, into two military zones, preventing any possible combination of the western and eastern tribes, and thus effecting piecemeal the reduction of the whole region.

Most of the eastern section is comprised under the general name of Daghestán, a Turko-Persian compound meaning "Highlands." This region, which is politically included in Trans-Caucasia, is considerably lower and far more irregular than the western section. Here a uniform elevation of 10,000 to 12,000 feet is maintained north of the Rion basin, and above this there tower, besides Elburz (18,526 feet), five other snowy peaks, all higher than Mont Blanc — the Koshtan-tau (17,096), Dikh-tau (16,925), Kazbek (16,546), Ushba and Aghish, or Adish-tau, each considerably over 16,000 feet. Here also the mean altitude is so great that for 100 miles between the sources of the Kuban and Adai-Kokh there are no passes lower than 10,000 feet. Even the Mamisson, about the head of the Rion near the Zikari ridge, is still 9390 feet; but east of this point openings are found from 6000 to 9000 feet high.

West of Elburz the western section assumes the character of a coast range skirting the Black Sea from the mouth of the Rion to the Taman peninsula. For some distance beyond Elburz it retains a great altitude, with snowy peaks such as the Marukh, the Juman-tau, and the Oshten, rising far above the snow-line. On the coast the incline is also continued for a great depth below the surface, where depths of over 12,000 feet occur close in shore. But beyond Pitzunda the coast range falls rapidly towards the Idokopaz hills near the port of Novo-Rossiisk, after which the chain merges through a few low scattered hills in the alluvial plains of the Kuban delta.

The eastern section, or Daghestán, although crowned

by no peaks equal to those of the Western Caucasus, exceeds it not only in breadth but also in its mean elevation. From Mount Borbalo (11,100 feet) at its western extremity, the main range and the Andi ridge diverge right and left towards the Apsheron peninsula and the Terek delta, thus enclosing, with the sea for a base, the

SOUTH SIDE OF MOUNT ELBURZ, AND ASAN GLACIER.

triangular space entirely occupied by the irregular masses and upland valleys of Daghestán.

This intricate highland system, which still awaits thorough exploration, culminates with the Shebulos-mta (14,970 feet) in the Andi range, which also contains several other "mta" or peaks, such as the Kachu and Diklos-mta, considerably over 13,000 feet. The highest elevations are the Bazarjusi (14,910), the Tkhfan-dagh (13,950), the Vitziri (12,910), the Sari-dagh (12,160), and the Baba-dagh (12,080). Within the triangular space two crests, the Shah-dagh and the Shalbuz-dagh, exceed 14,000 feet, lying towards its south-eastern extremity, where it falls down to the hills of the Baku district in the Apsheron peninsula.

The northern and southern slopes of these ranges differ greatly in their general aspect. The descent towards the Rion and Kura is everywhere far more abrupt than towards the Manich depression. Here the fall is broken first by a succession of nearly parallel ridges, and then by a series of upland limestone terraces sloping gently towards the steppe, but often presenting nearly vertical walls, 3000 feet high towards the central range. This range consists mainly of crystalline schists, and it is remarkable that the same formation prevails in the transverse Mesk ridge, connecting the Great with the Little Caucasus. On both sides of the higher schists the chief rocks are eocene and other old limestones, which disappear northwards beneath the pliocene formations of the steppes.

Porphyries and other igneous rocks abound in the higher regions, where Elburz was probably a still active volcano down to the close of the tertiary period, during which the Euxine and Caspian were connected by the Manich Strait. Its rival, the mighty Kazbek, overlooking the Dariel Gorge, together with the more northern crests, is also of volcanic origin. Underground forces are even still at work, not only in the mud volcanoes and slumbering fires at both ends of the range, but also in the numerous hot springs and naphtha wells which occur on both sides, but especially in the Terek and Kura valleys about Vladikavkaz and Tiflis. The Lower Kura and Aras valleys are, moreover, still subject to violent earthquakes, while clear traces of continuous upheaval on the Euxine coast, and of oscillations of level on the Caspian side, are visible at Sukhum-Kaleh, about Baku and elsewhere. On the other hand, there is a remarkable absence of large waterfalls and alpine lakes; the great reservoirs, which formerly studded the plains on both sides, having been drained since the glacial epoch.

The Little or Anti-Caucasus presents in almost every respect the most decided contrast to the great northern barrier. Instead of one sharply-defined system, rising somewhat rapidly from the plains, and with a single main axis running throughout in a given direction, we have here rather a rugged plateau formation, intersected by irregular masses, with axes running in all directions. So ill defined is the whole system that it has no natural southern limits at all. It rises abruptly from the Rion and Kura valleys, but towards the south merges everywhere imperceptibly in the Lazistán and Armenian highlands. Hence the Little Caucasus forms the true north-western scarp of the Iranian tableland, with which the Great Caucasus is connected only by the narrow Mesk ridge intersecting the Rion-Kura valley.

But although narrow, the connection is complete; for the Mesk ridge, which maintains an elevation of 8000 feet, is continued south-westwards by the Ajara or Akhaltzikh range into Lazistán. This range, which skirts the Black Sea within a mile of the coast, rises gradually towards the Turkish frontier, where it culminates with the Karch-shall (11,410 feet), south-east of Batúm.

The Ajara range is separated eastwards by the Upper Kura valley from the Akhalkalaki plateau, which is limited southwards by the Kars-chai valley beyond Lake Chaldir. This extremely irregular plateau, which has a mean elevation of 8000 feet, thus forms the water-parting between the Upper Kura and Aras basins. It is a rugged, bleak region, which seems to have been formerly flooded

by an extensive lacustrine basin, of which the sole remnants are the Chaldir and other smaller lakes, draining some to the Kura, some to the Araxis, while others are mere brackish tarns or marshes, without any outflow. Eastwards the plateau is limited by a double line of volcanic peaks, culminating in Mount Samsar (11,115 feet), with a crater nearly two miles long; and the Great and Little Abul (11,000), with their two cones springing, like those of Ararat, from a common base.

Southwards the Akhalkalaki plateau passes through rough transitions to the desolate tableland of Erivan and Kars, which has a mean elevation of nearly 3500 feet above the sea, and which continues to form the parting line between the Kura and Aras north and south. Here the great features are the basin of Lake Gok-cha and the mighty Ala-göz, lying about midway between the lake and the recently-annexed fortress of Kars. Farther south the plateau has been politically extended across the Aras valley to include Ararat, which lies on the borders of the Russian, Turkish, and Persian frontiers.

Ararat and Ala-göz.

Although thus rising in apparently isolated grandeur at the converging point of three empires, Ararat really forms the eastern and culminating point of the volcanic range, which here forms the water-parting between the Aras and Murad-chai, or eastern branch of the Euphrates. Its seemingly isolated position and imposing appearance have from the remotest times encircled it with a mysterious halo of legends and traditions. A number of places in the vicinity still betray in their very names the traces of the Noachian tradition. Thus the village of Aghurri lying on its slope means "he planted the vine;" Nakhichevan, the spot where the patriarch is said to have reached the valley, is interpreted, "here he first descended," and Erivan itself indicates the place where he permanently settled. His grave is shown at Nakhichevan, and is held in great veneration both by the Armenians and Tatars. But while these Biblical traditions survived, the Biblical name itself of the mountain was forgotten in the neighbourhood. The Armenians know it only as the Masis Lern—that is, the "Grand or Sublime Mountain"—and the Tatars and Turks as the Agri-dagh, or "Steep Mountain," the Persians alone calling it the Kóh-i-nuh, or "Noah's Mount."

Viewed from Nakhichevan, Ararat presents the appearance of a single cone bounding the horizon towards the north-west. But it really consists of two separate cones known as the Great and Little Ararat, resting on a common base, and separated by a deep intervening depression. The higher cone consists itself of a double peak, and the whole mass, with its projecting spurs, covers a space of over 370 square miles between Erivan and the frontier Turkish town of Bayazid.

The Armenians have a firm conviction that the summit is altogether inaccessible, and received with absolute incredulity the statement that Parrot had for the first time succeeded in scaling the highest peak in 1829. Subsequently, Bryce's solitary ascent of this mountain forms a striking episode in the history of travel.

On the northern slope there is a vast chasm where formerly stood the Convent of St. James and the village of Aghurri, both of which were overwhelmed by a terrific earthquake in the year 1840.[1] The upper portion of the chasm is filled by one glacier, while another glacier occupies a narrow channel towards the north-east.

[1] Some attribute the catastrophe to a landslip or avalanche, scme to an earthquake, and others to the reopening of an old crater above the Convent of St. James.

Owing to the slight moisture there are no large forest trees on Ararat, which, however, is clothed with vegetation to an altitude of over 11,000 feet. Pasturage extends thence to nearly 13,000 feet, beyond which an alpine flora struggles up to 14,200, which marks the snow-line. Its fauna is also very poor, including on the higher grounds little beyond a mountain goat, a species of hare, and the polecat.

The chief elevations are — Summit of Great Ararat 16,916 feet, Little Ararat 12,840, connecting ridge 8780, Upper Aras Valley 2800. This valley, in which are situated Erivan and Echmiadzin, political and spiritual capitals of Armenia, separates Ararat from its northern rival, the magnificent Ala-göz, a truncated volcanic cone 13,436 feet high. With its advanced spurs, the Ala-göz, or "Motley Mountain," so named from the various colours of its pumice, scoriæ, obsidian rocks, and foliage, covers a wider area than its southern rival.

3. *Hydrography: The Kalaus, Terek, Kuma, Ingur, Rion, Kura, and Aras Rivers—Lake Gok-cha.*

The Great Caucasus forms a clearly-defined water-parting between the Terek, Kuma, Kalaus, and Kuban, flowing from its northern slopes towards the former Ponto-Caspian Manich Strait and the Ingur, Rion, and Kura, draining from its southern slopes to the Black and Caspian Seas. South of the Rion-Kura depression the hydrographic system is far more intricate, comprising, besides an inland drainage, represented chiefly by Lake Gok-cha, the farthest sources of the Aras, Euphrates, and Chorukh, which flow east to the Caspian, south to the Persian Gulf, and north-west to the Euxine.

Since the disappearance of the Ponto-Caspian Strait the Terek and Kuma find their way eastwards to the

Caspian, the Kuban westwards to the Azov and Black Seas, while the Kalaus, a true steppe river, reaches the Manich only when swollen by the melting of the snows in spring. The Kalaus has its farthest head-streams in the advanced spurs of the Caucasus above Stavropol, and joins the Manich exactly at the water-parting between the Euxine and Caspian, 25 feet above sea-level. Its waters are thus divided into two channels, flowing during the floods one through the West Manich to the Azov Sea, the other through the East Manich to the Caspian.

The Terek, rising in a cirque 8000 feet above the sea, at the northern foot of the Kazbek, sweeps round through the Dariel Gorge and by Vladikavkaz northwards nearly to the 44th parallel. Above the Malka, its largest affluent, the discharge is over 17,000 cubic feet per second, and such a quantity of alluvia is washed down that the delta is encroaching on the Caspian at the rate of about 40 yards annually. Fishing hamlets which early in the present century stood on the coast are now 10 or 12 miles from the sea, and Baer asserts that the Terek is contributing even more than the Volga to the filling up of the Caspian. The waters brought down are doubtless considerable; but these are rapidly evaporating, while the sedimentary matter remains continually accumulating.

A combined system of canalisation and drainage has brought several hundred thousand acres under cultivation about the Lower Terek.

The Kuma has its source nearly under the meridian of Elburz, and pursues a uniform north-easterly course towards the Caspian, which it occasionally enters through several small channels at a point between the Terek and Volga deltas. Although a considerable stream on emerging from the hills, it gradually contracts as it approaches the coast, the diminished volume being due to evaporation, to the absence of any affluents during a sluggish course of 150 miles through the steppe, and to the irrigation-works of the Kalmuk and Tatar stockbreeders on both sides of its banks. From these combined causes the Kuma is often entirely exhausted within 50 or 60 miles of the Caspian. Formerly the discharge was much greater, as indicated by the old channels and dried-up watercourses, some joining the Manich, some reaching the coast at Serebrakovskaya, and all still occasionally flooded.

The Kuban is the only river flowing to the Euxine basin from the northern slopes of the Caucasus. Its farthest head-stream rises on the west side of Elburz, whence it flows north-west and west to its delta below Yekaterinodar. Here the main branch continues to flow westwards through the Taman peninsula to the Black Sea, while a considerable quantity of water is diverted through several smaller channels northwards to the Sea of Azov. During the spring, summer, and autumn floods the Taman is often swollen to the proportion of a large stream, from 300 to 400 yards wide and over 10 feet deep. But at other times it nowhere exceeds 4 feet, while the Azov channels sometimes run dry. It is ascended during the floods for some miles by the Kerch steamers, but it is permanently navigable only for flat-bottomed craft. The mean discharge is estimated at 40,000 cubic feet per second.

Two important rivers reach the Black Sea from the northern slopes of the Western Caucasus. These are the Ingur and Rion (Phasis), whose basins are completely enclosed north, east, and south by the Great Caucasus, the Mesk, and Ajara ranges. All the head-streams of the Ingur lie on the southern slopes of the Adish-tau and Ushba, two of the highest peaks in the Caucasus, whence it flows through mountain gorges and upland valleys down to the coast near Redut-Kaleh.

The Rion also and its chief tributary the Kvirila, rise at great elevations, the former at the Pasis-mta near Mount Garibolo, the latter at the Mamisson Pass (9520 feet). But the Kvirila soon reaches the Mingrelian plains, where it joins the left bank of the Rion below Kutais. The joint stream enters the Euxine at Poti close to the large Palaiostom lake or lagoon with which it was formerly connected. Although now cut off both from the sea and the Rion, its partly marine fauna shows that this now fresh-water lake at one time communicated with the sea, thus forming the "Old Mouth" of the Rion, as still indicated by its Greek name.[1] Formerly navigable for nearly 100 miles, the Rion has now scarcely 2 feet at low water, and even during the floods from January to June is ascended by small craft only for 30 miles from its mouth.

Both the Rion and the Ingur, whose basins comprise the ancient Colchis of the Greeks,[2] flow through romantic upland regions. The magnificent gorges of the Ingur, with their steep granite walls, are 800 to 1000 feet high in some places, and often clothed with a luxuriant sub-tropical vegetation. Their upland valleys are accessible from the Poti-Tiflis railway running up the Rion valley to Kutais in Imeria (Imeritia).

South of the Great Caucasus the Kura (Cyrus) and Aras (Araxes), by far the largest rivers in Caucasia, now flow through one mouth to the Caspian. Yet they belong mainly to two distinct basins; for although both rise in the Armenian highlands, the Aras remains an Armenian river for most of its course, whereas the Kura soon emerges in the Georgian plains and receives all its

[1] From *παλαιός*, old, and *στόμα*, mouth.

[2] It is remarkable that the Pasis-mta, or "Pasis Peak," source of the Rion, still preserves the old name *Phasis* (read *P'hasis*), by which the Rion was known to the ancients.

large affluents from the Great Caucasus. Hence the Aras is historically the Armenian, the Kura the Georgian river pre-eminently, and even within the historic period both reached the Caspian through independent mouths.[1] The mingling of their waters throughout their lower course is comparatively recent.

The Kura has its source in the Kizil-Gyaduk, 10,340 feet above the sea, whence it flows along the east base of the Arsiani and Ajara ranges north-eastwards to about the 42d parallel, where it receives several small feeders from the east slopes of the Mesk water-parting. It now trends eastward to Mtzkhet and Tiflis, whence it pursues a south-easterly course through a continuously broadening valley to the Caspian below the Apsheron peninsula. Above the plains of Tiflis it is almost a mountain torrent, rushing through a succession of wild gorges and rapids, in one of which it descends nearly 750 feet in a distance of 15 miles. Soon after receiving the Yora and Alazan, its great tributaries from the north, it becomes navigable for vessels drawing four feet for a distance of 450 miles through the Karabagh and Mugan steppes to its mouth. Its lower course is one of the most productive fishing grounds in the world, teeming as it does with enormous quantities of sturgeon and white fish.

The discharge of the united Kura and Aras rises from nearly 7000 cubic feet per second in winter to 25,000 in summer. Much of this water might easily be applied to irrigating the now arid but formerly productive Mugan and Karabagh steppes. Like most of the other Caucasian rivers, the Kura is continually encroaching on the sea, its delta having advanced over 50 square miles between the years 1830 and 1860.

The Aras has even a longer and a much more wind-

[1] Strabo tells us expressly that in his time the Kura and Aras entered the sea through separate mouths.

ing course than the Kura. Rising south of Erzerum, at the foot of the Bingol-dagh (see p. 54), it flows for some miles through Turkish territory north-east to the recently-advanced Russian frontier. Here it turns eastwards to the Erivan plain north of Ararat, whence it sweeps in a semicircle mostly between the Russian and Persian empires round to its confluence with the Kura.

Of the three great closed basins of the Armenian highlands, Lake Gok-cha is the smallest. Yet it fills a vast triangular cavity 540 square miles in extent, at an elevation of 6400 feet above the sea on the plateau, nearly midway between the Aras and Kura valleys. It has an extreme depth of 250 feet, and when swollen by the melting snows from the surrounding hills discharges its surplus waters through the Zanga towards the Aras in the Erivan district. On an islet at its north-west extremity lies the historical Convent of Sevan in one of the most desolate spots on the globe. At the opposite end the horizon is bounded by the huge volcanic mass of the extinct Ala-Polarim volcano, whose lava-streams descend in two channels to the edge of the lake.

4. *Natural and Political Divisions: Cis-Caucasia—The Northern Steppes and Slopes of the Caucasus; Trans-Caucasia—Colchis; Georgia; Russian Armenia.*

The old historical divisions of Caucasia—Georgia, Lesghistán, Imeria, Mingrelia, Kabardia, Abkhasia, Circassia—have been completely swept away under the new order of things. They were based almost exclusively on ethnical considerations, and in some instances, as in the case of the Circassians and Abkhasians, the very races themselves have all but disappeared which gave rise to these distinctions. No regard has even been had for the great historic nations of this region, and the ancient kingdoms of Georgia and Armenia are now in official language replaced by the Russian Governments of Tiflis and Erivan.

Our survey must therefore follow the great natural divisions, which are in the north the steppe-lands stretching from the Manich depression to the foot of the hills, and the zone of fertile and inhabited uplands between the steppe and the alpine regions of the main range; in the south the Rion-Kura and the Aras basins.

Travellers approaching the Great Caucasus from the north, after crossing the Manich depression are not for a long time sensible of any marked change of scene. The boundless level steppes of Southern Russia are still continued without any perceptible break south-eastwards far into the Ponto-Caspian isthmus. From the borders of the Government of Saratov and the Don Kossak domain right away to the advanced spurs of the Great Caucasus the land remains almost perfectly flat, broken only here and there by a few low hills or ridges.

Yet this extensive region, comprising fully half of all Caucasia, is divided into two natural sections, each with its special physical features. First come the lowlands stretching southwards to the Kuban, Malka, and Terek rivers, and following the line of their course seawards. This is a true steppe land, interrupted only here and there by a few deep furrows. It is marked by an almost total absence of timber beyond a few small plantations in the neighbourhood of Stavropol. There is also a great lack of moisture, most of the streams running quite dry in summer, while the lakes in the northern districts are mostly brackish. Here the great evil is the want of water for irrigation purposes. The black loamy soil is naturally highly productive, yielding heavy crops of cereals and rich pasturage whenever the rainfall is sufficiently abundant. But the summers are mostly rainless, the long droughts and great heat reducing the steppe vegetation to a fine dust, which forms dense clouds sometimes covering the whole horizon, and wafted to great distances by the winds.

The second section stretches along the foot of the main range between the Euxine and the Caspian for a distance of over 450 miles, but contracting in some places to a width of from 20 to 30 miles. Here the luxuriant growth of grasses and the genial climate remind the traveller of the Mississippi prairies. The steppe now soon merges in a boundless park-land, bordered southwards by the mighty central range with its frowning granite and basalt crags and glittering glaciers, northwards by broad rivers, east and west by two seas. Yet even here agriculture has been but slightly developed, and apart from a few patches of cultivated land, the whole region is still virgin soil. It produces when tilled magnificent crops, and the grass grows to a height of 5 or 6 feet.

At either extremity of the range are the two remarkable naphtha-producing peninsulas of Taman and Apsheron. Of its other natural products very little is yet known, although the presence of silver, lead, and copper has long been placed beyond doubt. Many of the hillmen cast their bullets from the lead or copper they pick up on the surface, and near Elburz there is an abundance of common pyrites, which they utilise in the manufacture of their gunpowder. Granites, magnificent green and red porphyries, various-coloured marbles, and rock-crystal, exist in great quantities, while extremely copious mineral waters of every description and coal combined with the inexhaustible naphtha springs promise to prove a future source of permanent wealth to the country.

Trans-Caucasia, comprising the Rion-Kura and Aras

basins, possesses even greater economic importance than the northern division. Here is the great commercial route between the two seas leading from Poti through Tiflis to Baku. Here is the historic domain of the Georgian nation, by far the most important of all the Caucasian races, and Tiflis the former capital of the Georgian States has naturally been selected as the political centre of the new Russian Government of Caucasia. Here all the streams are perennial, while the Kura-Aras alone discharges a far greater volume than that of all the northern rivers combined. Here are some of the most productive fisheries and one of the finest wine-growing countries in the world. Here amongst other mineral treasures is Mount Kulpi, a prodigious mass of rock-salt in the Upper Aras valley, the salt-mines of which are in some places over 200 feet thick. Although almost continuously worked since prehistoric times, as shown by the implements frequently picked up dating from the stone age, these mines show no sign of exhaustion, and the Armenians have a tradition that Noah drew his supplies from this source. The present average yield is about 16,000 tons yearly, although the workings are carried on in the most primitive fashion.

The Rion basin, the Colchis of the ancients, has been famous from the remotest times for its surprising fertility and resources of every kind. The legendary Argonautic expedition was fabled to have been fitted out by the Greeks to recover from this region the golden fleece, emblem of boundless wealth. It is completely enclosed by an amphitheatre of hills sweeping round from Sukhum-Kaleh to Batúm, and now crossed by the Poti-Tiflis railway at the Suram Pass in the Mesk range some 3000 feet above the sea.

This pass leads directly down to the ancient kingdom of Georgia, comprising the greater part of the Kura basin.

The depression through which this river flows may be regarded as a dried-up fiord or inlet of the Caspian, which formerly penetrated between the Great Caucasus and the Armenian highlands across the southern portion of the Ponto-Caspian isthmus westwards to the Mesk and Anjara ranges. The lower section of this basin, comprising the Mugan and Karabagh steppes, is now mostly waste land. But the traces of ancient canals, and the ruins of many villages, caravansarais, and even towns, show that it was once highly cultivated and thickly inhabited.[1] It is now visited only by the Tatar nomads, in spring, when the rainfall produces herbage.

Higher up rice was formerly cultivated along the banks of the Kura above the Alazan confluence. But the raids of neighbouring Lesghian marauders caused the irrigation works to be abandoned, and during the present century the Karayazi steppe between the Kura and Yora has reverted to a state of nature. An attempt, however, has now been made by the construction of the "Mary Canal" to bring this tract once more under cultivation. The recently-executed surveys also show that over 5,000,000 acres in the Lower Kura valley might again be easily rendered productive. Much of the soil consists of a rich black loam, and many of the old canals might be restored and extended at a moderate outlay.

At the same time the whole of this region is notoriously malarious, and farther north the Baku coast district is still subject to violent earthquakes. The centre of the seismic action seems to be the town of Shemakha, which was nearly destroyed in the seventeenth century,

[1] Some of these canals were nearly 100 miles long, and on one of them stood the great city of Bilgan, destroyed by Jenghis Khan. When Timur restored the canal Bilgan rose again from its ruins and continued to flourish till the close of the seventeenth century. But the subsequent wars with the Daghestán hillmen again caused the works to be abandoned, and Bilgan again disappeared.

and again suffered so severely in 1859 that the administration has since been transferred to Baku.

The Aras basin, comprising that portion of Armenia which is now included in Russian territory, differs in no respect from the Armenian highlands politically belonging to Turkey and Persia. Here also the most salient characteristics are the long ranges intersecting the plateau in every direction, and dividing it into a number of upland arid steppes, sparsely peopled and almost treeless. The hills falling abruptly towards the Rion basin are here and there thinly wooded; but the Tràletes and their spurs overlooking the Kura valley often present nothing but bare rocky surfaces for miles together.

Immediately north-west of Alexandrapol rises the extensive Chaldyr plateau. North of it flows the Upper Kura, here intersecting the Armenian frontier hills, and forming a natural approach to the frontier district of Ardahan between Kars and Batúm. This recently-annexed district also consists of a treeless tableland about 4500 feet above the sea, and enclosed north and south by almost inaccessible hills. Southwards stretches the rugged plateau of Kars, east of which the Arpa-chai valley leads down to the Upper Aras, which here flows through the plains of Erivan between Ararat and Ala-göz. This is almost the only comparatively low and well-watered level tract in the whole of the Armenian highlands, and here is, so to say, the focus of the Armenian nation, where are centred all its most hallowed associations. The highway approaching it from the north-west is fringed with the ruins of Ani, Vardzia, and other ancient cities, recalling the former greatness of the land.

But a far more sacred spot is the venerable Convent of Echmiadzin at the southern foot of Ala-göz, residence of the Armenian "Katholicos," who rules with a plenitude of spiritual jurisdiction over the two millions of

Gregorian Christians scattered over the continent from the Bosphorus to the Ganges.

5. *Climate: Rainfall.*

No other region of the same extent presents so great a diversity of climate as Caucasia. This is due partly to its peculiar position between two inland seas, at the southern verge of the Russian steppes and at the north-western edge of the Iranian plateau, partly to the extreme deviations in the general relief of the land ranging from the low-lying Mingrelian plains to the Elburz Peak nearly 19,000 feet above the sea.

Being exposed to the northern winds sweeping over the Russian steppes, Cis-Caucasia is both drier and colder than the southern slopes. Hence many of the rivers here run out during the summer heats before reaching the coast, and are ice-bound in winter; whereas in Trans-Caucasia all the streams are perennial, and frozen only in exceptionally hard seasons. For analogous reasons the western section, receiving the moist and relatively warm atmospheric currents from the Euxine, enjoys a far higher winter temperature and greater abundance of moisture than the eastern slopes facing the Caspian and arid Turkoman deserts. Here the contrast between the rainless and sultry Mugan and Karabagh steppes of the Lower Kura and the moist and moderately hot Rion basin is very striking. In general, the rainfall is three times heavier on the western slopes than on the Central Caucasus, and from eight to ten times more copious than on the east side of the Daghestán ranges. So little moisture is brought from the Caspian that at times no rain falls for six months together in the Lower Kura basin.

Although the extremes of heat and cold are much greater in the Caucasus than in the Alps and Pyrenees, the mean annual temperature is much the same in all these highland regions. Thus, while Caucasia and Switzerland have a common mean, the temperature varies in the latter about 18°, in the former as much as 25° or 26° between winter and summer. Hence in its extremes the Caucasian climate resembles the Asiatic, in its general mean the European, so that the region is a land of transition in its climatic as well as in other features.

6. *Flora and Fauna.*

This transitional character is especially conspicuous in its vegetable and animal kingdoms. In some respects Caucasia seems to be a land of dispersion, where certain vegetable species, such as the peach, apricot, cherry, and other stone-fruit trees, became differentiated, and thence distributed east and west over the two continents. Plants of this sort are found in such variety and abundance on both sides of the main range, but especially in the Rion basin, that they may be regarded as the typical vegetable order of this region.

The southern limits of Trans-Caucasia, lying under the same parallels as Central Italy, are the natural home of the laurel, orange, citron, vine, and mulberry. The vine arrives at great perfection, especially in the Georgian province of Kakhetia, which is famous for its fiery vintages. The plant has in recent years suffered from the ravages of the oidium; but large quantities are still produced of a very full-bodied wine, which is now largely used for improving the flavour of inferior sorts. The vine, like the stone-fruits, is probably indigenous in Caucasia.

Heavy crops of rice, maize, wheat, and other cereals of excellent quality, are raised on all the lowland tracts on both sides of the Great Caucasus, wherever water can be obtained in sufficient abundance. But the abandonment of the old irrigation works, especially in the Middle and Lower Kura valley, and the increasing dryness on the north side, have reduced to barren wastes extensive districts where these crops were formerly widely cultivated. The fruits of Southern Russia, such as the pear, plum, cherry, and walnut, flourish on the northern slopes and along the banks of the Terek, Kuma, and Kuban.

In the profound and precipitous mountain gorges of the central range, where a solitary sunbeam seldom penetrates, not a blade of grass will grow. But emerging from these abysses, we sometimes fancy ourselves transported to the Alpine valleys of Switzerland, with their luxuriant pastures, rich woodlands, and foaming mountain torrents. Here the forest zone stretches along both sides of the Great Caucasus for a distance of 500 miles, with a breadth varying from 10 to 20 miles. On the heights grow the maple, lime, ash, fir, pine, beech, and larch; farther down the oak, chestnut, several species of poplar, the plantain, box, and walnut. In the valleys thrive most southern fruits, as well as the loveliest flowering shrubs; and in the more favoured spots the cotton and olive. Conspicuous amongst the flowering shrubs is the *Azalea Pontica,* one of the glories of the vegetable world, rivalling the Himalayan rhododendron in the richness, variety, and splendour of its blossom. It reaches an elevation of 6000 feet, where the deep-red autumn tints of its foliage offer a surprising contrast to the sombre green of the surrounding conifers. A species of tea grows wild on the southern slopes of the Mingrelian highlands, and on the Caspian seaboard the Tatars raise crops of madder and saffron. In the hot moisture-charged atmosphere of Abkhasia and Mingrelia the vegetation is marvellously luxuriant; but man and nature alike are here still in a wild state. Wheat and rice no doubt flourish in the valleys, but the natives themselves grow nothing but millet, some barley, and maize. As a rule, all these cereals reach a higher elevation than in the Alps. Barley is cultivated by the Osses in the central ranges up to 8000 feet, while wheat thrives at 6500, and maize at 3000, in all the sheltered southern valleys.

The flora of Russian Armenia, a land apparently of diminished moisture, is incomparably poorer than that of Caucasia proper. The few forest tracts consist chiefly of oaks, beeches, aspens, and especially poplars. But here the characteristic plant is the curious nölbönd, a magnificent species of elm, with enormous leafy branches, through which the solar rays never penetrate. This highly-ornamental tree is absolutely unknown beyond the limits of the Aras basin. Of cultivated plants the chief, besides cereals, including rice, are the apricot, cotton, sesame, and the vine, which in some places yields a highly-flavoured wine, somewhat like Madeira.

The Caucasian fauna, like that of similar highland regions elsewhere, is far less varied than its flora. The tiger sometimes ventures across the Persian frontier, and the leopard and hyena are also met in the Lower Kura and Aras basins. The lowland thickets and sedgy river-banks are the favoured haunts of the wild boar, and the Abkhasian and Mingrelian forests are still infested by the panther, wolf, lynx, and bear. The Caucasian bear, not a very formidable species, reaches no higher than about 5000 feet, above which a few herds of the bison or wisant, wrongly identified with the aurochs, still linger in the upland forests on the slopes of Elburz. Still higher up the chamois and *túr*, a species of ibex, frequent the alpine valleys along the central range. A wider range is enjoyed by the marten, blue fox, squirrel, hare, fish otter, and some other wild animals of smaller size. On the whole, game is abundant, but is found chiefly in the low-lying and unhealthy wooded tracts along the northern slopes of the Caucasus (*Clive Phillips-Wolley*).

Pre-eminent amongst the domestic animals are the horned cattle of the Ingur and Rion basins. Here there are two fine breeds, one small and active, the other of magnificent size and symmetrical proportions, sprung originally from Ukranian stock. The horses, mules, asses, goats, and other domestic animals of this region, are all alike noted for their fine proportions and good qualities.

In general the Great Caucasus may be regarded as a parting-line between the European and Asiatic vegetable and animal kingdoms, the Cis-Caucasian flora and fauna being more allied to the western, those of Trans-Caucasia to the eastern continent.

7. *Inhabitants: Varied Ethnical and Linguistic Elements—Tabulated Scheme of the Caucasian Aborigines—The Georgians, Mingrelians, Imerians, Circassians, Abkhasians, Chechenzes, Lesghians, Osses; Non-Caucasian Intruding Races.*

Caucasia is inhabited by a highland population, comprising a multiplicity of distinct ethnical elements elsewhere almost without a parallel. The most varied tribes, speaking fundamentally distinct languages, here dwell in the closest proximity, hemmed in on the north by the Russian Slavs, southwards by the Armenian, Kurdish, and Persian Iranians. But in comparatively recent times all these, besides the Tatars and other alien races, have penetrated into the Kura, Terek, Kuma, and Kuban basins.

The popular view is, that we have in the Caucasus the remnants or fragments of the peoples who have from time to time been driven into these recesses from the surrounding lands, or who have passed through these highlands during the ceaseless flow of prehistoric and subsequent migration from Asia to Europe.

But this view was combated by Professor Virchow at the Archæological Congress held at Tiflis in the autumn of 1881, and will be farther dealt with in the Appendix to this volume. Here it will suffice to remark that the Caucasus could not have been a highway of migration when the ice-fields descended much lower than at present, and that the numerous Caucasian languages, with the single exception of the Ossetian, have no kind of affinity with those elsewhere current.

Partly on geographical, but mainly on linguistic grounds, all the Caucasian races are here grouped rather than classified in four great divisions as under:—

I. SOUTHERN DIVISION.

(*Kartveli Stock.*)

Georgian	East of Mesk range to Tiflis district	1,150,000
Imerian, Rachan	Imeria (Imeritia)	
Mingrelian, Gurian, Lechgum	Mingrelia	
Laz	Lazistán	
Svan	Upper Ingur and Tskhenis valleys	
Pshav, Khevsur	Sources of Alazan and Yora	

II. WESTERN DIVISION.

Cherkess	Ubych, Shapsuch, Dshiget	Left bank Kuban	138,000
Abkhasian		Coast of Euxine, N. of Ingur River	
Kabard		N. and E. of Elburz	

III. EASTERN DIVISION.

Chechenz	Ingush, Galgai, Kist, Tush, Karabulak	Right bank Upper and Middle Terek	164,000

Lesghian	Avar, Kazi-Kumyksh, Andi, Dargo, Dido, Duodez, Ude, Kubachi, Kurini	Daghestán	517,000

IV. CENTRAL DIVISION.

Oss or Ossetian	Both slopes of Great Caucasus about Kazbek	110,000

The Georgians, Mingrelians, and Imerians.

None of the Caucasian people except the Georgians possess any historic importance. Direct descendants of the old Iberians,[1] they still form the bulk of the population in the Governments of Tiflis and Kutais, and although essentially a lowland race, bear a marked resemblance to the Imerians and other highland members of the Kartvelian (Kartalinian) family. The term Grusia or Georgia,[2] does not occur till mediæval times, when it came into use after the north-eastern division of Kakhetia became detached from the old Kartvelian kingdom. Since the annexation to Russia this old national name, traditionally traced to a Kartlos, grandson of Noah, has again come into favour.

The several branches of the Kartvelian stock afford a striking illustration of the often-repeated remark, that the less favoured the land the more industrious and intelligent are its inhabitants. The magnificent race occupying Lower Mingrelia, who would be amply rewarded by the

[1] The present type and even the head-dress are absolutely identical with those of the statuettes found in the numerous graves dating from classic times scattered over the land.

[2] The name *Georgia*, of which *Grusia* is merely the Russian form, has been referred to the national saint, George, by whom they were early converted from Paganism to Christianity

labour bestowed on their fertile soil, are a hopelessly indolent and poverty-stricken people. Allowance should doubtless be made for the enervating climate; but everywhere the southern are far outstripped in energy and enterprise by the northern tribes.

The best conditions of existence are found at about an elevation of 4000 feet above sea-level. Here the vine still flourishes, sericulture is possible, maize and millet yield good returns, and wheat prevails on heavy soils. Here industry meets with a fair reward. The natives, although not grouped together in large centres of population, are still found in more compact masses than in the Mingrelian lowlands, where the agricultural villages are replaced by solitary farmsteads. Here rice, cotton, and sub-tropical fruits might easily be cultivated; but nothing is done, and the people remain poor. Much more prosperous are their northern kinsmen, who are compelled by the less favourable conditions to work for their living, and are not free from anxiety for the winter season. Their dwellings are also more substantially built, their cattle require to be housed, the vine must be carefully dressed and pruned, the silkworm needs constant attention, the wooded slopes must be cleared, the ground requires hoe and spade to provide sufficient for the sustenance of the family. Hence the Imerian is a better agriculturist than the lowland Mingrelian, and the higher we ascend the more industrious become the kindred tribes.

Although speaking one of the harshest languages in Caucasia, where a surprisingly harsh phonetic system is the rule, the Georgian race is distinguished by a passionate love of song and music. In the home, in the tavern, of which they are unfortunately constant visitors, in the market-place, at all their feasts and social gatherings, the Kartvelians are perpetually singing or shouting to the accompaniment of their tambourines, their balalaïkas, and other stringed instruments. Even their daily occupations and their field operations are relieved by a concert of voices, whose cadence is adapted to the movement of their various pursuits.

CIRCASSIAN.

The Circassians and Abkhasians.

Although their domain has been largely encroached upon by Tatars from the east, Armenians from the south, and Slavs from the north, the Georgians still constitute the most compact and homogeneous nationality in Caucasia. But more typical representatives of the Caucasian races are, or rather were, the Cherkesses or Circassians, formerly the most powerful and warlike of all the Western nations.

Their domain seems to have at one time extended round the Euxine seaboard, as far as the Strait of Kerch. But they were for centuries confined by the advancing Little Russians to the left or southern bank of the Kuban, and now since their final reduction in 1864, after a heroic resistance maintained for generations, nearly all their lands have been occupied by the Great Russians. A few scattered groups still cling to their ancient homes along the course of the Kuban and its affluents; but the great bulk of the nation withdrew after the conquest into Turkish territory, and isolated Cherkess communities are now found dispersed over Armenia, Asia Minor, Syria, and the Balkan peninsula. Here they have acquired an unenviable notoriety for lawless and turbulent habits. But the national character should rather be studied in the mountain fastnesses, where the race was moulded.

A similar fate has overtaken the kindred and neighbouring Abkhasians, of whom 20,000 migrated to Turkey after the late Russian war. Their territory is now reduced to a narrow tract on the coast of the Black Sea, north of the Ingur basin, where they are hemmed in by the Kartvelians, Tatars, and Great Russians. The Tatars, who are here isolated, separate the Abkhasians from their remote kinsmen, the Kabardians of the Central Caucasus.

The eastern division, whose most representative members are the Chechenzes and Lesghian Avars, have also been encroached upon, especially by the Nogai and Kumik Tatars, who have long been settled on the Caspian coast, south of the Terek, and more recently by the Russians, who have penetrated into the Chechenz territory as far south as Vladikavkaz.

The Chechenzes, Lesghians, and Osses.

The Southern (Kartvelian) and Western divisions,

amid much physical diversity, are at least characterised by linguistic unity. For the original identity of all the numerous Kartvelian dialects on the one hand, and on the other of the Cherkess, Abkhasian, and Kabardian, is an accepted conclusion of comparative philology. But the utmost ingenuity of specialists, who have devoted a lifetime to the study, has hitherto failed to introduce much order into the Babel of tongues still current amongst the innumerable tribes of the Eastern division. Here there are at least five stock languages, probably more, which can be affiliated neither to each other nor to any other known forms of speech.

Even the Chechenzes, by far the most important nation in the Eastern Caucasus, are split up into some twenty different groups, each with a distinct language. They occupy the whole of West Daghestán, between the Osses and the Avars, and long maintained a hopeless struggle under Khazi-Molla and Shamyl (Samuel) against the Russians. Even more fanatical Muhammadans than the Western Cherkesses, they fought with the dauntless valour inspired by religious enthusiasm and a passionate love of freedom. Since their reduction in 1859 large numbers migrated to Turkish Armenia, where most of them perished of want and hardships of every description.

Rivalling the Cherkesses in valour and physical beauty, the Chechenzes surpass them in generosity and self-respect. Their love of finery amidst the squalor of their wretched highland villages is very remarkable. Men and women dressed in rich flowing garments, worn with admirable grace, are often met residing in damp and gloomy underground hovels, or in huts formed of interwoven branches, or of huge stones thrown loosely together.

The Osses or Ossetians, who call themselves Iron, constitute the fourth division of the Caucasian races. But while the other three may be regarded as indigenous, the Osses are certainly intruders of Aryan stock. They occupy the most central part of the Great Caucasus along both slopes of the Kazbek, where they are conterminous with various tribes of all the other divisions—Kartvelians on the south-west and south-east; Chechenzes on the north-east; Kabards on the north-west. Of fair complexion, robust, of a somewhat heavy and sluggish temperament, and lacking the graceful carriage of the other highlanders, these Ossetians seem to resemble the Germans more than any other branch of the Aryan family. Yet their language belongs to the Iranian group, and the national name of Iron has been accepted as an argument in favour of their Persian origin.

Non-Caucasian Intruding Races.

Besides the already-mentioned Russian Slavs, Nogais, and other Tatars, the more important intruding peoples are the Armenians, who have advanced from the Aras to the Kura basin; the Kurds, some of whom have penetrated north to the Rion basin; the Greeks, numerous in the district west of Tiflis; the Tats and Talyshes of the Baku district, akin to the Iranian Tajiks; lastly, a German colony from Würtemberg settled in a few isolated communities in the Kura basin east of Tiflis.

These German colonists present a most remarkable anthropological problem.[1]

There is also a very old Jewish element in several parts of Caucasia, but nearly everywhere assimilated in speech and habits to the surrounding peoples. A number of places known by the name of Jut-Kend, or "Jewish Town," are now occupied by communities claiming Tatar descent, and the Jews of the Baku district have adopted the Persian garb and speech. These latter are said to

[1] See *Mémoire sur l'Ethnographie de la Perse*, p. 14.

have arrived from Persia during the time of the Sassanides, and are by some writers supposed to be descended from the Israelites, who were removed to Persia after the first destruction of the Temple by Salmanazar. This view seems confirmed by the family names still current amongst them, which belong to the period of the Judges, and which have been elsewhere obsolete for over 2000 years.

The prevailing religions in Caucasia are Christianity and Muhammadanism, with almost everywhere a substratum of the old pagan superstitions. Thus the Khevsurs, who belong to an extremely interesting group of tribes clustered round Mount Borbalo, about the sources of the Aragva, Yora, and Alazan, have developed a form of Christianity of a somewhat peculiar type. Having been followers of the Prophet before they were Christians, and heathens before their conversion to Islám, they keep the Friday with their Moslem neighbours, and the Sunday with the Georgians, intermingling the worship of trees and of the spirits of earth and air with more orthodox rites. Their chief deity seems to be the God of War, and they also do homage to the Mother of the Earth, the Archangel of Property, the Angel of the Oak, and many other lesser gods and angels. Yet they are very proud of their Christianity. The Khevsurs are also probably the only people in the world who still wear armour.

8. *Topography: Stavropol — Vladikavkaz — Derbent — Baku—Poti—Tiflis—Kars—Batúm—Erivan—Alexandrapol—Shusha—Nakhichevan.*

In Western Caucasia there are no towns of any size. Yekaterinodar, capital of the Kuban province above the delta of that river, is an important agricultural centre, much frequented during the autumn fairs, when produce to the amount of over £320,000 is usually disposed of.

Yeisk on the Sea of Azov is a thriving seaport and fishing station, and although founded so recently as 1848 has already become the largest place on the whole Caucasian seaboard. Taman, which gives its name to the peninsula, is a mere village, and Sukhum-Kaleh on the Abkhasian coast, although possessing a safe and deep harbour, is noted for its dolphin fishery.

Stavropol and Vladikavkaz.

Stavropol, capital of the government of like name, lies on the verge of the steppe about 2000 feet above the sea, and at some distance to the east of the south-eastern line of railway, whose present terminus is Vladikavkaz. This place, which is the capital of the Terek territory, lies almost in the heart of Central Caucasia, at the northern entrance of the Dariel Gorge, through which the great military road leads south to Tiflis. Although since the pacification of Caucasia it has lost its former strategic importance, Vladikavkaz has continued to flourish as a commercial emporium, and its commanding central position, 2300 feet above the sea, midway between the two seas, marks it out as the future capital of Cis-Caucasia. In the whole of this region its only rival is Tiflis, over which it possesses the great advantage of a genial and healthy climate. Piatigorsk, about equidistant from Stavropol and Vladikavkaz, is the chief centre of the Caucasian watering-places, which are nowhere surpassed for variety, copiousness, and health-giving properties. But although supplied with grand hotels, promenades, pleasure-grounds, and other attractions, it has hitherto failed to tempt many visitors from the West.

Derbent and Baku.

On the Caspian coast, Derbent, in lat. 42°, occupies a

peculiar position between the spurs of the Daghestán hills and the sea, completely guarding the narrow defile on the great historic route along this coast. It covers a long,

DERBENT.

narrow strip, enclosed by walls running from the citadel of Narin-Kaleh eastwards to the sea. The line of fortified works is continued over the hills for some distance westwards, and is traditionally supposed to have formerly stretched right across the isthmus from the Caspian to the Euxine. But a much more important place is Baku at the neck of the Apsheron peninsula, centre of the most productive naphtha district in Asia. This trade, which employs quite a little fleet of coasting vessels and steamers on the Caspian, yielded over 16,000,000 cwts. in 1883, and this quantity might easily be doubled with proper appliances. Close to Baku over 700 naphtha wells have already been sunk, none of which show any signs of exhaustion.[1] Here is a famous shrine of the Persian fire-worshippers, which is directly fed with inflammable gases from the subterraneous fires.

In the Rion - Kura depression the chief places are the port of Poti, Kutais, and Tiflis, all connected by the line of railway which at present stops at Tiflis, but which must eventually be carried down the Kura valley to the Caspian.

Poti and Tiflis.

In spite of its exposed and shallow roadstead and its pestilential climate, Poti had rapidly progressed before the late Russo-Turkish war. It still retains some importance as the terminus on the coast of the only railway south of the Caucasus. But since the acquisition of Batúm in 1878 its shipping has been greatly reduced. Whenever the railway is continued round the coast to Batúm, Poti will probably be abandoned to its swamps and mosquitoes.

By far the most important city in Caucasia is the capital, Tiflis, which lies on the left bank of the Kura, a little south of Mtzkhet, the ancient capital of the Georgian kingdom. It is a half European, half Asiatic town, consisting of a Russian quarter with some ambitious buildings

[1] In September 1881 the Krassilnikoff wells took fire, and continued to burn with intense fury for several days.

in the modern style, a clean and picturesque German suburb, and a Persian district with a decidedly Eastern appearance. The most prominent feature of the place is the fine open "Golavinsky Prospect," of which any pro-

TIFLIS.

vincial town in Europe might be proud. Less inviting is the quarter containing the Armenian bazaar, although historically interesting as recalling the time when Tiflis still acknowledged the authority of the Shah. The old castle of the Georgian princes, a reminiscence of a still earlier period, now lies in ruins on a hill rising precipitously above a wild romantic stream. Tiflis has been described as a city of contrasts. Cairo alone presents a similar mingling of Oriental poetry and decay with some of the humble types of European society.

Kars, Batúm, Erivan, and Alexandrapol.

South of the Rion basin Russia acquired in 1878 as the prizes of victory, besides the frontier town of Ardahan, the much-coveted seaport of Batúm, and the formidable stronghold of Kars. A straight line drawn from Batúm to Erivan, capital of Russian Armenia, will pass through Ardahan, leaving Kars a little to the right. Regarding the great natural strength and strategic importance of Kars there never could be any question. But the descriptions current in Europe of Batúm vary somewhat. It lies some 30 miles south-west of Poti, and its harbour, formed by the delta of the Chorukh advancing westwards, is completely sheltered, about 60 feet deep, and capacious enough to accommodate twelve large vessels. But this harbour is being gradually encroached upon by the very delta to which it owes its existence, a danger, however, which may easily be remedied by giving the river a better scour, or connecting it by canal directly with the port. When these works are carried out, and the railway, already commenced, completed through Uzurgeti to Poti, Batúm must become the commercial emporium as well as the naval station of Trans-Caucasia. It was declared a free port by the Berlin Congress of 1878, but this has not prevented the Russians from converting it into a second Sevastopol.

Erivan, capital of Russian Armenia, stands on the Zanga, which flows intermittently from Lake Gok-cha, and is here diverted into innumerable little irrigation canals before reaching the Aras. It occupies an important strategic position at the entrance to the route leading over the Gok-cha plateau to the Kura basin and Tiflis. But the climate, with its sudden changes of temperature, malaria, and dust-storms, is one of the worst in Caucasia. Hence Erivan has always remained a small place, and has already been outstripped by the comparatively new fortified town of Alexandrapol, now the largest place in the Aras basin. This stronghold was founded in 1837 on the then Turkish frontier. The possession of these places not only renders the Russian position impregnable in the Aras basin, but gives that power complete command of the head-waters of the Euphrates. The road through Ardahan to Batúm is extremely rugged; but the works now in progress will soon render it an easy military route from the Black Sea to the Turkish frontier.

Shusha and Nakhichevan.

East of Erivan the only places in the Aras basin calling for mention are Shusha and the ancient town of Nakhichevan. Shusha lies in the heart of the plateau near the water-parting between the Aras and Kura, some miles above their confluence. It is the largest place in the Yelizavetpol Government; but standing 3500 feet above the sea on an exposed terrace, its climate is excessively severe in winter and correspondingly hot in summer. Nakhichevan lies close to the left bank of the Aras near the Persian frontier. It enjoys the distinction in Armenian tradition of being the oldest city in the world,[1] founded in fact by Noah himself after planting the vine on the slopes of Ararat. But it is now chiefly

[1] Nakhichevan, of which the classic form was Naxuana, is explained to mean "The First Abode," *i.e.* of man after the deluge.

inhabited by Tatars, while its finest monuments, including the old palace gateway and the "Tower of the Khans," date from the Persian epoch.

9. *Highways of Communication.*

All the routes from Southern Russia through Rostov, Astrakhan, and other points, converge at Vladikavkaz on the north side of the Central Caucasus. From this place the great military and commercial highway leads through the magnificent Dariel Gorge and right under the Kazbek down to Tiflis on the south side of the main range. Here the roads again diverge in all directions—west through Gori, over the Suram Pass of the Mesk range, down the Rion valley, and through Kutais to Poti; south-east down the Kura valley through Yelizavetpol to Baku on the one hand, and on the other to Lenkoran on the Caspian near the Persian frontier; south direct to Alexandrapol and Kars, and over the Gok-cha plateau to Erivan, Echmiadzin, and Ararat. From Poti the road skirts the coast to Batúm, whence the military route leads over the Lazistán and Arsiani ranges through Ardahan to Kars, down the Arpa-chai valley through Ani to Erivan, and down the Aras valley to Nakhichevan and Julfa on the Persian frontier. From Kars and Julfa two short roads, crossing the Turkish and Persian frontiers, strike the great trade route from Trebizond to Tabriz at Erzerum and Khoi.

The old and difficult coast road from the Taman peninsula to Poti, followed by trade and warlike expeditions from the time of Mithridates down to the Middle Ages, and which had been abandoned, will be reopened for traffic. The coast route from Baku through Derbent to the Terek and Kuma basins is little used, although easier than the Black Sea road.

The Caucasian railway system is still limited to two completed lines. Of these the longest and most important forms a south-eastern extension of the Russian system, running from Rostov at the head of the Azov Sea through the Kossak stanitzas of Yekaterinovskaya and Kavkaskaya to Georgiyevsk for Piatigorsk, and so on across the Upper Terek to Vladikavkaz, its present terminus. The second runs from Poti through Kutais and over the Suram Pass down to Mtzkhet and Tiflis. The short coast line from Batúm to Poti will also soon be finished.

Several projects have been proposed for connecting both sides of the Caucasus. Of these the most practicable seems to be a line from Vladikavkaz along the military route through the Dariel Pass to Tiflis. Another will ere long be carried round the east coast through Derbent to Baku facing the Mikhailevsk terminus of the new Turkestán line. This will give direct railway communication from the heart of the empire to within "a measurable distance" of Herat, interrupted only at the narrowest part of the Caspian. A first section of this line from Yelizavetgrad to Petrovsk, north of Derbent, has been commenced, and the line from Tiflis to Baku, completing the junction of the two seas, has also been taken in hand.

10. *Administration: Results of Russian Rule—Armenian Politics—Administrative Divisions.*

The brief period during which Caucasia has formed an integral part of the Russian Empire has already produced great social and ethnical changes. A succession of able and energetic administrators have gradually succeeded in stamping out the last spark of independence amongst the highland tribes, and this mountainous region no longer forms a weak point in the colossal empire.

This result has been brought about by very simple

means—a steady but determined pursuance of purely practical ends. The administration has everywhere respected the local customs and usages, restricting itself to the maintenance of the preservation of social order. After the heroic defenders of their highland fastnesses were sufficiently reduced by military operations and by wholesale expulsion, efficacious military steps were taken to place the supreme authority beyond the reach of attack from the survivors, while interfering as little as possible with their local affairs. The several tribes were allowed to retain their primitive usages and social institutions, the judicial functions were carried out more in a paternal than in a bureaucratic spirit, and the administration was, wherever possible, entrusted to natives.

The authorities have from the first entirely abstained from interfering with the peculiar religious views of the people, so that it is scarcely correct to assert that the Russian rule has mainly resulted in the extinction of national life.

It is remarkable, in this connection, that the opposition of the Armenians was long directed rather against the Russians than the Turks. But since the acquisition of Erivan and Kars Russia has begun to exercise a growing influence over the Armenian people. In the neigh bourhood of Erivan is the Convent of Echmiadzin, the residence of the Armenian Patriarch, whose spiritual authority is absolute over the whole race wherever settled.

But the rising generation of Armenians has shown itself less submissive, in secular matters, to the authority of its spiritual guides. A political party has been formed in Constantinople, inspired by the modern revolutionary spirit, which has undertaken to quicken the slumbering sentiment of nationality, and direct the efforts of the people towards independence. Since the late political changes the Armenians have also made their voice to be heard, determined not to be overlooked in the midst of the innovations which promise to fundamentally modify the social condition of the Eastern races.

Meanwhile Russian Armenia forms a simple division of Caucasia, which constitutes a single administrative government under a viceroy or lieutenant-general responsible only to the Czar. To this government is also attached the Trans-Caspian territory, recently extended nearly to the Afghan frontier. The Caucasian portion of this region is subdivided into fourteen separate administrations, variously named governments, provinces, territories, or circles, as in the subjoined tabulated scheme of the several administrations, with their areas, populations, and chief towns. The populations are estimated for 1881, and the areas are reduced from the data published by the Statistical Commission of the Caucasus for the Caucasian section of the Russian Geographical Society.

11. *Statistics.*

Areas and Populations.

	Area in sq. miles.	Estimated Pop. 1881.
Cis-Caucasia	88,700	1,960,000
Daghestán	11,400	500,000
Trans-Caucasia	75,900	3,250,000
Conquests, 1878	10,000	236,000
	186,000	5,946,000

Cis-Caucasia.

Administration.	Area in sq. miles.	Pop.	Chief Towns.	Pop.
Stavropol (Government)	27,500	500,000	Stavropol	29,600
			Piatigorsk	13,600
			Belaglina	11,200
			Praskoveya	8,000
			Alexandrovskaya	7,400
Terek (Province)	23,700	600,000	Vladikavkaz	20,600
			Kizlar	9,200
			Grozniy	8,500
			Mozdok	8,400
Kuban (Province)	37,500	860,000	Yekaterinodar	32,500
			Yeisk	28,500
			Maïkop	22,600
			Temruk	11,200
			Novo-Petrovskaya	7,000
Total Cis-Caucasia,	88,700	1,960,000		

TRANS-CAUCASIA.

Administration.	Area in sq. miles.	Pop.	Chief Towns.	Pop.
Black Sea (Circle)	2,250	23,000	Anapa	5,200
Sukhum (Division)	3,000	85,000	Sukhum-Kaleh	4,000
Kutais (Government)	8,500	570,000	Kutais	13,000
			Zugdidi	5,500
Tiflis (Government)	16,300	770,000	Tiflis	105,000
			Akhaltzikh	13,300
			Signakh	9,250
			Telav	7,000
Sakatal (Circle)	1,600	80,000	Tali	6,000
Erivan (Government)	11,100	568,000	Alexandropol	20,500
			Erivan	12,500
			Nakhichevan	7,000
Yelizavetpol (Government)	17,500	594,000	Shusha	24,600
			Nukha	21,000
			Yelizavetpol	18,500
Baku (Government)	15,650	560,000	Baku	14,600
			Kuba	11,300
Daghestán	11,400	500,000	Derbent	13,800
			Gubden	5,600
			Akhti	5,600
Batúm Kars (Provinces)	10,000	236,000	Batúm	6,000
			Kars	10,000
Total Trans-Caucasia	97,300	3,986,000		
Cis-Caucasia	88,700	1,960,000		
Total Caucasia	186,000	5,946,000		

INHABITANTS OF CAUCASIA GROUPED ACCORDING TO RACES AND RELIGIONS.

Kartvelian Stock	Georgians, Imerians, Mingrelians	Christians, Greek rite	1,150,000
	Svans, Pshavs, Khevsurs	Nominal Christians	
	Lazis	Sunnis	20,000
West Caucasian Stock	Cherkesses, Abkhasians	Sunnis	138,000
	Kabards	Christians	
East Caucasian Stock	Chechenzes, Lesghians[1]	Sunnis	164,000
Semite Stock	Jews and Assyrians[2]	?	30,000

[1] All the Lesghians are Sunnis except the Dido tribe of the Upper Koïsu valley, who are said to be "Devil-worshippers."

[2] These are introduced on the authority of N. von Seidlitz, Director of the Statistical Office at Tiflis.

Slav Stock	Great Russians, Little Russians, Bulgarians, Bohemians	Orthodox Christians and Dissenters	1,500,000
Iranian Stock	Osses	Nominal Christians	110,000
	Armenians	Gregorian Christians	720,000
	Kurds	Sunnis	380,000
	Tats, Talishes	Shiahs	
	Greeks	Christians, Greek rite	50,000
	Germans	Lutherans	
Mongolo-Tatar Stock	Tatars, Nogais, Turkomans, Kumiks, Kirghiz[1], Turks	Sunnis mostly; some Tatars are Shiahs	1,550,000
	Kalmuks[1]	Buddhists	

MISCELLANEOUS.

Mean density of population per square mile, 80.
Christians, 3,560,000; Muhammadans, 2,220,000.

Land under vineyards	220,000 acres.
„ „ tobacco	10,000 „
Average yield of cotton	10,000 cwts.
„ „ silver	800 lbs.
„ „ lead	1,800 cwts.
„ „ copper	2,250 „
„ „ alum	130 tons.
„ „ salt	25,000 „
„ „ coal	6,000 „
Imports and exports (1879)	£2,200,000
Receipts (1878)	3,000,000
Expenditure „	11,000,000
Deficit „	8,000,000
Ordinary deficit	4,000,000

Chief source of revenue, alcohol, one-third of the whole.

For 1883 the total area and population for the "Caucasus," including "northern Caucasia" and "Trans-Caucasia," were set down thus:—

Square miles, 182,505; Population, 6,534,853.

[1] The Kirghiz and Kalmuks occupy a large area in the Lower Volga district, whence some of them reach southwards beyond the Manich depression along the right bank of the Kalaus and as far as the left bank of the Kuma.

THE CAUCASUS.

SCALE OF ENGLISH MILES 50 0 50 100 150 200

London: Edward Stanford.

Stanford's Geogl Estabt

CHAPTER X.

RUSSIAN TURKESTÁN.

1. *Boundaries—Extent—Area—Nomenclature.*

HERE we enter a region in which geographical, ethnographic, and political conditions are still far from being reconciled. Although somewhat simplified by the recent progress of Russian arms in West Central Asia, the very nomenclature is still in a confused state, and few even of the leading authorities are of accord as to the exact meaning of such common expressions as Turkestán or Central Asia. The Russians themselves often designate as Central Asia the second great administrative division of their Asiatic possessions, which is mainly comprised within the Aralo-Caspian depression. But this expression is misguiding in a geographical sense. To the portion of this division directly administered by the Governor-General, whose headquarters are at Tashkent, they give the still more questionable name of Eastern Turkestán—the true Eastern Turkestán, if there be any, lying beyond his jurisdiction in the Chinese province of Kashgaria. The confusion of nomenclature is increased by the distribution of the land, portions of which are attached either to the European governments of Orenburg and Perm, or to the administration of Caucasia, while the south-eastern section beyond the Upper Oxus belongs politically to Afghanistán, and thus forms part of the British political system.

This vast region is mostly comprised within the limits of the Aralo-Caspian basin, stretching west and east between the Caspian Sea and the Central Asiatic highlands, limited southwards by the scarp of the Iranian plateau, merging northwards in the west Siberian steppe. The south-eastern boundary line along the Afghan frontier was accurately laid down by agreement with the British Government in 1885. The south-western boundary line along the Persian frontier has been greatly extended by the reduction of the Akhal Tekke Turkomans in 1881. The rectified frontier line, as determined by the agreement with Persia in 1882, now runs from the south-east corner of the Caspian up the Atrak valley to Chat at the junction of the Sumbar, thence eastwards along the water-parting to a point south-east of Askabad.

Russian Turkestán is bordered on the west by the Caspian, the Ural river and mountains, on the east by the Pamir plateau, the Tian-Shan and Ala-tau ranges separating it from the Chinese Empire, northwards by the low ridge crossing the Kirghiz steppes about the 51st parallel, and forming the water-parting between the Aralo-Caspian and Ob basins. But here again the administrative overrides the geographic division, for a large portion of West Siberia beyond this natural boundary is now attached to the Turkestán Government.

Including this tract, which is alone about 400,000 square miles in extent, Russian Turkestán has an extreme length from the Caspian to Lake Issik-kul of 1400 miles west and east, with a breadth of nearly 1000 north and south, a total area of about 1,600,000 square miles, and a population of 6,500,000.

2. *Relief of the Land: The Great Pamir—Humboldt's Bolor Range—The Kizil-art and Alai Ranges—The Tian-shan and the Ala-tau Highlands—The Mugojar Hills—The Turkestán Depression—The Dried-up Central Asiatic Mediterranean—The Turkestán Deserts.*

Western Turkestán is commonly supposed everywhere to consist of vast low-lying sandy or saline plains. But nothing could be more opposed to the actual conditions, for the relief of the land here presents absolutely greater contrasts than are elsewhere found on the surface of the globe. The misconception is due to the failure to distinguish between the Aralo-Caspian depression and the Aralo-Caspian basin. The basin—that is, the whole area of drainage—consists of about even parts highlands and lowlands; and while the lowlands fall in the Caspian Sea as much as 85 feet below sea-level, the highlands in the culminating points of the Tian-shan and Great Pamir rise to 25,000 feet above sea-level. In no other region are such vast differences of relative level to the surface of the sea brought into such close juxtaposition.

The highlands, which lie mainly in the east, consist substantially of the Pamir and Tian-shan systems, with the Alai and other sections, all converging westwards between the Tarim and Aralo-Caspian depressions.

The Great Pamir or Bam-i-Dunya—that is, the "Roof of the World," as it has been not inaptly termed—forms the nucleus of the whole Central Asiatic highland system. Here converge the Hindu-Kush and Himalayas from the south-west and south-east, the Kuen-lun from the east, the Tian-shan from the north-east, while the plateau itself merges westwards in the snowy highlands and icefields about the sources of the Zarafshan, between the Oxus and Jaxartes valleys. Although still but imperfectly explored, its main features are now sufficiently ascertained, and we know that it consists of a vast plateau formation, some 30,000 square miles in extent, with a mean elevation of at least 12,000 feet, culminating in the east with the Tagharma (25,500 feet). Its southern limits seem to be marked by the ridge connecting the Karakorum with the Hindu-Kush, and forming the water-parting between the Upper Oxus and Indus basins. Northwards its limits are better defined by the Alai and Trans-Alai ranges skirting the south side of Ferghanah (Khokand), and forming the water-parting between the Zarafshan and Jaxartes valleys.

In this region Alexander von Humboldt imagined an isolated chain of the Bolor, Bilaur or Belut-tagh, running north and south between Eastern and Western Turkestán, and forming the axis of the whole continental system. But Major Biddulph, while showing that such a range has no existence, traces the term Bolor itself to an old state of that name, whose chief place is Iskardo. "There can be little doubt that it is to Iskardo we must look for the centre of the ancient kingdom of Bolor, as suggested by General Cunningham. In Gilgit, Hunza, Nagyr, and all the valleys to the westward, the name Iskardo is almost unknown, and the place is called 'Palor,' 'Balors,' and 'Balornts.'"[1] But it will probably be some time before Balor as a mountain range will disappear from the maps. Its place ought to be taken by the Kizil-art mountains, which, according to the explorers Hayward and Shaw, form the eastern limits of the Pamir towards Kashgaria.[2]

This range seems to run north-west and south-east

[1] *The Tribes of the Hindu-Kush*, p. 146.

[2] The Kizil-art has now been identified with the Tsung-ling, or "Onion Mountains," of the Chinese, so called from the quantity of garlic growing on their slopes.

between the Kashgar-tagh and the Karakorum at a mean elevation of 20,000 feet, culminating in Mount Tagharma (Taghalma), estimated at 25,500 feet. But the whole of these highlands, with their altitudes, directions, and nomenclature, still form one of the most obscure chapters in Central Asiatic geography. At the same time the abrupt fall towards the Kashgarian depression renders it very probable that the numerous Pamir lakelets, where they have an outflow, drain westwards to the head-streams of the Oxus, and that the Kizil-art forms the true water-parting in this direction.

In July 1876 Kostenko crossed the Trans-Alai, and was thus the first European to enter the Great Pamir from the north. From the Kizil-art Pass in the Trans-Alai he commanded a complete view of this region; and after for the first time visiting the romantic Lake Kara-kul, one of the sources of the Oxus, he undertook an expedition as far as the little Lake Ran-kul, towards the Kashgar frontier. Between these two lakes rises Mount Us-bel (15,600 feet), which forms the water-parting between the Oxus and the Tarim. From its summit a grand prospect is presented of an easterly range towering far above the snow-line, and shutting off the valley of the River Kashgar. This range, obviously Hayward's Kizil-art, Kostenko proposes to call the Konstantinov Mountains, assigning them an elevation of from 25,000 to 26,000 feet. The Trans-Alai itself he describes as an alpine chain 10,000 to 12,000 feet high, forming the northern boundary of the Pamir, which stretches thence southwards, and which is crossed in every direction by ridges, all rising above the snow-line, and dividing it into a number of smaller "Pamirs," or upland plains.

The whole region is destitute of trees or shrubs, and even the grass grows only in isolated patches along the banks of the streams and lakes. Here, however, it affords some of the very finest pasture in the world to the flocks of the Kara-Kirghiz nomads, who visit the Pamir during the summer season.

The hills consist of a soft stone, so that the passes are less abrupt and easier to cross, while the tracks are everywhere tolerably good. The streams also are seldom very rapid, and the soil consists either of grit or sandy loam, dotted over with salt or brackish tarns, which when dried up are covered with an incrustation of dazzling white magnesia.

It is difficult to assign a beginning or an end to the Tian-shan, or "Celestial Mountains," which separate the Tarim from the Issik-kul and Ili basins south and north, and stretch thence eastwards to about 120 miles east of Hami (Khamil) in 95° E. longitude. At this point, which may be taken as their eastern limit, the Tian-shan consist of a single ridge. But the whole system continually expands westwards, developing two or more lateral and parallel ridges, and in the extreme west ramifying into several distinct branches, which spread out like a fan far into the Turkestán lowlands. Of these branches the south-westernmost are the Alai and Trans-Alai, which stretch in parallel lines for 240 miles along the northern edge of the Pamir, down to the Turkestán plains. They are separated from the Tian-shan proper by the Kog-art[1] and Terek-davan[1] passes, but their diorite and granite formations show that they belong none the less to that system. The Alai, or Kichi-Alai, rises to over 18,000 feet in the east, and is crossed by several passes, of which the lowest is the Isfairan (12,000). From this point a view is afforded of the "Kaufmann" Peak, over 25,000 feet,

[1] In the Tian-shan, *art*, *davan*, *bel*, and *kutal*, are the general names for passes. The *art* is a high and dangerous gap, the *davan* a difficult rocky defile, the *bel* a low easy pass, the *kutal* a wide opening between the hills.

one of the very highest, if not the highest, in the whole Tian-shan system.

The Alai and Trans-Alai merge westwards in the snowy plateau and ice-fields, about the sources of the Zarafshan, whence the Shchurovsky and other enormous glaciers descend towards the surrounding upland valleys.

North of Ferghana, the most important western branch of the Tian-shan, are the Alexander Mountains, which run at an elevation of 15,000 feet from the closed basin of Lake Issik-kul along the northern edge of the Narin (Upper Jaxartes) valley, and are continued by the Aksai, Talas-tau, and Kara-tau (6000 feet) north-westwards between the Middle Jaxartes and the Chui valleys. Here the culminating crests are the Hamish (15,550) in the Alexander chain, the Kara-bura (11,000) in the Talas-tau, and the Min-jilke (7000) in the Kara-tau.

West of Hami the eastern section of the Tian-shan soon attains an elevation of 8000 to 10,000 feet. Between Hami and Barkul it is crossed by the Kosheti-davan Pass, over 9000 feet. Farther on there occurs a profound gap or break of continuity, through which the historical route leads from the Gobi desert through Turfan north to Urumtsi and the Ili basin. West of this defile the central section, under the name of the Katun range, rises far above the snow-line, and attains an elevation of at least 16,000 feet. Here there are no passes, and enormous glaciers have recently been discovered by Regel about the sources of the Kash, which flows from these highlands westwards to the Ili valley.

From its eastern extremity to this point the Tian-shan may be regarded as consisting substantially of a single range. But here are developed as many as four parallel snowy ridges, enclosing the two extensive dried-up basins known as the Great and Little Yulduz, or "Stars," 7000 feet above sea-level.

South of the Great Yulduz basin the main southern section runs under diverse names westwards to Lake Issik-kul. Here the chief sections are the Muz-art-tau, crossed by the historical Muz-art Pass (11,000), and the Khan-tengri (24,000), the most imposing and dominant mass in the whole Tian-shan system. Although exceeded in height by the Trans-Alai peaks, the Khan-tengri contains far more numerous glaciers, ice-fields, and snowy crests. Some of the glaciers in the little-known highlands west of the Muz-art rival the Aletch of the Valais, and from this pass the main range runs for over 60 miles westwards at a mean altitude of nearly 17,000 feet. Here all the peaks are higher than Mount Blanc by over 3000 feet, and yet are dominated southwards by the magnificent Khan-tengri or Kara-göl-bas.

West of the Yulduz basins the extensive dried-up marine basin of the Tian-shan-pelu rises westwards towards the plateau occupied by Lake Sairam, which is skirted north and south by the Zungarian Ala-tau and the Boro-khoro chains. Facing the Boro-khoro is the imposing Nian-shan (Temurlik) range, which rises abruptly above the south side of the Kulja plains, between the rivers Tekes and Kegen, east and west. North of the Narin valley the main range takes the name of the Ala-tau Terskei—that is, the "Shady Ala-tau"—to distinguish it from the Ala-tau Kungei, or "Sunny Ala-tau," which skirts the opposite side of Lake Issik-kul.[1] It culminates

[1] As much confusion is caused by the numerous "Ala-tau" or "mottled" ranges in this region, it may be well to explain that the Zungarian Ala-tau runs north of the Ili valley, over against the Tarbagatai chain, the Trans-Ili and Kungei Ala-tau lie south of the Ili and north of Lake Issik-kul, where they are separated by the valleys of the Great Kebin and Chilik Rivers; the Ala-tau Terskei runs south of Issik-kul, between it and the right bank of the Narin (Upper Sir) River; while a fifth Ala-tau running west from Issik-kul has now been renamed the Alexander range by the Russians. It may be added, that the hitherto

with the Ugus-bas (17,750 feet), and is crossed by the Barskaun Pass (12,000 feet) near the source of the Narin. The Terskei Ala-tau is continued westwards by parallel ridges, which enclose Lake Son-kul, and effect a junction north-westwards with the Alexander chain.

Thus is completed the vast system of the Tian-shan, which is about 1500 miles long east and west, with an average width of nearly 250 miles, and a total area of 400,000 square miles, or rather more than that of all the European highland systems taken together.

The long-suspected presence of still active volcanoes in the Tian-shan, and especially towards the Kulja frontier, has at last been settled in a negative sense. General Kolpakovsky, Governor of Semirechinsk, in the autumn of 1881 explored Mount Bia-shan, twelve miles north-east of the city of Kulja, in the Ailak highlands, and discovered that the fires which have been burning there from time immemorial are not volcanic, but proceed from ignited coal-beds. The caves in the side of the mountain emit smoke and sulphurous gases.

From the Mugojar ridge, running from the Urals to the head of the River Emba, the low hills which form the water-parting between the Ob and Aralo-Caspian basins may be traced across the Kirghiz steppes about the 48th parallel eastwards to Lake Balkhash. The Mugojar is nowhere more than 600 feet high, while the water-parting falls at one or two points to 220 feet. But this water-parting is continued eastwards along the north side of Lake Balkhash by the Denghiz-tau, or "Sea Range," which near Sergiopol merges in the Tarbagatai highlands. From this point the Tarbagatai—that is, "Marmot Moun-

little-known section of the Ala-tau north-east of the Sir valley, about the sources of the Talas and Arys, has been quite recently explored by Col. Ivanoff. He describes it in the *Isvestia* of the Russian Geographical Society (xvii. 3).

tains"—stretch in two main sections under the 47th parallel between Lakes Zaisan and Ala-kul eastwards to the Upper Irtish valley. Above the left bank of this river rise the snowy Tas-tau (9850 feet), and Muz-tau or Sauru (11,320 feet), culminating point of the whole system. West of the Muz-tau the range is crossed by the Khabar-assu Pass, which, notwithstanding its great height (7628 feet), has always been more frequented than any other, especially by traders between Siberia and the Ili basin. The Tarbagatai has a mean altitude of 6000 feet, but few of its crests rise above the snow-line, which here falls to about 9000 feet.

The Turkestán lowlands, which stretch from the Caspian and Ural River to the foot of the Central Asiatic highlands, possess no bold natural limits towards the north, where they merge imperceptibly with the West Siberian steppe. Southwards they are limited by the western continuation of the Hindu-Kush as far as the Hari-rúd valley, and beyond that point by the North Khorasán highlands as far as the Caspian. By far the greater part of this region is occupied by the Aralo-Caspian depression, which is the most extensive on the globe; for it properly includes the plains of south-east Europe, which drain through the Volga to the common basin of the Caspian. In this great inland sea the depression reaches its lowest level of 85 feet below the Mediterranean. Eastwards the Caspian is separated by the extensive Ust Urt plateau from the Aral Sea, which is the next largest reservoir in Asia.

The Chink, or eastern edge of the Ust Urt plateau, is 500 feet above sea-level along the west coast of the Aral Sea, but falls to 210 farther south towards the Sary Kamish lakes. This is probably little more than the mean elevation of the Turkestán lowlands, which are now known to have at one time formed part of a vast inland

KULJA, THE RUSSO-CHINESE FRONTIER, &c.

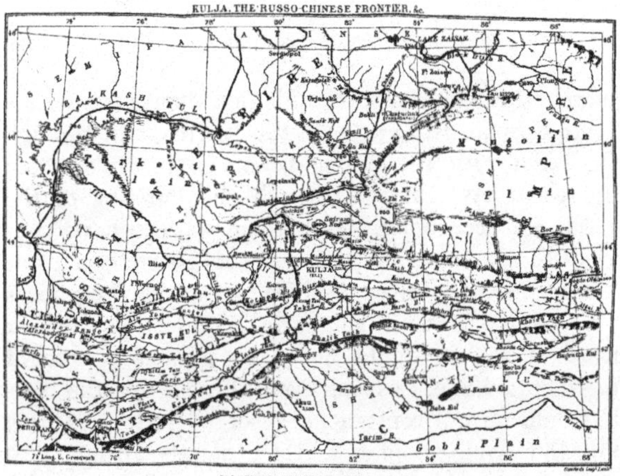

London: Edward Stanford, 26 & 27, Cockspur St. Charing Cross, S.W.

sea communicating through the Manich depression across the Ponto-Caspian isthmus with the Euxine, and through the Ob basin with the Arctic Ocean. This inland sea is supposed to have escaped through the bursting of the Bosphorus, an occurrence which some writers have connected with the legendary deluge of Deucalion about 1530 B.C. But however this be, were the Bosphorus again to be closed to a height of 220 feet, the former condition of things would again be gradually brought about. The Euxine, Caspian, Aral, if not also the Balkhash, would form a continuous sheet of water, draining through the Aralo-Caspian and Ob water-parting (220 feet at its lowest point) to the Arctic Ocean.

North of the Aral Sea the plains and steppes form a vast lacustrine region, dotted over with lakes or tarns with no outflow, and fed by intermittent steppe rivers. Some of these, such as the Chui and Sari-su, formerly reached the Aral Sea through the Jaxartes, while others found their way either directly or through the Emba to the Caspian. Lakes Balkhash and Kara-kul, fed by the Ili and Talas respectively, must have also communicated through the Chui with the Aral Sea, which, it will be seen farther on, itself drained through the Uzboi, an old bed of the Oxus, to the Caspian. Balkhash, again, is known to have formerly stretched much farther east than at present, forming a continual sheet of water with the Sassik-kul and the numerous other lakes strewn over the so-called "Zungarian Strait," which is supposed to have connected the Turkestán and Mongolian mediterraneans through the depression between the Ala-tau and Tarbagatai ranges. At the time when these inland seas were thus connected, their elevation must have been far more than 500 feet, for the present level of Balkhash is 514 feet. Consequently the inland waters may have at that time communicated over the Bosphorus barrier with the Atlantic, as well as over the present Aralo-Caspian and Ob water-parting with the Arctic. When, through the process of desiccation continually going on in Asia, the mediterraneans were reduced to about 220 feet, the northern and southern outflows were probably arrested, and then the pressure of such a prodigious body of water would help to account for the bursting of ,the Bosphorus and gradual draining of the Han Hai, or "Western Sea," as the Asiatic Mediterranean is called in the Chinese Chronicles.

The Turkestán lowlands proper are known by various names, such as the Kara Kum or "Black Sands," north of the Aral Sea; the Kizil Kum or "Red Sands," between the Oxus and Jaxartes; the Ak Kum or "White Sands," between the Alexander range and the Chui River; and the Khwarezm or Turkoman desert, between the Oxus and Caspian. But all alike present the same monotonous and desolate aspect, in which a dull brown is the prevailing colour. "The gloom of the West Turkestán steppes, which first impresses one so forcibly in the Karakum deserts north of Aral, seems surpassed by the sadness of Kizil-kum near the south-east corner of the lake."[1]

3. *Hydrography: The Rivers Oxus, Zarafshan, Murgh-ab, and Sir-darya—The Aral Sea—Lakes Balkhash and Issik-kul.*

All the streams of the Aralo-Caspian basin have their natural outlet in the Caspian Sea, the lowest part of the Turkestán depression. But none of them now reach that outlet except the River Emba flowing from the Mugojar hills south-westwards to the north coast, which it reaches after a course of 250 miles through the Kirghiz steppe. All the rest either run dry in the sands like the Tajand

[1] Major Herbert Wood's *Shores of Lake Aral*, p. 337.

and Murgh-ab from the Iranian plateau, the Zarafshan, Chui, and Talas, from the Tian-shan highlands, the Sari from the Ob water-parting, or else, like the Oxus and Jaxartes, are at present absorbed in Lake Aral. But as Aral stands about 160 feet above the Mediterranean, consequently 245 above the Caspian, its proper outflow should also be to the Caspian, with which, in fact, it formerly communicated. Hence the Oxus and Jaxartes must be regarded as rivers also arrested on their course to their natural outlet in the Caspian.

The Oxus—the Jihun of the Arabs and Amu-darya of the Persians—is the Vak-shu of Hindu writers, a term itself most probably derived from the Ak-su, or "White Water," of the indigenous Kara-Kirghiz nomads. It collects all the drainage of the Great Pamir through two main head-streams, the Panja, or southern, rising in Lake Victoria (13,900 feet), discovered in 1838 by Wood, the Ak-su (Murgh-ab), or northern, flowing apparently from the still smaller lake Barkai Yasin (13,100 feet), and receiving the outflow of Lake Kara-kul above the junction. The united stream flows first westwards towards Balkh, before reaching which place it gradually trends round to the north-west, and retains this direction for the rest of its course to the south coast of the Aral Sea. During its upper course it receives the Surkh-ab from the Trans-Alai and Karateghin Mountains, besides numerous other feeders on both its banks from the Hindu-Kush and Bokhara highlands, but none lower down for a distance of 700 miles, or about half of its entire length. At Kilif its bed is narrowed to 350 yards by the advanced spurs of the Hissar hills, but in the plains it broadens to a mean breadth of 800 yards with a depth of 20 feet and a velocity of over 5 miles during the floods which last from May to October. At Pitnyak, about the head of the Khivan irrigation works, the discharge is about 125,000 cubic feet per second, which is nearly equal to that of the Nile. But fully half of this volume is absorbed in supplying the irrigation canals of the Khivan Oasis, by which over 4000 square miles of marvellously fertile alluvial land are kept under cultivation. The sedimentary matter yearly brought down exceeds 16,000,000 tons, some of which maintains the productiveness of the soil, while much of it forms shifting banks and bars in the Taldik and Yani-su, the two main channels through which the Oxus enters the Aral Sea.

These channels present the usual outlines of a delta. But the triangular space thus formed is not a true delta, as it consists, not of alluvial deposits, but of much older formations, through which the river has cut its way to the Aral. Separate little deltas, however, have been developed at the Taldik and Yani-su mouths, where vessels drawing over four feet are already excluded by the accumulating deposits. Nevertheless the Oxus has been regularly navigated by small craft since 1875, when the Petrovsky steamer, drawing about three feet six inches, forced its way up the Yani-su and so-called Kuvan-Jerma, or "New Cut," to Nukus, at the head of the false delta.

The tendency of the Oxus, like that of the great Siberian rivers, is to press continually on its right or east bank. The consequence of this tendency, which is due to the rotation of the earth round its axis from west to east, is that the stream has been gradually deflected from the Kungrad channel, navigable in the seventeenth century, but now dried up, eastwards to the Taldik channel, now slowly disappearing, and thence to the Yani-su, or "New River," which thus at present receives the main discharge into the Aral. But in former times a far greater deflection took place. For it is now ascertained beyond doubt that no less than twice during the historic period the

Oxus has oscillated between the Caspian and Aral Seas. In the time of Strabo it was a sort of eastern continuation of the Ḳura water-highway, affording a continuous trade route from Georgia across the Caspian and the Khwarezm desert, under the 39th parallel, to Charjui, and so on to Baktra (Balkh) under the Hindu-Kush. Its course across the desert in this direction seems to be still indicated by the Igdy and other wells dotted over the plains in a line with its former bed, which reached the Caspian not through the Atrak, as was at one time supposed, but most probably directly through the depression between the Great and Little Balkan hills.

Later on Edrisi and other early Arab and Túrki writers find the Oxus flowing, as at present, to the Aral. But in the fourteenth century it had been again diverted to the Caspian; this time, however, through a fresh bed, known as the Uzboi. This bed is supposed to have run from near Nukus westwards to the Sari-Kamish steppe lakes,[1] and thence southwards to the Igdy wells, and so on along the original bed between the Balkans to the Caspian, close to Mikhailovsk, the seaward terminus of the Trans-Caspian railway, opened for traffic by the Russians in 1881.

The hopes till recently entertained by the Russians of restoring the Oxus to the Uzboi channel have now been abandoned for the apparently more feasible project of connecting the Aral and Caspian through the bed of the steppe river Chagan, round the northern edge of the Ust Urt plateau. The Chagan is only 65 miles from the Aral, and probably at a somewhat lower level, while it seems to have formerly reached the Caspian at the bay of Chuché-bas, through the deep Arys valley. The Aral

[1] Guédroich, however, who carefully surveyed the Oxus delta and the Sari-Kamish lakes in 1880, found no indications that the Oxus ever flowed through the Sari-Kamish basin.

being 243 feet above the Caspian, if its waters can be brought to the Chagan, the connection will be established.

On the other hand, all these ambitious engineering projects are confronted with the fact that Turkestán, like most of Central Asia, comes within the area of desiccation that has been in progress since the remotest times. The Zarafshan and Murgh-ab, the two great former tributaries of the Middle Oxus on its right and left banks, are now both absorbed partly in the sands, partly in irrigation rills before reaching it. The Zarafshan, or "Gold Distributor," rises in the Alai range, at the foot of a stupendous glacier (9000 feet), which is still 30 miles long, but which Mushketoff recently traced for 33 miles below its present limits. In this magnificent ice-world the Zarafshan is fed by no less than thirteen secondary glaciers, beyond which it receives the outflow of the romantic Lake Iskander, at an elevation of 7000 feet above the sea. Thence it flows westwards down to the Bokhara plains, being first tapped by the Russians at Samarkand, and then diverted into a thousand irrigating canals by the Bokhariots. Thus 60 miles before reaching the Oxus, the "Gold Distributor" is completely used up in bringing nearly 2,250,000 acres under cultivation.

The Murgh-ab (or "water-fowl river") also, which flows from the Garjistán Mountains in North Afghanistán, runs dry in the sands, after supplying the numerous irrigating rills of the Merv Oasis. A similar fate overtakes the Tajand, as already described at p. 195.

The Jaxartes — Sihun of the Arabs, Sir-darya (or "head-river") of the Persians, twin sister of the Oxus— has its source in the very heart of the Tian-shan. Here the Narin, as its upper course is called, has its chief head-stream at the foot of the Petrov glacier in the Ak-shiirak hills, whence it flows at first westwards through the former Khanate of Khokand, and present Russian province of Ferghana. Beyond Khojend it turns abruptly south-westwards, thenceforth running parallel with the Oxus to the north-east end of Lake Aral, which it reaches after a course of about 2000 miles. Through the unexplored Kapchegai defile the Narin falls nearly 3300 feet in a space of 46 miles, and passes through two other romantic gorges before reaching the Ferghana plains, where it becomes the Sir

CROSSING THE SIR-DARYA.

below Namangan. Here much of its waters is absorbed in irrigating some of the most fertile tracts in Turkestán; but enough remains to send down a mean discharge of about 90,000 cubic feet per second to the Aral.

Like the Oxus, the Sir has frequently shifted its lower course. But it can never have reached the Caspian directly, as has been asserted, but only through the Oxus, with which it formerly communicated through the Yani-darya. This channel branches off from the main stream 7 miles below Peroffsky, but although occasionally flushed during the floods, it now never gets beyond Lake Kukcha-denghiz, 180 miles from the Sir, and 60 from the Oxus delta. Below Peroffsky the main stream enters the Aral through one large branch and several small channels, forming a shifting marshy delta, haunted by an astonishing number of wolves, wild boars, deer.

With a bar covered sometimes by scarcely three feet of water, the Sir cannot be regarded as a navigable river, although the light craft of the Russian flotilla have contrived to reach Peroffsky, and return to the Aral without getting embedded in the shifting sandbanks.

The Aral—that is, "The Inland Lake—" has a present area of perhaps 26,000 square miles. But it was formerly far more extensive, the water-marks on the Chink and many other indications showing that its level was at least 200 feet higher than at present. Yet it still stands 243 feet above the Caspian, and consequently 158 above the Mediterranean. It can scarcely be shown that in historic times the Aral has ever been deprived of both of its great influents, on which its existence entirely depends. It is abundantly evident that but for them the lake would disappear in a few years. Even as it is, a slow process of desiccation is steadily going on, by which its size has in a short time been reduced by 1400 square miles.[1]

Under the Chink the Aral is about 225 feet deep; but it shoals continually eastwards, until it becomes merely a flooded swamp along its east and south-east coasts. Hence the mean can scarcely be more than 40 feet, which would give a volume only eleven times greater than Lake Geneva, while exceeding that basin 116 times in area.

[1] It is remarkable that neither the Greeks nor Marco Polo make any mention of the Aral.

Although the Turkestán depression was at one time covered by a vast marine basin, the Aral Sea cannot be regarded, like the Caspian, as a relic of that period. In its brackish waters, no doubt, both a fresh and salt water fauna exist; but the former greatly predominates, while the latter is also common to the Caspian. Seals have been spoken of by Pallas and others, which might point to an independent connection with the Arctic Ocean, at least if of a different species from those of the Caspian. But later observers have made it evident that none of these cetacea are found in the Aral basin, which must on the whole be regarded as a sort of intermittent steppe reservoir, oscillating between the conditions of a lake and a mere morass, according to the vagaries of its great feeders, the Oxus and Sir.

A similar statement might almost apply to Lake Balkhash, if regard be exclusively paid to the fact that this basin also has in recent times diminished enormously in size. But other considerations show that it could at no time have been exhausted, so that it may be regarded as a true remnant of the old marine basin.

Even within the historic period this lake, which is at present 514 feet above the sea, spread out westwards to three or four times its actual area, while stretching nearly 250 miles towards the east, where it absorbed the now isolated Sassik, Jalanash, and Ala Lakes, south of the Tarbagatai range. Even still its area, which is somewhat fluctuating along its low-lying south coast, cannot be less than 8500 square miles, with a total length of 330 miles and a circumference of 880 miles. But the depth nowhere exceeds 56 feet, so that it has only twice the volume of Lake Geneva, which it surpasses in extent thirty-six times. Its waters, which are very brackish, abound in fish, although generally frozen over from December to April. Besides the Ili, it receives several affluents on its south coast, which is over 450 miles long Such a quantity of alluvia is brought down by these streams from the Zungarian Ala-tau that the lake seems to be slowly filling in.

Of the upland lakes by far the largest is Issik-kul, which lies at an elevation of 5300 feet in the heart of the Tian-shan between the Ala-tau Kungei and the Ala-tau Terskei. But notwithstanding its great altitude this lake belongs still to the Aralo-Caspian basin; for it drains intermittently through the Kutemaldi to the Chui, which formerly reached the Sir at some point below Peroffsky. Like the steppe lakes, the Issik-kul has greatly diminished in size, as is evident from the water-marks fully 200 feet above its present level. Yet it has still an area of about 2300 square miles. Its blue and somewhat brackish waters abound in fish, and are overlooked from the east by the Khan-tangri, the giant of the Tian-shan.

4. *Natural and Political Divisions: Uralsk — The Dáman-i-kóh — Khiva — Bokhara — Ferghana — The Kirghiz Steppes — Semirechinsk.*

As already remarked, the Russian administrative divisions are not based upon ethnical and physical considerations. Thus the region beyond the Caspian, which now includes all the Turkoman country together with the Merv oasis, is attached, under the name of the Trans-Caspian Territory, to the Government of Caucasia. Farther north the province of Uralsk, east of the Ural River, includes portions of the districts of Uralsk, Gurief, and Kalmikov, west of that river, while the Nikolayevsk district of the adjoining province of Turgai lies within the limits of the Ob basin. On the other hand, the Sari-Suisk district of the Akmolinsk Province, West Siberia, encroaches across the Ob water-parting on North Turkestán.

The eastern section of Turkestán between the Aral Sea and the Chinese frontier is divided politically between Russian territory and the still nominally independent Khanates of Bokhara and Khiva. The Russian portion, collectively called "Eastern Turkestán" in official documents, comprises several natural divisions, which to some extent coincide with the administrative provinces and circles. Thus the province of Semirechinsk, or "Land of the Seven Streams," lies mainly between the Zungarian and Trans-Ili Ala-tau, and Lake Balkhash; the province of Ferghana, the former Khanate of Khokand, comprises the upland valley of the Narin (Sir); while the circles of Zarafshan and Amu Daria embrace the tracts watered by those rivers mainly between Bokhara and Khiva.

Uralsk includes all the land between the Ural and the Obshchy Syrt, and between the Caspian and Aral north of the Ust Urt plateau. Here the West Siberian grassy steppes are continued southwards to within 100 miles of the Caspian, the intervening space being occupied by saline wastes. These are succeeded by argillaceous desert tracts, which extend eastwards between the grassy steppe and the north coast of the Aral Sea to the Kara-kum sands. The same argillaceous formation prevails throughout the Ust Urt plateau, and along the east coast of the Caspian southwards to the old bed of the Oxus.

But the Dáman-i-kóh,[1] or northern skirt of the Khorasán range, is fringed by an almost continuous strip of fertile tracts or rich pasture lands stretching between the hills and the desert from the Little Balkans to the Tajand valley. Here is the domain of the Akhal Tekke Turkomans, who were reduced in 1881 by the Russians.

[1] Dáman-i-kóh is sometimes used by English writers in the sense of a mountain range. But it simply means the skirt of any mountain range, and is especially applicable to such ranges as slope somewhat rapidly down to the plains.

This district, which is dominated by the Kerawul (5000 feet), the Kúh-Giffan (7770), the Kúh-Bughun (8000), and other lofty crests of the Kuren and Kopet ranges, now forms by far the most important division of the Russian Trans-Caspian territory. It is already traversed by a line of railway running from Mikhailovsk on the Caspian through Kizil-Arvat to Bami in the direction of Askabad and Sarakhs on the Tajand. Abundant fuel for the locomotives on this line is supplied by the rich naphtha wells on the coast of the Caspian and the islands close to the Mikhailovsk terminus. A horse tramway now also runs from the Bala-Ishem station for 40 miles to the centre of the oil-producing districts.

Khiva itself properly belongs to the same sandy formation, and is only rendered fertile by the alluvia of the Oxus combined with a fine system of irrigation developed on an extensive scale. Hence the difficulty of assigning any definite limits to the khanate, which is almost everywhere encircled by desert and steppe lands, except on the north and east, where it is bounded by the Aral Sea and Oxus. Even here the Lower Oxus flows for some distance through wastes of shifting sands, saline marshes, or morasses overgrown with sedge and reeds.

In the centre of this inhospitable region lies the Oasis of Khiva, which lies mainly between the towns of Pitnyak and Kungrad near the now dried-up bay of Aibughir at the south-west end of the Aral Sea. But the settled population is mainly grouped along the left bank of the Oxus as far north as Khojeili over against the advanced Russian fort of Nukus at the head of the false delta. Here there is a rapid flow, by which the intricate system of irrigation works is greatly facilitated. From Pitnyak the land is seen to be covered by a complete network of canals fed entirely by the Oxus. To guard against the danger of excessive floods the Khivans have constructed a dam or level all along the left bank of the river, and through the pipes laid across this dam the flow of water to the irrigating rills is regulated. Works have also been erected to raise the water to the higher grounds; yet notwithstanding the skill and labour spent on these works, they have failed to render productive much more than one-third of the whole deltaic region, which has a total area of some 5500 square miles.

The Oxus is also the main stream of the southern Khanate of Bokhara, which, like Khiva, is still suffered to enjoy a certain show of political independence. It has, however, been deprived of the eastern and more important section of the country, which now forms the Russian circle of the Zarafshan, and includes the city of Samarkand.

Northwards, also, the two khanates are separated by the intervening Russian territory, which here reaches the Oxus across the Batkak and Khalata sands. Between the Russian station of Petro-Alexandrovsk on the Oxus, and Ak-kamysh farther up, the country, though thinly peopled, is well cultivated. But from that point the road crosses a sandy steppe at a considerable elevation above the river, but here and there falling to the level of the stream, where it forms little projecting oases covered with "jidda" and other steppe plants called "tugai" by the natives.

Farther south the land is being continually encroached upon by the shifting sands of the steppe. From the fortress of Ustin to Kara-kul there stretches a sandy waste for a distance of 15 miles, which is dotted over with the ruins of abandoned buildings and other silent witnesses of better days.[1] Even within the last thirty years several flourishing towns in this region have been

[1] The vast field of ruins stretching from Shahri Gulgula through Shahri Saman to Termes on the Oxus was for the first time explored by M. Bonvalet during the summer of 1881.

swallowed up in the sands, drifting like ceaseless billows from the north. Year after year the sand-storms encroach upon the last cultivated plots, and the Kara-kul district between Bokhara and the Oxus already presents a hopelessly desolate aspect. Besides the diminution of moisture, the more immediate cause of this widespread ruin is the increasing absorption of the Zarafshan waters for irrigation purposes in the Samarkand district.

Between Bokhara and Samarkand the land is in some places admirably cultivated. The circle of Kermin in the Miankal valley even surpasses the district of Bokhara, and the country becomes more flourishing as we approach the Zarafshan district. Along the broad valley of this river there stretches an almost uninterrupted chain of townships between the capital and Katti-kurgan. But this flourishing tract is hemmed in by the Kizil-kum sands advancing southwards to the Zarafshan, and scarcely arrested by the isolated ridges of argillaceous schists and igneous rocks here rising to a height of about 1000 feet. Vambery describes this region as a boundless sandy ocean.

The Narin, or Upper Sir, separated from the Zarafshan valley by the Alai ranges, flows through the heart of the Khanate of Khokand, now the Russian province of Ferghana. This upland region is entirely enclosed on three sides by the lofty mountain barriers of the Western Tian-shan, and open only towards the west, where the Sir escapes down to the Turkestán lowlands. Here the western spurs of the Tian-shan fall in a series of terraces to the plains, and send down numerous mountain torrents, which, however, are mostly absorbed in irrigation works before reaching the Sir. To this abundant supply of water Ferghana is indebted for an exuberant fertility nowhere surpassed in Central Asia. But lower down nearly all trace of vegetation disappears, and the vast region stretching from Ferghana to Lake Aral, between the Sir and Oxus, is mostly occupied by the Kizil-kum sands. Nevertheless there are still several fertile tracts in the strip of land stretching from the right bank of the Sir to the Ak-sai and Kara-tau, the last spurs of the Tian-shan projecting north-westwards between the Kizil-kum and Ak-kum deserts. Here are even several considerable towns, including Tashkent, the present centre of Russian authority in Turkestán.

The region about the Lower Sir everywhere presents the appearance of a land that had once been under water. The saline and clayey soil towards the Aral Sea has been rendered arable by the skilful irrigation works of its former and present rulers. But under the cloudless summer skies the land beyond the reach of the irrigating channels has the aspect of a wilderness strewn with salt, and producing nothing beyond a few prickly plants. The reedy morasses about the delta are infested by dense clouds of mosquitoes, the plague of the Russian sailors navigating these waters.

North of the Aral and Caspian the sands merge everywhere in the grassy Kirghiz steppes, which, with a mean elevation of about 300 feet, consist mainly of vast rolling or gently sloping tracts. The steppe is, however, occasionally intersected by broad and deep furrows. But scarcely a tree or shrub is anywhere visible, the whole region presenting the aspect of a boundless sea, whose rolling billows have suddenly become solidified. The only relief to this monotonous picture are the Mugojar hills, a continuation of the Ural range, nowhere exceeding 1000 feet in height. The hilly portion of the steppe consists everywhere of feldspar and porphyry, with which are often associated lead, copper, silver, and occasionally even gold. Between Aral and Lake Balkhash, the steppe is in many places strewn with lakes and tarns, often strung together like pearls on a string, but everywhere showing the same tendency as the Aral and Balkhash themselves to disappear. All have long been closed basins, nor do the Sari, Chui, or any other of the steppe rivers, now ever reach either the Aral or the Sir.

Beyond the water-parting of the Chui and Ili begins the Semirechinski-Krai, or "Land of the Seven Streams," consisting of the steppe which here stretches at a mean elevation of about 1000 feet between Balkhash and the Zungarian Ala-tau north and south, and between the Lower Ili and Lepsa rivers west and east. Although thus partly severed from the Central Asiatic highlands, this region is connected with them through the deep valley of the Ili leading to Kulja, recently restored by the Russians to China.

The seven streams whence the land derives its name are the Lepsa with the Baskan, the Ak-su with the Sarkan, the Biyen and the Karatal with the Kok-su. The Lepsa, Ili, and Karatal alone reach Balkhash throughout the year, all the others either losing themselves in the sands, or discharging their waters into the lake only during the floods. They all rise in the snowy Ala-tau, and flow through fertile upland valleys before reaching the open plains. These plains are dotted over with brackish lagoons, and their vegetation resembles that of the Aralo-Caspian depression, while the cultivated and well-watered upland tracts recall the lowlands of West Siberia and the East of Europe.

5. *Climate: The "Fever Wind."*

Notwithstanding the extraordinary difference in the relief of the land, the climate of the Aralo-Caspian basin is everywhere characterised by a remarkable uniformity. It is distinctly continental in its main features, intense heat being followed by equally intense cold, while great dryness prevails over the whole area. This general uniformity is largely due to the low elevation of the Aralo-Caspian and Ob water-parting, which offers no obstacle to the full play of the northern winds from the Arctic Ocean across the Turkestán depression. Thus the difference in elevation between this depression and the Tian-shan and Pamir uplands is to a great extent neutralised in winter, when the glass falls to 30° or 40° below freezing point in the Kizil-kum sands 300 feet, as well as on the Pamir plateau 14,000 feet, above the sea. The chief difference between the uplands and lowlands is the excessive rarefaction of the atmosphere on the great tablelands, due to their enormous elevation above sea-level. A greater quantity of moisture also falls on the highlands, the rain-bearing clouds from the south-west being first arrested by the Pamir and Tian-shan uplands. But even here the rainfall is much slighter than in the European highlands; while at times whole years pass without any rain falling on the Kara-kum and Kizil-kum sands. Such are altogether the climatic conditions that most of the Aralo-Caspian basin would be uninhabitable but for the moisture deposited mostly in the form of snow on the elevated lands. From this reserve the great rivers are fed, which in their turn supply the irrigation works on which the Bokhara, Khiva, Merv, and other oases, are dependent.

Along the Lower Oxus, which is usually frozen for four or five weeks only, summer begins in April and lasts occasionally into November. The long hot and dry season is rendered still more oppressive by the dust-storms prevalent throughout the lowlands. But more dreaded still is the "tebbad," or "fever-wind," to which the Kizil-kum and other desert tracts are exposed. At its approach the camels of the caravan utter loud moaning sounds, and cowering to the ground stretch their long necks flat on the parched land, or seek to bury their heads in the hot sands.

Behind them crouch the terrified guides and travellers, while over them the pestiferous blast sweeps with a dull soughing. The whole caravan is soon covered with a layer of hot sand, which falls like a shower of fiery sparks.

6. *Flora and Fauna: The Saxaul—Mosquitoes and Locusts—The Turkoman horse.*

Vegetation is represented in the wilderness chiefly by the saxaul, the jidda or wild olive, the poplar, and other hardy or prickly plants, which have invaded this domain from all quarters since the subsidence of the waters. As a rule, the flora of the Iranian tableland advancing northwards has prevailed over that of Siberia, and "it is interesting to observe how all those plants gradually adapt themselves to the changed conditions of soil and climate in the steppe. To resist the wind they acquire a more pliant stem, or present a smaller surface to its fury by dropping their foliage. To diminish the evaporation their bark becomes a veritable carapace, and their pith becomes mingled with saline substances. They clothe themselves with hairs and thorns, distilling gums and oils, whereby the evaporation is still further reduced. Thus are able to flourish far from running waters such plants as the saksaul, which, though perfectly leafless, produces both flowers and fruits. So close is its grain that it sinks in water, and emits sparks when struck with the axe."[1]

In the oases the cultivated plants, both cereals and fruits, are noted especially for their great abundance and excellent quality. In Khiva wheat will yield sixty, rice seventy, and the *jugara* as much as three hundredfold. This latter grain takes the place of oats, and its stalk that of hay for horses and cattle. Barley, lentils, and

[1] *E. Reclus*, vi. p. 196.

pease are also cultivated, besides cotton, hemp, *kunshut* (an oil-yielding fruit), madder, flax, and tobacco. The lack of grazing grounds is here obviated by the lucerne clover, which is mown three times and yields excellent fodder. But the special glory of this oasis are its fruits, which are remarkable for their fine flavour. Here flourish choice apples, pears, plums, apricots, peaches, the grape, the pomegranate, and, above all, the melon. Of trees, the poplar, naruan, and elm are grown for their timber, and the mulberry for the silkworm.

The mulberry, flax, and maize form the staple produce of Ferghana, which also grows fine crops of wheat, rice, sorghum, maize, cotton, and tobacco. In Semirechinsk fertile and well-watered upland valleys are succeeded by grassy steppes stretching away to the low-lying swampy shores of Balkhash. Higher up a splendid wooded zone clothes the slopes of the Zungarian Ala-tau, between 4500 and 8500 feet. In the forests of the Central Tian-shan the prevailing trees are the mountain ash and the spruce (*Picea Schrenkiana*), which are now supplanting the apple and apricot. The spruce "attains a height of 70 to 80 feet, with a thickness of stem 2, 3, and often 4 feet in diameter. It grows very much in the sugar-loaf shape, its thick branches hardly projecting from the general mass, so that the whole tree has the appearance of having been cropped by a barber" (Morgan's *Prejevalsky's Lob-nor*, p. 40). The spruce grows as far as 8000 feet and upwards above sea-level. These varied advantages, combined with a healthy climate, have attracted numerous Russian settlers to Semirechinsk, which has become a chief centre of Slav culture in Central Asia. The elevated lands with their abundance of water and timber supply the necessary essentials for the development of social culture. The low, flat steppes, with their arid wastes, here, as elsewhere, arrest the progress of civilisation. These waterless and treeless tracts admit of nothing but a nomad existence.

In the Turkestán lowlands a characteristic feature of animal life are the scorpions, lizards, snakes, and other reptiles, with which all the fissures in the ground are alive. But the special plague of these regions are the mosquitoes and locust. "Mosquitoes are serious evils in many other parts of the world, and stories have been told of seamen driven to jump overboard, and so to commit suicide on this account, in the Rangoon River. But it may be doubted whether any more exquisite torture can be suffered than that inflicted by the mosquitoes of the Lower Amu" (*Major H. Wood*). The same traveller, speaking of the prodigious quantities of locusts which swept over the Khivan Oasis in July 1874, remarks that "one of such clouds was estimated to measure 15 miles in length by 2 miles in breadth, and to have a depth of half a mile. . . . It might be inferred that the Khivan Oasis must be exposed to great danger from this plague; but in practice the large number of small birds in the planted groves of the khanate seem to afford a sufficient safeguard against their ravages."

Large beasts of prey, such as the wild boar, tiger, ounce, wolf, haunt the thickets of the swampy Oxus and Aral deltas. But the open desert is frequented only by swift gregarious animals, like the wild ass and gazelle.

The camel, horse, sheep, and cattle, are everywhere the prevailing domestic animals. In Russian Turkestán camels are not so numerous as might be supposed, scarcely amounting to 400,000, as compared with over 1,600,000 horses, 1,160,000 cattle, 1,350,000 sheep. The Kirghiz horses are a hardy, active breed, which traverse distances of 40 to 50 miles at a stretch. But a far finer animal is the Turkoman horse, which possesses many of the best points of the Arab and English with some excellent qualities peculiar to itself. "Well do these noble animals deserve all the care that is lavished on them, for in courage, speed, and endurance combined, they stand at the head of the equine race. It is probable that the race dates back, like our own thoroughbred, to the Arab. But the race is now distinct; and, besides being much larger, they far excel the Arabs both in speed and endurance. . . . In appearance they more nearly resemble the English race-horse than any other type, and average about the same height."[1]

7. *Inhabitants: Table of the Turkestán Races.—The Usbegs—Kara-Kalpaks—Kara-Kirghiz—Kirghiz-Kazaks—Turkomans—Tajiks—Sarts—Galchas—Russians.*

The Aralo-Caspian basin is commonly supposed to be the exclusive home of the Túrki race, from whom this region takes the name of Turkestán, or "Land of the Túrk." But the statement is true only of the unarable but still inhabitable grassy upland and lowland steppes, whose first occupants seem to have been the nomad tribes of Túrki stock, by whom they are still mainly inhabited. On the other hand, the arable tracts, especially in Khiva, Bokhara, and Ferghana, have apparently from prehistoric times been the joint home of men of Túrki and Iranian blood. Here an incessant intermingling of the two races has been going on for ages, resulting in a profound modification of both types, now represented by every shade intermediate between the two extremes. A third element, entirely distinct from the Túrki, and allied to but not identical with the Iranian, is found in almost exclusive possession of the productive upland valleys of Ferghana, the Zarafshan, and the Oxus. To these high-

[1] *Clouds in the East*, p. 214.

landers Ch. de Ujfalvy has given the collective name of Galcha, and these Galchas, whose true position seems to be intermediate between the Iranic and Indic branches of the Aryan family, are obviously allied to the Wakhis, Badakhshis, Siah-Posh Kafirs, Chaganis, and other highland races holding the upland valleys on both sides of the Hindu-Kush.

To the primitive Galcha, Iranian, and Túrki stocks are therefore reducible all the varieties of mankind from time immemorial in possession of the Aralo-Caspian basin, as in the subjoined scheme:—

I. Túrki Stock.

Usbegs	Kungrad	Bokhara, Ferghana, Khiva	2,000,000
	Naiman		
	Kipchak		
	Jalair		
	Andijani		
Kara-Kalpaks	Baymakli	S. and S.E. shores Aral Sea mainly	50,000
	Khandelki		
	Achamayli		
	Ingakli		
	Shaku		
	Ontoturuk		
Kirghiz-Kazaks	Great Horde (Ulu-Yuz)	Steppes between Lake Balkhash and Lower Volga	450,000
	Middle Horde (Urta-Yuz)		1,100,000
	Little Horde (Kachi-Yuz)		1,000,000
	Inner Horde (Bukeyevskaya)		200,000
Kara-Kirghiz (Buruts)	Right Section ("On")	Tian-shan and Pamir	300,000
	Left Section ("Sol")		100,000
Turkomans	Tekke	Ust Urt, Khwarezm, Dáman-i-kóh, left bank Middle Oxus	?600,000
	Goklan		
	Yomud		
	Sarik		
	Salor		
	Kara		
	Ali-Eli		
	Ersari		
	Chaudor		

II. Iranic Stock.

Tajiks	Khiva, Bokhara, Ferghana	?1,000,000
"Sarts"		
Persians		

III. Galcha Stock.

Maghians . Kshtuts . Falghars . Machas . Fangs . Yagnaubs . Karateghins .	Ferghana, Zarafshan, and Karateghin highlands, Upper Oxus valleys	? 250,000

Of all the Túrki peoples of Central Asia the Usbegs are by far the most civilised. The great majority in the two khanates have long abandoned the nomad life, and are now the chief agricultural element in Khiva and Bokhara. The term itself is rather political than ethnical, being the collective name of the numerous Mongolo-Tatar tribes, who became the dominant people in this region after the dissolution of Jenghiz Khan's empire. But although the ruling race, the Usbegs are intellectually inferior to the conquered Tajiks, with whom they now live in social harmony, and through alliances with whom they have become largely assimilated in appearance to the Iranian type. Many are settled in all the large towns, where they are partly occupied with trade, and furnish the principal contingent to the army. Besides the Osmanli, the Usbegs are the only Túrki people who possess a written language and a literature. Sultan Baber, founder of the so-called Moghul Empire in India, who belonged to this race, composed his well-known Memoirs in the Jaghatai, which is still the standard literary language of all the Central Asiatic Túrki peoples. Like the Tajiks, the Usbegs are all Sunnis.

The Kara-Kalpaks—that is, "Black Caps"—formerly a widespread branch of the Túrki family, now occupy a restricted area round the shores of Lake Aral between the mouths of the Oxus and the Yani-darya. They are a harmless, but feeble and somewhat sluggish race, evidently in process of extinction or absorption by their former oppressors, the Usbegs of Khiva.

The Kara-Kirghiz, or "Black Kirghiz," and the Kazaks, or Kirghiz-Kazaks, represent respectively the highland and lowland nomad elements all along the northern and eastern border-lands of the Aralo-Caspian basin. The Kazaks, who are by far the most numerous

USBEG WOMAN.

of the two, have never accepted the name of Kirghiz, which has been imposed upon them by the Russians to distinguish them from their own Kossaks.[1] Hence it would be more correct to speak of a Kazak than of a

[1] The unaccented *o* in Russian being pronounced like *a*, Kossak and Kazak have very much the same sound in the mouth of a Russian.

Kirghiz steppe, the Kazaks being here the exclusive element, whereas the Kara-Kirghiz dwell in the Tian-shan highlands and Great Pamir. At the same time there is no substantial difference between the two peoples.

The Kara-Kirghiz, who are the "Buruts" of the Chinese and Kalmaks, live partly in Zungaria and Turkestán, partly in the Western Altai, in the hilly

KIRGHIZ.

districts about the source of the Sir and its tributaries, in the Alexander range, in the highlands about Issik-kul and southwards to the sources of the Oxus on the Pamir. They speak an almost pure Túrki dialect, and their two great sections, *On* or "Right" and *Sol* or "Left," are again subdivided into numerous tribes and septs. North of the Sir, their grazing grounds are limited northwards by the Kazaks, while stretching southwards to the Hindu-

Kush. Their camping-grounds in the Tian-shan are here and there overlapped by the lands of the warlike Galchas. The northern Kara-Kirghiz have no common bond of union, nor any kind of political organisation. Even the lesser tribes are often split up into independent hordes still living in a state of constant feud amongst themselves. Thus all their energies have been wasted in internecine strife, or else in chronic hostilities with the Kazaks. Hence in spite of their personal courage they have at various periods been easily subdued by the Chinese, or by the Khans of Khokand. In recent years all the right section and many of the left have accepted Russian supremacy. Some years ago a small number penetrated to the Sarikia pasturages on the Karatash River near Sendshu, the southernmost point ever reached by the Kara-Kirghiz.

The Kazaks occupy a somewhat intermediate position between the Túrki and Mongolian races, possessing many physical traits in common with the latter, while still speaking a pure Túrki dialect. Their four hordes, all now subject to Russia, occupy a vast domain stretching from the Lower Volga to Zungaria, and from beyond the Aralo-Caspian and Ob water-parting southwards to the Aral Sea.

The Kara-Kirghiz is of a sullen, rude, and fierce temperament, but he is more straightforward and good-natured than the Kazak. Both are Muhammadans in little more than the name, without mollahs, mosques, or fanaticism, their whole religion being limited to a few simple rites, strongly tinged with the traditions of the old Shamanist cult. The Kirghiz cultivate more land than the Kazaks, but both are essentially stock-breeders, living mainly on the produce of their herds. Their chief drink is *kumiss*, fermented mare's milk, which is preserved in skins, and largely consumed throughout the spring, summer, and autumn. Kumiss is very wholesome, and a specific against all consumptive diseases.[1]

The monotony of nomad existence was formerly relieved by tribal warfare, and the so-called "barantas" or marauding expeditions generally directed against the encampments of their neighbours. The attacks were usually made towards dawn, when man and dog alike, wearied with the night-watch, were buried in sleep.

The Kazaks have all been gradually induced to acknowledge the sovereignty of the Czar. The Russian Government has been the more anxious to effect their submission, that all the caravan tracks between the Caspian and Altai highlands traverse their domain. Towards the south-east a few of the Kara-Kirghiz tribes still maintain their independence beyond Lakes Balkhash and Zaisan, within the Chinese frontier. But some of these have of late years shown a tendency to migrate westwards into Russian territory.

The Turkoman nomads have now also tendered their submission to the "White Czar." Even those of the Merv Oasis, struck with the irresistible power of the Russian arms, and flattered by the courteous reception accorded in the spring of 1881 to their chief, Tikma Sirdar, by the St. Petersburg authorities, were at last induced to accept the proffered protectorate of Russia in the year 1882. The example of the Merv tribes was soon followed by those of Yuletan, 35 miles farther south, and in 1885 those of Penjdeh, the last independent section of the race, were incorporated in the Russian empire.

Throughout the historic period the Turkomans or

[1] In "Koumiss . . . and its uses in the treatment and cure of pulmonary consumption" (*Blackwood*, 1881), Dr. G. L. Carrick fully establishes its claim to be regarded as a sovereign remedy for all affections of the chest and lungs.

Turkmenians seem to have been a plundering nomad race, never at any time united under an organised political system. They are divided into "Khalks" or tribes, each comprising several "tayfe" or hordes, who are again grouped in a number of "tir" or septs. Of all the Khalks the most friendly and civilised are the Goklans, who have long been settled in the Persian province of

TURKOMAN ENCAMPMENT.

Astrabád. Although recognising no general leader, except, perhaps, for a short time in great emergencies, the Turkomans do not live in a normal state of anarchy. Offences against their own unwritten code are even rarer amongst them than amongst other Muhammadan peoples. Everything is regulated by the all-powerful "dab" or "custom," religion exercising but a slight influence.

The various tribes formerly lived in mutual hostility to each other, showing little fear of the Persian Government, but great respect for the Russian power. To their individual tribes they remain true to the last, and even little children, five or six years old, know exactly the tayfe and tir to which they belong. The Turkomans are distinguished from other Asiatics by a bold, penetrating glance, developed by the dangers surrounding the "Alamans," or marauding excursions, to which they have been addicted from the remotest times. These alamans were preconcerted affairs, in which every precaution was taken against failure. The attack usually took place about midnight or at sunrise, and was generally successful. The Persian caravans were constantly taken by surprise, all who showed any resistance being cut down, and the rest carried off into slavery. But since the predominance of Russia in Turkestán the Khivan and Bokhara slave-markets have been closed. This was the first blow given to the Turkoman power, which was almost reduced to national bankruptcy by the stoppage of a traffic on which its very existence depended. Then came the massacre of the Yomud Turkomans, followed in 1881 by the crushing defeat of the Akhal Tekkes at Geok tepe.

The Tajiks are to be distinguished both from the Sarts and Persians. They are the original Iranian element settled in all the arable lands throughout Turkestán from the remotest times. It is probable that the diffusion of the Iranian race in this region, regarded as the peculiar home of the Túrki peoples, was brought about by the extension of the old Persian empire through Margiana (Merv) and Baktriana (Balkh) to Sogdiana (Bokhara). The Tajiks, direct descendants of those early Iranian settlers, differ in many respects from the Persians, who have in comparatively recent times settled in most of the Turkestán towns and oases. Many of them are descendants of those carried into captivity by the Turkomans, and sold in the Bokhara and Khivan slave-markets. They are everywhere the most industrious and intelligent section of the community, and are occupied chiefly in the cultivation of the land.

Sart is a term which has given much trouble to ethnologists, who have sought for a distinct race in a social distinction. The word seems to have originally meant "dealer" or "trader," and never had any ethnical value at all. It is applied to the settled in opposition to the nomad element in Turkestán, irrespective of race or nationality.

The Galchas, by De Ujfalvy at first described as "Tajik highlanders," but afterwards by him separated from that connection, present some obscure problems which still await solution. Most of them have been assimilated in speech to the Tajiks and Persians. But the Yagnaubs speak a distinct Aryan language, which will probably be found to be allied to the Wakhi and Siah-Posh of the Hindu-Kush.

The Russian Slavs, who are the latest intruders into this region, threaten ultimately to absorb all the others, at least in the settled and cultivated districts. They already form a continuous cordon stretching from the Urals round the northern and north-eastern borders of Turkestán beyond Barnaul and Semipalatinsk, and have also founded agricultural settlements in Semirechinsk, the Issik-kul uplands, and Ferghana. In the Aralo-Caspian basin their political status and higher culture give them a preponderance out of all proportion with their numbers, and this preponderance must go on increasing according as the Russian authority becomes more consolidated.

8. *Topography : Mikhailovsk — Kala Kaushid Khan (Merv) — Khiva — Urgenj — Bokhara — Samarkand — Tashkent — Khokand — Namangan — Charjui — Verniy.*

A region consisting mainly of sandy wastes, grassy steppes, bleak upland plateaux and highlands, cannot contain many large centres of population. In an area of nearly a million square miles there are only four towns with over 50,000 inhabitants, and these are all concentrated in the eastern districts of Russian Turkestán and Bokhara. Elsewhere, except in the Khivan Oasis, there is not a single place attaining to the dignity of a town. A few forts and military stations are scattered over the Russian circles of the Sir and Oxus; but the monotony of the Kirghiz steppes and Western Turkestán deserts is unrelieved by a solitary hamlet. But in the recently-organised Dáman-i-kóh of the Akhal Tekke Turkomans, a few places along the new line of railway have already acquired some importance. On the Caspian the military station of Krasnovodsk has superseded Chikislar near the mouth of the Atrak, and will itself soon be superseded by Mikhailovsk, the neighbouring seaward terminus of the Trans-Caspian railway. Beyond this point the chief stations are Kizil-Arvat, Bami, and Askabad, which last, nearly midway between Krasnovodsk and Herat, seems already to have superseded Bami as the political capital of the Trans-Caspian territory.

East of Askabad the sandy wastes beyond the Tajand valley are broken only by the Merv Oasis, where the famous historical city of Merv has ceased to exist since its destruction in 1784 by the Amir of Bokhara. It is now a mere collection of mud huts, and even as a strategical point has been replaced by the neighbouring fort of Kala Kaushid Khan, "which is protected by the Murgh-ab

River on two sides, being built in the loop of the river. It is about 2½ miles long and 1¼ wide. The Tekke have most wonderful confidence in the strength of the place, which will contain, they say, 50,000 alajaks, or Turkoman tents. It was commenced in 1860, and the Tekke have worked at it by fits and starts ever since. When the Persians now speak of Maur, or Merv, they mean Kala Kaushid Khan. The Turkomans themselves never speak of Maur as a town; when they use the term at all they mean the district where Merv was formerly situated. . . . The portion of the country fit for cultivation is about 90 miles long, and extends to about 11 miles on the east side of the river. The ground is very fertile, and produces melons and water-melons in plenty and of great excellence. . . . A canal which formerly existed, and which led from the Tajand River near Sarakhs to Kacha-kum (within 20 miles of the Merv Oasis), could be easily reconstructed. In 1860 the Persian general employed his army in damming up the Tajand and turning it into the bed of the old canal." [1]

Khiva, capital of the khanate to which it gives its name, lies near the head of the irrigation works at some distance from the left bank of the Oxus. It is intersected by two artificial canals, and surrounded by a mud wall four miles in circumference and 10 feet high. The palace of the Khan in the interior of the city, besides the houses of the officials and some religious buildings, are all protected in the same way, the whole forming a sort of inner town and citadel with three gates, and defended by twenty guns. The outer town contains a large bazaar and the summer palace of the Khan; but the whole place has a population (1874) of scarcely 5000.

By far the largest place in Khiva is Urgenj, near the

[1] Col. Stewart in *Proceedings of the Royal Geographical Society*, eptember 1881.

capital, a fortified town with a wall mounting several guns. But these and other little strongholds are completely overshadowed by the Russian fortress of Nukus, conveniently erected at the head of the delta on the Russian side of the river. From Nukus and Petro-Alexandrovsk, facing the capital, the whole khanate could be at any time occupied in four-and-twenty hours.

Near the point where the Zarafshan runs dry in the ever-encroaching sands, stands Bokhara, capital of the khanate of like name. But sufficient water still remains to supply the magnificent gardens, cotton, jugara, and other plantations, for which the surrounding district has long been famous. The great feature of the city is its well-stocked bazaar, whose vast size is a constant surprise to the stranger. Here all the shops and caravansarais are gorged with Russian and "Kábuli"—that is, English and Indian—wares. No less astonishing is the number of colleges, schools, mosques, graveyards, and "saints" of all orders. Yet Bokhara "the Noble" has fallen far below its former greatness, and in the 50 years between 1830 and 1880 its population has been reduced from 140,000 to 70,000, of whom about two-thirds are Tajiks.

The cause of this rapid decay is the gradual loss of water from the Upper Zarafshan, which is being drawn off in ever-increasing quantities by the Russians for the irrigation works of Samarkand. This renowned metropolis of Timur lies near the left bank of the river due east of Bokhara, 2154 feet above the sea, on a western spur of the Alai range, which here merges gradually in the Turkestán lowlands. The plains terminate east of the city, which, notwithstanding the splendour of its ancient buildings, now differs little from other Central Asiatic towns. Here we have the same belt of blooming gardens and orchards encircling the same confused mass of narrow gloomy streets, mud hovels, and crumbling walls, the whole pervaded by the same oppressive stillness, broken only in the vicinity of the great bazaar. Yet here are some of the grandest monuments of Islám, dating mostly from the time of Timur, and including several magnificent colleges, and the Shah-Zindeh, the most sumptuous mosque in Central Asia. The old palace of the Emir has been converted into a hospital by the Russians, whose administrative and military officers occupy a large portion of the ancient citadel. The same cause that is hastening the doom of Bokhara is furthering the prosperity of Samarkand, the population of which has increased between 1834 and 1880 from 8000 to upwards of 30,000.

It is probable that to this place will ultimately be removed the headquarters of the administration in Russian Turkestán, which are at present centred in Tashkent.[1] Next to Tiflis, Tashkent is the largest city in Asiatic Russia, and its population, which rose from 86,000 in 1874 to 100,000 in 1880, already nearly equals that of the Georgian capital. But beyond its size it presents few points of interest. Like most of the large towns in the Sir valley, it lies at some distance from the main stream, on the Chirchik, a small tributary flowing from the Aksai-tagh, and in a healthy district 1400 feet above the sea. In proportion to its population, Tashkent covers a very large space, being nearly 8 miles long and 4 broad. The Russian quarter has already a population of over 5000; but the great bulk of the inhabitants are Tajiks, who also form the chief element in Khokand, Namangan, Andijan, Marghilan, and the other large towns in Ferghana. Khokand, capital of the former khanate of like name, scarcely deserves the title of the

[1] The natives always say *Tashkand*, but the Russian pronunciation *Tashkent* seems to have gradually established itself in the West.

"Delightful," which has been conferred on it; for goître is here so prevalent that the Russians were compelled to transfer the centre of administration to Tashkent, lower down the Sir valley. Yet its bazaar is still one of the best stocked in Central Asia, and does a considerable trade in local produce and European wares. On the right or opposite side of the Sir, and near the Narin confluence, lies Namangan, the next largest place in Ferghana. It occupies the centre of a rich oasis at some distance from the river, and is the chief mart for the flocks of the Kara-Kirghiz nomads from the surrounding upland steppes. In the neighbourhood are some rich naphtha wells and coal-beds.

Besides Nukus and Petro-Alexandrovsk at the northern and southern extremities of the Khivan Oasis, Charjui higher up the Oxus occupies a position of great importance at the point where the river is crossed by the caravan route from Bokhara to the Merv Oasis.

Verniy (Vernöe), the old Almati, although a Russian town only since 1867, has already acquired importance as the capital of Semirechinsk. It lies near the southern base of the Trans-Ilian Ala-tau, 2430 feet above the sea, nearly midway between Lake Issik-kul and the left bank of the Ili. Although the centre of the Russian agricultural settlements in this region, its trade is still mostly in the hands of the Chinese dealers from Kulja, towards which it is the most advanced Russian outpost. Verniy is the mart for the Russian copper ware, which is distributed from this point over Central Asia and Mongolia.

9. *Highways of Communication.*

From Orenburg, the present terminus of the Russian railway system towards Central Asia, the northern postal and trade route passes through Orsk and Turgai across the Kirghiz steppes to Verniy, and thence up the Ili valley across the Chinese frontier to Kulja. The former Kossak stanitzas, established to curb the Kirghiz nomads, have now become so many postal stations along this line, which will some day be replaced by the " Great Northern Asiatic Railway."

At Orsk the south-eastern postal route branches off across the Kirghiz steppe through Kara-Bulak and Irghîz to Kasalinsk, thence following the Sir valley through Peroffsky and Yasi to Tashkent. " The post-stations, which have been built along the route crossing these desolate regions, afford excellent accommodation for travellers, and wells have been dug along the whole distance, though it is true that the water in many of them is not of good quality " (*Major Herbert Wood*).

From Tashkent the great historical military and trade route leads by Chinaz across the Sir, through the Jilanuti defile, over the Kara-tau to Samarkand, and thence down the Zarafshan valley to Bokhara. West of the Jilanuti pass stands the so-called " Gate of Tamarlane," a pyramidal slaty rock covered with Persian inscriptions, and marking the site of many a fierce struggle for the possession of the Zarafshan and Sir valleys.

Two parallel routes run from Samarkand and Bokhara through Karshi and Koja Sali across the Oxus southwards to Balkh, while a third leads from Bokhara across the Oxus at Charjui to Merv, and up the Murgh-ab to Herat. An alternative line runs from Merv by Sarakhs on the Persian frontier, up the Tajand valley to Herat.

From Khiva several tracks radiate across the Kwarezm desert southwards. But there appears to be only one recognised highway, which follows the right bank of the Oxus to Charjui for Bokhara. The desert tracks are :—
1. The Orta Yolu, nearly by the Usboi, or old bed of the Oxus, between the Great and Little Balkans, to the south-

east corner of the Caspian; 2. The Tekke Yolu, west of and parallel to the previous, to Kizil-Arvat and the Atrak valley; 3. The Hazaresp, through the Dara-gez district to Kuchan for Mashhad; 4. Direct to Merv for Sarakhs, Herat, and Mashhad.

From Ferghana to Kashgaria the route leads through Osh to Gulcha (4140 feet) and Sufi Kurgan (40° N. lat., 73° 30′ E. long), whence two roads run over the passes of Terek (12,500 feet) and Shart (13,000), which again unite at the outpost of Irkeshtam. Two other roads also run from Sufi Kurgan over the passes of Archat (11,500) and Taldyk (11,800), which unite on the Alai, and lead thence through the Khizil-art gorge, and passes (14,000) over the Trans-Alai to the Pamir. The Pamir itself is crossed in all directions by easy tracks, some of which would present no difficulties to the passage of large armies and artillery. The Alai passes, formerly supposed to have been closed from September till late in spring, were shown by Severtzof in 1877 to be free of snow at heights of 13,000 feet till the end of October.

For the distances of the main routes between Turkestán and the Iranian plateau, see p. 441.

Since the reduction of the Akhal Tekke Turkomans in 1881 the locomotive has penetrated into Central Asia. The Trans-Caspian line starting from Mikhailovsk runs along the Dáman-i-kóh south-eastwards to Kizil-Arvat and Bami, and the works are in progress to Askabad, the future capital of the Russian Trans-Caspian territory. The first goods train laden with Russian merchandise from Moscow, steamed into the Kizil-Arvat Station on 3d October 1881. But the traffic here is slight, and the Persian and Indian trade routes cannot be seriously affected by this line until it is opened to Sarakhs for Mashhad, and up the Tajand valley to Herat for Afghanistán. Meantime its strategical importance in view of future complications along the Iranian frontier is too obvious to call for any comment.

10. *Administration : Resources—Products—Trade.*

The administration of Russian Turkestán is of a purely military character. The Governor-General, or Yarin-padishah—that is, "Half King," as the natives call him—has his headquarters in Tashkent. Appointed by the Czar, to the Czar alone he is responsible for the exercise of the supreme civil and military functions centred in his person. He even enjoys the privilege of entering into diplomatic relations with the neighbouring States, an arrangement by which negotiations entered into in Central Asia may be confirmed or revoked by the Emperor according to circumstances. His jurisdiction embraces the Siberian provinces of Turgai, Akmolinsk, Semi-palatinsk, and Semirechinsk; but the extreme west and south comprise the Trans-Caspian territory attached to the Government of Caucasia.

The governors of the various provinces and circles are appointed by the Minister of War, and assisted by provincial councils chosen by the Governor-General. These governors are directly responsible for the revenue and maintenance of order in their several districts. All religions are tolerated, and the tribal usages of the Kirghiz and other nomads respected as far as consistent with the general interests of the State. The towns appoint their own magistrates, who, however, may be removed at the pleasure of the Governor-General.

Public instruction does not seem to have been as yet undertaken by the new masters of the land. The only education that receives any encouragement is the harmless reading of the Koran, as taught in the Medresseh or colleges attached to the mosques. In the whole of Turkestán there are scarcely 5000 Muhammadan children receiving regular instruction.

The chief source of revenue is the land tax, and the chief source of expenditure the army, which averages about 30,000 men. As all the supplies have to be brought from Russia, this item alone absorbs the whole of the revenue, so that there is a normal deficit amounting in some years to £1,500,000, or over £2,000,000.

Of the products of this region perhaps the most important next to live stock is cotton, of which there are two varieties. It has already become indispensable to the Russian manufacturers, and Vambery declares that it is of better quality than the Indian, Persian, or Egyptian, if not quite equal to the American. The best description is grown in Khiva, which is the chief area of the cotton cultivation in Central Asia. Sericulture, originally introduced by the Chinese, has also been long established in Turkestán, and especially in the eastern districts. Important articles of export are, further, the black lamb's wool, known in Europe as "Astrakhan;" the Turkoman horses now supplying splendid remounts for the Russian cavalry; wool, hides, dyes, cereals, and fruits.

The local retail trade has of late years been greatly developed, and the remotest nomad hamlets are now supplied with all kinds of wares from the bazaars of the central marts. Hence the importation of Russian and English goods has largely increased, somewhat to the detriment of the native industries. The Russians have the advantage over their English rivals in being first in the field, and more carefully studying the tastes of the Eastern nations. Their goods are also exempt from the heavy duties imposed by the Russian Government on English and Indian wares. The rapid organisation of the Trans-Caspian Territory will now enable them to send their manufactures direct to the Mashhad and Herat bazaars.

11. *Statistics.*

Areas and Populations.

	Area in sq. miles.	Pop. 1884.
Sir-darya (Province)	182,000	1,110,000
Amu-darya (Circle)	41,400	222,000
Zarafshan (Circle)	9,800	352,000
Ferghana (Province)	84,300	808,000
Semirechinsk (Province)	157,000	686,000
Uralsk east of the Ural (Province)	120,000	525,000
Turgay, with Hak Aral (Province)	150,700	327,000
Akmolinsk (Sari-Suisk District)	92,400	463,000
Trans-Caspian Territory[1]	146,000	206,000
	933,600	3,632,000
Bokhara (Khanate)	100,000	2,000,000
Khiva (Khanate)	23,000	700,000
Total Russian Turkestán and Dependencies	1,056,600	6,332,000
Merv Oasis	1,000	250,000
Unclaimed Desert and Pamir	500,000	...
Total Aralo-Caspian Basin and Dependencies	1,557,600	6,582,000

Inhabitants of Russian Turkestán classed according to Races and Religions.

Usbegs	Sunnis.
Kara-Kalpaks	Sunnis.
Kara-Kirghiz	Sunnis.
Kirghiz-Kazaks	Sunnis.
Turkomans	Sunnis.
Tajiks and Persians	Sunnis and Shiahs.
Galchas	Sunnis and Fire-worshippers.
Slavs	Christians.
Kurumas	Sunnis.
Turuks	Sunnis.
Mazang, settled Gipsies	Sunnis mostly.
Luli, nomad Gipsies	Pagans.

Chief Towns in Aralo-Caspian Basin.

Tashkent	100,000	Samarkand	32,000
Bokhara	70,000	Urgenj	30,000
Khokand	60,000	Khojent	29,000
Namangan	50,000	Karshi	25,000
Marghilan	40,000	Andijan	20,000
Shehr-i-Sebs	35,000	Osh	20,000

[1] Including the new Akhal Tekke district, 15,000 square miles; population, 250,000.

AGRICULTURAL RETURNS, RUSSIAN TURKESTÂN.

Provinces.	Under Crops. Acres.	Pasture. Acres.	Waste. Acres.	Total. Acres.
Semirechinsk . .	2,356,000	50,000,000	50,000,000	102,356,000
Sir-darya . . .	984,000	50,000,000	68,512,000	119,496,000
Ferghana . . .	1,650,000	8,250,000	8,525,000	18,425,000
Zarafshan . .	626,000	3,625,000	2,497,000	6,784,000
Amu-darya . .	126,000	3,625,000	19,949,000	23,690,000

LIVE STOCK, RUSSIAN TURKESTÂN.

Provinces.	Camels.	Horses.	Cattle.	Sheep.
Semirechinsk . .	97,412	892,007	523,200	6,296,000
Sir-darya . . .	242,130	395,563	293,550	3,183,000
Zarafshan . . .	1,248	51,991	84,463	283,000
Ferghana . . .	38,294	213,760	220,717	1,260,000
Amu-darya . .	11,267	48,000	38,070	329,000
	390,351	1,601,311	1,160,000	11,351,000

Army—Peace footing 30,000; war footing, 80,000.	
Revenue (mean) . . .	£450,000
Expenditure (mean) . . .	£1,400,000
Deficit (mean) . . .	£900,000
Cotton crop (mean) . . .	50,000 tons.
Silk (Bokhara) . . .	2,500,000 lbs.
Live Stock	£18,000,000
Wool exported to Russia (mean) .	£90,000

DISTANCES.

	Miles.
Chikishlar to Kizil-Arvat .	200
Ashurada to Kizil-Arvat .	221
Mikhailovsk to Kizil-Arvat, by rail	146½
Kizil-Arvat to Sarakhs .	320
Sarakhs to Herat . .	210
Herat to Merv Oasis . .	240
Merv to Sarakhs . . .	70
Merv to Charjui . . .	140
Charjui to Bokhara . .	90
Bokhara to Samarkand .	140
Bokhara to Balkh . .	290
Samarkand to Balkh . .	220
Bokhara to Tashkent . .	320
Samarkand to Tashkent .	180
Tashkent to Khokand . .	120
Orsk to Fort Karabulak .	120
Orsk to Irgiz . . .	240
Orsk to Kasalinsk . .	500

CARAVANS.

	Days.
Orenburg to Tashkent	50 to 60
Namangan to Semipalatinsk .	40
Bokhara to Herat .	25 to 30
Bokhara to Samarkand .	6
Samarkand to Khokand .	6
Khokand to Ush . .	4

CHAPTER XI.

SIBERIA.

1. *Boundaries—Extent—Area.*

A REASON analogous to that which awards the Caucasus to Asia gives the Urals to Europe. For the west Asiatic Mediterranean, which was formerly connected through the Ponto-Caspian Strait with the Euxine, also communicated with the Arctic Ocean over the low ridge forming the present water-parting between the Ob and Aralo-Caspian basins. This ridge transversely crosses the deep furrow stretching northwards along the Tobol valley to the Ob, through which the inter-continental strait flowed between the inland marine basin and the Arctic. Hence the Urals were at this time entirely cut off from the Asiatic continent, of which they form the present north-western boundary.

From this point Siberia stretches uninterruptedly eastwards across 130 degrees of the meridian to the Pacific Ocean. Its northern boundary is formed by the Arctic Ocean, whence it extends across 30 degrees of latitude southwards to China and Turkestán, forming its southern frontiers. But these southern frontiers are in many places extremely vague, and at some points purely conventional. Towards Turkestán the natural line follows the Aralo-Caspian and Ob water-parting, between 48° and 51° N. lat., which at its lowest elevation rises a few feet only above the surrounding Kirghiz steppe. Farther east the frontier towards China generally follows the line of the Altai from the Irtish valley to the Upper Amur valley, running thence along the course of that river to the confluence of the Usuri. Here it is deflected southwards along the Usuri valley, and beyond Lake Kenka to the Sea of Japan, about 43° N. lat. below Victoria or Peter the Great Bay. At this point the Russian territory thus impinges on the north-east frontier of Korea, and shuts off Chinese Manchuria from the Pacific seaboard.

South of the Upper Irtish valley the Russo-Chinese frontier line still remains to be definitely fixed. Here the political boundary running north and south has to be drawn *across* the highlands and depressions, which mostly run east and west. The line from the south-western extremity of the Altai across the Irtish valley at Lake Zaisan, and thence over the Tarbagatai range down to the Emil (Churtu) valley, seems never to have been definitely settled, and its settlement has recently been again postponed to some future period. But the line which thence follows the crest of the Zungarian Ala-tau down to and across the Ili valley to the Trans-Ilian Ala-tau was at least temporarily determined by the treaty ratified on 19th August 1881, in virtue of which Russia restored to China the province of Kulja, held by the Czar's troops during the troubles in the neighbouring districts. This line is drawn from the Boro-khoro hills along the Khorgos River down to the right bank of the Ili and thence across the valley to the Tengri-khan, culminating point of the Tian-shan. It thus leaves all the broad upper portion of the Ili valley—that is, Kulja proper—to China, Russia merely reserving a strip of land.

Lying mainly between 46°-78° N. lat., and 60°-190° E. long., and thus occupying the whole of North Asia, Siberia stretches from Orsk for over 4200 miles northeastwards to Cape Vostochni on Bering Strait, and from

Cape Severo (Chelyuskin) for about 2000 miles southwards to the Tarbagatai range, with a total area estimated at over 4,600,000 square miles, and a population of 4,500,000, or rather less than one to the square mile.

2. *Relief of the Land: The Altai, Sayan, Ergik-Targak, Yablonovoi, Stanovoi, Sikhota-alin, and Kamchatka Ranges.*

A region of such vast extent is naturally of very diversified configuration. Thus, while the south-western portion is exclusively a lowland country, considerable highland tracts are comprised in the southern and eastern sections. These highlands, often comprehensively spoken of as the Altai system, begin properly north of Lake Zaisan and the Upper Irtish valley, by which their westernmost extremity is clearly separated from the Tarbagatai range. On this account the Tarbagatai, although usually included in the Siberian mountain systems, has here been regarded rather as the northernmost extension of the Tian-shan. Its true position is that of a water-parting between the Arctic and the Central Asiatic closed basins. For it sends down streams northwards to the Irtish, flowing to the Frozen Ocean through the Ob, southwards to Lakes Ala and Sassik, which formerly communicated westwards with Lake Balkhash, eastwards with the Ebi-nor and the Mongolian Mediterranean.

From the Irtish valley the Altai, or "Gold Mountains," stretch mainly north-eastwards through the Sayan range to the Daurian Alps, and thence beyond the Baikal basin under diverse names, such as the Yablonovoi and Stanovoi, to the volcanic masses filling the greater part of Kamchatka, and through the Chukchi domain, gradually falling towards the north-easternmost extremity of the continent at East Cape. But it will be seen that the system is by no means continuous, being not only broken up into distinct sections by the deep gorges of the Upper Yenisei and Selenga Rivers, but merging round the Sea of Okhotsk in a moderately elevated plateau, where high ranges are figured on most of our maps.

Even the western section—that is, the Altai proper—is not so much a distinct mountain range as an aggregate of more or less detached chains running in various directions between the upper Irtish and Yenisei valleys. South of these valleys the main direction is rather west and east, but north of them the normal direction is north and south, while the whole system inclines towards the north-east. The portion to which the term Altai is more specially applied, and which scarcely comprises more than one-fourth of the whole western section, stretches from the River Bukhtarma, an affluent of the Irtish on its right bank, and from the Smeinogorsk, or "Snake" mountain, north-eastwards to the romantic Lake Altyn (Teletzkoïe) and to the Chulishman River, joining the lake from the east. Here the Altai is crossed by the much-frequented Suok Pass leading from Siberia to Mongolia. But this eastern limit of the chain is somewhat conventional, for east of the pass the system is continued by the Sayan range with no perceptible interruption to the Upper Yenisei valley.

The whole range has a mean altitude of perhaps 5000 feet, with numerous crests from 6000 to 10,000 feet, culminating in the Bieluka, or "White" mountain, whose twin peaks rise to 11,100 feet. The term "Great Altai," commonly applied to the little-known chains penetrating across the Chinese frontier into Mongolia, belongs rather to the Bieluka chain, which encloses the Kobdo plateau on the west, and several peaks of which rise above the snow-line. Hence recent Russian explorers now designate as the "Little Altai" the "Great Altai" of most geographers.

The western section—that is, the Kolyman or Russian Altai—abounds in ores, and encloses the romantic little Lake Kolyman, whose rugged granite banks are here and there clothed with fine timber. Elsewhere the numerous and rapid streams, the varied forms and colours of the hills, impart great variety to the scenery of the Altai. Between the detached chains there everywhere stretch extensive upland plains covered with snow or morasses, and intersected here and there by low rocky ridges or granite masses.

The southern spurs also consist largely of granites with crystalline schists and a hornblende porphyry, presenting fantastic, bare, and rugged outlines. Here lies the famous mining region of the Altai, which forms part of the Imperial domain, and has altogether an area of perhaps 200,000 square miles. The works at Serianovsk yield gold, silver, copper, lead, and tin. In the Smeinogorsk district the matrix of the metalliferous ores is augite porphyry, varied with schists and huge masses of auriferous quartz. The mines are worked exclusively after the German method, but water and horse power have not yet been supplemented by steam.

Beyond the wooded Sayan section the system is continued across the Bei-kem, or Western Yenisei valley, by the Ergik-Targak and other ridges rising here and there above the snow-line, and crossed by passes over 7000 feet high leading from Siberia to Mongolia. The Ergik-Targak on the Chinese frontier has an altitude of at least 10,000 feet, and is connected with the Baikal uplands by snowy masses which have been only recently explored. Conspicuous amongst them is the Munkusardik, or "Silver Mount," covered with ice-fields, and first ascended by Radde in 1859. This pyramidal mass forms an important water-parting between the great western and eastern branches of the Yenisei, and in the neighbourhood are the vast deposits of graphite discovered about thirty years ago by Alibert.

Beyond the Baikal region the plateau prevails over the strictly highland formation. Here the "Great Divide" between the Lena and Amur, or rather between the Arctic and Pacific Oceans, consists not of a distinct mountain range, as laid down on the maps, but of a vast tableland, contracting gradually north-eastwards towards the Chukchi peninsula, and intersected by a number of moderately-elevated ridges running mostly parallel, or at slightly-diverging angles. Hence Russian geographers now propose to substitute the expression *Stanovoi Vodorazdyel*, or "Main Parting Line," for *Stanovoi Khrebet*, or "Main Dorsal Range," hitherto applied to this upland system. The Russo-Chinese frontier line was doubtless laid down by diplomatists along the crest of the Stanovoi; but from the first this frontier line was purely fictitious, and has been altogether dispensed with since the Russians have established themselves along the left bank of the Amur.

The plateau has a total length of about 2400 miles between Transbaikalia and Bering Strait, and the loftiest and best-defined range in the whole system is the Yablonovoi, or "Apple" range, running south of Baikal, near the Chinese frontier, and culminating southwards with the massive Sokhondo or Chokhondo (8370 feet). The upper crests are composed of granitic and palæozoic rocks, which nowhere reach the line of perpetual snow. The range is easily crossed by the road from Lake Baikal to Chita. East of the Yablonovoi stretches the Daurian steppe, which has been compared to a fragment of the Gobi desert transplanted to Russian territory. It was formerly crossed by a wall attributed to Jenghis Khan, and is separated by extensive pine forests and the River

Onon from the Nyerchinsk steppe, which extends thence eastwards to the Argun valley.

North of the Amur the Yablonovoi section of the Stanovoi runs between the Rivers Aldan and Zeya, at a mean altitude of 7000 feet, beyond which the Aldan or Jugjur ridge falls to little over 3000 feet. Yet here the formations are most varied, comprising granites, porphyry, gneiss, underlying palæozoic, and even jurassic rocks, followed by coal measures towards Verkhoyansk, and basalts and trachytes near the Sea of Okhotsk. At this point the water-parting is deflected so far to the east that the head-streams of the Aldan have their source within a short distance of the Pacific, whence the gold, silver, lead, and iron ores of the neighbouring Aldan hills might easily be procured. A little farther north the "Captain" rises west of Okhotsk to a height of 4360 feet, and although falling short of the snow-line, it overlooks deep valleys filled with masses of perpetual snow and ice.

East of the Stanovoi proper a wooded range, variously known as the "Little Khingan," the Bureya or Daûss-alin, runs at a mean height of 2500 feet from the Amur north-east to the south coast of the Sea of Okhotsk, and culminates with the Lagar-aûl, 3450 feet. Still farther south and east the Maritime Province is traversed in its entire length by the Sikhota-alin, or so-called "Manchurian Mountains," which really consist of an extensive plateau intersected by innumerable ridges, with a mean elevation of scarcely 3000 feet. But notwithstanding their low elevation, these ridges are of very difficult access, so that but few passes lead from the Usuri valley across the plateau to the coast. Communication, however, is effected southwards through the depression of Lake Kenka, whence an easy pass leads down to the Suifun coast stream. Although commonly supposed to be of igneous origin, the Sikhota-alin seems to be mainly a sandstone formation. It culminates with Mount Golaya, 5550 feet.

The Stanovoi water-parting is still continued north-eastwards to the Bering Strait by the low straggling eminences traversing the Chukchi peninsula, and separating the head-streams of the Kolima flowing to the Arctic from those of the Anadir flowing to the Bering Sea. Here the continental system, nowhere more than 5000 feet high, falls to about 2000 as it approaches the coast.

But farther south Kamchatka is occupied by a totally different formation, belonging in its igneous character rather to the oceanic than to the continental system. The peninsula is traversed in its entire length by a chain of lofty burning mountains, fourteen still active volcanoes rising close to the east coast, amongst which is the Klyuchevskaya Sopka (16,000 feet), the highest active volcano in Asia. This igneous system is one of the grandest instances of a connected series in the world; yet it forms merely a link in the endless chain which stretches from Alaska through the Aleutian Islands, Kamchatka, the Kuriles, and Japan, to the Philippines and the Eastern Archipelago. The Kuriles are thoroughly igneous, and contain from eight to ten still active volcanoes.

3. *Hydrography: The Ob, Yenisei, Lena, Yana, Indigirka, Anadir, and Amur Rivers — Lakes Baikal and Kenka.*

Siberia presents the most extensive, but economically perhaps the least serviceable water system of any country in the old world. The land has a general inclination towards the north, so that all the great rivers flowing from the southern highlands pursue a normal and nearly parallel northerly course to the Arctic Ocean. But most of the large tributaries flow rather north-west and north-east to the left and right banks of the main streams, thus affording an almost uninterrupted water highway from the Urals to the Pacific, as well as from the Southern highlands to the Arctic. From the River Ural to Yakutsk, a distance of 6000 miles, this magnificent waterway is broken only by two short portages between the Ob and Yenisei, and between the Yenisei and Lena respectively. The whole country is in this way covered with a network of rivers, affording altogether some 30,000 miles of navigable waters. Unfortunately all these rivers are ice-bound for the greater part of the year, while the estuaries are open only for about ten or twelve weeks during the warm season. Even the Amur, which is the only great river draining from the southern watershed to the warmer Pacific seaboard, is blocked for six months at a time. Hence, notwithstanding the repeated efforts that have been made, especially since Nordenskjöld's successful expedition round the north-east passage, to open up a trade with Siberia through these arteries, it is not probable that the markets of the world will be affected by the agricultural produce from Northern Asia.

All the countless streams from the Urals and Southern highlands are collected and discharged into the Arctic mainly through seven independent channels, which, going eastwards, are the Ob, Yenisei, Khatanga, Olenek, Lena, Indigirka, and Kolima. In the same way those flowing towards the Pacific are grouped in two systems only, those of the Anadir and Amur.

Although not the longest of the rivers, the Ob drains the greatest extent of country, and with its tributaries affords the longest stretch of navigable water highways. The basin, merging eastwards almost imperceptibly with that of the Yenisei, bordering westwards and southwards on those of the Volga, Ural, and Aralo-Caspian, and penetrating south-eastwards far into the Mongolian plateau, has a total area of over 1,400,000 square miles. At Troitsk, where all the great affluents are gathered into one channel, the main stream, still nearly 700 miles from its estuary, has a width of no less than three miles.

The great head-streams are the Tobol, with the Tavda from the Urals, the Ishim from the Aralo-Caspian water-parting, the Irtish from the Kobdo plateau, and the Ob with the Katun, Biya, and Tom and Ket, from the northern slopes of the Altai. The Tobol and Ishim are collected on its left bank by the Irtish, which ought to be regarded as the true upper course of the main stream; for the Urungu, its farthest head-stream, has its source on the Kobdo plateau, south of the Russian Altai, whence it flows first to Lake Ulungur, or Kizil-Bash. From this lake there seems to be an intermittent surface discharge, and a perennial underground outflow to the Black Irtish, a torrent from the snowy upland valleys on the west side of the Chinese Altai. This connection of Ulungur with the Black Irtish is not shown on our maps, but there can be little doubt of its existence. At a certain point below the level of the lake the volume of the Irtish, without receiving any visible influent, is suddenly increased from about 640 to 1900 cubic feet per second. Whence this great access except from the neighbouring reservoir? Mattus-sovski, who visited the district in 1870, could detect no apparent connection, but he ascertained on the other hand that there are no intervening elevations between the lake and the river.

After receiving the Kaljir from Lake Marka, the Black Irtish, which even in Chinese territory is already a considerable stream, 500 feet broad and 10 feet deep, enters the east end of Lake Zaisan, a vast steppe lake on the Russo-Chinese frontier, 60 miles long and over 25 deep, with a mean area of 720 square miles. This lake, which abounds in fish, yielding over 1,500,000 lbs.

annually, has already been visited by a steamer, and it is now proposed to establish regular steam communication between Tiumen and the Black Irtish, a total distance of about 1000 miles.

From the west end of Lake Zaisan the river emerges as the White Irtish, and thenceforth pursues a somewhat winding north-westerly course between the Kirghiz and Baraba steppes, west and east to its junction with the Ob, about 1900 miles from its farthest source. After its junction with the Bukhtarma it pierces the western spurs of the Altai through the wild Gust-Kamenogorsk defile, here falling from 1300 to 1150 feet above sea-level, and from this point to the Ob it receives probably over a thousand tributaries, of which by far the largest are the Ishim and Tobol, both on its left bank. But the large Lake Chany, as well as most others of the Baraba steppe, have become closed basins, no longer sending even intermittent discharges to the Irtish.

The Baraba steppe is skirted eastwards by the Ob, which is formed by the junction of the Katun and Buja from the Altai, and which is joined by the Irtish 300 miles below Tobolsk. With a fall of scarcely 300 feet from the advanced spurs of the Altai to its estuary, the Ob pursues an extremely sluggish course, expanding here and there into broad steppe lakes, and occasionally almost undecided whether to flow west to the Irtish, east to the Yenisei, or in an independent channel north to the Arctic. Varying in breadth from half a mile to two miles, it expands during the spring floods into a great inland sea, which, even above Tomsk, is so broad that the opposite banks are quite invisible. After receiving the Tom and Chulim, it is joined near Narim by the Ket, which in some respects, though not the largest, is its most important tributary. For the Ket is navigable for no less than 600 miles towards the Yenisei, with which it is now proposed to connect it by means of a canal $2\frac{1}{3}$ miles long from Lake Kosovskoie across the portage to the Kas flowing to the Yenisei below Yeniseisk.

Below the Irtish confluence the Lower Ob flows to its estuary beyond Obdorsk, in two separate channels, known as the Great and Little Ob, the latter of which is most available for up-stream traffic, the former for craft going seawards. Both branches, which are everywhere connected by innumerable intermediate channels and backwaters, enter the gulf in a joint stream about two miles wide, and from 40 to 90 feet deep. The gulf or fiord runs first east and then north for over 480 miles beyond the Arctic Circle. This great water highway, which is navigable throughout nearly the whole of its course of over 3400 miles, was thrown open to the trade of the world by the expedition of Dahl, who reached the Ob from the Kara Sea in 1877. With the tributaries, there is a total navigable highway of perhaps 9000 miles, which has hitherto been but slightly utilised by steamers.

Next in importance to the Ob is the Yenisei basin, which occupies nearly the whole of Central Siberia between the Ob and the Lena water systems. Like the Ob and most other great Asiatic rivers, the Yenisei has its farthest sources on the central plateau behind the enclosing mountain barriers, through which it forces its way seawards. Of its two great branches, the Yenisei proper and the Selenga-Angara, the former is developed on Chinese territory, about 4000 feet above the sea, between the Sayan and the Tanu-ola mountains, by the junction of the Bei-kem and Khua-kem, from the Ergik-Targak frontier range and the northern slopes of the Tanu-ola. Flowing first west to a point where it finds an outlet in the Western Sayan range, the united stream here enters Russian territory, through which it henceforth pursues a northerly course, mainly between the 88th and 94th meridians, to its estuary in the Arctic Ocean.

The eastern branch also rises within the Chinese frontier, whence the Selenga, its farthest head-stream, flows from the north-eastern slopes of the Tanu-ola eastwards

CROSSING THE YENISEI.

to a point where it receives the outflow of Lake Kosso (Kosso-gol) from the north. This romantic basin, fed by the glaciers of Munku-sardik at its northern extremity, stretches north and south for a distance of about 70 miles, with an area of about 1300 square miles. In its centre is the islet of Dalai-kui, which the Mongolian

Buddhists hold in special reverence as the "Navel of the World."

Beyond this point the Selenga sweeps round to its junction with the Orkhon, which rises in the Gobi itself, near the ruins of Karakorum, the old Mongolian capital. The united stream flows thence between Kiakhta and the Khamardaban range, north and west to the east coast of the great highland Lake Baikal, by far the largest fresh-water reservoir in Asia. This lake, which fills two enormously deep fissures in the plateau at a present elevation of 1363 feet above the sea, seems to have formerly communicated directly with the Irkut valley. But its present outflow is through the Angara, which forces its way over a series of romantic gorges to the right bank of the Yenisei. The lower section of the Angara, where it trends west, takes the name of the Upper Tunguska in contradistinction to the "Stony" and "Lower" Tunguska, which join the main stream farther down.

Baikal, the Dalai-nor, or "Holy Sea," of the Mongolians, has a mean depth of 850 feet, sinking in some places to 4500, or considerably over 3000 below sea-level. Owing to this prodigious depth, its volume with an area of scarcely 14,000 square miles, is more than double that of Lake Michigan, which has an area of 23,000 square miles, but a mean depth of only 300 feet. Recent soundings have revealed a rocky ridge about 3350 feet high, dividing the lake into two secondary but now united basins, at a point where there is a depth of scarcely more than 200 feet. These surveys also show that Baikal was formerly far more extensive than at present. Within comparatively recent times its level has fallen about 20 feet; but at some more remote epoch it was high enough to drain through the Irkut to the Yenisei through a channel distinct from that of the Angara. Its waters, which are remarkable for their great transparency, revealing objects at a depth of 40 to 50 feet, are frozen from December to May to a thickness of 4 or 5 feet. Yet such is the fury of the winter gales that its icy fetters are constantly broken, thus affording fresh supplies of air to the salmon, sturgeon, and other fishes with which the lake abounds. Amongst its fauna is a species of seal, in appearance exactly resembling the Spitzbergen *Phoca fœtida.*

In summer the communication along the shores of the lake, which is 360 miles long, with a mean breadth of 35 miles, is kept open by a steamer, affording an opportunity of visiting the lovely island of Olkhon, famous for its alpine roses. The north-west coast contains some very grand scenery, the rocky granite masses being here in many places clothed with larch and pine forests from their summits to the water's edge. The shores also abound in hot springs, which are associated with still active underground agencies and frequent earthquakes. Near the hot springs is a flourishing Russian settlement, where rye, barley, and potatoes are successfully cultivated. During the long winter months the peasantry occupy their time in pursuing the sable, squirrel, and other fur-bearing animals, which, however, here, as elsewhere in Siberia, are rapidly disappearing.

M. Chersky's recent explorations (1881) show that the rocks on the west side of the lake belong to three different epochs—pre-Silurian (Laurentian ?), Silurian, and Jurassic. This naturalist also confirms the view that Baikal forms, as above stated, two distinct longitudinal cavities, connected together by a central ridge.

The Angara, which has a discharge of at least 105,000 cubic feet per second, flows for 40 miles below its junction with the Oka through a series of rapids between sheer rocky walls, rising in some places 600 feet above the surface. Here the average fall is about two feet in the mile; but the steamers now plying on these waters pass safely over the rapids. As the Selenga is also accessible to light craft as far as the Orkhon junction, there is an uninterrupted navigable highway of 2700 miles from Kiakhta through Lake Baikal and down the Angara and Yenisei to the Arctic Ocean.

Nor is the navigation entirely interrupted even by the rapids over which the Upper Yenisei descends from the Mongolian plateau through the Sayan range down to the Siberian plains. Even above the Angara junction the current is very gentle, though nowhere quite so sluggish as that of the Ob, the elevation at Krasnoyarsk being 530 or 200 feet more than that of the Ob under the same parallel. At Yeniseisk below the Angara confluence it is still 230 feet above sea-level. Here it has a mean width of 6000 feet, expanding in the floods to upwards of 4 miles, with a rise of about 40 feet.

Through the Lower Tunguska, which has a total length of 1620 miles, with a breadth of over half a mile at the confluence, the Yenesei approaches near Kirensk to within 14 miles of the Lena; and as the portage between the Ob and Yenisei (see p. 450) is only $2\frac{1}{3}$ miles long, the navigable water highway from the Urals across Siberia nearly to the Pacific seaboard is only interrupted by two breaks of less than 17 miles altogether. This highway is also of far greater economic importance than those of the main streams flowing northward to the Frozen Ocean. Lying mainly between the 58th and 60th parallels, it is open for a far longer period of the year than those of the great arteries running for several degrees beyond the Arctic Circle towards the North Pole The navigation of the Lower Yenisei for 300 miles above its estuary, where it expands to 30 or 40 miles during the floods, is rendered extremely dangerous by the northern gales, here sweeping with great fury over the tundra and against the current. During some seasons it is also blocked from the end of August till the first days of the following July, leaving scarcely six weeks of open navigation in the whole year. Yet since the discovery of the convenient harbour of Dicksonhavn in its estuary by Nordenskjöld, several trips have been made to the Yenisei by English and Scandinavian skippers, who have returned with cargoes of grain and other produce brought down from the Minusinsk steppe in the Upper Yenisei valley. The rafts and light craft engaged in this traffic are broken up for fuel or timber after discharging at Dicksonhavn.

The Yenisei drains a total area of about 1,180,000 square miles, and with its tributaries has a navigable water-way of not less than 5000 miles. The western branch, or Yenisei proper, is nearly 2600, the eastern (Selenga-Angara) 3300 miles from their farthest sources to the common estuary, which is separated from that of the Ob only by a comparatively small peninsula, 300 miles broad at its widest part.

Of all the great North Asiatic streams the Lena alone has its source on the seaward slope of the mountain range enclosing the central plateau. But the Upper Lena, which rises on the hills skirting the west coast of Lake Baikal, and which for some distance flows parallel with the Angara, seems to have formerly communicated with that river through a now dried-up depression in the low water-parting between the two basins. It is joined at Vitimskaya by the Vitim which flows from the east side of Lake Baikal round the elevated Vitim plateau north-westwards, and which, from its size and volume, might be regarded as the true upper course of the Lena. Below the confluence the main stream is deflected by the scarp of the Yakutsk tableland for hundreds of miles east-north-east to Yakutsk, where it again resumes its normal northerly course to its delta in the Arctic Ocean, over against the archipelago of New Siberia. Below Yakutsk the Lena is again nearly doubled in size by two great affluents, the Aldan from the Stanovoi uplands and the Viliui from the low water-parting between the Lower Yenisei and Lena basins. The vast basin of the Lena, draining a total area of 1,000,000 square miles, is thus enclosed by those of the Yenisei, Amur, and Indigirka, and presents a total water highway of perhaps 6000 miles, open, however, for only five or six weeks in the summer.

Expanding at the Aldan junction to 12 miles from bank to bank, the Lower Lena again contracts to 3 or 4 miles as it approaches its delta, which is nearly 9000 square miles in extent. Here the navigable channel is sometimes blocked throughout the whole summer by floes massed along this coast by the polar winds. Nevertheless the Norwegian Johannsen succeeded in ascending the Lena to Yakutsk in the steamer *Lena* for the first time in 1878. But it is doubtful whether any regular navigation can ever be established with this river, whose basin belongs entirely to the Arctic Ocean. Yet this region abounds in copper, iron, coal, lead, gold, silver, sulphur, salt, and other minerals. The Vitim sends down auriferous sands in large quantities, and the Lena is skirted for over 900 miles by coal measures, often cropping out above the surface. In some places the coal-beds, kindled by forest fires, have been burning for years, giving rise to the reports heard from time to time of still active volcanoes in the Lena basin.

Between the Yenisei and Lena basins two other large rivers—the Khatanga, 600 miles long, and the Olenek, double that length—enter the Frozen Ocean in separate channels. The Olenek is 6 miles wide and over

20 feet deep at its mouth; but both alike are practically useless as highways of trade.

This is also to a large extent true of the Yana, Indigirka, and Kolima, flowing east of the Lena from the Verkhoyansk range northwards to the Arctic, as well as of the Anadir, flowing from the southern watershed to the Bering Sea.

The Amur, whose basin is now politically divided between Russia and China, promises at no distant date to become more important economically than all the Siberian rivers taken collectively. It is formed by the junction of the Shilka and Argun, the former flowing from Mount Kentai in the Khan-ula range mostly through Russian territory, the latter from the south side of the same range, under various names, such as Kerulen, Lukin, etc., mostly through Chinese territory. The main stream is joined on its left bank by the Zeya at Blagevyeshchensk, above which point it is navigable by light craft for several hundred miles from its mouth in the Gulf of Tartary over against the island of Sakhalin. Below the Zeya its waters are swollen by the Bureya, also on its left, and by the Sungari and Usuri on its right bank. Between the Zeya and Usuri, a distance of 570 miles, it is accessible to ships drawing 8 feet, and thenceforth to deep-sea vessels, although a serious obstacle to navigation is offered by the bar at its mouth, with a depth of scarcely 13 feet.

On its seaward course the Amur has to force its way through the Khingan range, separating the Mongolian plateau from Manchuria, and farther down through the Sikhota-alin, by which its lower course is deflected almost due north in a line with the Usuri. Between these two points it breaks through the barrier of the Little Khingan, below the Bureya confluence, and again through the rugged Chanyatin uplands. Besides the bar, the shifting sands and intricate channels, both in the river itself and in the shallow waters between its mouth and Sakhalin, offer great obstacles to its navigation, which is further

VALLEY OF THE AMUR.

closed by ice for six months in the year. Owing to these difficulties, the attention of the Russians was long directed towards the southern coast region, which they finally secured in 1860, and where they now possess the more or less convenient harbours of Castries Bay, Port Imperial, Olga Bay, America Bay, Victoria Bay (now by them renamed the Gulf of Peter the Great), and Possiet Harbour on the Korean frontier.

On the south-west frontier of this maritime province is the extensive but shallow Lake Kenka (properly Hanhai), which is 65 miles long and about 25 wide, with an area of 1200 square miles, but is nowhere more than 30 feet deep. This basin, which, notwithstanding its great extent, is a mere reservoir for the rainfall of the surrounding hills, drains through the Sungacha northwards to the Usuri and Amur.

4. *Natural and Political Divisions: West Siberia (Governments of Tobolsk and Tomsk)—The Tundra—East Siberia (Governments of Yeniseisk, Irkutsk, Transbaikalia) — Amur — Maritime Province — Islands—Sakhalin.*

The conflict between permanent physical conditions and arbitrary political groupings, so common throughout the Russian Empire, is mostly restricted in North Asia to West Siberia. Some portions of this region, which is mainly comprised within the limits of the Ob basin, are attached to the Government of Turkestán, while other portions are included in the European Governments of Perm and Orenburg. To European Russia by a curious fiction are also attributed the Governments of Akmolinsk and Semipalatinsk, besides a part of Turgai, so that Tobolsk and Tomsk are the only West Siberian Governments which are not politically encroached upon. Tobolsk is limited by the eastern slopes of the Urals, while

Tomsk comprises the Altai highlands and the upper waters of the Ob and Irtish.

West Siberia.

Seen from the eastern slopes of the metalliferous Ural range these lowlands seem to stretch away eastwards, like a limitless ocean plain. For some distance beyond Tiumen the land presents the aspect of a heath varied with a few plantations of sickly firs. This is succeeded by the steppe, which as far as the Irtish is diversified with birch and brushwood thickets, interspersed with extensive swamps and shallow basins. But east of the Irtish the country assumes the character of a true steppe, a boundless grassy plain, here and there relieved by a few bushes on the distant horizon. Beyond Omsk it presents the appearance of a prairie with rolling hills, covered with short grass, which affords pasturage for the herds and flocks of the Kirghiz nomads. But the steppe is distinguished from the prairie especially by the numerous lakes, some of considerable size, frequented by wild swans, ducks, and other waterfowl. Towards Semipalatinsk, a feature of the scenery are the square graves of the Kirghiz-Kazaks, made of the trunks of trees, and looking at a distance like little houses or log-huts.

South of this place run the Arkat hills, bare granite masses of grotesque form and rugged aspect, rising to a height of 1200 feet in picturesque outline above the surrounding treeless plains. At Sergiopol the snowy crests of the Tarbagatai range come into view. North of this point the slopes of the Ala-tau afford good pasturage to magnificent herds of oxen, camels, horses, and fat-tailed sheep. Then follow the salt steppes, where animals and wayfarers sink at every step through the saline efflorescence covering the surface. But in the Ala-kul lacustrine district the ground is overgrown with an extensive dense jungle of reeds, affording in winter a slight shelter to the Kirghiz from the fierce snow-storms. A striking contrast to these monotonous wastes is presented by the magnificent scenery of the upland valleys and alpine lakes on the slopes of the Ala-tau.

Proceeding along the Ob valley north from Tobolsk, the impenetrable primeval woodlands are succeeded by dreary bottomless swamps, the true tundra, stretching without a break north and north-eastwards to the Ob and Yenisei estuaries. For eight long winter months the frozen ground is here covered with snow, the glass often falls to 45° below freezing point, and the cold converts the breath of animals into icy hoar frost. Birds on the wing often fall dead from the skies, the panes of glass start in their sashes, the hardened soil splits into wide and deep fissures, and the very ice on the lagoons bursts asunder. Here the fierce storm often rages for twenty-four hours at a time, during which man and beast remain patiently buried in the snow, as the only means of sheltering themselves from its fury. The heavens are perpetually overcast with dull leaden clouds, the atmosphere is raw and humid, the long gloomy nights are relieved only at intervals by the magnificent phenomenon of the Northern Lights.

Yet in summer the tundra can present even an inviting aspect. Nordenskjöld, who explored the lower course of the Yenisei in the August of 1875, in preparation for his famous expedition by the north-east passage, denies that the tundra presents the aspect of a dreary ice-bound waste, relieved here and there only by a stunted growth of sickly vegetation. Such, according to him, is its aspect at one point only, on the Yenisei, the vegetation being elsewhere, and especially in the islands of the river, of a surprisingly luxuriant character. The fertility of the soil, the boundless extent of the meadow lands, and the abundance of pasturage render the tundra a splendid grazing ground.

Farther south, between Turukansk and Yeniseisk, where the country is overgrown with extensive woodlands of great age, and succeeded near Krasnoyarsk by extensive plains covered with a thick layer of black mould, the tundra merges in a region fully as productive as the most favoured tracts in Scandinavia. Here the natural richness of the soil, combined with the abundance of fish in the rivers, the sparse population, and the absence of markets, renders provisions of all kinds fabulously cheap. The Rev. H. Lansdell, who visited this region in 1879, was offered "live ducks for five farthings each, large fish, called *yass,* for 1½d. a pair, and pike for a farthing each. Milk cost 2½d. a bottle, but young calves in remote villages could be purchased for 6d. each. The belt of rich black earth in the region immediately north of the Altai lets for 3½d. per acre, and from it wheat may be purchased for about one-twentieth its cost in England. Still farther north, in the forest region, rich in excellent timber and fur-bearing animals, meat was bought up wholesale in 1877 at less than a halfpenny per pound; whilst in the tundras the rivers are so full of fish that one of the ordinary difficulties of the natives is to avoid breaking their nets with the weight of the draught. The fish are frozen and sent more than 2000 miles to St. Petersburg, where a very moderate price realises for the fisherman a profit of nearly 100 per cent."[1]

East Siberia.

The course of the Yenisei marks the boundary line between West and East Siberia. Beyond this line the

[1] Paper in *Proceedings of Royal Geographical Society,* October 1880.

plains are far more diversified, by hills, ridges, and even hilly plateaux often deflecting the course of the streams east and west, whereas in the Ob basin all run north, north-west, and north-east. In fact in East Siberia true lowland plains of great extent are comparatively rare. Even in the Taimyr peninsula, between the Yenisei and Khatanga estuaries, the coast ranges are said to attain elevations of from 3000 to 4000 feet, while the Kharaûlakh hills between the Lower Lena and the Yana rise in some places to a height of 1300 feet. Farther east and south occur the vast elevated plateaux of Yakutsk, Transbaikalia, the Daurian and Nierchinsk steppes, the Stanovoi uplands, the Amur basin and Kamchatka, filling most of north-east Asia, and reducing the lowland formation to a relatively small area. Here also the Verkhoyansk water-parting runs at an elevation of from 5000 to 6000 feet from the Lower Lena right across to the north-east coast ranges, thus completely separating the head-streams of the Yana, Indigirka, and Kolima from the Lena basin. Speaking generally, while the mean altitude of the southern Altai mountain system falls gradually north-eastwards to the Stanovoi plateaux, the mean altitude of the northern region rises gradually from the Yenisei eastwards to the Bering Sea, here culminating in the Kamchatka peninsula. Even between the Yenisei and Lena basins there is a plateau of palæozoic formation.

In East Siberia, although Kamchatka disappears from the administrative nomenclature, the political divisions otherwise mostly follow the natural lie of the land. Thus the vast Governments of Yeniseisk, Irkutsk, and Yakutsk are mainly comprised in the Yenisei and Lena river basins, while the provinces of Transbaikalia and the Amur correspond with two physically distinct regions, the first stretching from Lake Baikal to the Argun-Shilka confluence, the second comprising the region between the

Stanovoi water-parting and the left bank of the Amur. Even the south-eastern divisions of the Usuri territory and the Maritime Province answer to two natural divisions, the first comprised between the right bank of the Usuri and the Sikhota-Alin highlands, the second including the strip of coast land between the eastern slopes of these highlands and the Sea of Japan.

To these governments are attached the islands and groups of islands lying off their respective coasts. Of these the largest are the desolate and uninhabited New Siberian or Liakhov Archipelago north of the Lena delta,[1] the Bear Islands north of the Kolima estuary, Wrangel Land, discovered in 1849 by Kellet on the spot previously indicated by Wrangel, and for the first time circumnavigated in the summer of 1881 by the crew of the American steamer *Rodgers*, not in the vessel itself, as has been stated, but in the small boats belonging to it;[2] lastly Sakhalin, whose northern extremity is almost connected with the mainland near the mouth of the Amur.

Formerly held jointly by Russia and Japan, Sakhalin[3] (Saghalien) or Karasto was by the treaty of 1875 ceded by the latter power to Russia in exchange for the Kurile Archipelago. It has an area of no less than 25,000 square miles, and stretches for 550 miles north and south,

[1] A grant of 14,000 roubles has been made by the Government for two Polar observing stations at the Lena delta and on one of the islands of New Siberia during the year 1882.

[2] Reports have recently been current respecting the formal occupation of this island both by the United States and the Dominion Government. But whatever be the claim of Canada to its possession, the United States are understood to be barred by treaty engagements from occupying any Arctic lands west of Bering Strait. The island appears to be about 60 miles long and quite uninhabitable.

[3] This word is a corruption of the Manchu "Sakhalan anda Khanda"—that is, "Rock of the Amur Estuary"—applied originally to an islet in the mouth of the Amur, and afterwards, by mistake, extended to the island now known as "Sakhalin."

with a breadth varying from 15 to 80 miles. It is traversed in its entire length by parallel ridges, of which the loftiest and most continuous is the west coast range with crests from 3000 to nearly 5000 feet, culminating with La Martinière (Ktönspal) Peak about the centre, 4860 feet. From this chain, which nowhere reaches the snow-line, a few streams flow for short distances mainly southwards to the coast. Amongst the products the most important is coal, which, although inferior to the English, still commands higher prices than that of Japan or Australia. There are also extensive forests of valuable timber, frequented by numerous fur-bearing animals. The climate and soil are unfavourable to agriculture, but vegetables may be grown and stock-breeding carried on in some sheltered districts. Latterly the Russians have used this island chiefly as a convict settlement for political prisoners, who are sent thither by the sea route through the Suez Canal. The chief stations are Dui on the west coast, and Mauka Cove towards the south-west extremity. The latter, of which a full account has recently been communicated by Captain Anderson to the British Hydrographic Office, has been chosen as the headquarters of a company which has just obtained from the Government the monopoly of the trade in fish, bêche de mer (trepang), edible seaweed, and other local produce, for a term of ten years.

5. *Climate: Region of Intensest Cold.*

Amid much diversity, natural in such a wide area, the Siberian may on the whole be taken as the most essentially continental climate on the globe. Here the maximum of cold is reached, not in the Yakutsk district, as is commonly supposed, but in Verkhoyansk, on the Upper Yana, just within the Arctic Circle. Here the glass

usually falls to 49° C. below freezing point in January, the mean in Yakutsk being 42° or 43°. "Within the isothermal of −40° C., a temperature at which the quicksilver freezes, Verkhoyansk alone is included for the whole period from November to February, Yakutsk for December and January only, and Ustyansk (at the mouth of the Yana) for January only, while Tolstoy Noss (at the mouth of the Yenisei) lies beyond this isothermal."[1] It also appears that in these very places the glass rises from 28° to over 38° C., or occasionally as high as 102° F. in July and August, which is about the normal summer heat of most lands lying about the equator. No other region can show such amazing extremes as these, consequently the claim of North-East Siberia to the possession of the most typical continental climate is established. It would seem to be at once colder than the North Pole,[2] and hotter than many uplands under the equator, a condition due to the combination of more cold and heat producing causes than occur elsewhere in the northern hemisphere.

But while the intense heat lasts only for a few weeks, the intense cold prevails for many months, the two extremes being separated by short intervals of broken spring and autumn weather. The result is that in the course of ages the ground has gradually become permanently frozen in many parts of the tundra from about 2 feet below the surface to depths of from 100 to 300 feet, and perhaps even more. It might be supposed that the great elevation of the Altai regions would have the effect of neutralising the difference of latitude, thus

[1] *Verhandlungen der Ges. für Erdkunde zu Berlin*, July 1881, p. 275.

[2] The two poles of greatest cold in the old and new worlds oscillate about Verkhoyansk and Cockburn Bay, Adelaide Peninsula (*H. W. Klutschek*, of Schwatka's Expedition).

rendering the southern highlands as cold as the northern lowlands. But this is far from being the case, and as we proceed southwards the normal temperature rises steadily. The Russians compare the climate of some places in these latitudes with that of Italy, and there can be no doubt that many of the Altai and Amur districts are favoured by a genial healthy climate suitable for the development of agriculture.

The prevailing winter winds are from the south and south-west between the Urals and the Yenisei, but in the Lena basin from the north-west. These icy north-west gales blow steadily for months together, and are felt far beyond the limits of Siberia in the Japanese waters, in the Amur basin, and on the Mongolian plateaux. In summer the rarefaction of the atmosphere causes Arctic breezes to prevail along the western seaboard. But farther east these are succeeded by moist south-east winds from the Pacific, and to this cause the regions east of Lake Baikal are indebted for their abundant rainfall. On the coast lands the mean exceeds 40 inches, falling westwards to 10 at Yakutsk and 8 at Kiakhta. In winter the snows are much lighter in the east than in the west, and the Lower Lena and Yana basins, where the cold is intensest, are remarkable for their clear blue winter skies.

6. *Flora and Fauna: The Argali, Marmot, and Lemming—Extinct Mammalia.*

In North Asia the northern limits of timber, while following the coast-line, scarcely anywhere reach the Arctic seaboard. Long before reaching their actual limits, the few stunted larches straggling northwards assume strange distorted forms, trailing rather along the surface than shooting upwards, and often presenting the aspect of withered branches or dead trunks of trees. Nevertheless, in his expedition down the Obi in 1880, Khandachefsky discovered a magnificent forest of large cedars and larch in the valley of the River Nadym, at a point where Petermann places the extreme northern limit of the forest zone.

This forest zone, or "taiga," consisting mainly of species common to Europe, stretches almost uninterruptedly right across the continent, merging everywhere northwards in the tundra. Here the vegetation consists almost exclusively of mosses, lichens, and grasses. South of the forest zone the Ob basin is occupied by the steppe, which in some respects resembles the tundra, both presenting the same cheerless, monotonous aspect, and absence of timber. Even the same species are often found in both, such differences as exist being caused by deficient moisture in the steppe, and deficient heat in the tundra. Sometimes particular species of reindeer and other mosses predominate in the tundra, imparting a pale white or a dull yellow aspect to the scenery for miles and miles along the lower reaches of the great rivers.

In the taiga[1] the prevailing trees are the larch, birch, alder, cedar, and a noble species of pine peculiar to Siberia, which shoots up to a height of nearly 100 feet, with a slender stem seldom exceeding a foot in diameter. But perhaps the most characteristic plants are those producing berries in great variety and abundance. These uncultivated fruits supply food to man and beast, and quantities are preserved for use during the winter.

In the remote volcanic peninsula of Kamchatka, the

[1] This term is used somewhat differently in different parts of Siberia. In the Altai it means the wooded uplands abounding in fur-bearing animals; in the north it is applied to the zone of uninhabited woodland tracts bordering on the mossy tundra, which stretches thence to the Arctic seaboard.

banks of the inland waters are decked with a clothing of grass, growing with an almost tropical luxuriance, interspersed with bright flowers, alpine rose-bushes, the cinquefoil, and the rare Kamchadale lily. The poplar and the birch grow in clusters on the lowlands, while the slopes of the hills are covered with the sombre foliage of extensive pine-forests.

The prevailing humidity of the Pacific seaboard has also favoured the development of a magnificent vegetation in the Amur basin, where the flora, especially in the islands and along the river-banks, abounds in endless varieties of leafy shrubs and undergrowths. Here also the conifers, oak, elm, ash, walnut, cork-tree, maple, and linden, often attain majestic proportions. On the Usuri the ginseng is largely cultivated for the market of China, where this plant fetches its weight in gold, and where it is supposed to be a sovereign remedy against all disorders. The wild vine in some places yields a good grape, and the grasses flourish with astonishing luxuriance; for in this more favoured region the flora includes plants peculiar to the cold, temperate, and warm zones.

This is also largely true of the animal kingdom. In Manchuria the tiger reaches his northernmost limit, and is here associated with the panther, lynx, glutton, and wolf. There are two species of bear, and a transition is effected to the fauna of Siberia proper by the sable, black and red fox, marten, ermine, and other fur-bearing animals. The cedar-groves are here enlivened especially by a species of dark-gray squirrel, whose skin is much prized, fetching large prices on the spot. The ruminants are represented chiefly by the deer, elk, roe, and musk-deer, while there are over 200 species of birds. The Amur, and especially the Usuri River, with Lake Kenka, are incredibly rich in fishes, including the sturgeon, salmon, carp, sterlet, and many other varieties.

This teeming animal life has elsewhere its counterpart in the prodigious multitudes of marmots and other species of small rodents inhabiting vast tracts from the Tarbagatai right away to Kamchatka and the extreme north-eastern Chukchi lands. The ground in many places is honeycombed with the galleries and subterranean townships of these pretty little troglodytes, who may at times be seen mounting guard in interminable lines on the hillocks at the entrance of their dwellings, suddenly disappearing at the least sound, and as suddenly reappearing to ascertain the cause of the alarm.

Equally abundant in the north-east are the lemmings. The line of march of these migratory rodents often stretches for miles across the plains between their winter quarters and summer camping-grounds. In Kamchatka a lasting alliance has been struck between them and the natives. Whenever the latter are driven by distress to draw from the supplies of their provident little friends during their absence on some distant expedition, they are always careful to replace the stores in more prosperous times. It is also said that, to guard against similar plunder by other less scrupulous marauders, the lemmings conceal their underground granaries with poisonous herbs.[1]

Most of the fur-bearing animals have disappeared from these north-eastern regions, causing many of the old hunting stations to be abandoned. But in Kamchatka the trappers still obtain from 6000 to 8000 sable skins for the Russian market. In other parts of Siberia many species of these animals are also becoming extinct. But here their destruction is often due as much to the destruction of the taiga as to the skill of the trapper.

In West Siberia birds are very numerous, and here amongst the more characteristic species are the golden eagle, the white-throated alpine lark, and the gray-headed

[1] So, at least, Krasheninnikov was informed by the natives.

wagtail. In this region the Arkat hills still afford a refuge to the Argali (*Ovis ammon*), a magnificent mountain sheep, with enormous thick and twisted horns over three feet long. This species, whose original home seems to be the Central Asiatic plateau, resembles in its habits the steinbok and chamois, and, owing to its extraordinary speed and velocity, is very difficult to bring down.

Of domestic animals, the most useful are the fat-tailed sheep and camel in the steppe, the reindeer in the tundra, and the yak in the Upper Yenisei basin, where it reaches its northernmost limits.

Siberia was in former epochs the home of a large species of rhinoceros and of the mammoth, some specimens of which have in recent times been found preserved by the ice in an almost perfect state.[1] Vast quantities of fossil ivory from these animals, amounting at one time to 40,000 lbs. yearly, have been obtained in the archipelago of New Siberia, and parts of the Arctic seaboard.

7. *Inhabitants: Table of the Siberian Races—The Buriats and Mongolian Buddhism—The Tunguses, Yakuts, Koriaks, Kamchadales, Ostiaks, and Shamanism—The Samoyedes and Voguls.*

Excluding the Bashkirs, who dwell chiefly west of the Ural River, and the Kara-Kirghiz and Kirghiz-Kazaks, whose camping-grounds lie chiefly south of the Aralo-Caspian and Ob water-parting, the native inhabitants of Siberia scarcely number 750,000 altogether. And even these, few as they are, seem to be mostly in a process of more or less rapid extinction or absorption in the advancing Slav element. Certainly Siberia belongs henceforth to the Russians, in the same sense that Australia

[1] A rhinoceros by Pallas in 1771 on the Vitin; a mammoth by Adams in 1799 on the Lena, and others during the present century.

has become a new home of the English race. They have already occupied a continuous broad zone stretching from Europe across West Siberia and along the southern highlands to Lake Baikal, and thence through Transbaikalia and down the Amur and up the Usuri to the Pacific seaboard. Here they are firmly established at Nikolayevsk and Vladivostok, the extreme northern and southern points of the Maritime Province. They have also occupied both banks of the Yenisei throughout its entire course, most of the Ob and Irtish, the Lena down to Yakutsk uninterruptedly, besides numerous detached stations on the Lower Lena, in Kamchatka, Sakhalin, and elsewhere. Large portions of the really arable lands are thus already held by Russian agricultural colonies, and great efforts are now being made by the Government to direct the migrations of the peasantry from Europe to the Amur basin. Certain tracts, such as the distinctly steppe region of West Siberia, will doubtless continue to remain in the hands of the natives, for they are uninhabitable except by nomad tribes. But all the broad lands available for cultivation will be occupied by the Slav race.

None of the natives have any vitality except the Yakuts of the Lena basin and the Kirghiz of the West Siberian steppes. The Ostiaks of the Ob basin and the Yukaghirs of the Indigirka and Kolima Rivers are actually dying out, and a similar fate threatens to overtake the Giliaks of Sakhalin and opposite mainland, as well as the Samoyedes of the Lower Ob and Yenisei.

With the exception of the still unclassified "Hyperborean" group, all the aborigines belong to various branches of the Mongolo-Tatar ethnical and linguistic family. From the subjoined table of these races the Bashkirs and Kirghiz are omitted, the main sections of these races being included in Europe and the Aralo-Caspian basin respectively:—

Mongolian Stock.

Races.		Religions.	Population.
Kalmuks	Zungars	Buddhists and Shamanists	20,000
	Targuts		
	Khoshods		
	Turbets		
	Chorasses		
	Teletzes		
Buriats (East Branch)	Kudara	Shamanists, Buddhists, and Christians	250,000
	Selenga		
	Khorinsk		
	Barguzin		
Buriats (West Branch)	Tunka		
	Verkolensk		
	Olkhon		
	Kuda		
	Ida		
	Balagansk		
	Alarsk		

Manchu Stock.

Tunguses	Lamuts	Shamanists and nominal Christians	80,000
	Oroches		
	Orokhos		
	Chapogirs		
	Golds		
	Manegrs		
	Manguns		
	Samagirs		
	Ngatkons		
	Nigidals		
	Negdas		
	Tazi		
	Olenes		

Finnic Stock.

Samoyedes	Yuraks	Shamanists and nominal Christians	35,000
	Tagurs		
	Abators		
	Koibals		
	Soyots		
	Motors		
	Karagasses		
	Kamastes		
	Tagvis		
Ugrians	Ostiaks	Shamanists	25,000
	Voguls	Nominal Christians	4,500
Mixed Finno-Tatars	Darkhats	Buddhists	15,000
	Soyons		
	Assan	Shamanists	5,000
	Arinzi		
	Kotti		

TŪRKI STOCK.

Races.	Religions.	Population.
Yakuts	Christians and Shamanists	200,000
Red Tatars	Mostly Christians	80,000
Black Tatars	Mostly Christians	
Teleuts	Sunnis	
Kumandes	Christians	

UNCLASSIFIED SUB-ARCTIC RACES.

Races.	Religions.	Population.
Koriaks	Pagans	5,000
Chukchis	Pagans and nominal Christians	12,000
Yukaghirs	Shamanists	1,600
Kamchadales	Nominal Christians	3,000
Ankali, extinct?	Pagans.	...
Giliaks	Pagans.	5,000
Ainos	Pagans.	3,000
Eskimos	Pagans.	500

SLAV STOCK.

Races.	Religions.	Population.
Great Russians	Orthodox	4,500,000
Little Russians	Orthodox	
Poles	Roman Catholics mostly	

SUNDRIES.

Races.	Religions.	Population.
Chinese	Buddhists.	10,000
Manchus	Buddhists.	
Koreans	Buddhists.	7,000
Japanese	Buddhists.	400
		5,262,000

The Mongolian race is in Siberia best represented not by the Kalmuks, but by the less known Buriats, who have been long settled on both sides of Lake Baikal. Previous to the Russian conquest all were still addicted to the old Shamanist religion of Siberia. But towards the close of the seventeenth century those dwelling east of Lake Baikal adopted Buddhism, while most of the others conformed to the Orthodox Church. Like most Mongolian peoples, the Buriats are of a decidedly phlegmatic temperament, betraying such an inborn disinclination to work that they often need the stimulus of actual hunger to exert themselves in any way. They are stolid, reserved, sullen, and uncourteous to strangers. Through the Russians they have acquired a passionate love of drink and tobacco, and children eight or nine years old are now often met with Chinese pipes in their mouths. The Buriats are in other respects a harmless, peace-loving people. Amongst them murder is rare, and highway robbery unknown, although they are still prone to acts of petty theft. Formerly nomads and stock-breeders, they have recently become successful agriculturists, and also show a marked capacity for industrial pursuits, often proving more skilful than their Russian teachers.

Beneath an outward show of Buddhism and Christianity, the Buriats, like so many other Siberian peoples, are still at heart genuine Shamanists. The Shamanistic cult, which is based entirely on oral tradition, and which is little removed from nature-worship, was formerly universally diffused throughout Siberia. But it could scarcely hope long to resist the attacks of the Buddhist propaganda, supported as this was by a zealous priesthood and a rich religious literature.

Of the Buddhist Sacerdotal order there are three degrees in Siberia, the two first alone bearing the title of Lamas.

Notwithstanding their ignorance, the Lamas have betrayed a fanatical zeal in the cause of Buddhism, everywhere suppressing Shamanistic practices, and even successfully resisting the spread of Russian Christianity amongst the aborigines of East Siberia.

Conterminous on the north with the Buriats are the Tunguses, who occupy an enormous domain in East Siberia, stretching from the Yenisei to the Pacific seaboard, and at two points reaching northwards to the Frozen Ocean. In the Lena basin this domain is largely encroached upon by the Yakuts; but the coast lands from the Amur nearly to the Arctic Circle are still almost exclusively held by the various divisions of the widespread Tungus family. The Tunguses contrast most favourably not only with the sluggish Buriats, but with all the other races of Siberia. Travellers are never wearied of extolling their many admirable qualities; and there can be no doubt that they are one of the very noblest types of mankind. They are cheerful under the most depressing circumstances, persevering, open-hearted, trustworthy, modest yet self-reliant, a fearless race of hunters, born amidst the gloom of their dense pine-forests, exposed from the cradle to every danger from wild beasts, cold, and hunger. Want and hardships of every kind they endure with surprising fortitude, and nothing can induce them to take service under the Russians, or quit their solitary woodlands, where they cheerfully face the long and harsh winters, when the snow-storm often rages for days together.

The dress of the Tunguses is picturesque, and even elegant, especially when contrasted with the coarse and slovenly garb of the Buriats. "Surprising resemblances in the designs of the materials seem to show that the Tunguses must at one time have maintained constant intercourse with Japan." [1]

Most of the Tunguses in the Baikal district have been baptized; but Russian orthodoxy has scarcely penetrated below the surface. They look on the rites of the Church as mere formalities, practising them only under compulsion, or in the presence of the Russians. When engaged in the chase, or remote from the European settlements, they are still true nature-worshippers.

Hemmed in and continually encroached upon, especially by the Russians and Yakuts, the Tungus race seems destined to ultimate extinction as a distinct nationality.

[1] *Reclus*, vi. p. 359.

The domain of the Yakuts, who are the most energetic and versatile of all the Siberian peoples, lies mainly on both sides of the Middle and Lower Lena, with isolated settlements on the left bank of the Lower Indigirka and Upper Kolima. This is the north-easternmost point reached by the Túrki race, of which the Yakuts are a distant branch. During their migrations eastwards the Yakuts have become largely intermingled especially with the Tunguses. Their Túrki type has thus become so profoundly modified, that their original kinship with the Western Túrki peoples, from whom they are separated by a vast interval, is now attested chiefly by their speech. At the same time, there is perhaps some exaggeration in the oft-repeated statement that the Lena Yakuts and the Osmanli of Stambul can easily converse together.

While all the other aborigines of Siberia seem to be dying out, the Yakuts are actually increasing in numbers.[1] They have been not inaptly described by Wrangel as "men of iron," and more inured to cold than perhaps any other people in the world. Their territory includes both Yakutsk and Verkhoyansk comprised within the zone of intensest cold in the old world. Yet they seem to be almost indifferent to the rigours of a climate where the glass falls in winter to nearly 50° below freezing point. In a temperature of −32° R., Kennan met them airily arrayed in nothing but a short shirt and a sheepskin, lounging about, joking or gossiping, as if they were enjoying the balmy summer zephyrs of some favoured temperate zone. They are at the same time extremely industrious, skilful artisans and agriculturists, and probably the most intelligent traders in North Asia. From their preternatural cleverness in driving a bargain the less quick-witted Russians have named them the "Jews of Siberia,"

[1] From about 50,000 in the beginning of the century to 200,000 in 1880.

and, unless it be the Chinese, they certainly yield in this respect to no other Asiatic people.

In their greater frugality the Chinese have also the advantage over the Yakut, who, with all his inherent energy and powers of endurance, seldom works except under pressure of actual want.

Although mostly baptized, the Yakuts are no better Christians than the Tunguses and the other "converts" to Russian orthodoxy in Siberia. Beneath the outward parade of Christianity they are not merely Shamanists but true nature-worshippers at heart. With many curious rites they conjure the powers of nature, filling mountain, stream, and valley with many good and evil spirits, whose numbers have been increased by additions from the Calendar and pandemonium of the Russians. Above all there is doubtless a supreme being; but he is too far off to hear their prayers, or too good to need their supplications. It is the evil ones who require to be propitiated, as no harm can come from the good spirits. The two principles of good and evil took part in the creation, the former making the earth small and level, the latter coming and tearing it up enraged, whence the hills and the valleys. And the valleys became river-beds, and the great lakes and seas gathered round about the high mountains.

Beyond and partly overlapping the Yakuts and Tunguses is the interesting group of "Hyperboreans," filling the Chukchi and Kamchatka peninsulas, and occupying a portion of Sakhalin and of the opposite mainland about the Lower Amur. Since Nordenskjöld's expedition round the north-east coast, the Chukchis, who give their name to the north-eastern peninsula, have again been the subject of much controversy. But there can be little doubt that W. H. Dall is right in affiliating them to the Koriaks, who probably form a connecting link between them and the Kamchadales. All these tribes, together with the neighbouring Yukagirs, would be readily grouped with the Mongolo-Tatar peoples, but for their speech, which differs in its structure fundamentally from the Ural-Altaic linguistic family. But it will be seen in Chapter XIII. (Japan), that the Ainos, if not also the Giliaks, stand on a different footing, and must be separated altogether from both connections.

The Koriaks, probably the parent stock of all the sub-Arctic races except the Ainos, possess many commendable qualities, and are especially noteworthy for their gentle and kindly disposition. A harsh word is never uttered against their women, and the children are treated with extreme tenderness. Hence all the more surprising seems the long prevalent practice of despatching their nearest kindred when enfeebled by age or infirmities. But this inconsistency is more apparent than real. The weak and aged are both alike incapable of performing the offices or undergoing the hardships inseparable from the nomad state. Hence to these untutored children of nature it seems a merciful act to spare them a lingering death by this means.

Essentially distinct from the Koriaks are the Lamuts, who dwell on the shores of the Sea of Okhotsk, and who are evidently a branch of the Tungus or Manchu family.

The Kamchadales, or aborigines, of the Kamchatka peninsula, differ both in speech and appearance from the neighbouring Koriaks. They are favourably spoken of by Kennan and other travellers who have visited them, but since the year 1780 they have been reduced to one-half their former number by disease, famine, and other troubles. In summer they spear the salmon as they ascend the stream, and cultivate a little rye, potatoes, and turnips, besides which they keep some cattle and barter their furs with the Russians for tea and sugar. The interior of their houses is scrupulously clean, the walls, roof, and floor being planked over with rough but spotless birch boards, while the windows are adorned with chintz curtains and the walls hung with American engravings; but the doors are so low that ingress has to be effected on all fours. All have long been Christians, and the little Byzantine church is never missing in the centre of the villages, which invariably stand amidst clumps of trees on the banks of streams abounding in fish. Some of the northern islands of the Kurile Archipelago are inhabited by the Kamchadales.

In West Siberia nearly all the still surviving aborigines are members of the widespread Finnish race. The Finns are supposed to have come originally from the Altai and Sayan highlands, where they are still represented by the Soyots. These Soyots, by many regarded as the parent stock of the race, occupy a considerable area about the head-waters of the Yenisei, on both sides of the Russo-Chinese frontier, reaching from the Tanu-ola range northwards to the Krasnoyarsk district, and from the Upper Yenisei eastwards to the Buriat domain about Lake Baikal. They are extremely skilful artisans, and seem to have inherited the arts of the so-called "Chudes," an extinct prehistoric race, traces of whose culture are still met in various parts of Siberia.

The other branches of the Finnish stock in Siberia are the Ostiaks, Samoyedes, and Voguls. The Ostiaks are scattered in isolated groups along the Ob basin northwards to the estuary, and eastwards to the Yenisei between Yeniseisk and Turukhansk. In this wide domain of some 400,000 square miles they scarcely number 25,000 souls altogether, and seem to be everywhere either rapidly dying out or becoming absorbed amongst the surrounding Russian settlers. Their old national organisation is completely broken up, and they have almost ceased to dwell in settled abodes since the destruction of their villages and strongholds by the early Russian invaders of West Siberia. They formerly paid tribute in peltries, but with the gradual disappearance of the fur-bearing animals the taxes have been raised in specie. Such is the depressed condition of the race, that it has been proposed to distribute the children amongst the Russian peasantry, and leave the adults to die out in the course of nature.

Even after death the prospects of the Ostiaks are not of the brightest; for although there is a "third world," where there are no more bodily ailments, they never reach this stage, but are confined to the "second world," a far less happy abode lying somewhere beyond the Ocean, away north of the Ob estuary. Their belief in Shamanism is still unshaken, and nowhere else does the wizard, or medicine-man, enjoy more influence than among the Ostiak tribes. The brave man may possess bodily strength, they say, but the Shaman has the words of wisdom. The strong man may hurl the dart, but its course is directed by the Shaman, through whom alone it hits or misses the mark.

A still more primitive people are the Samoyedes, whose territory lies mostly within the Arctic Circle from the head-waters of the Khatanga westwards to the Kanin peninsula. They are usually represented as dwelling on the Arctic seaboard; but the eastern or Tavgi branch do not appear to have anywhere quite reached the coast, which they hold to be the rightful domain of the "white bear people."

Some of the Samoyedes are baptized, but all alike are true Pagans, or idol-worshippers. Their gods are carnivorous, and are fond of raw flesh, which is accordingly thrust between their teeth at stated times. The Khatanga tribes keep entirely aloof; but some of the others have already become assimilated to the Russians, while others on the Ket and other eastern tributaries of the Ob are spoken of as "Ostiaks."

The eastern slopes of the Urals are occupied by the Voguls, who reach southwards to Yekaterinburg, and eastwards along the Konda valley nearly to the Lower Irtish. Reduced merely to a handful, the Voguls were formerly a powerful people, representing the primitive stock, whence came the Magyars of Hungary.

"Like many other Finnish peoples, the Voguls have their family *totems* tattooed on their heads, arms, and legs.

"The Voguls are probably the least sociable of the Siberian aborigines. In summer they live in isolated family groups; in winter they pitch their tents or build their huts far apart from each other. Even the family spirit seems but slightly developed. The hunter may have one or more wives according to his means, but the least disturbance dissolves the union, and the husband will then often live quite alone, accompanied only by his reindeer and dog."[1]

8. *Topography: Omsk—Tobolsk—Yekaterinburg—Tomsk—Beresov—Obdorsk—Smeinogorsk—Barnaul—Semipalatinsk—Krasnoyarsk—Irkutsk—Kiakhta—Vladivostok.*

The Siberian towns claim consideration rather for their prospective than their actual size and importance. Many hamlets, consisting of fifty or sixty log-huts, such as Turukhansk or Okhotsk, figure in large letters on the maps, either because they are the official centres of administration, or because they are the only stations or settlements occurring for hundreds of miles in the almost

[1] *Reclus*, vi. p. 340.

uninhabited regions of East Siberia. Excluding Yekaterinburg, which is comprised in the European Government of Perm, there are only three towns altogether with upwards of 20,000 inhabitants—Irkutsk, Omsk, and

OKHOTSK.

Tomsk ; and beside Yakutsk, there is one only in the whole of East Siberia, Krasnoyarsk, whose population exceeds 10,000. Yet some of these towns cover vast spaces, being laid out with broad, straggling streets, and low wooden houses in the midst of extensive plantations and waste grounds. Most of them are concentrated in the Ob basin, which alone contains about four-fifths of the entire population of North Asia. All, except Tiumen, are comparatively new, dating since the Russian conquest,[1] and none of them can boast of any "monuments" in the European sense. The dreary monotony of log-cabins and wooden huts is seldom relieved by anything beyond a few whitewashed houses and public buildings, such as the official quarters, barracks, convict prisons, and the like, all designed on a uniform plan, and imposing only because of their mean surroundings.

In the Ob basin the largest place is Omsk, capital of West Siberia, which takes its name from the Om, standing at the confluence of that river with the Irtish. Here the Government buildings are of brick, and besides some extensive wastes Omsk contains several churches, including one for the Kossaks, one for the Roman Catholics, and a third for the Protestants, besides a large mosque for the surrounding Bashkirs and Kirghiz. Lower down the Irtish, Tobolsk presents from a distance a really picturesque appearance, perched as it is on a bluff facing the junction of the Tobol from the north-west. From Tobolsk the steamers ascend the Tobol to the Tura, on which stands Tiumen, which claims to be the oldest place in Siberia. Above the Tura the Tobol is joined by the Isset from the Urals, about the head-waters of which stands Yekaterinburg, centre of the important mining industries of this region. Standing almost on the verge of the two continents, Yekaterinburg presents more the aspect of a European city than any other in Siberia. It is the pre-

[1] Even Sibir, capital of the Tatar kingdom, overthrown by Yermak in 1581, has been washed away, with the bank of the Tobol River on which it stood. But Tumen (Tiumen), which appears on Herbertstein's old map (1549), is still a flourishing place.

sent Asiatic terminus of the Russian railway system. Here are a meteorological observatory and the "Society of the Ural Naturalists," besides extensive porphyry, malachite, jasper, rock-crystal, and other ateliers, whose products adorn the palaces and museums of every city in Europe.

In this district are several large mining stations, the general *entrepôt* for which is Irbit, at the junction of the Rivers Irbit and Nitza. A mere village during most of the year, Irbit becomes the Siberian Nijni Novgorod during the annual fair, when it is visited by 15,000 to 20,000 dealers, whose wares exchange hands to the value of about £1,750,000.

In the eastern parts of the Ob basin the chief place is Tomsk, on the Tom above its junction with the Ob. Tomsk is the centre of a very large local trade, and some of its streets, with their bright and well-stocked shops, present quite a cheerful aspect. Here was laid, in 1880, the first stone of the Siberian University, which with its future botanic garden and other branches is destined to render Tomsk the intellectual centre of North Asia. Here begin the extensive gold-fields discovered in 1830, which, before the opening of the Californian El Doradoes, yielded more of the precious metal than the whole of America.

Some 640 miles below Tobolsk, on the verge of the forest and tundra zones, lies the little town of Beresov, on the banks of the Sosva, near its junction with the Ob.

In the Altai uplands the chief mining stations are Ziryanovsk, Smeinogorsk, and Verkhniy Pristen. At the latter place the ores are shipped on a peculiar kind of craft called "karabass," and floated down the Irtish to Ust-Bukhtarminsk. The road beyond Ust-Kamenogorsk to Smeinogorsk crosses a cultivated hilly district, dotted over with several large villages. Here is probably the oldest mine in the Altai region. But the operations have been lately suspended, and replaced by the smelting works for the ores brought down from Ziryanovsk, Rybinsk, and elsewhere. Not far from Smeinogorsk is the famous imperial stone-polishing establishment of Kolivan, where the finest porphyries, jaspers, and marbles are dressed, but only for such large objects as chimney-pieces, monumental vases, tables, slabs, and the like.

From Kolivan the way leads through the Baraba steppe north to Barnaul, the chief town in the Altai highlands. This place, which lies on the left bank of the Upper Ob, in an extremely fertile and flourishing district, possesses several scientific institutions and technical establishments.

South-west of Barnaul lies the busy agricultural town of Semipalatinsk, most of whose inhabitants are Siberian Tatars, distinct from the Kirghiz, partly Sunnis, partly Christians, with seven mosques and two churches. Here the broad, sandy streets, lined with low wooden houses, give the place the appearance of a city of dunes. The road leading thence southwards to Lake Balkhash passes by Sergiopol, whence a view is commanded of the snowy Tarbagatai range.

In the Yenisei basin the first place of any consequence crossed by the main route from Tomsk is Krasnoyarsk on the Upper Yenisei, in a fertile district, where the streams wash down auriferous sands. Below the junction of the Yenisei and Angara lies the little town of Yeniseisk, centre of a vast administration, and near the head of the Angara, where it emerges from Lake Baikal, stands the city of Irkutsk, capital of East Siberia. In some respects Irkutsk is the most important, as it is the largest, city in Siberia. Frequently wasted by fires, it has always risen rapidly from its ashes, and, thanks to its vital position on the trade and military route through Kiakhta to China, it must always remain an important place. Irkutsk is also a great centre of the fur trade, and amongst its public institutions are a gymnasium, library, theatre, and a flourishing geographical society.

In Transbaikalia the main highway leads up the Selenga valley to the Chinese border, where stand the well-known trading towns of Kiakhta on the Russian and Maimachin [1] on the Chinese frontier. Through these places pass the great tea, silk, and rhubarb caravans from China to Russia. But Kiakhta has lost much of this trade since the opening of the Suez Canal, which has developed a sea-borne traffic between the Chinese free ports and Odessa on the Black Sea.

In the north-eastern regions Verkhoyansk on the Upper Yana and within the Arctic Circle is noteworthy as perhaps the coldest place on the globe. But here the most considerable town is Yakutsk on the left bank of the Middle Lena above its junction with the Aldan. This is the proper capital of the Yakut nation, the most enterprising and prosperous of all the indigenous races.

There are no large towns in the Amur province, Blagoveshchensk, the capital, having scarcely 3000 inhabitants. But on the coast are the two important naval and trading stations of Nikolayevsk and Vladivostok, the former at the mouth of the Amur, the latter close to the Korean frontier on the Sea of Japan. But Nikolayevsk has lost much of its importance as the port of entry of the Amur, owing to its severe climate and the intricate navigation of the river, which is usually blocked with ice for six months in the year. Vladivostok—that is, "Ruler of the East"—was founded lower down the coast in 1860 to obviate these inconveniences. Considerable sums have already been spent on its docks, piers, arsenals, and fortifications, with the intention of making it the chief naval station on the Pacific seaboard.

[1] That is, *Mai-Mci-Chin* = "The Chinese Mart."

9. *Highways of Communication: The Trakt—Railway Projects.*

Few regions present fewer obstacles than Siberia to the general movement of the population. Doubtless many parts of the taiga are almost impenetrable, and the great rivers run mainly in the direction of the meridian. But beyond the limits of the dense forest zone the open steppe and boundless rolling plains stretch with little interruption from the Urals to the Pacific seaboard. Even of the great rivers, the Middle Amur flows east and west, while many of the tributaries of the others follow the same direction. A great navigable highway, broken only by two short portages between the Ob, Yenisei, and Lena basins, thus affords a natural line of communication across North Asia to within a short distance of the east coast of the Lena and Aldan, and quite to the coast by the Amur. This circumstance, combined with the sparseness of the population, explains the surprising rapidity with which the whole land was overrun by the Kossaks within twenty years of Yermak's second expedition, resulting in the capture of Sibir in 1581. Since then the country has been traversed with comparative ease in all directions by naturalists and scientific explorers, such as Gmelin (1733-42), Pallas (1771-2), Lesseps (1787-8), Wrangel (1821-3), Erman (1828-30), Castren (1842-3), Middendorff (1843), Radde (1855-9), Venyukov (1856), Krapotkin (1865-6), Finsch (1876), and many others. Yet the survey of the seaboard can scarcely be said to have been completed till Nordenskjöld successfully made the north-east passage in 1878-9.

From Yekaterinburg, the present Asiatic terminus of the Perm railway, the great caravan route runs across the Tobol and Ishim valleys to Omsk. Here one branch follows the Irtish valley southwards through Semipalatinsk and Sergiopol to Lake Balkhash and the Ili valley. But the great Siberian trunk line is continued from Omsk for 400 miles across the Baraba steppe to Kolivan on the Ob. Here it is deflected southwards to Tomsk, whence it runs due east as far as Krasnoyarsk on the Yenisei, and thence north-east to Irkutsk on the Angara. From Irkutsk the communication is maintained both by steamer across Lake Baikal and by land round its southern extremity to the Selenga valley, where the trade route runs southwards to Kiakhta on the Chinese frontier. Another line is continued eastwards across Transbaikalia and over the Yablonovoi range down to Chita on the Shilka River. Here the road mainly follows the course of the Shilka across the Nierchinsk steppe to the Amur and down the main stream through Blagoveshchensk to the Usuri confluence. Here it ramifies northwards along the Lower Amur to Nikolayevsk, southwards up the Usuri valley and over the Sikhota-alin coast range down to the Pacific at Valdivostok.

The *trakt*, as the great trunk line from Perm to Kiakhta is called, has been one of the chief instruments in developing trade and diffusing civilising influences throughout Siberia. It is traversed by long lines of waggons and sleighs, which will often make from 40 to 50 miles a day. Along the route the various halting stations have gradually grown into considerable centres of population, generally consisting of a single line of two-storied houses from one to two miles long.

From Tomsk the old route to Yeniseisk is continued north-eastwards to Yakutsk, whence one road branches northwards over the Verkhoyansk range down to the Yana valley at Verkhoyansk, while a second runs due east across the Aldan valley and over the Stanovoi plateau down to Okhotsk on the Sea of Okhotsk. From Verkhoyansk a track followed by Müller leads to Nijnekolimsk on the Arctic, and thence across the Chukchi country to the Gulf of Anadir, in the Bering Sea. Okhotsk also communicates round the head of the Sea of Okhotsk and down the Kamchatka peninsula with Petropavlovsk. This line was traversed by Lesseps in 1777-78, and again in 1865-66 by Kennan, who also explored the region between the Gulf of Anadir and the Sea of Okhotsk.

The railway from Perm over the Urals to Yekaterinburg is not yet connected with the European system. But it forms the first section of the great North Asiatic trunk line, which is intended ultimately to run from the Urals to Pekin, a distance of about 3500 miles. A second section of 225 miles from Yekaterinburg to Tiumen is now in progress, but the estimate for the whole line being about £80,000,000, its completion is a somewhat remote contingency. At the same time, there are few engineering difficulties to contend with, for several broad openings and moderately sloping depressions lead from the Aralo-Caspian, Ob, and Amur basins to that of the Hoang-ho. Of these the most serviceable are probably those stretching from Lake Balkhash between the Ala-tau and Tarbagatai ranges, and from Lake Saisan up the Black Irtish between the Tarbagatai and Western Altai to the Mongolian plateau. Farther east the Selenga valley gives access from Transbaikalia through Kiakhta to Urga, on the northern verge of the Gobi desert.

10. *Administration: Education—Industries.*

Excluding the portions attached to Europe and Turkestán, the whole of North Asia comprises the two great administrative divisions of West and East Siberia, whose capitals are Omsk and Irkutsk respectively. Each of these is subdivided into a number of governments and provinces, which in their turn are distributed into circles and districts. The military system still largely prevails in the Amur Government, which is divided into Kossak "regiments" and "battalions." But elsewhere the administration is mainly modelled on that of European Russia. The municipal, judicial, and ecclesiastical departments are all theoretically based on the same uniform plan. But owing to the vast distances and the difficulty of communicating in winter with the central authorities, the local officials and commanding officers enjoy almost absolute control in their several jurisdictions. In some places the natives hardly understand the existence of the "White Czar," and know of no higher power than the district magistrate. But in the more settled parts, and especially in the Tobolsk, Tomsk, and Irkutsk Governments, these functionaries are fully as responsible to the higher authorities as those of European Russia.

In some respects the people enjoy even greater personal liberty than in the west. In Siberia there are no nobles or specially privileged classes; serfdom never gained a footing in the land, and through the increasing traffic with California, ideas of freedom and independence, unknown to the Western Mujiks, have already penetrated into Siberian society. "As we advance eastwards," remarks a Russian writer, "the freer and more independent do we find life and opinions among us."

Education is still in a very rudimentary condition. In some places the Russian settlers have even forgotten their mother-tongue and become assimilated to the surrounding aborigines. In East Siberia, with a population of 1,500,000, there were only 283 schools, attended by 8610 scholars, in 1870, and in the whole of Siberia there were only two periodicals, a weekly and a monthly, in 1878. Nevertheless a certain intellectual life has been fostered in the larger centres of population. There are geographical and other learned societies in Irkutsk, Yekaterinburg, and elsewhere, and in 1880 the late Czar at last gave his consent to the establishment of a "Siberian University" in Tomsk, from which much is expected.

Meanwhile agriculture is in such a backward state that the crops scarcely yield sufficient for the local consumption. Stock-breeding, however, is conducted on a very extensive scale, and notwithstanding the ravages of the "Siberian plague," said to have first broken out in the Baraba steppe, vast herds of cattle and of horses, almost in a wild state, are bred on the rich pasturages of the southern plains and upland valleys. On the other hand, both the chase and the deep-sea fisheries have fallen off with the gradual disappearance of the fur-bearing animals and of the large cetacea from the northern waters. The mining industry also, although still of primary importance, has suffered by the competition of the Californian, Australian, and South African fields. Of other industries, perhaps the most important is distilling. Vast quantities of coarse spirits are produced from grain and potatoes, and retailed in the taverns which abound in all the towns, and especially in the mining districts. Returns of these various resources will be found in the subjoined tables.

11. *Statistics.*

Areas and Populations.

Administrations.	Area in sq. miles.	Population, 1870-77.
ATTACHED TO EUROPE AND TURKESTÁN.		
Asiatic portion of Government of Perm .	54,050	1,105,361
Asiatic portion of Government of Orenburg	62,060	627,120
Nikolayevsk (Government of Turgai) . .	33,990	?90,000
WEST SIBERIA.		
Tobolsk (Government)	588,660	1,086,848
Tomsk (Government)	344,950	838,756

EAST SIBERIA.

Administrations.	Area in sq. miles.	Population, 1870-77.
Yeniseisk (Government)	1,028,540	396,783
Irkutsk (Government)	322,290	358,629
Yakutsk (Province)	1,553,430	238,067
Transbaikalia (Province)	85,800	430,780
Amur (Province)	112,840	28,589
Maritime Province	?	50,510
Usuri (Territory)	?	
	5,215,110	5,251,443

APPROXIMATE POPULATION OF SIBERIA ACCORDING TO RACES (1880).

Mongol Stock.		*Finnish Stock.*	
Tunguses	80,000	Voguls	4,500
Buriats	250,000	Samoyedes	25,000
Kalmuks	20,000	Ostiaks	25,000
Chinese, Manchus	10,000	Soyots	8,000
Koreans	3,000		
Sub-Arctic Races.		*Slav Stock.*	
Chukchis	12,000	Great Russians, Little Russians, Poles	4,500,000
Koriaks	5,000		
Kamchadales	3,000		
Giliaks	5,000		
Ainos	3,000		
Yukagirs	1,600		
Túrki Stock.		*Sundries.*	
Yakuts	200,000	Gipsies	5,000
Tatars	80,000	Dolgans	500

CHIEF TOWNS IN SIBERIA.

Irkutsk	32,000	Semipalatinsk	10,140
Omsk	30,553	Smeinogorsk	10,000
Tomsk	31,552	Vladivostok	9,000
Yekaterinburg	25,133	Kiakhta	9,000
Tobolsk	20,132	Yeniseisk	7,140
Barnaul	14,000	Biisk	6,350
Krasnoyarsk	14,000	Ishim	5,800
Tiumen	13,144		

GROWTH OF POPULATION IN SIBERIA.

1796	1,193,145	1870-73	3,340,362
1816	1,540,424	1884	4,093,535
1869	3,327,627		

Yield of the Altai Mines from 1745 to 1860—3,568,750 lbs., valued at £10,000,000.

PRODUCT OF THE ALTAI MINES IN 1876.

Silver	25,250 lbs.
Gold	2,665 ,,
Copper	1,380,000 ,,
Nickel	64,050 ,,
Pig and Cast Iron . . .	1,730,750 ,,
Total value	£350,000

Yield of the Siberian Gold Mines from 1726 to 1880 .		£120,000,000
Present average annual yield of same . . .		£1,200,000
Yield of the Transbaikal Silver Mines to 1880 . .	7,500,000 lbs., valued at	£20,000,000
Present average annual yield of same . . .		£200,000
Gold-seekers in East Siberia (1877) . . .		51,272
Annual yield of Iron in Siberia and Urals . .		492,000 tons.
Yearly export of Furs from Siberia . . .		£600,000

Factories in Siberia (1876), 1100 ; hands employed, 4000 ; yield, £1,000,000.

EXILES TO SIBERIA FROM 1823 TO 1858.

Men, 238,482 ; Women and Children, 42,844 ; Women and Children voluntarily accompanying their friends, 23,285 : total, 304,618.
Total during the last 250 years, about 1,000,000.
Present yearly average, 8000 to 9000.

PUBLIC INSTRUCTION IN SIBERIA EXCLUSIVE OF THE URALS.

Elementary Schools (1876), 600.
Attendance—14,000 Boys ; 2200 Girls : Total, 16,200.
Higher Schools, 96
Attendance, 3800.

THE GREAT RIVERS OF SIBERIA.

	Length. Miles.	Area of Drainage. Sq. miles.	Nav. Waters. Miles.
Ob-Irtish . . .	3400	1,420,000	9000
Yenisei-Angara . .	3300	1,180,000	5000
Lena-Vitim . . .	3280	1,000,000	6000
Amur-Argun . . .	3066	800,000	6000

SECTION D.

EASTERN ASIA: BUDDHIST STATES.

CHAPTER XII.

CHINESE EMPIRE.

1. *Boundaries—Extent—Area.*

BY the expression Chinese Empire will here be understood all the lands, either absolutely administered from Peking, or indirectly forming part of the Chinese political system. In this chapter will therefore be treated not only China proper, Mongolia, Manchuria, Kulja, and Kashgaria, but also Tibet and Korea, which, notwithstanding a certain more or less real autonomy, are practically controlled in all their foreign relations by Chinese diplomacy.

The region thus defined, besides the political unity derived from this circumstance, is further united by the bonds of race and religion. For the vast majority of its inhabitants belong to various modified forms of the Mongolic type, and constitute various branches of the Buddhist religious world.

Compared with the other great states of the world, China takes the foremost rank in respect of population, while in extent yielding only to England and Russia. It occupies the whole of Central and East Asia, the Indo-Chinese peninsula alone excepted. For by the expression Central Asia should properly be understood the great continental tablelands confined north and south by the Altai and Himalayan mountain systems, and stretching from the Pamir—that is, from the converging point of these systems—eastwards to China proper. On these tablelands of the Pamir, Tibet, and Mongolia rise the great continental rivers—Oxus, Sir, Indus, Brahmaputra, Yang-tse, Hoang-ho, Amur, Ob, and Yenisei—which flow west, south, east, and north, to the Aralo-Caspian basin, the Indian, Pacific, and Arctic Oceans. Excluding the Pamir, which is at present a sort of neutral land between the three empires of British India, Russia, and China, all converging at this point, these central plateaux constituting the true heart of the continent, and determining its great water systems, form politically an integral portion of the Chinese Empire. Consequently to China alone belongs Central Asia, although the expression has found a place in the official language of Russian beaurocracy.

Thus comprising Central and East Asia, the Chinese Empire is almost everywhere clearly delimited, on the north and north-west by Asiatic Russia, on the south and south-west by British India, on the south-east by Indo-China, and on the east by the Pacific Ocean. From the Kizil-art, the water-parting of the Oxus and Tarim basins, about 75° E. long., it stretches across 53 degrees of longitude for a total distance of 3000 miles to the east coast of Korea in 128° E. long.; and from the great northern bend of the Amur on the Siberian frontier across 34 degrees of latitude for 2400 miles southwards to the island of Hainan. Within these limits it has a total area roughly estimated at 4,500,000 square miles, with a population of perhaps 350,000,000.

An estimate based on official returns for 1842 gave 405,000,000 for China proper, to which probably 10,000,000 should be added for the rest of the empire. But since then enormous losses were caused not only by the wars of the Taipings and Muhammadan Dungans in the south and north-west, but also by the inundations and shiftings of the Hoang-ho, and the terrible famine by which the northern provinces have been wasted in recent years.

2. *Relief of the Land: The Kuen-lun Mountain System—The Nan-shan, Khingan, and Nan-ling Ranges—The Cross Ridges—Plateaux and Depressions.*

The great frontier mountain systems of the Himalayas and Altai, enclosing the central plateaux on the south and north, and converging westwards round the Pamir, as well as the Tian-shan, lying partly in Russian and partly in Chinese territory, have been described in previous chapters. Our information with regard to the internal systems, especially in the western regions, is still extremely defective. In a general way it may be stated that in the west—that is, in Tibet and Mongolia—the great ranges run mainly west and east, and assume somewhat the character of bold escarpments to the great central tablelands, which stretch at different elevations from the Himalayas northwards to the Altai. But in the east—that is, in China proper—the direction is rather north-east and south-west, and even north and south. Here also the tendency is, especially on the Tibetan and Indo-Chinese frontiers, to broaden out into extensive and irregular highland regions, in which the general direction of the main ridges is indicated by the course of the great rivers flowing from the Tibetan plateau to the Chinese and Indo-Chinese seaboards.

Thus we have in the west the great Kuen-lun range breaking away from the Karakorum, and running under diverse names, such as the Tuguz-daban (9000 to 10,000 feet), the Altyn (13,500), and Kilien-shan (Nan-shan

(13,600), mainly west and east along the northern edge of the Tibetan plateau as far as the Chinese frontier, about the Tsaidam and Kuku-nor district.

The Altyn-tagh section of the system was one of the most surprising results of Prejevalsky's expedition in 1877 to the Lob-nor. For this snowy range rises abruptly to an elevation of 13,000 to 14,000 feet within 120 miles of the lake, where it was formerly supposed that the sands extended for several degrees of latitude southwards to the scarp of the Tibetan plateau. The discovery is of great importance as helping to explain many hitherto unintelligible passages in Chinese records in connection with the wars and migrations of the Huns and Mongols.

The Tsaidam plains are skirted on the south by the parallel Shuga (15,600) and Burkhan-Buddha (15,800) ranges, which seem to branch off from the Altyn south-eastwards, and are continued far into China proper, between the Hoang-ho and Yang-tse valleys, under the names of the Hsi-king-shan, Tsing-ling, Funiu-shan, and Mu-ling. German orographists group these ranges collectively as the Eastern Kuen-lun, of which they regard the Burkhan-Buddha[1] as the central, and the Tuguz-daban and Altyn as the western section. The Kilien-shan, or Nan-shan, thus sinks to the position of a subordinate northern offshoot of the Kuen-lun system, which stretches from the Kizil-art (Eastern Pamir range), with many interruptions, across North Tibet and Central China, for 2700 miles eastwards to the Lower Yang-tse-kiang.

The Nan-shan is again continued north-eastwards, partly along the great wall of China, through the snowy

[1] The Burkhan-Buddha, which, notwithstanding its great elevation, nowhere reaches the snow-line, runs for 130 miles along the northern edge of the Tibetan plateau, and forms the southern limit of the marshy Tsaidam steppe.

Kuliang and Liang-chu, the Ala-shan[1] (11,600), the Khara-Narin-ula, In-shan,[2] Munni-ula (9000), Sirung Bulik, Suma-hada, Shara-hada, and other parallel ridges, to the head-waters of the Lohan, at the converging point of Mongolia, Manchuria, and China, north of Peking. Here the system is gradually contracted till it forms a junction with the volcanic Great Khingan range, which runs between Mongolia and Manchuria, due north to the Amur, near the confluence of the Argun and Shilka.

East of the Khingan, Manchuria is occupied in the north by the Ilykhuri-alin and Duss-alin, skirting the right bank of the Amur, and by the Shanyan-alin (10,000 to 11,000 feet), forming the frontier towards Korea. This peninsula is mountainous throughout, and especially on the east side, where a coast-range, here and there flanked by parallel inner ridges, forms a southern extension of the Sikhota-alin coast-range of the Russian Maritime Province. Geologically the Kuen-lun, or at least its western section, is of far older date than the Himalayas. The prevailing rocks are syenitic gneiss and more recent triassic formations, whereas in the southern range is comprised the whole series between the palæozoic and eocene deposits. Hence the Kuen-lun, rather than the Himalayas, must be regarded as the eastern extension of the Hindu-Kush, and the true backbone of the continent in this direction.

In China proper the provinces of Shan-si and

[1] The Ala-shan mountains rise abruptly above the left bank of the Hoang-ho, and run for 150 miles north and south between Kan-su and the Ala-shan country, Mongolia. They culminate with Mount Bugutin (11,600 feet), but nowhere reach the snow-line.

[2] "The natives do not know this name, and have their own names for different parts of the range. In a wider sense the term In-shan applies to all the mountains from the northern bend of the Hoang-ho, through the Chakar country to the confines of Manchuria" (Prejevalsky's *Mongolia*, i. 153).

Pechili are traversed by the Siwe-shan, Man-tu-shan, Tao-tsu-shan, running south-west and north-east from the Hoang-ho. Parallel with this system are the Utai-shan, Luyen-shan, Mian-shan, Niao-ling, and other ridges, filling the whole of Shan-si, and continued beyond the Hoang-ho by the Ming-shan and Sung-shan through North Ho-nan. Here a junction is effected with the Funiu-shan section of the Eastern Kuen-lun, which at this point forms the boundary between the provinces of Ho-nan and Hu-pe.

South of the Lower Yang-tse the whole of south-east China is occupied by extensive and nearly parallel chains, such as the Shi-shan (16,000 feet), the Ja-ling, Ta-yu-ling, Yung-nien-ling, Tung-lo-ling,[1] and Pu-ling, whose normal direction is also from south-west to north-east. This system merges through the Nan-ling range with the southern and south-western highlands of Kwang-si and Yunnan on the Indo-Chinese frontier.[2] Here we enter one of the least-known regions on the globe, on which some little light has recently been thrown by Gill, Riley, Desgodins, and Baber. But its thorough exploration is needed to solve the many obscure questions connected with the sources and water-partings of the Yang-tse, Min, Mekhong, Salwin, Irawady, all of which flow for long distances in close proximity through the narrow longitudinal valleys formed by the Cross Ranges stretching from the Yunnan highlands along the Tibeto-Chinese frontier between the Brahmaputra and Yang-tse basins northwards towards the Kuen-lun system.

[1] The Tung-lo-ling, an easterly section of the Nan-ling, separates Hunan and Kiang-si from Kwang-tung, but does not form a true water-parting between the Yang-tse and Canton basins, for it is pierced by a stream rising north of it and flowing south to the Kiang or North River of Canton. It is crossed by the important Che-ling (1200 feet) and Mei-ling passes.

[2] To the whole of this south-eastern system, which in the Ping-ya-shan rises above the snow-line, Richthofen gives the collective name of Nan-shan.

These "Cross Ridges," as Blakiston calls them, are obviously in a geological sense an eastern extension of the Tibetan plateau itself, which has here been cut up into parallel chains running mainly north and south. The beds of the running waters, to which the Ranges would seem to owe their existence, lie still at elevations of from 8000 to 10,000 feet above the sea, and the great trade route from Lassa through Batang to West China maintains a normal elevation of no less than 12,000, with occasional passes nearly 17,000 feet high. The ranges between the River Kinsha (Yang-tse) and its tributaries, the Yalung and Min, rise above the snow-line which Gill here fixes at from 14,000 to 15,000 feet. The Nenda, or "Holy Mountain," east of the Upper Kinsha valley under the parallel of Batang, is some 20,000 feet high, and sends down glaciers to all the surrounding valleys. Farther east the snowy peaks of Surong, running north-west and south-east, are nearly if not quite as high, while above a parallel chain east of the Yalùng River, the Jara, or "King of Mountains," commands all the surrounding heights by 5000 feet. This chain is continued northwards to the Bayan-Khara system, where some of the crests may possibly rival those of the Himalayas themselves. Amongst them are the Ngomi-shan, ascended in 1879 by Riley; the Siwelung-shan, or "Snow Dragon;" the "Seven Nails," supposed by Gill to have an altitude of 19,000 to 20,000 feet.

Between the Tarbagatai and Zungarian Ala-tau lies the depression of the so-called "Zungarian Strait," through which access is gained from Turkestán along the Sassik-kul, Ala-kul, Ebi-nor, Sir-nor, and other eastern extensions of the Balkhash lacustrine system, into the Mongolian plateau. In the same way between the Barluk-Orkochuk and Little Altai lies the valley of the Black Irtish, which again gives access to North Mongolia through Lake Ulungur and River Urungu. For the valley of this river sweeps round the south-eastern extremity of the Little Altai to the Kobdo plateau, where it has its source. The Kobdo plateau itself stretches from the Little Altai beyond Lake Ubsa to the Tanu-ola range, by which it is separated from the valleys of the Upper Yenisei and Selenga. From this point the North Mongolian plateau is broken by no other prominent range, until we reach the Great Khingan, by which, as already stated, it is separated from Manchuria.

From this rapid survey it appears that China proper is by no means a vast lowland plain formed by the alluvia of the twin Rivers Hoang-ho and Yang-tse. It is a distinctly highland region almost everywhere occupied by vast mountain systems, except along the lower courses of the great streams and on the east coast. And even here the lowland formation is broken in the upland peninsula of Shan-tung, projecting seawards between the old and new channels of the Hoang-ho, and culminating at its extremity in the Kuan-in-shan (2900 feet).

It also appears that the great Central Asiatic plateau consists in reality of several distinct sections differing enormously in elevation and extent from each other. These sections are grouped round the central basin of the Tarim, which is in fact rather a depression than a plateau, falling to little over 1600 feet above sea-level. South of it the land rises in successive stages from 3000 to 6000, 10,000, and 15,000 feet, the probable mean altitude of the Tibetan plateau, at once the most elevated and extensive on the globe. Above this vast tableland the intersecting ranges attain altitudes of from 20,000 to 25,000 feet, culminating in the southern scarp of the Himalayas with peaks ranging from 26,000 to 29,000, the highest summits on the surface of the earth.

North of the Tarim basin the land also rises in

terraces of 3000, 6000, and 15,000 feet, here culminating with the Tengri-khan (25,000), central and highest point of the Tian-shan. Beyond the Tian-shan the ground again falls gradually to about 1500 feet in the Zungarian depression (Tian-shan Pe-lu), north of which it attains a height of 7000 or 8000 feet in the Kobdo plateau. This elevation is maintained in North Mongolia eastwards to the head-waters of the Amur. But in the central parts the Gobi desert stretches from Lob-nor at a mean height of probably not more than 3000 feet to the Khingan range. Lastly, the closed basin of the Kuku-nor between the Nan-shan and Burkhan-Buddha ranges stands at an altitude of not less than 10,500 feet above sea-level.

3. *Hydrography : Inland Drainage, Lob-nor and Ili Basins—Seaward Drainage, The Hoang-ho, Yang-tse-kiang, Pei-ho, Liao-ho, and Si-kiang Basins—Lakes Kuku, Dangra-yum, Tengri, Pangkong, and Palti.*

Those water systems of Central and East Asia which are altogether comprised within the limits of the Chinese Empire are few in number, and seldom of great extent. Excluding the already-described Amur basin, now shared between China and Russia, not more than five large rivers find their way in independent channels to the Pacific coast, and of these two only, the Hoang-ho and Yang-tse, attain the proportions of great continental streams. The inland drainage also, apart from the numerous small lacustrine closed basins of Tibet, is represented chiefly by the Ili flowing beyond the frontier to Lake Balkhash, by the Ike-aral and Ubsa-nor basins of the Kobdo plateau and the Lob-nor of Kashgaria.

Of these inland systems that of the Lob-nor is by far

the most extensive. This lake, the true position of which was first determined by Prejevalsky in 1877, receives through the Tarim River nearly the whole drainage of the region variously known as "Chinese Turkestan," "Eastern Turkestán," Kashgaria or Jety-shahr. Here the surrounding Tian-shan, Kizil-art, Karakorum, and Kuen-lun, send down numerous streams, including the Ak-sai, Ugen-daria, Shah-yar, Kashgar, Yarkand, and Yurung-kash (Khotan), all of which are collected by the Tarim and carried through a still imperfectly-explored course to the Kara-buran, or west end of Lake Lob. The Tarim also receives from the north the discharge of Lake Bagrach (Bostang-nor) through the Koncheh-daria (Kaidu-gol), which forces a passage through the intervening Kuruk-tagh ridge. But the Cherchen-daria, rising in the Tuguz-daban (Western Kuen-lun), flows through the sands intermittently from the south directly to the lake at the Tarim confluence.

At the Ugen-daria junction the Tarim, which the natives call the Yarkand-daria from its chief head-stream, is from 350 to 400 feet wide, with a depth of 20 feet. Lower down it throws off the Kiok-ala-daria, which, after a course of 75 miles in an independent channel 150 feet wide, again joins the main stream 60 miles above its mouth. Much of its water is drawn off to irrigate the surrounding "tara," or fields, whence the name of *Tarim*, now applied to the river itself.

The lake consists of two sections, Kara-buran, about 18 miles long, and Kara-kurchin, or Chon-kul, 50 to 60 miles long, and nowhere more than 12 wide. Both are connected by the channel of the Tarim, and seem to be little more than 3 or 4 feet deep, except at the junction, where the Tarim is 14 feet deep and 125 wide, with a velocity of 170 feet per minute. The whole basin is little more than a flooded morass, choked with reeds, and gradually disappearing eastwards in saline marshes. But the lake itself was found by Prejevalsky to be fresh, and well stocked with carp and marena (*Coregonus marœna*). "The whole of Lob-nor is equally shallow, only here and there occur occasional pools, 10 to 13 feet deep. . . . But the fact that almost all the lakes of Central Asia show signs of desiccation is well known" (Prejevalsky's *Lob-nor*, p. 100).

The term Lob-nor is applied by the natives to the whole course of the Lower Tarim, the lake itself generally taking the name of Chon-kul, or "Great Lake."

The Ili is formed at the head of the Kulja valley by the junction of the Tekes and Kunges. From this point it flows through Kulja westwards beyond the Russo-Chinese frontier to Iliysk, where it trends north-westwards to its delta at the south-eastern extremity of Lake Balkhash. The Ili is thus partly a Russian and partly a Chinese river, and its valley forms one of the weak strategical points of the Chinese Empire, for the upper course of the Kunges leads beyond the Narat Pass (9800 feet) between the Odon-kura and Katun-daba, spurs of the Tian-shan, eastwards to the Mongolian plateau. From the sources of the Tekes in the Muzart (11,600 feet) to its mouth, the Ili has a total length of 750 miles, of which about 450 are navigable to a point 50 miles above the town of Kulja.

Apart from the already-described Brahmaputra, the seaward drainage of the Chinese Empire is mainly represented by the Hoang-ho and Yang-tse-kiang ("Yellow" and "Blue" rivers), which jointly drain an area of probably not less than 1,360,000 square miles. Both differ from the other great continental streams, inasmuch as they have their farthest sources not merely within the scarp or near the edge, but in the very heart of the great central plateau. The Hoang-ho is supposed to rise in the springs known as the Sing-su-hai, or "Starry Sea," on the Odon-tala plain in the Tangut country, south of the Burkhan-Buddha, or Central Kuen-lun range. It is remarkable not only for its extremely circuitous course, but also for its tendency to break through formidable water-partings, and to shift its channel from epoch to epoch. It flows first north-east and east along the southern base of the Tsi-shi-shan ridge, beyond which it sweeps suddenly round west and north to the Burkhan-Buddha, through which it forces a passage, as if determined to enter Kuku-nor. But here its course is deflected east to Lan-chew-fu in Kan-su, close under the great wall of China. At this point it makes a tremendous bend along the east slope of the Ala-shan and south foot of the In-shan round the Ordos peninsula, north-east and south to the Funiu-shan, or Eastern Kuen-lun range.[1] Failing to break through this barrier into the Yang-tse basin, its course is turned at Tung-kwan abruptly east to Lung-men-kow in Ho-nan.

At this point begin the extraordinary shiftings of its lower course, which for their vast extent and destructive character are altogether elsewhere unparalleled. "In all our ordinary maps the Hoang-ho enters the sea in lat. 34° south of the great peninsula of Shan-tung.[2] This was its true course down to some thirty years ago (1853), and for six centuries before that. But in the earliest times of which the Chinese have record, the Hoang-ho discharged into the Gulf of Pechili—that is, north of Shan-tung and its mountains—and it continued to do so, though with sundry variations of precise course,

[1] Prejevalsky, however, who again visited the Upper Hoang-ho in 1880, throws some doubt upon the reality of this sudden bend, as usually represented on the maps.

[2] Even Petermann's map, issued in 1880, still shows the Hoang-ho flowing from Kai-fong due east to the Yellow Sea, instead of north-east to the Gulf of Pechili.

till the thirteenth century. Before the latter period the river had occasionally thrown off minor branches to the south of Shan-tung, but it then changed its course boldly to the latter direction, and so continued till our time. The tendency to break towards the old northern discharge had long existed, and was resisted by a vast and elaborate series of embankments. These gave way partially in

SCENE ON THE HOANG-HO.

1851; following floods enlarged the breach, and in 1853 the river resumed its ancient course, the plains of Pechili, and now enters the gulf of that name in lat. 38°" (*Col. Yule*).

The whole of this region from the Yang-tse along the Grand Canal across the old, and down to the new course of the Hoang-ho, has recently been explored by the English engineer G. J. Morrison. "For some little distance down the new course seems to have resisted all attempts to confine it within reasonable bounds. It has overflowed the low-lying country, and presents the appearance of a lake with numerous shoals and channels between. None of these had more than two feet of water throughout their entire length, and this part of the river can hardly be said to be navigable. After about 25 miles, however, there is a change for the better. Embankments have been built along both sides, which protect the country from floods. The authorities have also attempted to confine the river by planting trees. But the river has eaten away the banks, in some places leaving the trees growing, in other cases carrying them off and depositing them in heaps elsewhere, thus forming serious and dangerous obstructions in the river. At one point we passed through what appeared to be more like a flooded plantation than a river, and although the current of at least 4 miles per hour was with us, it took about two hours to go 2 miles."[1]

In proportion to its length, estimated at from 2500 to 2600 miles, the Hoang-ho receives fewer large tributaries than perhaps any other river in the world. Its upper course is joined by the Ta-tung-ho flowing from the southern slopes of the Nan-shan across the Kuku-nor eastwards to its left bank. This river, which flows through a very mountainous region, has not yet been thoroughly explored, although crossed at several points by Prejevalsky during his excursion to Kuku-nor.

This great lake, although approached from the south by the Hoang and on the north by the Ta-tung, is none the less a closed basin, standing 10,500 feet above sea-level in the Tangut country, which is a sort of debatable land between China, Mongolia, and Tibet. The lake,

[1] Paper in *Proceedings of Royal Geographical Society*, March 1880, p. 148.

which has the form of an ellipse over 200 miles in circumference and 2300 square miles in extent, is very salt, and of an exquisite dark-blue colour, compared by the Mongolians to blue silk. The shores are flat and shelving, but towards the west is a rocky island with a temple inhabited by ten lamas, who have no means of communicating with the mainland during the summer. But in winter pilgrims cross over the ice with presents of provisions for the hermits.

Although now a closed basin, the Kuku-nor seems to have formerly communicated westwards with a long-vanished lake, the largest in all Tibet, which filled the whole of the Tsaidam plain between the Nan-shan and Burkhan-Buddha north and south. This swampy region is now traversed by several streams, the chief of which is the Bayan-gol ("Rich River") or Tsaidam, which flows for some 300 miles in a north-westerly direction, at last losing itself in the Dabsun-nor marshes. Here is a depression between the Altyn and Nan-shan, through which the former Lake Tsaidam must have sent its superfluous waters north-west to the Lob-nor basin. The gradual isolation of all these basins affords one of the most striking illustrations of the process of desiccation that has been going on throughout Central Asia from the remotest times. First the Kuku-nor fails to reach the Tsaidam; then the Tsaidam ceases to communicate with Lob-nor and ultimately dries up, while Lob-nor sinks to a mere reedy morass some 3 or 4 feet deep, and the thickly-peopled plains of the Tarim basin are converted into the Takla Makán sandy waste.

Some 70 miles south of the Shuga range the Hoang-ho basin is separated from that of the Muriu-ussu, or Upper Yang-tse-kiang, by the Baian-kara-ula ("Rich Black Mountains"), which under various names runs about 450 miles east and west without anywhere reaching the snow-line. But towards the north-west this range is connected with the Kuen-lun by the snowy Gurbu-naiji, whence flows the Napchitai-ulan-Muren, a chief head-stream of the Yang-tse. The confluence of these rivers in 94° E. long. and 34° 50′ N. lat., marks the limit of Prejevalsky's expedition to the Kuku-nor in 1871-73, when he was obliged to retrace his steps, unable either to reach Lassa or the actual source of the Yang-tse. At the confluence he found this river already 750 feet broad, and flowing in a bed over a mile wide, which, "as our guide assured us, is entirely covered with water during the rainy season in summer, when it sometimes even overflows the banks" (ii. 221). The first ford, even after the subsidence in autumn, lies 20 miles above the confluence, and the actual source of the main stream is said to be in the Tang-la mountains, a ten days' march farther up. From this point the Muriu-ussu, or "Winding Water," flows east and south to its junction with its great tributary, the Min, at Siu-chau-fu, and the Chinese, who regard the Min as the true head-stream, call the other branch Kinsha-kiang ("Gold-sand River"), from its source to the confluence.

The windings of the Yang-tse are no less remarkable and somewhat analogous to those of the Hoang-ho. The upper courses of both are separated only by the narrow Bayan-khara water-parting, and their waters have at various epochs been intermingled in a common delta, or at least connected by numerous natural and artificial channels in the coast province of Kiang-su. But while the normal direction of each is from west to east, in their middle course they are deflected for hundreds of miles north and south respectively, the Hoang sweeping round the Ordos peninsula to the foot of the In-shan mountains, the Yang-tse penetrating southwards far into the Yunnan highlands. Here their channels diverge as much as

15 degrees of latitude (26°-41° N. lat.), enclosing an intervening space fully 1000 miles long, and stretching from South Mongolia nearly to the frontiers of Burma. From the source to the mouth of the Yang-tse the distance in a straight line is scarcely more than 1800 miles, but owing to these astonishing meanderings its total length cannot be much less than 2800 miles, with a drainage estimated by Blakiston at 750,000 square miles.

From the junction of the Min on the Yunnan frontier, where the Kin-sha becomes the Yang-tse, the main stream flows with many windings, mainly east-north-east, through the great provinces of Se-chuen, Hu-pe, Ngan-whei, and Kiang-su, to its delta in the Tung-hai, or Eastern Sea, over against the southern extremity of Japan.

Throughout the greater part of its course the Yang-tse is fed by a vast number of tributaries, some of great size, and jointly affording a navigable water highway of not less than 12,000 miles. Its upper course is joined by the Yalung (Yarlung) or Niachu, which flows from the Bayan-khara slopes parallel with the Muriu-ussu. Lower down comes the Min-shan (Min) or Wen from the north, which at low water is navigable for 200 miles to Sintsin-hien, nearly 2000 miles from the sea. But below the junction the main stream is obstructed by several rapids in its passage through the "Cross Ranges" down to the central plains. From Pingshan to the coast, a distance of 1760 miles, the total fall is about 1500 feet (Blakiston), but very unevenly distributed. The descent is very rapid in the romantic hilly region on the Se-chuen and Hu-pe frontier, where several magnificent gorges follow in quick succession. Here the stream is contracted to a few hundred yards, and rushes in some places at the rate of 10 or 11 miles per hour through its deep rocky bed.

Between Kwei-chow and Ichang the chief "tan" or rapids extend over 100 miles; but below the stupendous Lon-kan and Mi-tan defiles the hills suddenly recede, the great river broadens out to a uniform width of 2600 feet, and a depth of 20 to 30 feet at low water.

Throughout its lower course the Yang-tse is lined on both sides, but especially on its right bank, by numerous shallow lakes or reservoirs, which during the floods are

YANG-TSE-KIANG RAPIDS.

filled up by the overflow of the main stream and its tributaries. These lakes, of which the largest are the Tung-ting-hu in Hunan and the Po-yang-hu in Kiang-si, thus serve to regulate the inundations and prevent the widespread ruin often produced by this cause in the Hoang-ho basin. The Tung-ting has an area of at least 2000 square miles, and receives the Yuan-kiang, Lo-kiang (Tse-kiang), and Heng-kiang, and other rivers from the south, which collectively drain an area of 80,000 square miles in the province of Hunan. From this basin the two adjacent riverain provinces are respectively named Hu-pe and Hunan—that is, "North of the Lake" and "South of the Lake."

Below the Tung-ting the Yang-tse receives its great affluent, the Han-kiang, flowing from the Tsing-ling (Eastern Kuen-lun) through Shensi and Hu-pe south-eastwards to its left bank at Hankow. Although obstructed by rapids at some points, the Han is throughout navigable, and in summer accessible to steamers for a distance of 600 miles. It flows through a magnificent region, well watered, extremely fertile and healthy, and abounding in vegetable and mineral resources of all kinds. But in its lower course it flows in a bed higher than the surrounding lands, so that during the floods the whole country from the confluence to Lake Tung-ting is sometimes transformed into a vast inland sea.

Beyond Hankow the great Lake Po-yang plays the same part for the province of Kiang-si that the Tung-ting does for Hunan. Its chief influent is the Kia-kiang, which flows from the Tung-lo-ling range through Kiang-si to a large delta on the south-west side of the lake. This basin, which has an area of 1800 square miles, is studded with islands, and although in many places covered with forests of reeds, its northern section is very deep and skirted by lofty hills. The wooded headlands, inlets, and islets are everywhere interspersed with towns, villas, towers, pagodas, crowning every eminence, and rendering this one of the most charming regions in China.

Below the Po-yang the Yang-tse flows in a majestic stream through the hilly and flourishing province of Ngan-whei, and by the great cities of Ngan-king and Nan-king to its delta, which fills a large part of the province of Kiang-su. Here it has a mean discharge of perhaps 735,000 cubic feet per second, ranking in this respect, according to Guppy,[1] next after the Amazons, Congo, and La Plata. Its basin occupies nearly one half of China proper, comprising some of the richest lands in the world, with boundless material resources of every

THE GOLDEN ISLAND, YANG-TSE-KIANG.

kind, and especially including vast coal-measures. The riverain provinces, where were sown the first seeds of Chinese culture, have for ages supported a teeming population, variously estimated at from 100 to 200 millions.

[1] *Nature*, 20th Sept. 1880.

The main stream itself and its navigable tributaries are everywhere crowded with many hundred thousand junks and boats of all sizes, on which a floating population of millions pass their whole lives. A fire which broke out at the port of Wuchang in 1850 destroyed 700 junks, several thousand smaller craft, and proved fatal to no less than 50,000 of this riverain population.

The tides penetrate for over 200 miles up the estuary, which in many places is over 300 feet deep and 60 miles wide at its mouth. Here, however, the navigation is much obstructed by numerous islands and shifting sands, with scarcely more than 14 feet at low water in the deepest channels. But at the flow, vessels drawing 18 to 20 feet easily pass up, and the danger most to be dreaded is perhaps the dense fogs often enveloping the whole estuary and neighbouring seaboard. At the same time, the sedimentary matter, brought down at the rate of perhaps 6300 millions cubic feet yearly, is constantly accumulating at the mouth, raising the sandbanks and enlarging the islands. Tsungming, the largest of these, formerly washed by the tides, now supports a population of 2,000,000 industrious peasantry and fishermen on an area of about 400 square miles.

Great changes have taken place in the lower course of the Yang-tse, which formerly threw off branches from various points above Nan-king eastwards to the large Lake Tai-hu, and thence southwards to Hang-chow bay. The whole of the Shang-hai peninsula thus formed a portion of the delta, which also extended through countless channels northwards to the Hoang-ho, when that river discharged into the Yellow Sea. Here the connection is still maintained even with the present channel of the Hoang-ho by means of the famous Imperial or Grand Canal, which runs through a district where land and water seem to be inextricably intermingled, mainly northwards from the Yang-tse across both the old and new beds of the Hoang to the Pei-ho.

The basin of the Pei-ho, although vastly inferior in extent to those of the Hoang and Yang-tse, derives a special importance from the fact that within its limits is situated Peking, capital of the empire. It is formed by a large number of streams, such as the Pei-ho, When-ho, Tsu-ho, Hu-to-ho, Chang-ho, and Wei-ho, which flow mostly in independent channels to within a comparatively short distance of the coast. Here they all converge at Tien-tsin, whence the united stream flows in a broad navigable channel to the Gulf of Pechili. It is probable that the various branches of this river system formerly found their way in separate channels to the coast; for all the lowland Pechili plains form a marine basin, which has been slowly filled in with the alluvia brought down by these rivers from the Shansi and Mongolian frontier highlands. They are still subject to extensive inundations, covering an area of about 6000 square miles, and representing, as it were, the former extension of the Gulf of Pechili towards Peking and the surrounding hills. These inundations are caused by the swollen waters of the head-streams, which, being unable to find room in the common channel below Tientsin, overspread its banks far and wide, and present a continuous sheet of water stretching from the gulf inland nearly to Pao-ting. Amid this waste of waters nothing is visible except the towns and villages perched on eminences dotted over the plains; the river banks are washed away, the streams shift their course, the crops are destroyed, and the people left a prey to famine and disease. Yet in an area of scarcely 60,000 square miles the province of Pechili mainly comprised in the Pei-ho basin, was estimated to have a population of close on 37,000,000 in the year 1842. At present it probably amounts to less than half that number. The inhabitants have been forced to emigrate in hundreds of thousands to Manchuria and Mongolia, owing to the disastrous floodings of the Pei-ho.

According to Guppy, the Pei-ho drains an area of some 56,000 square miles, has a mean discharge of 7500 cubic feet per second, and brings down to the Gulf of Pechili about 80,000,000 cubic feet of alluvial matter every year.

In the neighbouring province of Manchuria the drainage is mainly through the Sungari and Usuri to the Amur. But the southern portion between the Khingan range and Korea drains through the Liao-ho southwards to the Liao-tung Bay, which is a northern extension of the Gulf of Pechili. In its upper course the Liao takes the name of Sira-muran-pira (Shara-muren), which with its chief affluent, the Lohan-pira, rises in the highland district where the Mongolian escarpments merge in the Great Khingan range. The Liao pursues a tortuous semicircular course from these uplands round to its mouth at the head of Liao-tung Bay. It is navigable during the floods only in its lower course within the Chinese province of Shing-king; but since the Russian occupation of all the seaports north of Korea this river has acquired great importance as the only seaward outlet of Manchuria.

In the extreme south-east of China proper the Nanling with its eastern extensions forms the water-parting, broken at one point between the Yang-tse and the Si-kiang, or "West River." The Si-kiang basin thus occupies nearly the whole of Kwang-si and Kwang-tung, besides parts of Yunnan and Kwei-chow, and also encroaches beyond the Tung-lo-ling range (see p. 503) on Hunan in the Yang-tse basin. Here the Pe-kiang ("North River of Canton") flows from the north side of the range due south to its junction with the main stream west of Canton. This river has its farthest head-streams in the Yunnan and Kwei-chow highlands, whence it flows due east to its delta and estuary in the China Sea.

The North River between Lo-chang and Shao-chow "has some shallows, which would be impediments to steam navigation; but from Shao-chow the only troubles would

A CHINESE JUNK.

be the freshets, which cause the water to rise quickly, and the river to run swiftly; but as far as I could learn there is nothing to prevent regular steamboat traffic being carried on as far as Shao-chow" (*G. J. Morrison*). Thus affording direct communication between the Yang-tse basin and the southern provinces, the Pe-kiang has at all times occupied a position of primary importance as a military and trade route. In this respect it ranks far before the Si-kiang itself, which has yet a navigable course of nearly 900 miles. Below its junction with the Koli-kiang it penetrates through a series of magnificent gorges into the province of Kwang-tung, where it has a depth in some places of from 50 to 150 feet. The tides are felt for a distance of 180 miles inland, and at high water most of the countless channels and branches of the delta are navigable. In the delta, which has an area of over 3200 square miles, the chief channel is the east branch, known as the Canton River, which is joined from the east by the Tung-kiang. Below the confluence the stream is contracted to a narrow bed, commanded by Forts Humen—that is, "Tigers' Throats"—whence this part of the river takes the name of Bocca Tigris, or simply the Bogue. Immediately below the forts it broadens out to the estuary of the Shu-kiang, or "Pearl River," said to be so called from another stronghold, the Hai-chu, or "Pearl Fort," familiarly known as the Dutch Folly.[1] The network of channels in the delta is inhabited by an enormous floating population, estimated at about 250,000, most of whom pass their whole lives on the water.

Besides the lakes already mentioned in connection with the river systems, there are few bodies of still water in the empire anywhere except on the Tibetan plateau. Here, although mostly of small size, lakes are extremely numerous, and the lacustrine character of this region becomes more evident with every fresh exploration. More than half of the Khachi tableland is dotted over with closed basins, which are probably the remains of inland seas formerly draining through openings in the frontier ranges. In the west the largest would seem to

[1] That is, the "Dutch Fort," the English word *fort* being pronounced *foli* in "Pigeon English."

be the Ike-namur and Bakha-namur, which form the central basins of a lacustrine system, stretching for over 120 miles south-west and north-east. Many of these lakes were visited in 1874 by Nain-Singh, who found that they were the remains of far more extensive basins, some of which were already reduced to mere swamps or quagmires covered with a saline efflorescence. Most of the lakes are salt or brackish, while some are still perfectly fresh. The Dangra-yum, or "Mother Dangra," 86° N., 31° E., is 180 miles in circumference, and is dominated on the south by Mount Targot-yap, or "Father Targot."

East of the Dangra-yum several other lakes are said to discharge their waters northwards to the Chargut-tso, which is itself supposed to communicate with one of the great rivers flowing to the Indian Ocean. But better known is the Tengri-nor lying in the south-east corner of the Kachi plateau, within 60 miles of Lassa. It is 50 miles by 15 to 24, and of unknown depth. A favourite place of pilgrimage is the convent of Dorkia on its west side, commanding a superb view of its blue waters and of the surrounding snowy peaks. The Tengri is not a closed basin, as had been supposed, for it discharges through the Nak-chu, flowing from its north-west corner to the outlet of the Chargut-tso. North of it lies the Bul-tso, or "Borax lake," covering an area of 24 square miles, whence formerly came much of the so-called "Venetian" borax.

On the frontier of Tibet and Kashmir the three lakes, Pangkong, Mognalari, and Noh, form an almost continuous basin, stretching nearly west and east, but at different levels, and apparently in the line of an old watercourse. They wind along with a present depth of about 150 feet, and a total area of 210 square miles north-westwards, to a point where they formerly communicated through the little River Tankseh with the Shayok, a tributary of the Upper Indus. But with the gradual subsidence of their waters the emissary ceased to flow, and they now form a closed basin, saline in the Pangkong or lower section, but still fresh higher up.

Towards the south-east of Tibet lies Lake Palti (Yamdok-tso), usually described as ring-shaped, with a large island filling most of its basin, and rising 2300 feet above its surface. But the native explorer who visited it in 1875, ascertained that the supposed island is really a peninsula connected by a narrow tongue of land with the south side of the lake. The water, described by Manning (1811) as slightly brackish, this explorer found to be perfectly fresh.

4. *Natural and Political Divisions: Tibet—The Tarim Basin (Kashgaria) — Mongolia — Zungaria and Kulja—Manchuria—The Great Wall—The Gobi and West Mongolia—South-East Mongolia—Korea—China Proper—Islands: Hainan, Formosa, Macao, Hong-Kong.*

In the Chinese political system the great administrative divisions correspond to a large extent with the main physical regions. Tibet and Korea, off-lying members of the system in the extreme south-west and north-east, are marked off from the rest of the empire by their geographical position no less than by their political status. In the same way the remote western regions of Kashgaria and Zungaria—the former occupying the Tarim basin, the latter embracing the upper courses of the Ili and Irtish—are held by military tenure, while the nomads of the vast Mongolian plateau are ruled through their respective hereditary Khans. Since the conquest (1644) of China proper by the Manchus, the north-eastern region of Manchuria has been brought into more intimate relationship with the Hoang-ho and Yang-tse basins. Here also the great provinces are generally disposed in harmony with the respective areas of drainage. Thus Pechili is mainly comprised in the Pei-ho water system, while Hu-pe, Hunan, and Kiang-si are each occupied by the basins of large rivers flowing to either bank of the Yang-tse.

Tibet.

Tibet, taken in its widest sense, comprises the whole region between the Kuen-lun and Himalayas, which is at once the most elevated section of the Central Asiatic plateau and the loftiest tableland on the face of the globe. It forms a mass of irregular uplands sloping gradually eastwards, but nowhere falling below 10,000 or 12,000 feet above sea-level. Northwards, the Kuen-lun escarpment falls rapidly towards the low-lying plains of the Tarim basin, while on the west and south the massive sweep of the Himalayan system forms a natural barrier towards British India. The eastern boundary has been crossed only at a few points, but here also the "Cross Ridges," roughly answering to the Yung-ling of Chinese geographers, serve as the frontier line towards China proper.

The little-known northern division of Kachi is occupied in the west by the Hor of Túrki stock, in the east by the Sok of Mogul stock, and has hence been sometimes called Hor-Sok-pa. Here are the numerous chains of lakes discovered by Nain-Singh, which very possibly drain eastwards to one or more of the great rivers of Indo-China. Still farther south the high and apparently continuous Ninching-tangla ridge divides this basin from that of the San-po, or Upper Brahmaputra River.

Here is the true Bod-yul, or Land of the Bod—that is, of the Tibetan race; and here is situated Lassa, capital of Tibet, together with the more fertile and thickly-peopled portion of the country.

CAMPING-GROUND, WEST TIBET—THE KARAKORUM IN THE DISTANCE.

Bod-yul comprises seven provinces:[1]—1. Nari-

[1] Markham (*Tibet*) speaks of four only—Nari (Ari), U, Isang, and Kam, treating Hor as distinct, and omitting reference to Dok Thol and Monhuil.

Khorsum (chief town, Gartokh), embracing all Western Tibet as far as the Mariam La Pass (82° 30′ long.); 2. Dok Thol (chief town, Sarka-Jong), extending from the Mariam La Pass to the Kalha Pass (87° E.); 3. Chang (chief town, Shigatze), extending eastward to the Khamba La Pass; 4. U (chief town, Lassa) stretching thence eastward to the twelfth stage on the southern route from Lassa to Peking; 5. Monhuil, embracing the Tawang district between Bhutan and the Daphla (Lhopa) country; 6. Kham (chief town, Tsiamdo), between U and Se-chuen; 7. Hor-Sok, or Kachi, occupying the northern parts as far as the Kuen-lun, and over which the Central Government has scarcely a nominal jurisdiction.

The total area of these seven provinces has been roughly estimated at some 560,000 square miles; but with the north-eastern Kuku-nor district, Tibet covers a space of about 675,000 square miles, and 800,000 square miles if we include the districts in Kashmir and Se-chuen (West China) occupied by peoples of Bod stock.

The Tibetans are essentially a commercial people, to such an extent that most of the officials and head lamas of the monasteries are said to keep agents and carry on trade on their own account. There are also many Muhammadan and other foreign traders settled in Lassa, the great emporium of the country. From Northern China come silks, gold lace, precious stones, and carpets; from Mongolia and Kachi, leather, saddlery, sheep, horses, salt, and borax; from Se-chuen, tea, cotton goods, porcelain; from Tawang, Bhutan, and Sikkim, rice, indigo, coral, pearls, sugar, spices, and Indian wares; from Ladak and Kashmir, saffron, silk, and Indian produce.

Among the chief exports are gold and silver, productive mines of which are found in various parts. The richest gold-field is that of Thok Jalung, north-east of Gartokh, in 32° 30′ N. But the great staple is wool, of which vast quantities, and of the finest texture, might be produced on the boundless grassy plains and mountain slopes of Bod-yul. Tibetan musk is highly esteemed, but so great is the demand that it reaches the coast in a very adulterated state. Salt abounds everywhere, and is obtained chiefly by solar evaporation in shallow basins.

Of the imports specially important is the brick-tea, consisting of the coarser leaves and twigs, and described by Baber as the merest refuse. It is first pounded into moulds, and then broken into "bricks" 9 or 10 inches long by 7 wide and 3 thick, conveyed by mules and carriers over the lofty passes into the country. The annual import from Ta-chien-lu to Batang, the Tibetan emporium, is estimated at 10,000,000 lbs., valued at £160,000. The tea is apparently paid for mainly by Indian rupees; which have entered the country in vastly-increased quantities of late years, and are said to have now become the currency of Tibet.

Chinese cottons are also largely imported, but at Lassa are supplanted by Indian goods. Silks of bright colours fetch their weight in silver, and English woollen cloths are in great demand notwithstanding the competition of Russian goods. Desgodins saw numerous packets on their way to the salt-works, bearing the name of a Halifax maker, and he adds that scarlet is the favourite colour, although a good golden yellow would sell well. Mr Edgar reports that the demand for indigo is very great, and that the profit varies from 50 to 100 per cent.

Altogether it is evident that the Tibetan trade offers special advantages to English and Indian traders, were it possible to establish free communication. But ever since the war between China and Nepal in 1792 the passes to India have been closed. Even the native surveyors, sent forward disguised as traders, have had great difficulty in passing through. The Tibetans say that this policy is due to orders from Peking; but there is reason to believe that the real obstacle is at Lassa. Nevertheless, Tibet offers a good market for Indian teas, which would find a valuable exchange in wool, gold, borax, and other products.

The Tarim Basin (*Kashgaria*).

The basin of the Tarim River forms the western section of the relatively low Central Asiatic plateau, which the Chinese geographers styled the Han-hai, or "Dried-up Sea." From the Pamir to Lob-nor it stretches west and east about 900 miles, with a mean breadth of 500 between the Tian-shan and Kuen-lun ranges. Its prevailing character is that of a vast sandy plain enclosed north, west, and south by a horseshoe-shaped rampart of the loftiest and grandest mountains, rising in ridges of 18,000 to 20,000 feet, with peaks shooting up to 25,000 and even 28,000 feet. From the snows and glaciers of these highlands rush down the streams, which flow through the common bed of the Tarim to Lob-nor.

The open approach to the east afforded in former times easy means of access to migrating nomad tribes and military expeditions from China and Mongolia. Along the northern and southern edges of the basin there lie at intervals the remains of fertile oases, such as the Lop, Charchand (Cherchen), Kiria, Khotan on the southern route, and Hami (Khamil), Kuchar, Aksu on the north, some of which were of great extent and importance. But the sands driving before the winds in ceaseless billows from the eastern Gobi, have gradually encroached on the cultivated lands, swallowing up populous and flourishing cities, memorials of which are still found in the gold and silver ornaments, and even in the "bricks of tea" constantly exhumed at certain spots. Extensive ruins of cities are known to exist in the Lob district.

The whole basin comprises four natural divisions—highlands, lowlands, desert, and swamp or lacustrine tracts. The first include the plateaux and deep valleys of the encircling ranges, barren hill slopes, rich pastures on the more level portions, but with a general deficiency of vegetable and animal life. The lowlands comprise the strip of country intervening between the mountains and the desert. This is the only permanently-settled and cultivated portion of the country. Although the soil is naturally poor, it is extensively irrigated and brought under cultivation in the vicinity of the Khotan, Yarkand, Kashgar, and Aksu Rivers, where there are some exceptionally fertile tracts.

But most of the Tarim basin consists of an undulating plain of shifting sands, sloping gradually eastwards from 4000 to 2000 feet, and traversed by the various rivers flowing to the Tarim. The banks of these streams are fringed by strips of fir, poplar, willow, and tamarisk forest, interspersed with dense growths of tall reeds and grass. But all the rest is an inhospitable waste, with a deep coating of loose sandy or saline soil, impracticable alike to man and horse.

The swamps and lakes Lob, Bagrach, Yeshil-kul, and Karga, are formed by the overflow of the rivers, and are unhealthy tracts overgrown with reeds and swarming with waterfowl. "The poplar woods with their bare soil, covered only in autumn with fallen leaves, parched and shrivelled with the dry heat, withered branches and prostrate trees encumber the ground.

"But cheerless as these woods are, the neighbouring desert is even more dreary. Nothing can exceed the monotony of the scenery." (Prejevalsky, *Lob-nor*, p. 60.)

The whole of the Khotan district, and especially the neighbouring Kuen-lun mountains, abound in gold, silver, iron, lead, copper, antimony, salt, saltpetre, sulphur, soda, and coal. Gold and precious stones are chiefly found in the beds of streams, and the Kappa, Sorghak, and other auriferous districts of Khotan, are said to employ 7000 hands, with an annual yield of nearly 80 cwts. Its transport to India is generally a very lucrative venture, bringing in profits of from 20 to 24 per cent.

Silk is cultivated in Khotan, but notwithstanding its good quality, a defective method of reeling renders it of little use for the export trade. It is employed with wool and gold thread in the manufacture of the Khotan carpets, which are made from patterns usually handed down from master to pupil. A renowned product and former article of manufacture is jade, found only in Khotan and the northern Kuen-lun valleys. Here there is a plentiful supply, especially from the quarries in the Karakash valley and south of Khotan. It was carried far and wide in mediæval times, and jade implements have been picked up even in Western Europe. The Yu-moun, or "Jade Gate," in the Great Wall in north-west Kansu, seems to have been named from the jade caravans which passed that way to China. This interesting historical industry has been suspended since the expulsion of the Chinese from Kashgaria in 1864.

Mongolia—Gobi.

The swampy Lob-nor district offers little interruption to the sweep of sandy wastes which stretch continuously across 40 degrees of the meridian from below the cities of Kashgar and Yarkand eastwards to the Great Khingan range. The western section of this inhospitable wilderness as far as Lake Lob, takes the name of the Takla Makán desert; the eastern, thence to Manchuria, that of the Great Gobi or Shamo desert. But the whole forms essentially one geographic unit, and is by some writers spoken of simply as the Eastern and Western Gobi, or even as Eastern and Western Mongolia. For while the whole of this region forms on the one hand the true primeval home of the great Mongolian branch of the human family, it is, on the other, often difficult, at times even impossible, to say where Mongolia begins and Gobi ends.

Taken thus in its widest sense, Mongolia comprises the whole northern section of the Central Asiatic plateau between the Kuen-lun and Altai mountain systems. Towards the west the Tian-shan projects midway between these ranges to an unknown distance eastwards, thus dividing the western portion into a northern and southern region roughly indicated by the Chinese expressions Tian-shan Pe-lu, and Tian-shan Nan-lu—that is, the northern and southern Tian-shan routes. By the use of these terms the Chinese people showed from the earliest times a surprising appreciation of the disposition of the land in Western Mongolia. In their eyes the Tian-shan, itself mostly impassable, clearly indicated the routes to be followed in order to penetrate into the Western world. But whereas the Nan-lu led, so to say, to a cul-de-sac at the eastern foot of the Pamir, the Pe-lu gave direct access through more than one depression to the Aralo-Caspian basin. Hence the vast importance to China of the extreme north-western portion of Mongolia, now commonly but most inconveniently spoken of as Zungaria. For within this region are comprised all the natural openings which either through the Balkhash or the Irtish basins lead from Central to Western Asia.

Zungaria—The Great Wall.

The term Zungaria, unknown to the Chinese, derives from the Zungars, a branch of the Kalmuks, or Western

Mongolians, who suddenly acquired great power early in the eighteenth century. Their empire stretched east and west from Hami to Lake Balkhash, and they were strong enough to invade Tibet and sack its capital in the year 1717. But after a chequered history of some sixty years they fell as rapidly as they had risen above the political horizon. Their overthrow by the Chinese in 1757 was attended by the most frightful massacres, in which the whole nation perished, leaving behind it nothing but the name which Western writers still continue to apply to the region, at one time forming the centre of their power.

Zungaria, which is administratively connected with, but physically separated from, Kulja (Upper Ili valley), occupies the whole region between the Central Tian-shan and the Western Altai. It has no natural frontier towards Mongolia, with which it everywhere merges imperceptibly, and which it resembles in its main physical features. Towards the west it is not bounded so much as intersected by the Ektag-Altai, the Tarbagatai, and the Ala-tau, which with their eastern extensions run rather east and west than north and south. Thanks to this disposition of the ranges between the Altai and Tian-shan, the Central Asiatic tableland, elsewhere enclosed by continuous and mostly impassable mountain barriers, here opens through no less than three distinct depressions down to the Aralo-Caspian basin. Between the Ektag-Altai and the Tarbagatai lies the Upper or Black Irtish valley, continued right into Mongolia by the Urungu River (see p. 451), and nowhere rising more than 2500 feet above sea-level (*Sosnovsky*). But far deeper is the southern depression between the Tarbagatai and the Ala-tau, which is itself divided into two sections by the intermediate Barluk-Orkochuk ridge also running east and west. Between this ridge and the Saura, or eastern extension of the Tarbagatai, runs the second approach, which passes by the town of Chuguchak, and which, although less open, is more frequented than the others. Lastly, the third and southernmost passage is clearly marked by the Ayar-nor, Ebi-nor, and the undecided steppe rivers, all formerly presenting a continuous waterway communicating eastwards with the Central Asiatic mediterranean (Gobi), and connected westwards through Lakes Ala, Sassik, and others, with Lake Balkhash—that is, with the Aralo-Caspian basin.

The physical complexity is reflected in the ethnical confusion especially of the Ili valley, which has been the common battle-ground of rival races and conflicting creeds for ages. Kulja, as the Upper Ili valley is now called, is naturally by far the richest land in the empire beyond the limits of China proper, and has at times supported vast populations dwelling in numerous large cities and thriving towns scattered over its fertile and highly-cultivated plains. But the frequent revolts, first of Zungars, then of Dungan and Taranchi Muhammadans, in which momentary success on either side was invariably followed by wholesale extermination, have in recent times converted these magnificent lands into a howling wilderness. The victims of the successive Zungarian and Dungan insurrections, extending over more than a century, must be reckoned literally by millions, and the scene of desolation now presented by the ruined cities and wasted plains of unhappy Kulja baffles all description.

Kulja, which was temporarily occupied by Russia from 1871 to 1880, forms a triangular space some 26,000 square miles in extent, wedged into the very heart of the Central Tian-shan, and opening down the Ili valley towards Semirechinsk and Lake Balkhash. Its population has been reduced from over 1,000,000 to little more than 100,000 in 1880, and in the whole of

Zungaria, with an area, including Kulja, of 146,000 square miles, there are less than 500,000 inhabitants.

From Zungaria Mongolia proper stretches south of the Altai highlands eastwards to the Khingan range and almost to the gates of Pekin. North of the imperial capital the great commercial route is soon reached, which crosses the Gobi desert and Mongolia to the Siberian frontier town of Kiakhta. A two days' trip from Peking towards the Great Wall brings the traveller to the "fortified city" of Chang-piu-chao, which on a closer inspection proves to be a mere village surrounded by mud walls. A few hours beyond it lie the five mighty gates of the valley of the imperial tombs. In this sandy plain, enclosed by an amphitheatre of lofty mountains, stand the colossal tombs of thirteen Chinese emperors disposed in crescent form at the foot of the wooded hills. Farther on lies the wild and frowning gorge of Nang-kao, through which formerly flowed a rushing torrent, its narrow bed here confined between steep rocky banks. An interminable line of massive walls, flanked at intervals by turrets and battlements, is carried over the crests of the craggy heights, following snake-like all their sinuosities as far as the eye can reach. At the first glance it becomes evident that this is the Great Wall of China, and after penetrating farther into the rugged valley we perceive two parallel lines running close together over the summits of the rocky hills, and sharply defining their outlines against the horizon.

A little farther on rises the barrier of ramparts separating China from Mongolia. The buttresses and apertures of the bastions are somewhat out of repair, but no trace of decay or damage can be detected in the Great Wall, which rises suddenly to the right and left, broken at regular intervals by square towers, and, like a huge snake turned to stone, winding away over the summits of the highest ranges. Repeatedly repaired, rebuilt, and even altered in its general direction, little if any of Shi Hoang-ti's original structure now remains. But such as

THE GREAT WALL.

it is, with all its windings and the double and triple lines erected at certain points, it has a total length of 2000 miles, or one-twelfth of the circumference of the globe.

Manchuria.

The Great Wall is continued westwards across the northern bend of the Hoang-ho to the neighbourhood of Su-chau about 39° 30′ N., 99° E., and eastwards round to the Gulf of Liao-tung, thus completely enclosing China proper and part of Tibet from Mongolia and Manchuria. The hilly region of Manchuria stretches from Northern China northwards to the Amur, and from the Great Khingan range eastwards to Korea and the Usuri River.

The Sungari, which rivals the Amur itself in the volume of its waters, is the main artery of Chinese Manchuria. This province enjoys a healthy climate, with a fertile soil and great mineral wealth, so that it is quite capable of receiving the superabundant populations of North China. These industrious agricultural colonists have gradually migrated in such numbers to the Sungari valley, that the aboriginal Manchu tribes now form the minority of the population. Since 1864, the Rev. A. Williamson has made several important expeditions into this region.

In the year 1870 the Russian Archimandrite Palladius also traversed the country from Mukden to the Nonni.

At Sang-Sing the Sungari is joined from the right by the Hurka (Khurkha), whose banks are thickly peopled by Chinese settlers. From Ninguta on this river a road leads over the ridge separating the Hurka from the Siufun, a small coast stream flowing to Victoria (Peter the Great) Bay near Vladivostok.

Russian steamers now ply on the Sungari, and have even penetrated to Tsitsihar on the Nonni. They have also entered the Hurka, though this river is so shallow and rocky that Ninguta can be reached only in small boats. The frontier towards Korea is continued from the Shan-yan-alin range by a narrow strip of neutral and uninhabited territory southwards to the Yellow Sea. But the so-called wooden palisade traced on the maps from the eastern extremity of the Great Wall to the Upper Sungari has long ceased to exist, at least as a distinct boundary line. It never could have possessed any strategical importance, being quite incapable of defence, nor is it possible any longer to make out its general direction from the few straggling stumps of trees, which are now all that survives of the original stockade.

Western Manchuria has been far less explored than the other portions of the province. Here the Khingan range is crossed by passes 3800 feet high, leading to the Mongolian plateau. Nor do these mountains everywhere form an effective barrier between the two countries; for they have long been invaded by Mongolian tribes, which have encroached far beyond this natural barrier into South-West Manchuria. At the same time, both Mongolians and Manchus are being gradually absorbed or displaced by the Chinese immigrants, so that the whole of Manchuria may perhaps ere long become ethnically an integral part of China proper. The Khingan range must then resume its position as the natural frontier towards Mongolia.

For the fullest information on this borderland between the two countries we are indebted to the Russian astronomer Fritsche, who travelled in 1873 from Peking to the Russian frontier station of Staro-Zurukhaituyevsk on the Argun.

In eighteen days he reached Bei-lei-gu, beyond which point the Chinese begins to merge in the Mongolian population.

Soon after leaving He-shui, he reached the Shara-muren, here a turbid stream flowing between sandy banks about 1500 feet above sea-level.

About 47° 7′ N. and 118° E., crossing the Cholotu-davan at an altitude of 4000 feet, he reached the west side of the low border range, which here runs south-west and north-east in a rolling steppe, 3000 feet above the sea, but gradually sloping down to the River Argun at a level of 2000 feet.

Apart from the thinly-scattered "Yurtas," and the Chinese city of Khailar, along the whole route through the lands of the Barin, Ude-Michin, Khalka, and Solon Mongolians, between 43° and 50° N., the explorer passed only seven Lama monasteries, which in Mongolia seem to take the place of towns. The Mongolians of this region are governed by their own princes, the head-governor alone of the above-mentioned vassal-lands being a Manchu appointed from Peking and residing in Khailar. Here the Chinese traders are numerous, from this centre distributing their tea, tobacco, bread, saddles, yurts, and other wares, at little isolated stations scattered over the steppe.

The Gobi and West Mongolia.

West Mongolia proper, comprising the lowest plateau between the Altai and Tian-shan and the eastern section of the region between the Tian-shan and Kuen-lun, has a mean elevation of probably not more than 2000 feet. Farther east the waterless and treeless plains of Gobi stretch from the Tola, a head-stream of the Selenga, south-eastwards to the Darkhanola range, which rises to an elevation of 5000 feet. So far the land does not yet assume the aspect of a true desert, for the hill-sides are still overgrown with scrub from 2 to 3 feet high, and the plains covered with grassy tracts supporting numerous herds of cattle. But here begins the extensive depression which reaches to the Mandal Pass, 3700 feet high. At the Olong Baishing ruins the land falls to a still lower level, and here is seen the so-called "Rocky Girdle," a natural rampart of syenite stretching in a straight line east and west, and forming a clear landmark between North and Central Mongolia. South of this line begins the true desert of Gobi—the Shamo of the Chinese—the lowest points of which are found at Ergi, Ude, Durma, and Shabadurghuma. The higher grounds are in some places strewn with rubble and blocks of porphyry and jasper, besides chalcedony and carnelian interspersed with saline plants. The depression itself consists not so much of drift sand as of a sandy soil charged with alkalies, evidently the bed of a former marine basin, where still flourish the arundinaceæ and nearly all the species common to the Caspian Sea. South of Durma the land again rises to the level of the shores of this dried-up mediterranean, attaining at Tsagan-Balgasu an elevation of 4550 feet, a height corresponding exactly with that of the northern edge of the basin at Urga. The plateau attains its greatest elevation towards the east, where it is cut off from Manchuria and the plains of Pechili by the intervening Khingan range and the highlands, stretching thence south-westwards to the Inshan mountains.

The region stretching north of Ulia-sutai to the Kobdo plateau was explored by Mattussovski in 1870.

This traveller also visited Lake Ike Aral, one of the largest in West Mongolia. Here he ascertained that Lake Kirghiz in the north-east of Kobdo, although of small size, forms nevertheless the centre of the water system in this region, receiving the overflow of all the surrounding lakes and rivers.

Our knowledge of Mongolia has also been greatly enlarged by the remarkable journey undertaken in 1872 by the English traveller, Ney Elias, from Peking westwards to the Russian Altai. Beyond Kalgan he reached the Belgian missionary station of Si-yun-tse, where wheat, oats, millet, and especially the poppy, are cultivated. The poppy seemed to be the chief inducement for the Chinese to settle here, but no reliable data could be procured respecting the opium trade, which, notwithstanding the high duty, is said to form the most lucrative business in Mongolia. The route to Kwei-hwa-chang lies for over 140 miles across a somewhat hilly pasture-land, and about 40 miles farther on a pass 5900 feet high leads down to a valley whose soil consists of a brown-yellow loess, intersected by numerous clefts and fissures, often 30 feet deep. Beyond the hills these crevasses even serve as regular dwellings for the people. Kwei-hwa-chang consists of two towns, and enjoys an extensive trade in tea, flour, millet, and the wares in demand amongst the Mongolians. From this point the traveller visited Hokow on the Hoang-ho, a small but busy place near extensive beds of a hard, slaty coal.

From Kwei-hwa-chang two routes—a government road and a caravan track—lead to Ulia-sutai. But the Mongolian steppe presents little variety for the traveller either way. The general aspect of the desert consists of low hills with intervening valleys and plains, rather stony than sandy, here and there intersected by low, rocky ridges, and mostly destitute of vegetation. The best water is found near the hills, where it is always sweet, while that of the plains is often brackish.

On 8th October Elias reached the River Onghin, which, after a south-easterly course of about 100 miles, loses itself in the desert. Proceeding westwards along the slopes of the rugged red and gray granite Kangai hills, he reached the Tui and the Baitarik, the largest of the Kangai rivers. Here the country is wild and barren, although frequented by wild asses and ponies in herds of from twenty to thirty each. On 25th October he camped on the left bank of the Chagan-tokoi, which flows south-west and west parallel with the Sirke range. This range forms an important geographical feature of the land, some of its crests rising from 3000 to 4300 feet above the general level of the surrounding plains. To the north-west lie the hills whence flow the Ulia-sutai and the Buyanta, and which are crossed by a pass 7450 feet high. From Ulia-sutai the traveller made his way to Kobdo by the River Yabkan and Lake Ike-Aral. A pass over 9000 feet high leads from Kobdo to the Chinese frontier town of Suok, whence a second high but easy col in the Altai brings to the River Chu and the town of Biisk.

South-East Mongolia.

For the latest information regarding south-east Mongolia we are indebted to Col. Prejevalsky, who visited the Ordos and Ala-shan regions on his journey to Kuku-nor in 1871. Proceeding in a south-easterly direction from Kalgan, this intrepid explorer came upon the In-shan range, skirting the northern bend of the Hoang-ho. Even before leaving the Kiakhta caravan route, a change is perceptible in the aspect of the country. The hills become higher and more craggy, while grass becomes more scanty, the pasturages being succeeded still farther west by extensive waterless valleys, where the nomads are entirely dependent on the wells dug at intervals along the route. The highest ranges are the Shara-Hada and Suma-Hada, wild and rugged uplands, where the traveller discovered the wild ovis Argali in flocks of as many as fifteen together. Farther on the Muni-ula range, over 7000 feet high, forms with the Hoang-ho a well-defined landmark in the distribution of birds and mammalia. From these mountains the city of Bautu was reached, a large and busy but dirty place on the Hoang-ho, near the hill where the wife of Jenghiz-Khan is supposed to lie buried.

From Bautu the route lay across the Bagakhatun, southernmost and largest branch of the Hoang-ho, to the Ordos country, where the population is entirely confined to the Hoang-ho valley for about 70 miles west of Bautu. On the left bank of the river lies the Ala-shan region, mostly a dreary lifeless waste of shifting sands, destitute alike of vegetation, birds, and mammals. The small tracts, where the sand is mingled with the loam and alkalies, produce a scanty but peculiar vegetable growth. The Ala-shan range rises some twelve miles to the west of Din-yuang-ing, capital of the province and residence of a native "van" or prince. The range, about 140 or 150 miles long, rises everywhere abruptly above the Hoang-ho valley, and presents a decidedly alpine character, culminating southwards with Mount Bayan-Tsumbur, 10,600 feet high.

Farther north lies the domain of the Urutes, occupying all the country between the Ordos and the territory of the Chakhar and Khalkha Mongolians in Ala-shan. Here the land is undulating and even hilly, rising steadily to a height of 5900 feet, or 2300 above the Ala-shan plains and 2500 above the Hoang-ho valley.

It was during this journey that Prejevalsky witnessed the somewhat rare spectacle of a steppe fire near Lake Dalai, north of Kalgan. "Towards evening a small light was visible on the horizon, which in the course of two or three hours became a long line of fire advancing rapidly across the open plain. A solitary hill in the centre was soon enveloped in flames, and appeared like a great building burning above the rest. The lake resounded with the loud cries of startled birds, while all was still and quiet on the plain."[1]

On the same occasion this traveller paid a visit to the famous temple of Bathar Sheilun, in the Sirung Bulik mountains, a little north of Bautu. It is "picturesquely situated in the midst of wild rocky scenery, and regarded as one of the most important in South-East Mongolia. The gorgeous shrine is four stories high, and surrounded by a cluster of houses inhabited by 2000 lamas, whose numbers are increased in summer by the pilgrims who visit the temple to 7000, many coming from great distances. We ourselves saw near Lake Dalai a Mogul prince on his way to pray here. He had a large quantity of goods and chattels, and was followed by a train of several hundred sheep to supply him with provisions on the road."[2]

Korea.

The peninsula of Korea, projecting sonthwards between North China and Japan, must in some respects be regarded as an independent section of the Asiatic mainland. It stretches in a south-westerly direction from 42° 31′ to 34° 40′ N., and from 125° to 129° E., between the Yellow Sea and Sea of Japan west and east; while the northern frontier towards China and Russian Manchuria is marked by the course of the Yalu and Tuman rivers. With the numerous islets on the south and south-west coast, its superficial area is estimated at about 82,000 square miles, and most of this area is of a distinctly highland character. The surface rises continually eastwards, attaining in the east coast ranges an altitude of from 7000 to 8000 feet. Since 1882 several expeditions have been made to the interior, which consists mainly of wooded hills, with intervening fertile valleys,

[1] *Mongolia*, i. p. 108. [2] *Ibid.*, i. p. 155.

and in the north vast lava fields, in some places 40 miles long and 100 to 140 feet deep.

The ordinary native name of the country is Tsyo Syeun, with an alternative Keirin, whence the current form Korea.[1] Politically it constitutes an autonomous hereditary monarchy, divided into eight "tao," or provinces. But of all modern States it had till quite recently maintained the most exclusive isolation. The barriers of seclusion, however, were at last removed in the year 1882, when a treaty of amity and commerce was made with England, and since then with the United States, Germany, and other Western powers. The ports of Fu-San, Yuen-San, and Jen-Chuan are now open to the trade of the world, and much information regarding the resources of the country has already been collected by Carles, Douthwaite, and other recent travellers.

Above the bleak northern highlands towers the mighty Peh-tan-shen (Paik-tu-san), or "White-crested Mountain," forming a conspicuous landmark towards Manchuria, and whose height Chinese geographers have estimated at 20 li, or about 7 miles! From its western and eastern slopes flow the above-mentioned rivers Yalu and Tuman, the former to the Bay of Korea, a northern extension of the Yellow Sea, the latter north-east to the Sea of Japan. In the interior there are said to be 33 cities of the first, 28 of the second, and 70 of the third rank. Seul, or Kyung, capital of the kingdom, lies in a somewhat central position near the west coast, on the Han-kang, the chief river of the interior. After a north-westerly course of perhaps 150 miles through the province of Kiung-ki, this river forms below the capital an extensive delta interspersed with many rocky and wooded islets collectively known as the Prince Imperial Archipelago.

[1] But Keirin is itself a corruption of Korai.

Notwithstanding its great natural resources, the country is generally described as wretchedly poor, trade and agriculture in a very primitive state, the people of rude and simple manners. A census taken in the year 1793 gave a population of over 7,340,000, and the present estimate is under 7,000,000. The broad western valleys sloping seawards seem to be thickly peopled, the east side far less so, and the north very sparingly. Towards the Chinese frontier an artificial wilderness has been created by the Government as a protection against the warlike Manchurians, and for this purpose four large cities, besides many villages, are said to have been razed to the ground. Hence the broad zone of neutral and uninhabited land traced on the maps between Korea and Shing-king (Liao-tung).

Of the products one of the most useful is hemp, of which several varieties are cultivated and manufactured into a strong coarse material for the dress of the lower classes. There is, strictly speaking, no foreign trade, but ginseng and paper, the only exported articles, are either smuggled across the Chinese frontier, or brought to the fairs held at stated times along the border lands under the sanction of the authorities. This merchandise is also taken to China in considerable quantities by the suite of the embassy, which proceeds every year to Peking. The Korean paper, made of cotton and the inner bark of a species of mulberry, is very strong, and, like that of Japan, applied to a great variety of purposes.

The peninsula abounds in minerals, such as gold, silver, iron, copper, lead, and coal. But the State reserves to itself the exclusive monopoly of the mines. The natives are quite as skilful as the Japanese in the working of metals, often betraying much artistic taste in the designs. But the more useful arts are in a very backward state. Navigation is conducted on the rivers with flat-bottomed boats, on the coast with small and crazy-looking junks.

Regarding the political institutions and the details of the administration, our knowledge is extremely limited. The monarchy is known to be of an absolute type, modelled on that of China. But besides the royal family there are privileged classes and a hereditary aristocracy, an institution unknown in China. Amongst the nobles several parties have been formed, of which the State is compelled to take account. Although the crown is hereditary, the succession often gives rise to contentions, in which the magnates play an important part.

Officials are said to be mostly appointed, as in China, after a searching investigation, although they occasionally acquire office by purchase or the royal favour. The higher functionaries have almost unlimited control over the lives and property of those within their jurisdiction. The penal code is atrociously cruel, and the application of the bamboo of daily occurrence for the most trivial offences.

All are liable to military service, yet there is no standing army, and the bodyguard of the king are alone entitled to be regarded as soldiers in the ordinary sense of the term. Discipline and tactics are of course unknown; but the rural population, who are bound by a sort of villanage to the crown, are summoned at stated periods to the chief towns of the circles, where they serve as soldiers or armed police. Their weapons are the spear, bow and arrow, and matchlocks, although some really good firearms are manufactured in the capital. The guards wear helmets and breastplates, and in war long overcoats so thickly padded with cotton as to be proof against sword-cuts and musket-shots, but not against the rifle. But this uniform is so heavy that it prevents all free and rapid movements of the troops. A large army of Koreans would be almost helpless in the presence of a well-appointed company of Europeans.

The State religion, like so many other social features, resembles that of China. Both Buddhism and the Lao-tse doctrines are widely spread amongst the people, while the

KOREAN TYPE.

upper classes rest satisfied with the colourless moral teachings of Confucius. The Korean language, probably intermediate between the Mongolo-Tatar and Japanese, is written with a true alphabet of twenty-seven letters, and of uncertain origin. But this alphabet is held in slight esteem, being employed chiefly by women and children, while all the lettered classes are familiar with the Chinese ideographic system. By this means the Chinese and Koreans, although speaking totally distinct languages, are able to communicate their thoughts in writing. Thus the sign for *man*, read off as *yen* by the former, and *saram* by the latter, will convey to both alike the same concept of *man*, just as all Europeans, however they may pronounce them, attach the same value to the Arabic ciphers. Commercial relations between Korea and Japan have existed from the earliest times, and have recently been renewed.

China Proper.

Comprising most of the Hoang-ho and Yang-tse drainage, besides the Pei-ho, Kiang-si, and other smaller river basins, China proper occupies altogether rather more than one-third of the whole empire. But this smaller section is immeasurably the most important in respect of population, products, trade, industries, and material resources of every kind. To such an extent is this the case, that the loss of all the other great divisions would not appreciably diminish the status of China as one of the great powers of the world. Indeed it may be doubted whether its position would not be thereby strengthened. For ages the resources of the Central Government have been strained to the utmost in the endeavour to keep together an unwieldy, and, to some extent, an incongruous political system, the vast frontier line of which it is almost impossible to defend at all points. Were this frontier line contracted to the still broad limits of China proper, the gain in greater concentration alone would probably more than balance the loss of the vast sandy wastes of Mongolia and the bleak upland Tibetan plateau.

Within its natural limits China presents the form of an irregular circle, the landward and seaward semicircles of which are about equal in extent. The inner curve sweeping round from the head of the Yellow Sea to the Gulf of Tonkin, runs successively along the borders of the conterminous regions of Korea, Manchuria, Mongolia, Tibet, Burma, Siam, and Annam. The outer curve or coast line follows the Pacific seaboard through its entire course along the Gulf of Tonkin, the China Sea, Fukien Strait, the Eastern and Yellow Seas, the Gulfs of Pechili and Liao-tung, and Korea Bay.

Excluding the great islands of Hainan and Formosa, this compact mass of land stretches from 20° to 42° N. and from 98° to 122° E., or very nearly 1400 miles both ways, with a total area of about 1,556,000 square miles. From the Tibetan plateau the surface slopes uniformly eastwards, in which direction all the main streams drain to the Pacific. Of these the Yang-tse-kiang, traversing all the central provinces, and dividing the country into two nearly equal portions, is by far the most important artery of trade, and affords the easiest means of access to the interior. Hence this line has been followed by Cooper, Margary, Gill, M'Carthy, and most of the intrepid explorers who have in recent years traversed the land from the Pacific coast to Burma and British India. A good idea of its general features may be had by following in the footsteps of any of these travellers. Perhaps the most memorable and instructive journey, notwithstanding its tragic termination, was that undertaken in 1874 by the unfortunate Augustus Margary, who perished almost at the very goal of his ambition.

Lieutenant Margary followed the course of the Yang-tse to Hankow, whence he started in a native junk on 4th September 1874. In the province of Hu-pe, in recent times more than once wasted by marauders and the Taiping rebels, the level tracts along both sides of the river were planted with cotton and sesame. Here numerous rafts were met, and in Losfan an inquisitive and importunate crowd flocked round the stranger, who was the first European visitor to that place. Round about Losfan runs a crescent of dunes, beyond which stretches an interminable level plain, forming one of the great rice-fields of the empire.

On 20th September the expedition reached the island of Chun-shan at the entrance of the great Lake Tung-ting. Here the turbid Yang-tse was succeeded by the clear water of the shallow lagoon, across which a favourable breeze wafted the vessel to the River Yu-an. At Ni-hsien-tang, 36 miles from the mouth of this river, the hitherto bare or slimy banks gave place to fine grassy tracts and meadow lands, carefully-cultivated fields mostly planted with cotton, stately residences, and a general appearance of prosperity. Landing at Chang-te, the traveller crossed over to Tao-yuan-hsien, a large and flourishing city, formerly a chief centre of the earthenware industry. From this town the pottery is sent in large quantities to Chen-chu-fu, which has of late years been the most unruly and turbulent place in the province. Beyond Tao-yuan-hsien a picturesque highland region was reached, where the crests and slopes of the hills were clothed with pines, aged oaks, and palms. The stream now flowed in a narrow bed between bold and sheer rocky walls, while higher up the channel was obstructed by ledges, forming dangerous rapids.

At Chen-chi-hsien the river describes a great curve, sweeping for 60 miles southwards and the same distance northwards, after which it again resumes its normal westerly course. On 27th October Margary reached Chen-yuan-fu, a city enclosed by grand rocky heights. Here ended the wearisome journey by water, during which he was so ill that he lay nearly the whole time prostrate on his couch. Nevertheless he at once resumed the land journey in a palanquin, and soon after recovered his wonted health. At Ching-ping-hsien coal from the neighbouring pits was seen exposed for sale. As the capital of the province was neared, the towns became larger and the villages more numerous, while the cultivated tracts, chiefly under rice and tobacco, covered continually wider spaces. Beyond Kwei-ting-hsien the route lay mostly through narrow gorges, the grassy hills approaching so near that little room was left for tillage. Yet the road was here lined on both sides by close quickset hedges. On 5th November the traveller entered Kwei-yang, capital of Kwei-chow, which lies in a rolling plain planted with trees and encircled by hills. These hills are mostly isolated and clothed with a rich vegetation to their summits, which consist of level black rocks mostly crowned with splendid temples. At the end of the highway, and facing the city, stand a number of remarkable white marble triumphal arches, erected to the memory of devoted women.

Beyond this place stretched spacious arable plains, now overgrown with tall grasses, but still betraying many signs of former cultivation. The valleys extend mostly in the direction from west to east, wild flowers fringed the road-side, and in the hedges the wild tea plant was in full bloom. Very remarkable are the numerous isolated hills which rise to an average height of about 300 feet, in the Kwei-yang plain, in the districts beyond Ching-chi-hsien and away to the south. Kwei-chow, usually described as an inaccessible highland region, Margary was agreeably surprised to find diversified with many smiling plains, while the route lay chiefly over grassy hills and tracts of rolling steppe all the way to Hang-tai.

Here he at last entered a really highland region

where the road was often extremely precipitous, winding its way over hills rising to a height of 4000 feet and upwards. In this neighbourhood the River Metou marks the limits of the domain of the wild Miao-tse and Chung-chia tribes. Farther west the land under tillage becomes more extensive, the villages grow more numerous, and a limited trade is done in oranges and other local produce. The climate of Yunnan is colder than that of Kwei-chow, where the dwellings are not calculated to resist the effects of severe winters.

On 20th November the Yunnan pass was crossed, and the expedition reached the first frontier town of Yunnan. Between Yunnan-fu, capital of the province, and the Irawady valley, North Burma, the natural difficulties of the route offer very serious obstacles to the local trade. The whole of Yunnan is covered with mountains of an extremely rugged character. From the capital to Bhamo in Burma the track lies over ranges rising to 3500 or 4000 feet above the plain, which is itself about the same height above sea-level. The country is thinly peopled, and even the valleys are but partially cultivated. Here opium is produced in considerable quantities, one-third of all the crops apparently consisting of poppies. The natives are generally poor and wretchedly housed. The so-called "lekin," or stations, seem to be the most flourishing establishments. They are met everywhere, and are constantly used as residences, being usually the best and cleanest places in the towns.[1]

[1] Fresh particulars relating to Margary's death have been obtained by the Rev. H. Soltau and J. W. Stevenson, who have recently travelled from Burma to Yunnan in company with a caravan conveying goods to the Shan country. After the murder it appears that a proclamation was issued, which has made travelling in the interior of China much safer than it had ever been before. At Talifu they were well received by the natives, who were all eager to know when the English were coming to

As was the region traversed by Margary, so is China generally—a land almost everywhere presenting the most violent contrasts both physical and social, amidst an all-prevailing and undefinable sense of monotonous uniformity. We feel that this is still the home of the children of Han, a land of wealth and want, exposed to the most violent vicissitudes in the midst of an eternal stability, devastated by sudden outbursts of nature's pent-up forces or by the still more terrible display of human passion, but ever rising with fresh youth from the desolation of its ruined cities and wasted plains.

This surprising vitality is due partly to the inexhaustible fertility of the soil, partly to the character of the people. Not only is the "Chinese Mesopotamia" the richest granary in the world, but its productiveness is enhanced by the ingenuity of a laborious and patient race. Wheat, barley, tobacco, pulse, are the chief products of the northern provinces, where cold winters are succeeded by hot summers. Cotton, the sugar-cane, pepper, betel, spices, tropical fruits, and especially rice, which is the staple of food, are mainly cultivated in the central provinces, where the heats are excessive and the cold seasons accompanied by storms and a heavy rainfall. The chief rice-growing tracts form the heart of the country, the seat of the densest population, and the focus of commercial life.

The western hills, or rather their western slopes, are one of the few regions where the valuable medicinal rhubarb plant is indigenous, and here also the opium-producing poppy is largely grown. The taste for this

open up the trade with the coast. The farthest point reached was Laowatan within two days' journey of the Yang-tse, 828 miles from Bhamo and 1756 from Shanghai. The journey, which took 58 days altogether, lay mostly through pleasant scenery and amid a friendly, hospitable people.

narcotic is said to have been unknown in China during former times, although doubtless some other drugs must have been used. During recent times the consumption has increased to a large extent among the inhabitants of whole provinces. One-fourth of the people are now said to be opium-smokers, and the cultivation of the plant is now monopolised by the Chinese Government, which even appears to encourage its growth. The home-grown article comes chiefly from Mongolia and North Manchuria, and although some more southern provinces have of late years begun to take part in its cultivation, still the local supply is insufficient for the demand. Large quantities are imported from India for the use of the upper classes. The Indian Government levies high taxes on the exportation to the amount of £6,000,000 to £7,000,000 annually. This taxation has been much discussed lately, but it rests on the same foundation as the taxation levied by other Governments on spirits, and in principle it is defended as rendering dear an article which, if consumed temperately, may be harmless, but if consumed to excess will be injurious.

It is very remarkable that, notwithstanding the general deficiency of timber, the Chinese peasantry carefully clear the land of all its bush and forest growths. Although the great demand for wood and the want of more space for tillage may elsewhere account for this practice, such motives cannot explain the systematic burning of the hilly woodlands, more especially that the ground is not needed for stock-breeding, an occupation very little pursued in China. The Abbé David attributes this reckless destruction to the desire to get rid of wild beasts, such as the tiger and panther, which still infest many provinces. But whatever the cause, the consequences are most disastrous, leading to denudation of the surface soil in the uplands, and to constantly-increasing inundations in the lowlands. Unless the process be checked, it must ultimately reduce some of the richest lands to deserts, and bring about a state of things similar to that of Persia and parts of Central Italy, where the disappearance of forests on the uplands has been followed by the disappearance of agriculture on the lowlands.

A chief source of national wealth is the production of raw silk, which forms a staple of trade in a country where sericulture has from time immemorial been indigenous. At Shanghai silk ranks with tea and cotton as a staple export, while in Canton it takes the first place.

But next to agriculture the main resource of China lies in the ground itself, which harbours supplies of ores and coal sufficient some day to revolutionise the trade of the world. Although coal has long been used as an article of fuel, the attention of Europeans has but recently been directed to the vast coal measures of the great river basins. Some of the more important mines had already been visited by Kingsmill, Pumpelly, David, and other intelligent explorers. But for a more accurate knowledge of China's carboniferous deposits we are indebted mainly to Baron von Richthofen. South of the crystalline Peling (Tsin-ling-shan), over 11,000 feet high, which with the Funiu-shan forms the eastern extension of the Kuen-lun between the Hoang-ho and Yang-tse valleys, the measures, although less important than on the northern side, are still far from inconsiderable. The superficial area of the vast coal basin in Se-chuen alone is estimated by Richthofen at 100,000 square miles. This basin is enclosed on all sides by lofty ranges, and the deep channels excavated by the affluents of the Yang-tse everywhere show the fossil cropping out, while the working of the mines is facilitated by the streams themselves, which are navigable to the limits of the basin. Here the quality varies, the good bituminous beds of the north and west being elsewhere replaced by moderately good anthracites. Large deposits of excellent anthracite also occur in Yunnan, here associated with rich copper, tin, zinc, and lead ores.

But the old and true carboniferous formations lie in the more central province of Hunan, which basin may be compared with the rich Pennsylvanian deposits. It has an area of 16,000 square miles, stretching along both banks of the Tse-kiang (Lo-kiang) as far as Siang-tang, and contains excellent anthracite in the south, and bituminous coal in the north.

But in the northern provinces are found the most extensive and richest measures, spreading over 25 degrees of the meridian from the western deserts to the Yellow Sea. They belong mostly to the old carboniferous formations, and iron ores also abound in Shan-si, a chief centre of these deposits. Not only are the coal-fields extremely rich, but they might be worked under more favourable conditions than perhaps any others in the world. The exceptionally thick seams are mostly horizontal, spreading over a vast plateau 32,000 square miles in extent, and resting on a horizontal limestone foundation. This plateau of Tal-hal-shan rises from 2000 to 3000 feet above the surrounding plains, and in many places the seams have been denuded by the watercourses right down to the lower limestones. Here the anthracite is of the best quality, very thick and pure. Honan and Kan-su bordering on Shen-si are less rich in coal-fields, which, however, are also found in Pechili. Here the Government is now about to utilise the vast coal, copper, and iron deposits north and west of Tien-tsin, which were surveyed by Henderson a few years ago. The coal of this district is of prime quality.

Islands—Hainan, Formosa.

The Gulf of Tonkin is enclosed on the east by a peninsula projecting southwards, and separated only by a narrow channel from the mountainous but unhealthy island of Hainan. The interior, which contains productive gold-mines and valuable timbers, has been but slightly explored, being still mostly held by independent wild tribes. The Chinese hold possession of the coast only, on the north side of which is the large seaport of Kien-chow. The island, which seems to culminate with the snowy Ta-uchi-shan, has an area of over 16,000 square miles, with a population estimated at 1,500,000.

Farther north lies the great island of Taiwan or Formosa, separated from Fu-kien by the broad Fu-kien or Formosa strait. The west coast alone belongs to China, the east and the interior being still occupied by unconquered hill tribes. The island is 245 miles long north and south, with extreme breadth 76 miles, and total area 14,982 square miles.

Formosa—that is, the "Beautiful"—received this name from the early Spanish navigators, struck by the charming aspect of its magnificent wooded heights. A volcanic range with long extinct craters crosses the interior, merging westward in an extensive fertile plain watered by numerous limpid streams. This inviting region has attracted many industrious colonists from Fu-kien, who have brought most of the land under cultivation. But the eastern districts, where the hills reach down to the coast, afford little room for tillage, and have consequently been left to the wild native tribes of Malay stock. In the central range the culminating points are Mount Morrison (10,800 feet), Mount Sylvia (11,300), Ta-shan (12,000 ?), and an unnamed peak near Sylvia (12,800).

A trip along the west coast presents a panorama of

the loveliest scenery imaginable. Above the highly-cultivated lowlands, the advanced ridges rise to a height of 5000 feet, beyond which the horizon is bounded by

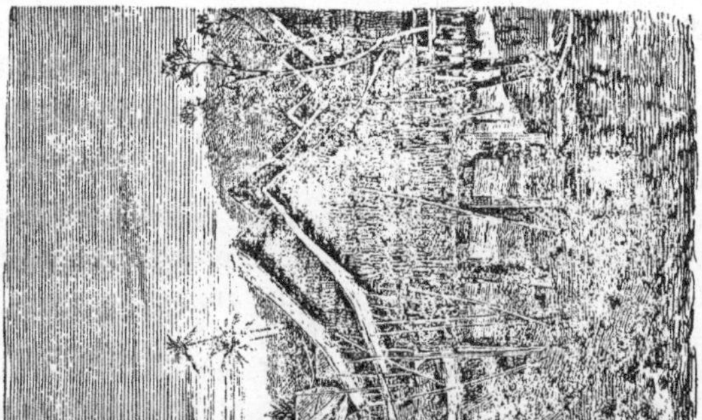

MALAY VILLAGE, FORMOSA.

the central range with a mean elevation of 10,000 feet. The outlines of these highlands are as fantastic as they are beautiful, domes and slender towers, curious jagged cliffs and steep rocky ramparts towering everywhere above the soft grassy slopes, and sending down mountain torrents, which merge lower down in sparkling silvery streams. Here and there the brighter hues are toned by the darker shadows of the native hamlets clustering in the rocky gorges. These villages are grouped in two classes, the independent and the "Peppohoan," which acknowledge the authority of the Chinese, intermarry with them and adopt their customs, while retaining their own Malay speech. These subject tribes serve to promote intercourse with the independent natives of the interior. They are a people of much promise, amongst whom Christianity seems to be gaining ground.

The wild tribes tattoo the face, and, according to Captain Box, construct elegant huts of bamboo and palm leaves. Over the doorway are often suspended, as trophies, the skulls of wild boars, deer, and apes, and a more than usually vainglorious savage made a display of a tuft of six pigtails detached with his own hand from the heads of his Chinese victims.

Amongst the natural products of Formosa, which is about the size of Sardinia and Corsica rolled into one, are sulphur, petroleum, coal, and camphor.

The whole eastern seaboard of Asia is fringed by a rampart of islands, and insular groups stretching from Kamchatka in a series of graceful curves southwards to the Philippines. The Kuriles and Japan are followed by Linshoten and Liu-kiu, reaching to Formosa, beyond which the Bashi and Babuyan groups complete the chain to Luzon. These islands have been compared to a line of fortifications, and the comparison is justified by the fact that from the Yellow Sea to Amoy the Chinese seaports are completely sheltered by them from the typhoons which elsewhere sweep the Eastern seas. Arrested by the rocky barrier of Formosa, these terrific tornadoes are diverted to the southern seaboard, where they often spend their fury on the exposed rocks of Hong-Kong and Macao.

Macao—Hong-Kong.

These two islets, at the eastern and western entrance of the Canton River, symbolise the setting star of Lusitania and the rising sun of England in the Eastern waters. Macao, for ever associated with the names of Camoens and Xavier, is all that now remains in these regions to the Portuguese, who first revealed to the West the water highway of the Eastern world, and whose influence was at one time paramount from Mozambique to Japan. And even for this rock and decayed seaport with the high-sounding name of Cidade do Santo Nome de Dios de Macao, they pay an annual tribute of £150 to the imperial exchequer; for the Chinese Government has never acknowledged the absolute right of Portugal to the possession of this little tongue of land at the southern extremity of the Si-kiang delta. In a space of about 12 square miles there is here massed a population of about 60,000, including 5000 Eurasians. But the monopoly of the trade with Europe enjoyed by Macao for three centuries disappeared with the opening of the Chinese Free Ports, and the traffic in slaves, disguised under the name of Coolie emigration to Peru and the West Indies, was finally suppressed by the Imperial Government in 1874. Since then Macao has continued to decline, and it is now chiefly noted for its gambling-houses. The sea-going vessels that formerly crowded its spacious harbour now either pass up the river to Whampoa for Canton, or else cross over to Victoria on the north side of Hong-Kong.

This barren granite and basalt mass, some 36 square miles in extent, had a population of about 2000 when it was occupied by the British in 1841. Now it is one of the great marts of Eastern trade, with magnificent quays, dockyards, arsenals, a yearly tonnage of 4,000,000, and exchanges already exceeding £12,000,000! The whole island is dotted over with large villages, suburban villas, public buildings, frowning battlements, and is overcrowded with a motley population of 250,000 Chinese, Hindus, Burmese, Malays, Polynesians, Europeans, and Americans. Formerly a very unhealthy place, Victoria has by a proper system of drainage become a sanitarium for the English residents in China. But it lies unfortunately within the limits of the cyclones, during one of which in 1874 over 1000 houses were blown down, 33 large vessels and several hundred junks wrecked, and many thousand lives destroyed.

Hong-Kong is a Crown colony, administered by a governor and legislative council. In 1881 this body voted 30,000 dollars for an observatory and time-ball to warn shipping against approaching storms.

5. *Climate: Prevailing Dryness—Steppe Storms—Typhoons.*

The climatic conditions are essentially different in Central and East Asia. Considerable uniformity prevails on the lofty Tibetan plateau and the less elevated plains of Mongolia, where the prevailing features are great extremes of heat and cold combined with excessive dryness. But the Hoang-ho and Yang-tse basins, being exposed to the soft moisture-bearing winds from the Pacific, enjoy a far more equable temperature. Here the winters are doubtless severe in the north, the summers hot in the south; but there are nowhere such extreme vicissitudes as are characteristic of strictly continental climates, while there is nearly everywhere a sufficient rainfall.

At the same time the copious rains would appear to extend much farther inland than is commonly supposed, embracing not only the region of the "Cross Ridges" on the Tibeto-Chinese frontier, but even the Ala-shan highlands of South-East Mongolia. Here Prejevalsky was more than once overtaken by tremendous downpours of quite a tropical character. On one occasion he tells us "the rainfall was so great that streams formed in every cleft and gorge, even falling from the precipitous cliffs, and uniting in the principal ravine, where our tent happened to be pitched, descended in an impetuous torrent with terrific roar and speed. Dull echoes high up in the mountains warned us of its approach, and in a few minutes the deep bed of our ravine was inundated with a turbid coffee-coloured stream carrying with it rocks and heaps of smaller fragments, while it dashed with such violence against the sides that the very ground trembled as though with the shock of an earthquake. Above the roar of the waters we could hear the clash of great boulders as they met in their headlong course."[1]

This was in the month of July, and in a district usually included within the rainless zone of the Gobi. But the explanation lies in the direction of the Indo-Chinese ranges and East Tibetan "Cross Ridges," which run south and north, or south-east and north-west, thus giving free access to the rain-clouds from the Indian Ocean and China Sea far into the interior of the continent. This orographic disposition also serves to account for the extraordinary number of large rivers which here take their rise, and which have in the course of ages cut up the East Tibetan plateau into so many independent watercourses. The mountains which are not upheavals, but the result of fluvial action, grew, so to say, hand in

[1] *Mongolia*, ii. p. 264.

hand with the rainfall, resulting in an intricate highland river system elsewhere unparalleled.

Farther west such is the extreme dryness, that for months together not a single snowflake will fall on the elevated Tibetan plateau. It is a remarkable fact that the snow-line descends much lower on the southern than on the northern slopes of the Himalayas, where Forsyth found the Cayley Pass quite free at an elevation of over 19,000 feet. Owing to this absence of moisture the passes between Kashmir and Yarkand are open throughout the year, and some of the more difficult passes are elsewhere lined with the withered or mummified bodies of yaks, horses, or sheep, which dry up where they fall without passing through the process of putrefaction. But notwithstanding the absence of snow the cold is not the less intense, and this, combined with the "mountain sickness," produced by the extreme rarefaction of the air, causes great sufferings to travellers and animals in winter. Even in summer the streams often freeze.

In the Tarim basin and West Mongolia the air is also extremely dry. Here there is scarcely any spring, intensely cold and late winters being followed almost immediately by equally intense heat, when the glass rises even in April to 93° F. in the shade (*Prejevalsky*).

In East Mongolia the spring is also cut up by late frosts, lasting even into May, when still water sometimes freezes an inch thick during the night. Here the sudden changes of temperature are very trying, and north-westerly gales prevail for weeks together, obscuring the sun's rays and filling the air with clouds of sand mixed with fine particles of salt from the saline marshes. Steppe storms rage at times with great violence, "during which even the camels accustomed to the desert would turn their backs to the storm and wait till its fury had abated."[1]

[1] Prejevalsky, *Mongolia*, i. p. 119.

The climate of Korea is healthy, although severe in the north, while resembling that of Japan in the south. In winter, however, snow and ice are common everywhere. Korea being washed on three sides by the sea, the rainfall is very heavy and the vegetation correspondingly vigorous.

Owing to the steady trade-winds from the Bay of Bengal and the monsoons from the Pacific, China proper enjoys both a more copious rainfall [1] and more regular seasons than Central Europe. Although it lies much farther south, reaching from about 40° N. into the tropical zone, the mean temperature of both regions differs but slightly. China, however, is at once colder and hotter than Europe. The dreaded typhoons visiting the Eastern seaboard are produced by the conflict of the south-westerly and south-easterly trade-winds which meet in the Pacific not far from the coast. The word typhoon, which has nothing more than a curious coincidence in sound and meaning with the Greek *τυφῶν*, is derived either from the Chinese *ta-fung* = great wind, or much more probably from the Formosan *tai-fung*, thus quaintly described in the *Tai-wan-fu-chih*, or native annals of Formosa—"The winds of our sea are very different from those of other seas. A fierce storm, which here blows, is called *keu;* but greater strength is possessed by the *tai.* The keu rises suddenly, as it also ceases suddenly, while the tai rages ceaselessly day and night. The keu blows in the season between February and May, the tai from June to September; in September follows the north wind (north-east monsoon). . . . The tai is accompanied by strong rain, uproots trees, blows down brick walls, unroofs houses, and rends rocks. Ships at anchor are dashed to pieces; but as soon as thunder is

[1] Mean rainfall on east coast, 40 inches; at Canton, 50; at Shanghai, 42; at Peking, 25.

heard the storm is over. When the north wind rises in the seventh month, a *tai-fung* is very likely at hand. To recognise a true tai we must observe its course. For the tai is a storm which blows from every quarter, and all the tai winds follow this law; but an ordinary storm rages in one and the same direction."

The form "tai-fung" and the description here given leave little doubt as to the origin of the term typhoon, which was first used by Pinto (1560), who derives it not from the Greek but from the Chinese.

6. *Flora and Fauna: Rhubarb and Ginseng—The Yak, Wild Camel, Birds of Passage.*

The great elevation of Central and West Tibet is unfavourable to the growth of trees, which are here represented chiefly by a few poplars and hardy fruit trees, in some places found at elevations of 12,000 and 13,000 feet. At these heights the breezy plateaux are often covered with coarse grasses, strong and sharp enough to pierce the boots of travellers and the hoofs of pack animals. Barley grows in sheltered spots as high as 15,000 feet, and many of the less exposed plains are covered with rich pasturage. Forests abound in the well-watered valleys of South-East Tibet, where a variety of the holly rivals the conifers in height, while far surpassing them in the richness of its foliage.

Equally rich in timber is the highland region between Kulja and the Tarim basin, where the spruce, larch, poplar, birch, apple, and apricot, are varied with an undergrowth of hawthorn, woodbine, brier, and wild hop. The poplar, aspen, peach, willow, birch, and some other trees, also flourish on the uplands of Eastern Mongolia. But in the Ala-shan and Kuku-nor districts the scanty vegetation is represented chiefly by a few dwarf elms, clumps of acacia, two or three species of shrub, and some flowering grasses. Even in the neighbouring province of Kansu, notwithstanding its rich and varied flora, forests occur only on the northern slopes of the southern ranges. Here the red birch, mountain ash, spruce, poplar, willow, and arboreous juniper, growing to a height of 20 feet, are interspersed with the wild rose, barberry, gooseberry, raspberry, currant, and other shrubs.

Indigenous in these regions is the medicinal rhubarb, the Shara-moto or "yellow tree" of the Mongolians, which grows 8 or 10 feet high, with a stalk nearly 2 inches thick, and leaves 3 feet by 2. This useful plant, which is found as high as 10,000 feet above the sea, is sent down the Hoang-ho for Peking and Tien-tsin, where it is shipped for Europe.[1]

Owing to its copious rainfall, Korea possesses an abundant vegetation, including rice, millet, maize, and other cereals, which are largely cultivated. Tobacco, apples, pears, apricots, pomegranates, hemp, and cotton, also flourish here. Of great commercial importance is Ginseng, a species of *pannax*, which is raised from the seed sown under sheds covered with pine bark. The roots arrive at maturity in five years, and are then collected and dried for the China market. Although less esteemed than that of Manchuria, the Korean variety fetches in Peking from £3 to £4 per pound.

The Chinese flora is extremely rich. Forests, in the European sense, are rare; but evergreens, flowering shrubs, and especially resinous plants, are found in great variety. Proceeding southwards, the transition is very gradual from the Manchurian to the tropical flora of Indo-China. Hence

[1] During his last expedition (1880) to Tibet, Prejevalsky found the rhubarb in the district beyond Gomi, on the Upper Hoang-ho, growing to a colossal size. One root taken by him "measured 16 inches in length, 12 in breadth, and 7 in thickness, and weighed 26 lbs."

in some of the central districts there is a remarkable intermingling of species belonging to different zones, the bamboo flourishing by the side of the oak, while wheat and maize crops are interspersed with paddy fields, sugar and cotton plantations. In general the cultivated species are everywhere encroaching on the wild flora, and in many districts little is seen except vegetables, fruit trees, sugar-cane, cotton, the poppy, mulberry, rice, the bamboo, and tea plant. The three last named are of vast economic importance, rice supplying the staple food of hundreds of millions, the bamboo yielding the chief material for the construction of their houses and furniture, while tea, forming their national beverage, is now also exported in ever-increasing quantities to England, Russia, and America.

Notwithstanding its scanty vegetation, Tibet, which by some naturalists is regarded as a chief centre of evolution for animal life, possesses a fauna of extraordinary richness. In the west, Nain Singh met the antelope in herds of as many as two thousand bounding over the plains. Here the yak, gazelle, wild goat, various species of sheep, wild ass, fox, jackal, wild dog, white wolf, and even a white bear, resembling the polar bear in appearance, are amongst the most characteristic animals. But the avifauna is poor, chiefly comprising the eagle, vulture, raven, but no singing birds. In East Tibet numerous herds of buffaloes are preyed upon by the wolf and panther. The musk-deer is found as high as 8500 feet, while the monkey and squirrel here form the transition to the Indian fauna. Amongst the birds are the pheasant and lark, which latter soars to a height of 15,000 feet.

The Tibetans have domesticated the yak, sheep, and horse, as beasts of burden. The sheep carries loads of 20 to 30 lbs. over the highest passes, and from the yak and Indian zebu, the dzo, a useful cross-breed, has been obtained, which however reverts after the fourth generation to the original types. There is a formidable species of watch-dog, which degenerates in India, but has been acclimatised in England. But of all the animals the most valuable is the goat, whose soft down (pashm), growing under the outer coating, supplies the material for the finest Kashmir shawls.

Amongst the wild animals of the Tarim basin and South Mongolia are the wild boar, tiger, antelope, and hare. The wild camel was formerly common in the Lob district and along the Altyn-tagh range, but is now almost entirely confined to the desert tracts east of the Tarim. Unlike the domestic species, the wild variety is remarkable for its sagacity and highly-developed sense of smell, sight, and hearing. It will climb the most inaccessible places, and when once it takes to flight will run for 20 or 30 miles at a stretch.

A notable feature of the Lob, Dalai, and other great steppe lakes, are the birds of passage which alight at these halting-places on their long journeys north and south in the spring and autumn.

Old writers speak of the rhinoceros, tapir, and elephant as formerly roaming over the plains of China. But the only large wild animals now found in the Yangtse and Hoang-ho basins are the tiger and panther, and even these seem to be disappearing. Nevertheless the fauna, especially of the wooded highlands towards the west, is very rich, especially in animals of the snake, salamander, and lizard orders. On the whole, the species common to Europe are few, including 146 out of 764 birds, 10 out of 200 mammals, and no fresh-water fish except the eel. There are several species of the monkey, some of which are met as far north as the neighbourhood of Peking.

Stockbreeding is not a favourite pursuit in China; hence cattle, sheep, and even horses, are comparatively rare, and of indifferent breeds. But buffaloes and swine are very common, as well as ducks, which, with fish, vegetables, and rice, form the chief articles of food. In many places the cormorant is trained to fish.

7. *Inhabitants: Table of Races in the Chinese Empire—The Chinese; Jews, Muhammadans, and Christians, in China—The Tibetans—Buddhism—The Mongolians.*

Within the limits of the Chinese Empire are probably comprised one-fourth of mankind, and of these at least nine-tenths are concentrated in China proper.[1] Korea and parts of Manchuria seem to be fairly well peopled. But most of the other outlying regions are, to a large extent, uninhabitable. In Tibet the population is mainly confined to the San-po basin, in Mongolia to the northern and eastern edges of the Gobi desert, in the Tarim basin to the middle courses of the Kashgar, Yarkand, and Aksu rivers. In China the great river valleys are amongst the most densely-peopled regions on the globe, and even the most inaccessible highlands on the Indo-Chinese and Tibetan frontiers are occupied by numerous hill tribes.

Apart from these hill tribes, whose affinities are still largely undetermined, all the inhabitants of the empire, except the few Iranians of Kashgaria and Zungaria, and Malays of Formosa, belong physically to various branches of the great Mongolo-Tatar family. But there are at least six fundamentally distinct linguistic groups, as shown in the subjoined table of all the races of the empire.

Most of the hill tribes are still nature-worshippers. Muhammadanism has been largely diffused throughout Kashgaria, Zungaria, West and North-West China. Christianity has secured a footing in various parts of China proper, and even in Mongolia and Manchuria. The Jews,

[1] Much light is thrown on the early history of these races by Demetrius Boulger's new *History of China.*

or "Blue Muhammadans," as the Chinese call them, were formerly numerous, but are now reduced to a few hundreds, mostly centred in Kai-fung-fu, capital of Honan. They claim descent from the tribe of Asser, and say they reached China in the Han dynasty (202 B.C.—264 A.D.) But they have forgotten their language, and the few "Aronists," or rabbis, who can still decipher the Pentateuch, pronounce the Hebrew words Chinese fashion, so that *Israel* becomes *Ye-se-lo-ni.* But nobody understands the text, and their traditions have got so confused that they believe Mecca and Medina to be their holy cities. But the great bulk of the inhabitants have long adopted the Buddhist tenets, variously modified according to the national temperament, usages, and traditions.

I. Mongoloid Races of Mongolo-Tatar Polysyllabic Speech.

Sharra or Eastern Mongolians.	Khalka: Tushetu, Tsi-tseng, Jasaktu, Sain-noin	N. Mongolia mainly.
	Uchumsin; Chakar; Genshikten; Barin; Kartsin; Jarot; Uniot; Sunni; Tumet; Kortsin; Durban; Urut; Naiman; Ahkhanar	S., E., and S.E. Mongolia.
	Ordos	N. bend of the Hoang-ho.
Eleuts (Kalmuks) or Western Mongolians.	Chorass; Turgut; Khoshot; Durbat	Zungaria, Kulja, N.W. Mongolia.
Urianhai		Upper Yenisei basin.
Sok-pa		N.E. Kachi (N.E. Tibet).
Taldi (?)		W. Kansu.
Tungus Family.	Manchus; Tungus	Manchuria.
	Solons; Sibos	Upper Ili valley, Kulja.
Túrki Family.	Taranchi; Kirghiz-Kazaks	Kulja.
	Kara-Kirghiz	Central Tian-shan.
	Kashgarians	Tarim basin, Kulja.
	Dolans	Kashgaria.
	Salars (Kara-Tanguts)	? About source of Yang-tse.
	Hor-pa	W. Kachi (N.W. Tibet).

II. Mongoloid Race of Korean Polysyllabic Speech.

Koreans	Korea.

III. Mongoloid Races of Tibetan Intermediate Speech.

Bod-pa (Tibetans proper)		San-po basin mainly.
Tanguts		Kansu, Kuku-nor, Tsaidam.
Drok-pa		Central Kachi, between Sok-pa and Hor-pa.
Chak-pa		Central Kachi, between Sok-pa and Hor-pa.
Cham-pa		East of Noh, Tibet.
Kham-pa		? Central lake region, Tibet.
Chang-pa		East of the Kham-pa.
Si-fan	Andoan	Tibeto-Chinese frontier from Kuku-nor to Yunnan.
	Tochu	
	Arru	
	Gyarung	
	Telu	
	Manyak	
	Melam	

IV. Mongoloid Races of Chinese Isolating Speech.

Chinese proper	N. and Central China.
Punti	Kwang-tung.
Hwui-chan	Kwang-tung.
Hakka	Kwang-tung, Fo-kien.
Hok-lo	Swatow district (Fo-kien).
Tungans	Kansu, Zungaria, Kulja.
Khambing	Kulja.
Chimpan (extinct ?)	Kulja.
Khatozun (extinct ?)	Kulja.

V. Highland Races of Undetermined Ethnical and Linguistic Affinities.

	Man-tse (I-jeu)	W. Se-chuen.
	Sumu	W. Se-chuen.
Miao-tse, or Nan-man group.	Pe-Lolo	S. bend of the Yang-tse. S. Se-chuen, N. Yunnan.
	Shu-Lolo	
	He-Lolo	
	Sen-Lolo	
	Chung	Kwei-chow uplands.
	Ngnchung	
	Tuman	
	Kilao	
	Kitao	
	Yao	Lipo district, S. side Nan-ling mts.
	Seng	Nan-ling mts.
	Tung	N. Kwang-tung.
	Lyssu	S.E. Tibet, between Lu-tse-kiang and Lan-tsan-kiang.

Mosso (civilised Lyssu)	N.W. Yunnan, S. of the Lyssu.
Lu-tse (Anong) . .	N. of the Lyssu.
Remepang . . .	E. of the Lyssu.
Pagni (Paï, Terong, or Ba-yul) . . .	W. of the Lu-tse.
Tsarong . . .	N. of the Lu-tse.
Ku-tse . . .	N. of the Remepang.
Diju	N. of the Ku-tse.
Jrupa	N. of the Diju.
Mu-ûa (Anampel) .	Upper Irawady, Burmese frontier.
Shutung . . .	W. Yunnan.
Shang-lai . . . Shuk-lai . . .	Island of Hainan.

VI. Malay Stock and Speech.

Pepukwan . . . Yukan . . .	Island of Formosa.

VII. Aryan Stock and Speech.

Tajiks . . .	Kashgaria, Kulja.
Kara-Kultsi (?) . .	Lower Tarim River.
Lobnorski or Kara-kurchin . . .	Lob-nor district.

Of the peoples comprised in this table scarcely any can pretend to claim unsullied lineage. In the Ili basin, where whole populations have been more than once extirpated by fire and sword, the Taranchi, Tungans, Solons, and other immigrants from Kashgaria, Kansu, and Manchuria, present the most varied types with almost every shade of transition between the fair and yellow stocks. The Taldi seem to be half-caste Chinese and Mongolians, the Salars a mixed Tibetan and Túrki people, the Kashgarians a curious blending of Túrki and Iranian, the Tanguts a still more remarkable fusion of Tibetan, Mongol, and Chinese elements. Of the original stock of the aborigines in the southern and south-western Chinese uplands, we know next to nothing; but we do know that these aborigines were already in possession of the Yang-tse basin long before the arrival of the children of Han. Here they were partly extirpated, partly absorbed, or driven into their present inaccessible fastnesses by the yellow intruders from the Tibetan plateau, who gradually re-settled the land. Hence the Chinese themselves can in no sense be regarded as a pure race, and the many diversely-modified forms of the Mongol type which they present must be attributed to the various interminglings that took place between them and the aborigines during the conquest and settlement of the "Middle Kingdom."

The Chinese.

Nevertheless the Chinese are the most important branch of the Mongolian family of mankind, far surpassing all the rest combined in numbers, wealth, and power. The great antiquity of their culture is reflected in the primitive state of their isolating speech, which is still destitute of declension, conjugation, or grammatical inflection of any kind. The same root is capable of representing all the parts of speech without further change, and in virtue of its position alone. Yet so subtle are the laws regulating the place of the word in the sentence, that the language not only amply suffices for ordinary intercourse, but has become an adequate instrument in the hands of the lawgiver, philosopher, historian, and poet. In their simple speech the Chinese have shown what great things may be accomplished by small means (*R. K. Douglas*).

Equally original and characteristic is the writing system, which is still mainly ideographic—that is, expresses not sound but thought. There is consequently no such thing as an alphabet, and the only approach to true phonetics are the 214 so-called "keys" or "tribunals" used in combination to indicate the pronunciation of unknown characters. There are practically as many symbols as there are words in the language, or 43,496 altogether; but of this number 13,000 are totally irrelevant, and for the expressions in ordinary literature about 4000 signs appear to suffice. The writings of Confucius and his disciples can even be read by the help of only 2500, and a knowledge of these will enable the student tolerably to understand all Chinese works on history and philosophy.[1]

Owing to their tenacious adherence to these and other primitive methods and traditions, the Chinese now find themselves far behind other nations in scientific attainments. Nevertheless, the patriarchal conception of the State as an enlarged family, and of the family as the State in miniature, has given rise to many excellent institutions and many charming features in the social and domestic life. Hence it would be a manifest mistake to regard the Chinese as a decrepit or hopelessly corrupt people on account of the present low state of their culture. Their standard of morality certainly differs widely from ours; but in the midst of much vice the nation has, so to say, been safeguarded by its extreme frugality and thrift. Drunkenness has hitherto been a rare phenomenon amongst them; but they have during recent times largely extended their habit of opium-smoking. They not only consume all the opium that can be obtained from India, but they grow it largely for themselves. The Indian opium is consumed by the upper classes, the indigenous or Chinese opium by the humbler classes. Whether opium-smoking is deleterious, even in moderation, is a question much discussed at present. Many authorities believe that it is not more harmful than spirituous liquors in European countries. Some even contend that it is less injurious than the other stimulants which, in some form or other, are used by nearly all nations.

[1] F. Ballhorn, *Grammatography* p. 32.

The status of woman is not so bad as is often supposed. Those of the lower classes have doubtless to work hard for the support of the family, whose ordinary diet is rice and cabbage. Still, their lot is not, perhaps, any harder than that of the same class elsewhere, more

CHINESE LADY, SHANGHAI.

especially as with the Chinese hard work has become a second nature. The husband has, in certain cases, the power of life and death over his helpmate, yet he seldom strikes her, although the reverse would seem to be far from rare.

A special occasion for rejoicing is the birth of a son;

but in the case of a daughter the midwife would find it as difficult to recover her fee as is so often the case in India.[1] That the birth of a son should be regarded as a propitious event is all the more natural that, according to universal usage, the son remains in the house and becomes the support of his parents in their old age, while the daughter either founds a new home or becomes a burden to the family. Old age is held in special veneration, and a man advanced in years is highly flattered by the inquiry after his "honourable teeth," the conventional phrase employed in asking people their age.

At the death of the Emperor the whole nation is thrown into mourning, the rites attending which are of a very stringent character. For a hundred days the court and people of rank wear white trimmed with white fur, this being the mourning colour in China. For the same period the men abstain from shaving, while the women lay aside the favourite ornamental head-dress. After this first term the garments assume a black or dark hue for the following twelve months. During this period no betrothals take place amongst the better classes, while for others the corresponding term is limited to a hundred days. All entertainments and public rejoicings are strictly suspended for a year.

The Chinese, if not the greatest, are among the oldest traders in the world, and during the course of ages all the complicated relations of buyer amd seller have been regulated by prescriptive usage. The legal rate of monthly interest on advances is fixed at 3 per cent, this high rate being partly explained by the great risk incurred by the lender.

Very beneficial are the many trading guilds, which

[1] "In midwifery cases a large fee will sometimes be paid spontaneously [if] the child be a boy, and no fee whatever if it be a girl" (Mrs. S. Heckford, late of Bhopal, *The Queen*, Nov. 19, 1881).

regulate current prices, advance money in cases of temporary embarrassment, and protect their members from the exactions of mandarins. The society defrays the funeral expenses of associates dying without means, and these guilds even form fire brigades, which are placed at the service of the general public.

The Chinese believe that the whole course of their lives, down to the smallest details, is predetermined by unalterable laws. The most ordinary events are referred to supernatural agencies.

On the other hand, the State religion, of which the Emperor is head, has neither a hierarchy nor ceremonials of any sort beyond a few symbolical rites observed by the Emperor and provincial governors on the New Year. Nothing further is prescribed except the study and contemplation of the moral precepts of Confucius and Lao-tse. Although the bulk of the people have adopted Buddhism, China has never forgotten the teachings of these two sages, of whom the former taught a system of practical ethics, rather than a religion, the latter a pantheistic mysticism analogous to that of the Hindu Patanjali.

Buddhism is encouraged by the State because it is found useful in imparting instruction to the people. But it is held in contempt by the learned, the indifferent, and the materialist. In the north-west and south-west Islám has at times made great progress, and the late uprising of the Muhammadan Panthays in the southern provinces, and of the Tungans in Zungaria, for a moment threatened the empire with ruin. Both movements were suppressed with wholesale massacre, in which millions perished, and extensive tracts were depopulated. Yet there are still probably 30,000,000 Muhammadans in China. They are called *Panthays* in Yunnan and the south generally; *Tungans* (Russ. Dungan) in Kansu, Zungaria, and the north generally, both terms of uncertain meaning and origin. But the Chinese confound them all under the general name of *Hoï-hoï*, a term formerly applied to the Uigur Tatars, from whom many doubtless are descended, especially in the north. But they call themselves *Kiao-mun*, or "Religious Folk," looking on their neighbours as a godless, impious people. Although all Sunnis, they are divided into two sects, the Azemi and the Shafieh, who, however, always forget their differences in the presence of the common enemy. M. de Thiersant predicts a great future for Islám in China.[1]

Small Christian communities have long been established all over the interior. France has assumed the protectorate of the Roman Catholic missions, at the same time displaying great zeal for the spread of the Catholic religion amongst the natives. Of the 500 European missionaries three-fourths are Frenchmen, and to the same nationality belong nearly all the "sisters" engaged at the stations. These stations stretch in an unbroken chain from the coast to the western frontiers. The French priests display remarkable zeal and self-devotion; they are compelled to assume the costume and conform to the customs of the people, and were formerly even obliged to renounce their native land for ever. They were not allowed to leave the country or give strangers any information regarding the interior. Thus alone was it found possible to allay the ever-watchful suspicion of the authorities.[2] These missions are said to be at present in a flourishing state, with native congregations of perhaps not less than 500,000 altogether.

The Tibetans—Buddhism.

According to a Chinese official estimate quoted

[1] *Le Mahométanisme en Chine*, etc., 1879.
[2] T. Cooper's *Travels*. London, 1871.

by Desgodins, the inhabitants of Tibet number about 4,000,000, and Klaproth estimated them at not more than 5,000,000. Such a sparse population in such a vast area is doubtless mainly due to the sterility and bleakness of the land. But the great number of celibate monks, combined with the custom of polyandry and the low tone of morality, are also contributing causes.

The Tibetans constitute a very distinct branch of the Mongol family, and are described by Huc as a people with small, contracted, black eyes, thin beard, high cheek-bones, flat nose, wide mouth, and thin lips. The skin of the upper classes is as white as that of Europeans, but the ordinary complexion is tawny. They are of middle height, and combine agility and suppleness with strength and energy. They have the reputation of being frank and generous, brave in war, extremely superstitious, and fond of display. They have domesticated the yak, they breed ponies, sheep, and goats in large numbers, cultivate such cereals as will ripen in their climate, work the precious metals, and are skilful weavers and potters.

Although Buddhism was not introduced till the seventh century A.D., Tibet has become the centre of the Buddhist world. A native king, founder of Lassa in 617, having married a Chinese princess, is said to have sent to India for the Buddhist Scripture, causing it to be translated into Tibetan with an alphabet derived from the Devanagari. In the fourteenth century the great reforming lama Tsong-kaba introduced many changes, prohibiting clerical marriages and necromancy, and organising frequent conferences of the priesthood. His followers were distinguished by a yellow dress and cap, and called Duk-pa, while the old unreformed party were called the Red Sect or Gelupka. At present the Red Sect prevails in Ladak, Bhutan, and Sikkim, the Yellow in Tibet proper (*Col. Yule*).

Gedun-tubpa, another reformer, arose soon after, and built the monastery at Teshu-lumbo in 1445, and it was in his person that the system of perpetual incarnation was supposed to begin. The sixth in succession of those incarnations, called Navang Lotsang, made himself master of all Tibet in the middle of the seventeenth century, and founded the Dalai and Teshu Lamaships as they now exist at Potala near Lassa and Teshu-lumbo respectively. There is a third incarnation called the Khutuktu, resident at Urga in Mongolia, and probably a fourth in the person of the Changay Lama or High Priest of Peking, besides a female incarnation—an abbess of a convent on the island in Lake Palti, referred to by Bogle and by Giorgi.[1]

The professed monks or clergy, subordinate to the holy lamas, are also called lamas, and are very numerous. They live in monasteries, some of vast extent, scattered not only over the inhabited valleys, but even in some of the wildest parts of Tibet. Their ritual consists mainly in the recitation and chanting of the *sutras*, or precepts and rules of discipline, to the sound of musical instruments. A characteristic feature is the prayer wheels, metal cylinders charged with rolls of prayers, which are kept revolving during the service and placed over streams to be turned by the current. They have been in use over 1000 years, being mentioned by the pilgrim Fa-Hian. Another peculiarity is the votive pile of stones frequently met with along the road-side, from a few feet to half a mile in length, and stuck over with flags inscribed with the mystic formula, *om mani padme hûm*, *i.e.* "O the Jewel in the Lotus! Amen!" These primeval six syllables, "among all prayers on earth, form that which is most abundantly recited, written, printed,

[1] In his great work, the *Alphabetum Tibetanum*, issued by the press of the Propaganda in the last century.

and even spun by machines, for the good of the faithful. They are the only prayers known to the ordinary Tibetans and Mongols; the first words the child learns to stammer, the last gasping utterance of the dying. The wanderer murmurs them on his way, the herdsman beside his cattle, the matron at her household tasks, the monk in all the stages of contemplation; they form at once a cry of battle and a shout of victory! They are to be read wherever the Lama Church has spread, upon banners, upon rocks, upon trees, upon walls, upon monuments of stone, upon household utensils, upon strips of paper, upon human skulls and skeletons! They form the utmost conception of all revelation, the path of rescue, and the gate of salvation!" (*Col. Yule*).

But the most extraordinary feature in the Tibetan Buddhist system is undoubtedly the external resemblance between its ritual and that of the Roman Catholic Church, a resemblance often extending to the minutest details. The priests of both hierarchies wear the tonsure together with flowing robes covered with gold embroidery. They fast and mortify the flesh, observe spiritual retreats, confess the faithful, intercede for them with the saints of heaven, make long pilgrimages to shrines where relics are devoutly preserved. "Celibacy is common to both, and under the shadow of church and temple alike, communities of men and women devote themselves entirely to a life of contemplation. Church and temple are in the same way furnished with high altar, candlesticks, reliquaries, holy water fonts, and belfries. The lama, like the priest and bishop, officiates with mitre and crozier, cope and dalmatica, salutes the altar, bends the knee before the relics, intones the service, recites the litanies, utters prayers in a language unknown to the congregation, solicits offerings for the repose of the faithful departed, heads the processions, pronounces blessings and exorcisms. Around him the choristers sway the incense-burner, and the devout tell their beads."[1]

The early missionaries were struck with the outward identity of the two rituals. Some have endeavoured to trace it to the early Christian Church of India, with which country Tibet has had direct relations since the seventh century. But from India Tibet derived not Christianity but Buddhism, and a more probable solution may perhaps be found in the pre-Christian Zoroastrian rites, spreading east and west from Irania, and influencing the religious thought of both regions during restless periods of transition. It is curious that the name of the Persian *mitre* should survive in Roman ecclesiastical nomenclature, while the object, variously modified, is still in use in the Latin and Greek churches as well as in the Tibetan and Mongolian Buddhist temples.

The salient features of Buddhism as originally constituted are threefold:—

1. *Socially*, Buddhism claims for itself a superiority over worldly power, holds that religion has a first claim upon all property, and forbids caste distinctions.

2. *Dogmatically*, it cannot be designated as theistic, as it deifies humanity and moral ideas.

3. *Ethically*, it teaches the vanity and emptiness of all mundane things, the transmigration of the soul and its ultimate absorption in *Nirvâna*.

But Buddha himself, like Confucius, was personally a philosopher, or expounder of an ethical code and a mirror of virtue, not professing to be a redeemer of fallen humanity, but declaring that man can work out his own redemption.

The Mongolians.

The Mongolians, once the terror of the world, and

[1] *Reclus*, vii. p. 80.

founders of ephemeral empires stretching from the Pacific seaboard to Central Europe, have ceased to possess any political cohesion since the destruction of the Zungar power in the second half of the last century. Their primeval home is undoubtedly the region still known as Mongolia, where every prominent feature of the land is associated with some national legend, where every mountain is a king, every lake or stream a divinity. During centuries of migrations and ceaseless military expeditions the race has become largely affected by Chinese, Túrki, Tibetan, Iranian, and other foreign elements, and the original type seems now best preserved in the Khalka branch occupying the whole of North Mongolia besides the Ala-shan district in the south.

The Mongolian is of middle height, robust and capable of enduring hardships that would kill an ordinary European. But these must be such as he is accustomed to; for although he will keep his seat on the camel for 15 hours at a stretch with the glass marking –20° F., a short walk across the steppe will completely subdue him with fatigue. He is always mounted, a skilful horseman, and extremely fond of racing, in which the whole encampment takes part. But in other respects the race has greatly degenerated, and under 200 years of Chinese government and lama influences it has even lost the personal courage which formerly enabled its warlike hosts to overrun continents. Vanquished, disorganised, and broken up into hostile factions, the nation has even acquired a sense of its helplessness. The independence, love of freedom, equity, and tolerance, which, fully as much as their martial spirit, formerly proclaimed their immense superiority over the surrounding Asiatic races, are now far less conspicuous than the degrading superstition, gluttony, indolence, filthy habits, and other vices attributed to them by all recent observers. "The gluttony of this people exceeds all description. A Mongol will eat several pounds of meat at one sitting, and some have been even said to devour an average-sized sheep in the course of twenty-four hours. But the most striking trait in their character is sloth. Their whole lives are passed in holiday-making, which harmonises with their pastoral pursuits. Their cattle are their only care, and even they do not cause them much trouble. The Mongol is so indolent that he will never walk any distance, no matter how short, if he can ride. His legs are bowed by constant equestrianism, and he grasps the saddle like a centaur. The wildest steppe-horse cannot unseat its Mongol rider.

"But the first thing which strikes the traveller in the life of the Mongol is his excessive dirtiness. He never washes his body, and very seldom his face and hands. His clothing swarms with parasites, which he amuses himself in killing in the most unceremonious way. The uncleanliness and dirt in which they live is partly attributable to their dislike, almost amounting to dread, of water or damp. Nothing will induce a Mongol to cross the smallest marsh where he might possibly wet his feet, and he carefully avoids pitching his yurta anywhere near damp ground.

"Lamaism, which has struck deep root in their midst, is represented by the *Kutukhtu* of Urga, ranking next to the Dalai Lama and Pan-tsin-Erdeni of Tibet. Besides the Kutukhtu there are over one hundred 'Gigens' or minor saints, who never die, but pass from one body to another. Their personal influence is unlimited. A prayer offered up to one of them, the touch of his garments, his benediction, are regarded in the light of the greatest blessings humanity can enjoy. But they are not to be had gratis. Every believer must bring his offering, which in some cases is very large. Lamaism is open to objection as attracting an undue part of the male population, and, by its unbounded influence, deprives the people of the power of rising in the intellectual scale.

"But although this religion has taken so strong a hold on them, superstitions are equally prevalent, and evil spirits and witchcraft beset the Mongol's path.

"The Mongols expose the bodies of their dead to be devoured by birds and beasts of prey, their lamas deciding in which direction the head should lie. Princes, gigens, and lamas of importance are interred or burned after death. Masses are said for the departed for forty days on payment of a sum of money."[1]

8. *Topography: Lassa—Yarkand—Kashgar—Hami—Urumchi—Kulja—Urga—Girin—Peking—Shanghai—The Treaty Ports.*

Outside of China proper, large centres of population are extremely rare. In Tibet the only real town is the capital, Lassa (Lhasa)—that is, "The Place of God"—which lies towards the south-east, on a plain surrounded by mountains and dotted over with large monasteries. It has a circumference of 2½ miles, in the centre of which stands a large temple containing images richly inlaid with gold and precious stones, and surrounded by bazaars with shops kept by Tibetan, Kashmiri, Nepalese, Chinese, and Muhammadan traders. Close by is the palace and monastery of Potala, residence of the Dalai Lama, encircled by eleven other monasteries, with an aggregate of 20,400 monks. The city itself has a population of scarcely more than 15,000, besides a garrison of 1000 Tibetans and 500 Chinese armed with flint guns.

When visited by the British Mission to Yakub Beg in 1874, Yarkand, the ancient capital of Kashgaria, was

[1] Prejevalsky, *Mongolia, passim.*

still the largest and wealthiest place in the Tarim basin. It stands on an open plain on the left bank of the Yarkand River, and is surrounded by strong walls 4 miles in circuit, beyond which lie several populous suburbs. Close by is the Chinese fort of Yang-shahr, or "New Town," residence of the governor, with his officials and garrison troops. The city had a population of some 20,000 in 1874, and was a busy place, with a staple industry of leather ware. Before Yakub Beg's revolt, Yarkand was a flourishing emporium, and apparently a much pleasanter place to live in than under the Muhammadan rule. "People bought and sold every day," said a citizen to Dr. Bellew, "and the weekly markets were much jollier affairs. There was no kazi, with his six satellites, to flog the people off to prayers and drive the women out of the streets, and nobody was bastinadoed for drinking spirits and eating forbidden meats. There were also musicians, acrobats, fortune-tellers, and story-tellers, who moved about and diverted the people."

Kashgar, the present seat of government, whence the province takes its European name, consists of the Kuhna-shahr, "or Old Town," on the Tumen River, and Yangi-shahr, "or New Town," on the plain, 5 miles farther south, with the Kizil River flowing between. The Yangi-shahr, built in 1838, is surrounded by lofty massive walls and a wide ditch, and in 1874 contained the Amir's residence with the garrison and military bazaars.

East of Kashgar follow successively Aksu, Karashar, Turfan, and Hami (Khami), all more or less important towns and stations on the old historical Tian-shan Nan-lu, or southern Tian-shan route between China and Samarkand. Aksu, the old Arpadil (Ardabil), is a very ancient place, and was formerly the centre of the Chinese trade and the limit of the trade privileges granted by the government to the Khokand Khan over the cities of the western division. Turfan, on the north-eastern verge of the Tarim basin, is sometimes called Kuhna Turfan in distinction from Ush-Turfan, lying farther west. In the district are the extensive ruins of an older Turfan, capital of the Uighur empire, visited in 1880 by Dr. Regel from Kulja. The highway leads east of Turfan to Hami at the junction of the route over the Tian-shan from Barkul. Lying in a fertile oasis on the skirt of the desert, and at the converging points of the Moslem and Buddhist worlds, Hami presents a curious mixture of Buddhist and Muhammadan monuments, including a fine temple and a magnificent mosque, dating from Shah Rukh's embassy to China (1420 A.D.)

In the Tian-shan Pelu, or northern Tian-shan route, (Zungaria) Barkul and Urumtsi correspond to Hami and Turfan of the southern route respectively. Their importance is due mainly to the fact that between them the Tian-shan range is crossed by only one difficult pass, so that they form vital points on the great trade and military route from Peking *via* Bautu and Hami to Kulja. Urumchi is the Bish-Bakil of mediæval times, and occupies a prominent place in the history of Central Asia during the last 600 years. It lies at the foot of the triple-crested Bogdu-ola (14,000 feet), and consists of an old town on the slope of the mountain, and a new or Manchu quarter lower down. In the neighbourhood are some hot sulphur springs, but Humboldt's view of the volcanic nature of the district has been denied by Severtzoff. Like Hami, Urumchi is a great centre of the trade between China and the West, and has a population of about 15,000 (*Regel*).

In the Ili valley the chief place is Kulja, a name which has recently been extended to the whole province. There are or were two towns of the name—the old or Taranchi town, lately the headquarters of the Russian administration, and the new or Manchu Kulja, 25 miles lower down, which was a thriving city of 75,000 inhabitants before the revolt of the Tungans, who exterminated its Chinese population, and laid the place in ashes. Kulja was visited in 1880 by Mr. E. Delmar Morgan, who found a Chinese Roman Catholic chapel there, with a layman officiating since the disappearance of its Italian pastor during the late troubles.

Urga, capital of Mongolia, and in Buddhist eyes second in sanctity only to Lassa, lies due south of Kiakhta, on the high road to Peking, in 47° 55′ N., 106° 41′ E. The large monastery of this city, visited by countless thousands from every part of the empire, is the residence of the Kutukhtu, Lama king, or High Priest of the Mongolians, who, like the Dalai Lama, is worshipped as a god incarnate. But the city itself, with its irregular and straggling houses and busy tent bazaar, has more the appearance of a nomad encampment than of a town, while the prevalent practice of exposing the dead converts the neighbourhood into a haceldama. Here the dead bodies are flung to the dogs and birds of prey.[1] But the Kutukhtu is very wealthy, owning 150,000 slaves, and receiving a constant flow of rich offerings from his pious votaries.

In Western Mongolia the chief place is Ulia-sutai, lying on the Kobdo plateau in a romantic district at the junction of the river of like name with the Bogdo. Ulia-sutai, which is the centre of the military administration, is rather a fortress than a town, being scarcely the third of a mile both ways, and enclosed by high palisades, with a gateway and four towers. Its 4000 inhabitants are nearly all Mongolian officials and troops. Business is transacted in a village about a mile off, where the Chinese traders barter their cotton goods, plush, tobacco, and

[1] Prejevalsky, *Mongolia*, i. p. 14.

leather, for sheep-skins, tallow, cattle, horses, buffalo horns, and peltry. This place was first visited in 1868 by the Russian consul Shishmarev, and again in 1872 by Ney Elias, who found the normal temperature in November six degrees below freezing point. In summer it rises to 110° and 120° F.

Kobdo, which gives its name to the plateau, lies in a wide plain destitute of vegetation, and consists of the government quarters, enclosed by a mud wall, and an open trading quarter. Its trade is said to be more considerable than that of Ulia-sutai, which it exceeds in population. Of its 6000 inhabitants, 3000 are Mongolians and over 1600 garrison troops.

Girin, the new capital of Manchuria, lies on the Upper Sungari, and is said to have a population of 120,000. Its streets are laid down with planks, and it has a large trade in lumber from the neighbouring wooded uplands. Other important places in Manchuria are Mukden, the old capital, near the Chinese frontier; Petuna (Sin-chung), below Girin, near the junction of the Nonni; Tsitsihar and Mergen, both in the Nonni valley; Aigun, on the right bank of the Amur, facing Blagoveshchensk, capital of Russian Manchuria; and Ninguta, a flourishing and well-built city on the Hurka (Khulkha), a tributary of the Sungari from the east.

The Korean towns, which are said to be very numerous, including over 30 of the first rank, are mostly unknown even by name. Seul, or Kyung, the capital, lies on the Hang-kang, a few miles above its mouth, on the west coast.

In China proper, large cities, some of great antiquity, are even more numerous than in India. There are at least 100 with populations of 100,000 and upwards, while two or three have certainly more than a million inhabitants, although the still higher estimates for Peking and some other places have proved to be gross exaggerations. At the same time, the Chinese cities, which reflect a uniform civilisation of a somewhat stereotyped character, are far from possessing the same antiquarian or architectural interest as those of India, where so many distinct cultures have left their impress on the land. Hence in China the towns seem to be nearly all cast in one mould, the same types of buildings, temples, towers, narrow irregular streets and rickety houses everywhere succeeding each other with oppressive monotony. What variety they may possess is due rather to the nature of the ground on which they stand than to the few slight departures from the standard models which may occur here and there. All are officially grouped in three classes, subordinate to each other. Those of the first rank, distinguished by the title of "fu," have under them several "chew," or cities of the second order, under which again come the "hien," or towns of the third rank.

Above all stands Peking, the present capital of the empire, which lies in the Pei-ho basin, about 40 miles south of the Great Wall, and 100 from the Gulf of Pechili. It consists of the Chinese or outer quarter and the Manchu or inner quarter, in the heart of which is the imperial palace. The Chinese quarter alone is accessible to foreigners, although an occasional view has been obtained of the Manchu, known also as the "imperial" or "forbidden" city, which seems to differ in few respects from the outer town. They form two parallelograms, so disposed that the shorter side of the Manchu faces the longer side of the Chinese city. A conspicuous object is the Temple of Heaven, which is visited in great state by the Emperor once a year, and never opened except on this occasion. With its various shrines, gardens, courts, moats, etc, it measures altogether two miles in circumference. Attached to it is the Temple of the Seasons, in which four wooden pillars support a cupola, the only structure of the kind said to exist in China (*Von Hübner*). This sanctuary is entered by the Emperor alone with the imperial princes and their suites.

WEST GATE, PEKING.

Other buildings are the great lamaserai or Buddhist temple of Yung-ho-kung, and that of Confucius ("Wen-Miao"), both in the north-east quarter of the Manchu city. In a little court of the latter are contained all the writings of the sage engraved on black marble tablets for the use of the *literati*. In the Manchu city are also four Roman Catholic churches, and on its east wall stands the famous old observatory with its globes, real masterpieces of science and art, executed by Chinese workers under the direction of the Jesuits. Of the four churches the most interesting is the Pei-tang, or North Church, with which is connected a seminary with a valuable library and a rich cabinet of curiosities. The number of Christians in the vicariate of Peking is at present 27,000, of whom 8000 reside in the capital, including many artisans and all the watchmakers in the place.

Peking produces on the whole an unfavourable impression, caused by the state of the streets, the dilapidated condition of the bridges, many built of marble blocks in the last century, by the neglected appearance of the canals, and the pictures of general decay which meet the eye. The population is far less than was at one time supposed, and is now generally estimated from 800,000 to 1,000,000, while one authority reduces it to about 500,000.

Much the same ruinous aspect is presented by Kiang-ning, better known as Nan-king, on the Lower Yang-tse, the former southern capital of the empire. A far more important, though scarcely a more inviting place, is the great seaport of Shanghai, a chief centre of British trade in the East, which is situated in a low-lying swampy district on the left bank of the Wu-sung, just above its mouth in the Yang-tse estuary. Of almost equal importance is the great city of Canton, which with its port of Whampoa occupies a favourable position at the head of the Si-kiang delta just above the Bocca Tigris (see p. 522), and midway between the confluences of the Pe-kiang and Tung-kiang from the north and east.

Shanghai and Canton are by far the most important of all the "treaty ports," which since the year 1842 have been successively thrown open to the trade of the world.

Besides Shanghai and Canton the list now includes Peking, Niu-chwang, Tien-tsin, Teng-chew, Chi-fu, Ching-kiang, Kiew-kiang, Hankow, Ning-po, Fu-chow, Amoy, Swatow, besides Ke-lung, Tan-shui, and Taiwan in Formosa, and Kien-chow in Hainan. To these must now be added the following, also declared free ports by the treaty concluded between England and China on 12th September 1876:—Sha-si, Ta-tung, Wu-hu, and I-chang on the Yang-tse, and Wen-chew and Pei-hai or Pakhoi on the coast. On the Yang-tse there have also been established six stations where steamers are permitted to stop for the purpose of landing and shipping goods and passengers.

9. *Highways of Communication.*

Roads as understood in the West can scarcely be said to exist anywhere in Central and East Asia. Even in China proper, settled as it has been by a practical and intelligent people for ages, the main highways are little more than beaten tracks left to take care of themselves. Morrison tells us that on one occasion in 1880 Mr. Wylie's carter preferred to deviate from the roadway into the neighbouring fields, where walls 18 inches high had to be crossed.

But such as they are, well-known routes have for thousands of years intersected the empire in all directions, marking the lines followed at all times by trade, at intervals by migrating hordes and conquering hosts, or occasionally by such illustrious explorers as the pilgrim Hwen Tsang, and the famous traveller Marco Polo.

China proper is intersected in every direction by 2000 imperial highways, which with the great number of navigable streams, and the extensive system of canalisation, render the country one of the richest in means of communication in the whole world. Unfortunately the State has neglected to keep either roads or canals in repair, or protect them from the wear and tear of time and weather, so that they are now partly impracticable. Morrison gives a deplorable account of the present condition of the Grand Canal on which Peking largely depends for its supplies. In 1880 some parts were in such a ruinous state that the boats could not pass through, and portages were formed at the sides for discharging and reshipping farther on.

In Tibet the chief road connecting Gartok and Lassa leads also to Rudok, Leh, and India, on the one hand, and on the other to Batang and China. This highway is fairly well maintained with "tarjums" or stations at intervals of 20 to 70 miles, where horses, yaks, and coolies are kept in readiness for the Government messengers and officials. The messengers are never allowed to stop except to eat and change horses. To prevent them from taking off their clothes, the breast fastening of the overcoat is sealed, no one being allowed to break the seal except the official to whom they are sent. At the end of their 800 miles' ride they look haggard and worn, their faces cracked, their eyes bloodshot and sunken. The entire distance is covered in about twenty-two days.

Next in importance is the northern route between Lassa and Peking, running by Nak-chu-ka to Kuku-nor, Sining, Mongolia, and the capital. This route is generally followed by the Kutukhtu of Urga on his pilgrimage to Lassa, and was also traversed by Huc and Gabet in 1845, and partly by Prejevalsky in 1872. From Nak-chu-ka two other routes are said to run north-west to Polu and across the Sok country to Lob-nor. In 1874 the Pundit followed a track from Lake Pang-gong and Noh across grassy plains and lacustrine valleys by Tengri-nor to Lassa. A path also leads from Noh near Rudok north to Khotan across the west end of the Kuen-lun and through Polu. Although at present less used than that over the Karakorum, this route may some day acquire great importance, as it seems to afford the most direct and easiest means of communication between India, Tibet, and Kashgaria, *via* Rudok, the Bogo-la Pass, Totling, and the Niti Pass.

But the most convenient routes at present between Tibet and India are—1. That by Lake Palti over the Pari-jong Pass, down the Chumbi valley, and over the Yelap Pass to the Tista valley and Darjiling, partly followed by Bogle, Turner, and Manning; 2. Nain Singh's route over the Karkang-la Pass to Tåwang, whence three roads lead to Odalguri and Dewangarhi in Assam.

But the whole line of the Himalayas is pierced by numerous passes giving access from Tibet to India, the chief of which are the Niti, between Garhwal and Gartok; the Kirong, from Katmandu; the Nilam route leading through the Butia Kosi Gorge, where the stream is crossed fifteen times, and where the path for one-third of a mile is formed by stone slabs from 9 to 18 inches wide, supported by iron pegs driven into the face of the cliff 1500 feet above the torrent; lastly, the Karkang-la, crossed by the Lassa-Tawang route.

The Nan-lu is connected with the Pe-lu, or corresponding northern route, by several passes leading north and south over the Tian-shan, between Zungaria and the Tarim basin. Of these the westernmost is the Muzart, or "Ice Pass," between Aksu and the Ili basin, which leads over a huge glacier close to the Tengri-khân, and which is extremely difficult, although kept open by gangs of labourers. Farther east are the routes between Turfan and Urumchi, recently explored by Dr. Regel, and between Hami and Barkul, beyond which the Tian-shan would seem to gradually merge in the Gobi.

Between the Pe-lu and Nan-lu, consequently in the heart of the Tian-shan, there is a route, followed in 1876 by Prejevalsky, leading from Kulja in the Ili valley over the Narat Pass (9800 feet) and across the Lesser Yulduz plateau to Lake Bagrach and Korla, and thence by the Tarim valley to Lob-nor and the Altyn-tagh range. Here the "Kalmuk Road" leads south to Tibet, while another track is said to run along the north foot of the Altyn-tagh to Sha-chan, and thence across the Gobi north to Ngasi-si-chan and Hami. Hami is also reached from Su-chow at the western extremity of the Great Wall by a path which forms a section of the great historical highway between China and the West.

At present the chief trade route between China and Russia runs farther east from Peking *via* Kalgan to Urga and Maimachin, close to Kiakhta on the Siberian frontier. South of this the more direct road runs from Hankow through North Kansu and across the Gobi to Hami, and thence by the Pe-lu route through the Zungarian depression, or the alternative Irtish valley, to Orenburg. The Russians are beginning to see that the future trade route must follow this line, which is practicable for carriages throughout the whole distance of 2580 miles from Zaisan to Hankow, except a section of 160 miles, which presents no difficulties to pack animals. It can be traversed in 140 days, whereas by the far more difficult Kiakhta road, which is 1800 miles longer, it takes 202 days to reach Peking.[1]

Meantime there are no railways in the empire. The short line, 9 miles long, opened by an English company in 1876, from Shanghai to Wusung on the Yang-tse estuary, had a brief career of sixteen months, when it was bought up and destroyed by the Government. Never-

[1] The whole of the route between Hankow and Zaisan has recently been traversed by Sosnovski.

theless Shanghai is at least traversed by tramways, while macadamised roads radiate from this centre of future progress for a few miles to the surrounding villas. A Danish company has also undertaken to construct a telegraphic line from Shanghai to Peking and Tien-tsin, which was completed in December 1881. Shanghai and Amoy are also in direct telegraphic communication through a submarine cable with Japan and the rest of the world.

10. *Administration: Patriarchal Government—Education—Cabinet—Maladministration—Army—Material Resources—Foreign Relations.*

The "Middle Kingdom," as the Chinese commonly designate it, takes its name either from its original central position in the Yang-tse basin, or from the idea that it lies in the centre of the universe. For here there are five cardinal points—the four common to the rest of mankind, and a fifth whence these radiate, and which is represented by the "Flowery Land." But this centre of the universe is still largely administered by primitive patriarchal institutions. The "Ta-tuang-ti," or "Great Emperor," is looked on as the father of his people, possessing unlimited authority over them. The fundamental laws of the empire are embodied in the first four books of Confucius (Con-fu-tse), and are based on the principle that the State ought to be ruled by the same laws that govern the private family. The sovereign is at once the head of the empire, of the State religion, and, so to say, of each domestic hearth. He is represented by mandarins of the highest rank in the eighteen "sang" or provinces, into which China proper is divided, and each of which enjoys a separate administration regulating every imaginable branch of the political and social system.

To say that the people are still held in leading strings

only expresses half the truth. They are so accustomed to State interference in the minutest details of private life, that they have lost, or rather have not yet developed, the sense of personal independence. Thus they think it quite natural that the State should separate man and wife, as is the case with the Chinese emigrants, and even with the traders in Mongolia, who are not allowed to take their families with them. And when the merchants in the remote frontier cities of Urumchi or Kulja forget their homes amid the amenities and distractions of a less austere social existence, the first impulse of the neglected wives and children is to apply for redress to the authorities, by whom the heads of families are often ordered back under escort.

Although there is a recognised State religion, public instruction is strictly secular. Education of a certain stereotyped order has received such a wide expansion that few are met incapable of at least reading and writing their own dialect. The lettered classes alone are acquainted with the mandarin literature, to the study of which they often devote their whole life. Yearly examinations are held in the provincial capitals for the first two literary degrees, while candidates for the two highest honours are examined in the imperial capital. The supreme degree requires a knowledge of all the written characters, as embodied in the extant literature. But so arduous is this task that few accomplish it early enough to leave home for the study of the physical sciences. Hence the majority are satisfied with committing to memory antiquated formulas and axioms of doubtful practical value. Each degree is denoted by a distinctive academical head-dress, and the various public appointments are awarded to the lettered classes in proportion to their literary qualifications.

Great improvements have, however, been recently introduced into the educational system, thanks especially to the action of Mr. Robert Hart, Inspector-General of Chinese Customs. A series of elementary scientific works are now being translated for circulation amongst all classes, and the *Tungwen*, or Foreign College of Peking, already embraces a scientific curriculum superadded to the study of European languages, which was its original aim. It has now 9 foreign professors, and 102 students, and among the subjects taught are chemistry, natural history, mathematics, physiology, and astronomy. A complete course lasts for eight years, and successful students are appointed to high offices in the various official departments.

The administration of State affairs is, in the first instance, entrusted to a Privy Council composed of four members—two of Manchu, and two of Chinese nationality. Two representatives of the "Kang-ling," or "Great College," take part in the sittings, their duty being to see that no measures are adopted at variance with the fundamental laws as contained in the writings of Confucius. The members of the Council and their associates are called "Ta-hyo-si," or Ministers of State, and control the ministerial functions. Amongst these are—1. The Ministry of the Civil Officials, which watches over the conduct of these officers; 2. The Ministry of Finance; 3. The Ministry of Customs and Ceremonies, whose province it is to see that the national laws and usages are observed by the people; 4. The Ministry of War; 5. Public Works; 6. The Supreme Military Tribunal.

Theoretically above all these departments is the "Tu-khe-yi-veng," or Board of Public Censors, consisting of forty or fifty members, with a Chinese and a Manchu President. According to ancient usage, each member of this Council has the right of personally directing the Emperor's attention to anything in the State calling for censure. At the deliberations of the various ministries one of the censors is always present, though not entitled to vote. Others visit the various provinces and inquire into the administration of the viceroys.

But with all this censuring the country continues subject to much maladministration. Under the ruling dynasty the public functionaries are understood to be often wanting in probity. In an absolute yet democratic system, in which knowledge and personal merit are the only legal avenues to distinction, a universally high moral tone can alone ensure respect for the laws and administration of justice. But these qualities began to lose their efficacy as soon as the State became tainted with corruption. In a patriarchal form of government, imposing on the authorities no check beyond the duty to act in a paternal spirit, the absence of morality had necessarily evil results for the nation, which thus became, to some extent at least, the victim of venality. Those political institutions of the *literati*, which, seen from a distance, are apt to excite the admiration of enthusiastic believers in the higher theories of a freedom and equality limited only by intelligence and morality, have here produced a depressing thraldom. The lower mandarins are chiefly occupied in raising money, either to redeem the pledges made to those of higher rank to whom they owe their appointments, or to procure still higher honours by fresh payments. All the authorities are surrounded by satellites attached to them as bodyguards, police agents, or executioners. These men are charged with the faults usually characteristic of pretorian guards. And it is doubtful whether the administration is conducted with the integrity and efficiency which would be expected in a European State.

The imperial forces consist of two distinct branches, one belonging to the nationality of the present Manchu dynasty, the other composed of Chinese and other races. The first, on which the chief reliance is placed, is organised in eight "banners," and supplies the garrisons of the large towns, fortresses, and other important positions throughout the empire. Hitherto their defective equipment and rudimentary instruction have rendered these troops quite incapable of contending with European forces greatly inferior in numbers. Recently, however, a marked improvement, especially in the artillery service, has been effected by the English instructors employed to train the men.

The second branch consists of about 800,000 Chinese and other races distributed over the empire, who reside with their families, support themselves by their own labour, and assemble only on special occasions in the capitals of their several provinces.

China also supports a small navy, chiefly employed in suppressing the pirates, who infest the numerous creeks and islets about the estuaries of the large rivers.

Enjoying the advantage of a vast seaboard, intersected by numerous streams, estuaries, and canals, abounding in cereals, tea, silk, cotton, flax, hemp, sugar-cane, indigo, tobacco, coal, and iron ores, China possesses perhaps the largest material resources of any country in the world. For the last thirty years the West has paid in cash fully three-fourths of the value of the exports in silk and tea alone. And if we consider the amazing industry of the Chinese, the energy and endurance of the working classes, their thrift and frugal lives, their respect for precedent, their love of peace and order, their cheerfulness under the greatest hardships, and further, their keen commercial spirit, we shall be able to form some idea of the enormous latent power possessed by this nation of some 350 or 400 million souls. It is understood that the Chinese are endeavouring to get into their own hands the trade heretofore conducted by foreigners.

Here we see a repetition of what is taking place in California. Until recently the chartering of steamers by natives was an unheard-of thing; now they not only charter but even own large ocean steamships, and are coming straight to the London market. Li-hung-chang, viceroy of Pechili, and founder of the "China Merchants' Steam Navigation Company," has just taken a step aimed at the total exclusion of European traders from the country. The Mei-fu steamer, belonging to this company, arrived in the Thames on December 5, 1881, with a cargo of tea and the Chinese managers of an association established for the purpose of opening direct relations between the two countries through native firms in China. A capital of £150,000 has already been paid up, and the undertaking is being encouraged in every way by the Chinese Government.

The weightiest commercial interests are represented by the English exchanges, which average about £42,000,000 yearly. Of growing importance also, although perhaps more in a political than a commercial sense, are the relations of Russia with the Celestial Empire. Both are border States, whose conterminous frontier extends for thousands of miles, and both have had relations in common for the last 200 years. In connection with these relations the Russians have given frequent proof of the tact they possess for dealing with Asiatics. Without any wars, and by means of a skilful diplomacy, they have secured enormous advantages, including the possession of vast and fertile tracts in the Amur basin and on the Pacific seaboard.

11. *Statistics.*

AREAS AND POPULATION.[1]

Provinces.	Area. English sq. miles.	Population.	Population per square mile.
Chili	58,949	17,937,000	304
Shantung	65,104	36,247,835	557
Shansi	56,268	12,211,453	221

[1] *Statesman's Year-Book* for 1886.

AREAS AND POPULATION.

Provinces.	Area. English sq. miles.	Population.	Population per square mile.
Honan	65,104	22,115,827	340
Kiangsu	44,500	20,905,171	470
Nganhwei	48,461	20,596,288	425
Kiangsi	72,176	24,534,118	340
Chêhkiang	39,150	11,588,692	296
Fukien, with Formosa	53,480	25,790,556	482
Hupeh	70,450	33,365,005	473
Hunan	74,320	21,002,604	282
Shensi	67,400	8,432,193	126
Kansu	86,608	5,411,188	62
Szechuen	166,800	67,712,897	406
Kwangtung, with Hainan	79,456	29,706,249	377
Kwangsi	78,250	5,151,327	65
Kweichow	64,554	7,669,181	118
Yunnan	107,969	11,721,576	108
Total	1,297,999	382,978,840	Aver. 234

According to official data referring to 1842 the population of the 18 provinces of China Proper was 413 millions.

Outlying Provinces.	Area in sq. miles.	Estimated Pop. (1842).
Korea	...	8,000,000
Manchuria	380,000	7,000,000
Mongolia	1,360,000	4,000,000
Tibet	600,000	6,000,000
Kuku-nor Tsaidam	120,000	150,000
Kashgaria	250,000	1,000,000
Zungaria	120,000	300,000
Kulja	26,000	140,000
Total	2,856,000	26,590,000

ESTIMATE OF POPULATION ACCORDING TO RACES.

Chinese proper	350,000,000
Si-fan, Man-tze, and other hill tribes	20,000,000
Koreans	7,000,000
Manchus	8,000,000
Bod-pa (Tibetans)	4,000,000
Mongolians and Kalmucks	4,000,000
Kashgarians	750,000
Tanguts	150,000
Sok-pa and Hor-pa	?100,000
Taranchi	40,000
Kirghiz	35,000
Tungans of Kulja	10,000
Total	394,085,000[1]

[1] See also a paper read before the Statistical Society, February 1885, by Sir Richard Temple.

Chief Towns.

China.

	Pop.
Canton	1,500,000
Peking	1,000,000
Siang-tan-fu	1,000,000
Singan-fu	1,000,000
Tien-tsin	930,000
Chingtu-fu	800,000
Hang-chew-fu	800,000
Hankow	700,000
Ching-cheng-fu	700,000
Shanghai (1881)	600,000
Fuchew-fu	600,000
Fuchen	600,000
Lanchew-fu	500,000
Fan-chan	500,000
Chao-hing	500,000
Lu-chew	500,000
Su-chew	500,000
Chang-chew	500,000
Yang-chew	360,000
Taiwan	355,000
Su-chew-fu	300,000
Hu-kow	300,000
Chang-cha	300,000
Taï-yuan-fu	250,000
Wei-hien	250,000
Liang-kiang	250,000
Teng-chew-fu	230,000
Kiung-chew (Hainan)	200,000
Yung-ping	200,000
Kalgan	200,000
Lanki	200,000
Tsinan-fu	200,000
Wu-chew	200,000
Chu-hing	200,000
Ching-kiang	170,000
Wen-chew	170,000
Victoria (Hong-Kong)	160,814
Tsing-chew	160,000
Ning-po	160,000
Pao-ting	150,000
Nan-king	130,000
Tung-kung	120,000
Che-fu (Yentai)	120,000
Tsongan	100,000
Hu-chew	100,000
Chih-lung	100,000
Tong-chew	100,000
Yuen-ching	90,000
Amoy	88,000
Lohui (Hainan)	80,000
Hanchung-fu	80,000
Tungkuang	70,000
Tsing-chew	70,000
Yuyao	65,000
Macao	60,000

Tibet.

Lassa	15,000
Shigatze and Tashi-lumpo	14,000
Chetang	13,000
Gyantze	12,000

Kashgaria.

Yarkand	60,000
Kashgar	50,000
Khotan	40,000
Sanju	35,000
Aksu	20,000
Kiria	15,000

Zungaria and Kulja.

Old Kulja	15,000
Suidum	4,000
Chuguchak	4,000

Mongolia.

Jehol (Chingte-fu)	40,000
Urga	30,000
Kuku-khoto	30,000
Dolon-nor	30,000
Paku	20,000
Urumchi	15,000
Turfan	10,000
Hada	10,000
Hami	6,000
Kobdo	3,000

Manchuria.

Mukden	180,000
Girin	120,000
Ninguta	60,000
Ajeho	40,000
Yingtze	40,000

The twenty-two ports to which England has access by treaty are, New-chang, Tien-tsin, Chefoo, Ichang, Hankow, Kiu-kiang, Wuhu, Nanking, Ching-kiang, Shanghai, Ning-po, Wen-chow, Foo-chow, Tamsui, Kelung, Taiwan, Takow, Amoy, Swatow, Canton, Kiung-chow, Pakhoi.

Trade Returns.

Year.	Imports.	Exports.
1868	£21,995,000	£20,234,000
1871	24,124,000	22,458,000
1875	20,340,000	20,673,000
1878	21,241,000	20,500,000
1884	20,010,000	18,468,000

Chief Imports (1884).

Opium	£7,205,000
Cotton and Cotton Goods . . .	3,100,000
Woollens	1,020,000
Metals	1,128,000

Chief Exports (1884).

Raw Silk and Silk Textures . . .	£6,430,000
Tea	8,230,000
Sugar	1,040,000
Straw braid	540,000

Average Trade with Britain and its Colonies.

92 per cent of all imports from.
74 per cent of all exports to.

Shipping (1884).

	Vessels.	Tonnage.
British	14,141	12,153,000
Chinese	4,625	2,994,000
German	1,758	940,000
American	2,381	2,382,000
French	48	94,000
Japanese	296	215,000

Exports to India	£150,000
Imports from India	14,600

Annual import of Brick Tea into Tibet from China £350,000.

Trade and Shipping of the Chief Chinese Seaports (1879).

	Imports.	Exports.	Tonnage.	Of which British.
Tien-tsin	£6,800,000	£471,000	531,000	195,000
Hankow . . .	£11,302,000		733,830	402,960
Kiukiang . . .	3,382,000		1,421,170	889,100
Shanghai . . .	19,126,000	17,423,000	3,062,682	1,309,500
Wen-chew . .	121,900		19,780	?
Fu-chew . . .	1,742,000	3,179,000	418,083	345,569
Amoy . . .	2,775,000	1,839,000	892,000	720,000
Swatow . . .	5,482,000		566,250	?
Canton and Whampoa	2,558,000	4,852,000	1,664,000	1,451,750
Victoria (Hong-Kong)	?		4,000,000	?

Gross Foreign Trade of the Nineteen Open Ports.

1870.	1880.
£35,250,000	£46,278,000

Revenue (Estimated for 1885).

Tax on provisions	£6,333,000
Land tax	5,703,000
Customs	4,752,000
Salt	1,583,336
Sale of titles and privileges	2,256,666
Miscellaneous	4,546,668
	£25,175,000

External debt (1885), £4,600,000.

Emigration.

To United States since 1820	224,050
To Peru, 1860 to 1874	86,700
To Singapore last 50 years	100,000
To Dutch East Indies last 50 years	320,000
To Australia last 50 years	130,000

Army.

Manchurea	70,000
Centre	50,000
Territorial	200,000
	320,000

But the territorial army can be raised in time of war to 600,000. There is also Tartar cavalry, and a force styled the Turkestan army. The total cost of the army is estimated at £15,000,000.

Chinese Navy (1885).

Armoured ships	3
Unarmoured cruisers	4
Gunboats	12

(Built in Europe.)

Squadrons of Foo-chow, Shanghai, and Canton include 40 unarmoured cruisers, corvettes, sloops, and gunboats, besides several torpedo boats.

Chinese Telegraphs (1885), 3000 miles of line.

Foreigners in China (1879).

English	1953	Russians	55
Americans	420	Austrians	38
Germans	384	Scandinavians	35
French	224	Dutch	24
Spaniards	163	Sundries	368
Japanese	81		
Danes	69	Total	3814

Note.—*Port Hamilton.*—This spacious and well-sheltered harbour was occupied by the English in 1885 (Douthwaite, *Notes on Korea*).

CHAPTER XIII.

JAPAN.

1. *Boundaries—Extent—Area—Name.*

In its widest sense the Japanese Empire comprises the three insular groups of the Kuriles, Nip-hon, and the Liu-kiu islands, skirting the east Asiatic seaboard from the southern extremity of Kamchatka to Formosa. It thus stretches in an almost continuous chain across 26 degrees of latitude from 50° N. nearly to the tropic of cancer for a total length of about 1800 miles, and consists altogether of about 4000 islands of all sizes. But excluding the southern Liu-kiu group, the sovereignty over which is still disputed by China, and the Northern Kuriles recently ceded by Russia in exchange for Sakhalin, Japan, as ordinarily understood, comprises the great Central Archipelago of Nip-hon, compared with which the two others are almost of no account.

In this narrower sense Japan consists of about 3850 rocky islets, and the four large islands of Yesso, Hon-do (Hon-shin), Kiu-shiu, and Shi-koku, lying between 30° 35′ to 45° 30′ N., and 129° to 146° E., with a total area of about 160,000 square miles, and a population (1880) of 35,925,313. It is washed east and south by the Pacific Ocean, and separated on the north by La Perouse Strait from Sakhalin, on the west by the Strait of Korea and the Sea of Japan from Korea and Russia. Between Yesso and Hon-do flows the narrow Sugaru (Sangar)

Strait, while Kiu-shiu is separated by the Bugo Channel, Suwo Sound, and Shimonoseki Strait, from Shi-koku and Hon-do. Lastly, between Shi-koku and Hon-do flows the marvellous Inland Sea (Seto-Uchi), which, with its countless islets, bluffs, headlands and inlets, its limpid waters and glorious sub-tropical vegetation, presents a varied panorama of almost unrivalled natural scenery. At the same time, the navigation of these waters is everywhere endangered by innumerable rocks, shallows, reefs, and intricate channels, while the more open Sea of Japan is swept by fierce cyclones accompanied with waterspouts, whirlpools, and rapid currents.

The term Japan, applied in various modified forms to this great archipelago, is not merely synonymous, but absolutely identical with the corresponding native name Nippon. The history of this word Nippon, which is by European writers wrongly restricted to the large central island of Hon-do, is extremely curious. The original Chinese form was Nit-pön, meaning the Land of the Rising Sun, the Orient, from *nit* = sun and *pön* = origin. The word was in this form adopted about the seventh century of the Christian era by the Japanese, who soon assimilated the *t* to the *p*, whence Nip-pon, Nip-hon, and even Nif-hon. But in China the *t* was first dropped, whence Ni-pön or Ni-pen, and the initial *N*, through Mongolic influence,[1] afterwards changed to *J*, whence Ji-pen, the form current in the time of Marco Polo, whose Venetian Zipangu derives directly from it, and is the parent of all the European varieties of the word *Japan*. This word was, as stated, from the first applied to the whole Archipelago, and not exclusively to the large island, for which the Japanese had no general name till that of Hon-do—that is, Original or Main Division—was introduced some

[1] During the Yen or Mongol dynasty (1260-1366), the mandarin or court language was greatly influenced by the Mongol phonetic system.

six years ago. Hence in our maps *Nip-pon* ought to be either altogether suppressed or extended to the whole group—that is, made synonymous with *Japan*, both being varieties of the common prototype *Nit-pön*.[1]

2. *Relief of the Land: Highlands—Volcanoes, Fuji-yama, Aso-san, Asama-yama.*

Japan forms a main link in the chain of volcanic islands which fringe the Asiatic seaboard from Java to Alaska. It is traversed in its entire length by irregular mountain masses of igneous origin, occupying most of the surface, and imparting to the Archipelago a distinctly highland character. In Yesso, the interior of which is little known, the ridges seem to be of moderate elevation, and here the narrow intervening valleys are traversed by numerous small streams which find their way in independent channels to the coast. The highest points are Shribetsi (7874 feet), in the south; Unabetsu (5039), in the north-east; Ofuyu (6000), on the west coast; Ishikari (7710); and Tokachi (8200), near the centre.

In Hon-do the main axis towards the middle of the island recedes somewhat from the east coast, where is developed an alluvial lowland district watered by numerous streams, and occupied by Tokio (Yedo), capital of the empire. But west and south of this district the hills attain their greatest elevation in Mounts Nantai (8195 feet), Asama (8260), Haku (9185), and the magnificent snow-capped cone of Fuji-yama (12,400), culminating point of the Archipelago. Fuji-yama, which rises in solitary grandeur some 70 miles south-west of Tokio, is visible in clear weather for a distance of nearly 100 miles. It has been quiescent since the year 1707; but although

[1] The expression Dai-Nippon—*i.e.* "Great Nippon"—also occurs, and answers to our "Great Britain," as applied to all the British Isles.

FUJI-YAMA.

To face page 610.

the highest, Fuji-yama is not the largest volcano in Japan. This honour is claimed by Aso-san, in Kiu-shiu, 20 miles from Kumamoto, the crater of which is said by Milne[1] to be 12 miles in diameter, and consequently larger than Maunaloa, hitherto supposed to be the largest in the world. It was visited in 1880 by Milne, who found it occupied by several villages.

The Asama-yama, which occupies a somewhat central position to the north-west of Tokio, was ascended in 1873 by the French traveller G. Bousquet. From its crater, 1000 feet across, this volcano emits constant volumes of smoke and vapour, and from its summit a magnificent prospect is commanded of the surrounding country.

To its numerous active volcanoes Japan is probably indebted for its immunity from the disastrous consequences of the earthquakes, which, although of almost daily occurrence, are seldom of a violent character.[2]

3. *Hydrography: Table of Rivers above 50 Miles long—Lake Biwa.*

The lofty range stretching southwards from Mount Asama forms the water-parting between the Pacific and the Sea of Japan. But owing to the disposition of the mountain system, covering probably nine-tenths of the whole surface, no room is left for the development of large rivers. Those that do exist bear somewhat the character of mountain torrents with very rapid courses,

[1] *Popular Science Review*, Dec. 1880.

[2] The Seismological Society for the study of these phenomena, recently founded at Yokohama, has just issued the first volume of its *Transactions*, containing a valuable paper by Professor Milne, the vice-president, on "The Earthquake which took place in the Tokio district on February 22, 1880," the severest and most destructive experienced since the opening ol the country to foreign intercourse (*Athenæum*, Nov. 26, 1881).

and liable to sudden and disastrous floodings in their lower reaches. Hence they are almost more damaging than beneficial even for irrigation purposes. To navigation they are not merely useless, but a positive hindrance, owing to the large quantities of sedimentary matter which they bring down, and with which some of the best harbours in the country have been gradually filled in. Such has especially been the fate of Osaka and Niigata harbours, formerly accessible to the largest vessels, but which can now be approached only by small craft. In Japan "a river-bed is a waste of sand, boulders, and shingle, through the middle of which, among sand-banks and shallows, the river proper takes its devious course. In the freshets, which occur to a greater or less extent every year, enormous volumes of water pour over these wastes, carrying sand and detritus down to the mouths, which are all obstructed by bars. Of these rivers the Shinano, being the biggest, is the most refractory, and has piled up a bar at its entrance through which there is only a passage 7 feet deep, which is perpetually shallowing."[1]

As little is known in Europe regarding the Japanese rivers beyond the names of a few of them, the subjoined table of all above 50 miles in length will be found useful:—

Name.	Source.	Outlet.	Length. Miles.
Shinano	E. Shinano	Niigata	180
Tone	N. Kôdsuke	Gulf of Tokio, and Pacific	170
Kitakami	N. of Rikuchiu	Ishinomaki and Murohama, E. coast Rikuzen	140
Ishikari	N. prov. Ishikari, Yesso	Ishikari, W. coast	130
Tenriu	Lake Sua	Pacific	120
Kiso	S.W. Shinano	Pacific	115
Sakata	S. of Usen	Sakata, W. coast Usen	110

[1] *Unbeaten Tracks*, i. p. 212.

Name.	Source.	Outlet.	Length. Miles.
Okuma .	S.W. of prov. Iwaka	Watari, E. coast Iwaki . . .	110
Noshiro .	W. of prov. Rikuchiu . . .	Noshiro, W. coast of Rikuchiu .	100
Akano .	Lake Inawashiro in prov. Iwashiro .	Near Niigata, W. coast of Echigo .	90
Sumida .	E. Musashi . .	Gulf of Tokio . .	90
Toshima .	S.E. of prov. Ugo .	Kubota, W. coast of Ugo . . .	70
Fujii . .	Kôshiu . . .	Pacific . . .	70
Yodo .	E. Iga . . .	Bay of Osaka . .	70
Baniu .	Yamanaka (Kôshiu)	Pacific . . .	60
Oi . .	N. Kôshiu . .	Pacific . . .	55

Of the few lakes none are of any size except Biwa, a magnificent sheet of water some 45 miles long, with a mean breadth of about 10 miles. Biwa, which is traversed by the River Yodo, lies within 8 miles of Kioto, the ancient capital of the Mikadoes, who usually spent the summer months with their suites on its romantic banks. It is closed north and west by lofty forest-covered mountains, and elsewhere skirted by an open, highly-cultivated country dotted over with numerous villages and tea-houses, the resort of pleasure-seekers from all parts. Its clear waters, which abound in fish, are enlivened by fleets of tiny craft, including probably one hundred small steamers always crowded with passengers.

4. *Natural and Political Divisions : Yesso, Hon-do, Kiu-shiu, Shi-koku.*

The four great islands, forming so many main natural divisions, were formerly divided into nine "do," or departments, which comprised as many as 85 "kuni," or provinces. But these old divisions have, since the revolution of 1868, been suppressed, and the empire is now, for administrative purposes, divided into three "fu" or urban districts, and 60 "ken" or "prefectures." Yesso, however, forms with the Kuriles a so-called "Hok 'kaido" or "Northern Sea Circuit," administered by the "Kaitakushi," or Colonisation Department. The aspect of this little-known island, which is considerably larger than Ireland, is described as peculiarly romantic. The southern peninsula, which was visited in 1878 by Miss Bird, everywhere abounds in magnificent woodland scenery, commanded towards the east by the volcano of Komono-taki, whose cone rises 3830 feet above a charming lake at its base. The crater has fallen in on the east side facing seawards, and it is everywhere intersected by yawning chasms emitting dense volumes of steam. Here the prospect over land and sea is superb, the middle distance and background being occupied by wooded hills, whose deep green foliage merges imperceptibly in a violet tint. A long chain of volcanoes stretches at the foot of the spectator, with the quiescent Mount Ussu bounding the distant horizon.

The shores of Volcano Bay are extremely rich in minerals, including gold, silver, lead, iron, rock oil, and coral. At Yesan, in the south-east corner of Yesso, there is a "solfatara" with sulphur works, and another at Mount Iwauno-bori, the central cone of a triple group of volcanoes. The interior of Yesso, which is said to be crossed by a series of mostly extinct volcanoes, still remains to be explored, the expeditions hitherto undertaken having been mainly restricted to the neighbourhood of Volcano Bay and the adjoining peninsula.

The great central island of Hon-do, with an area of 92,000 square miles, although thickly peopled and generally well cultivated, is by no means such a naturally rich land as is generally supposed. The soil is sandy and only moderately fertile, except in the Tokio district, which is covered with a rich black loam. Here the people are prosperous and comfortable, but elsewhere the rural districts have often a poverty-stricken air. The wretched houses serve to shelter man and beast alike, the villages are in some places indescribably filthy, and the natives clothed in rags or covered with loathsome sores. Miss Bird speaks of the painful sight presented by the importunate crowd pressing upon one another.

Yet "their industry is ceaseless, they have no Sabbaths, and only take a holiday when they have nothing to do. Their spade husbandry turns the country into one beautifully-kept garden, in which one might look vainly for a weed. They are economical and thrifty, and turn everything to useful account. They manure the ground heavily, understand the rotation of crops, and have little if anything to learn in the way of improved agricultural processes. The appearance of poverty may be produced by apathy regarding comforts to which they have not been accustomed. The dirt is preventible, and the causes of the prevalence of cutaneous diseases among children are not far to seek."[1]

A glance at the map of Japan will show that the three large islands in the south almost touch each other at certain points, where they are separated only by narrow channels. Of all the large islands, Shi-koku, with an area of 7000 square miles, is the least volcanic. Even in Kiu-shiu the sandstone formation prevails over the igneous rocks. Like all the other islands, it abounds in grand natural scenery, and especially romantic is its fiord-like and rugged iron-bound south coast. The narrow Shimonoseki Strait separating it from Hon-do, and giving access to the Inland Sea, affords the shortest and pleasantest route by steamer from Nagasaki on its west coast to Yokohama.

5. *Climate.*

Japan lies entirely within the temperate zone, and,

[1] *Unbeaten Tracks*, i. p. 166.

thanks to the proximity of the warm Pacific currents, it enjoys a far milder climate than the neighbouring Asiatic mainland lying under the same parallels of latitude. It may be described on the whole as equable and healthy, without any great extremes of temperature, and with a copious rainfall distributed over the whole year, but heavier in summer than winter. The summer heats are tempered by the cool northern sea-breezes prevalent during that season, while in winter the eastern seaboard is greatly influenced by the warm equatorial current, known as the Kuro Siwo or "Black Stream." The coasts washed by this current, which presents many remarkable analogies to the "Gulf Stream" of the Atlantic, enjoy an exceptionally mild climate, while the less-favoured northern sections are relatively colder than the parts of Europe crossed by the same parallels. In latitudes corresponding to those of Marseilles and Gibraltar it freezes hard in winter; and Yesso, lying between 42° to 46° N., is altogether much colder than Scotland, lying between 55° to 59°.

The southern islands are also naturally affected by their low latitude (30° to 34° N.), and here the glass rises normally in summer to 96° and 98° Fahr. in the shade, while in winter there is little snow except on the highest peaks. Between these extremes lies the great island of Hon-do, where the climate of the east coast is so genial and healthy that it has become the chief sanitarium for Europeans settled on the less salubrious and more relaxing Chinese seaboard. The temperature at Yokohama ranges between 20° and 90° F., averaging about 60°, with not more than eight or ten snowy days in winter, and a correspondingly short period of oppressive summer heats. But the west side of the island, being unaffected by the Black Stream, is much colder with an average snowfall of thirty-two days at Niigata. Here the "canals and rivers freeze, and even the rapid Shinano sometimes bears a horse. In January and February the snow lies 3 or 4 feet deep, a veil of clouds obscures the sky, people inhabit their upper rooms to get any daylight, pack-horse traffic is suspended, pedestrians go about with difficulty in rough snow-shoes, and for nearly six months the coast is unsuitable for navigation owing to the prevalence of strong, cold north-west winds. And all this in latitude 37° 55′ —three degrees south of Naples. . . . Europeans and their children thrive well in all parts of the Empire." [1]

6. *Flora and Fauna: Tea-culture—The Crows of Yesso.*

With an abundant rainfall, a moderately fertile soil, and a temperature ranging from the almost Siberian winters of Yesso to the tropical heats of Kiu-shiu, the vegetation of Japan could not fail to be one of the richest and most varied on the globe. Notwithstanding its long settlement and comparatively dense agricultural population, the forest area is still about four times more extensive than the portion brought under cultivation. This is due to the relief of the land in which the highland formation, ill suited for tillage, and here well adapted for the growth of timber, so greatly prevails over the plateau and lowland formations. Nevertheless sufficient space is left on the plains and on the more gently sloping hill-sides for the development of an extensive and skilful agricultural system yielding results of the most abundant and varied character. The vegetation of the lowlands, offering a striking contrast to that of the higher grounds, presents a magnificent flora, including many rare specimens which have already contributed to enrich the European botanical collections. The fruits are naturally of large size, and have been further developed by careful and

[1] *Unbeaten Tracks*, i. pp. 219 and 4.

systematic culture. Rice, which, with wheat, millet, fish, and vegetables, forms the staple of food, arrives at great perfection, one variety being fully equal to the best kinds produced in India or America. Wheat, barley, millet, and buckwheat, are also extensively grown on the more hilly districts above the level of the rice zone; but the crops are of inferior quality.

Next in importance to rice are the mulberry and tea plantations, which cover large tracts in almost every part of Hon-do and the southern islands. "The tea-plant was introduced from China about the beginning of the ninth century. . . . The plant is pollared to render it more branchy, and therefore more productive, and must be five years old before the leaves are gathered" (*Mossman*).

Of the forest trees the most prevalent are the cypress, firs, and other conifers, including the "umbrella pine" (*Koya Maki*), a magnificent variety often growing to a height of 100 feet, with a dense foliage of broad leaves in shape somewhat like an umbrella. Still larger are the cedars, which tower to a height of 200 feet, with trunks 18 feet in girth. But of all forest growths, the most remarkable is the Camphor Tree, which has a trunk over 50 feet round. The camphor is extracted from the stem and roots cut into small pieces, by a simple process of decoction. There is also a large species of oak yielding an edible acorn, which is boiled and much esteemed for its nutritious properties by the peasantry. But economically a far more important tree is the mulberry, of which there are several varieties. Besides supplying the natural food of the silkworm, the bark of this tree is also used in the manufacture of paper, cordage, and coarse dress materials. Yet even the mulberry must yield in importance to the bamboo, which is as indispensable in Japan as it is in China. It grows very rapidly in dense thickets to a height of 50 or 60 feet, and the uses to which it is put are endless. The framework of the houses and most of their contents, sails of junks, screens, mats, paper, pipes, walking-sticks, are amongst the innumerable objects made from this most useful of plants.

The wild mammalia of Japan includes no large animals except the bear, wolf, and wild boar. Amongst the smaller ones are the fox, red and black badger, the monkey, marten, otter, and squirrel, all closely allied to the corresponding species on the mainland, and pointing at a former connection of the Archipelago with the Asiatic continent (*Pumpelly*).

Very remarkable is the paucity of domestic animals, which are limited to the ox, the horse, two varieties of the dog, the cat, and poultry. Neither sheep, goats, nor the ass seem to be anywhere indigenous in these islands. Economically the silkworm is the most valuable animal, supplying the material of one of the staple industries. From the Japanese cocoons the deteriorated stock in Italy and other Western States has been largely renovated in recent years. " One has to walk warily in many villages lest one should crush the cocoons which are exposed upon mats, and look so temptingly like almond comfits." [1]

The avifauna includes few birds of bright plumage, and scarcely any songsters. But the eagle, hawk, heron, quail, stork, and pheasant, are numerous, while the crow forms a salient feature of the landscape, especially in Yesso. Here " there are millions of them, and in many places they break the silence of the silent land with a babel of noisy discords. They are everywhere, and have a degree of most unpardonable impertinence, mingled with a cunning and sagacity which almost put them on a level with man in some circumstances. Five of them were so impudent as to alight on two of my horses, and so be

[1] *Unbeaten Tracks*, i. p. 257.

ferried across the Yurapugawa. In the inn-garden at Mori I saw a dog eating a piece of carrion in the presence of several of these covetous birds. They evidently said a great deal to each other on the subject, and now and then one or two of them tried to pull the meat away from him, which he resented." [1] The Japanese is much larger and stronger than the European species, or about the size of our ravens, and fully a match for small dogs.

The fauna no less than the flora of Yesso differs in many respects from that of the neighbouring Hon-do. While conifers prevail in the latter, the forests in the former consist mainly of hard wood. The jays and woodpeckers are of different species; the Yesso birch grouse is not found in Hon-do, while the Hon-do ptarmigan and pheasants, as well as a sheep-faced antelope or goat, a monkey and a black bear, are all confined to the southern island. This, combined with many symptoms of upheaval, especially on the west and north coasts, points to a time when the narrow Sugaru Strait, now only 11 miles wide, formed a broad marine channel between Yesso and Hon-do.

7. *Inhabitants: Ainos and Japanese—The Shinto and Buddhist Religions—Christianity.*

From prehistoric times Japan has been exclusively occupied by the Ainos and Japanese, two races fundamentally differing from each other in physique, speech, and culture. The Japanese are obviously a branch of the Mongolian family, but a branch separated from the parent stock for a far longer period than is commonly supposed. The Ainos, whose affinities will be elsewhere discussed, are at present entirely confined to the northern island of

[1] *Unbeaten Tracks*, ii. p. 149.

Yesso, the southern districts of Sakhalin, and some of the neighbouring Kurile group. But there can be little doubt that they were at one time in possession of the whole of Nippon, or at all events of Hon-do, whence they were gradually driven northwards, or else partly exterminated or absorbed by the Japanese intruders from the mainland.

Milne concludes that in former times Ainos lived in the districts of Hon-do, where the kitchen middens and other archæological remains are now found, as they still live in Yesso, where similar remains are still found. And this is confirmed by tradition and history, according to which the present Japanese, on arriving in Nippon, "found it tenanted by Ebisu or barbarians, whom they recognise as the ancestors of the modern Ainos. Year by year the aborigines were driven step by step towards the north. About the year 800 they were struggling near Morioka, and by the year 1200 they seem to have been practically exterminated from Nippon, and those who remained or had taken refuge farther to the north in Yesso were completely subjugated."

The Ainos.

The traces of the partial fusion of the two races are most conspicuous in the northern districts of Hon-do, where the "Ebisu" held out longest and came most frequently into contact with the Japanese. They undoubtedly occupy, both socially and intellectually, a very low position, and some writers have even denied them the capacity of improvement. But Captain Blakiston and Miss Bird, both careful observers of the habits of this race, assure us that they possess many excellent qualities, and gladly take advantage of every opportunity to better themselves. Yet, like so many other primitive peoples, they seem incapable of enduring the contact of a higher culture, in the presence of which they slowly disappear. Blakiston, Brandt, Pumpelly, and others, are unanimous on this point; and if the process of extinction has lasted for perhaps thousands of years, the reason probably is because the Ainos found room to withdraw continually northwards, and thus escape destruction. They are now reduced to about 10,000 in Yesso, where alcohol, small-pox, and the gradual reclamation of the land, must eventually cause them to disappear. Further fusion with the ruling race, who regard them with infinite contempt, seems now no longer possible.

The Ainos are described by Miss Bird as "about the middle height, broad-chested, broad-shouldered, very strongly built, the arms and legs short and muscular, the hands and feet large. The bodies of many are covered with short, bristly hair. I have seen two boys whose backs are covered with fur as fine and soft as that of a cat. The foreheads are very high, broad, and prominent, and at first sight give one the impression of an unusual capacity for intellectual development. The nose is straight but short, the cheek-bone low, the eyebrows full, forming a straight line nearly across the face. The eyes are large, tolerably deep-set, and very beautiful, the colour a rich liquid brown, the expression singularly soft, the skin of an Italian olive tint, mostly thin, and light enough to show the changes of colour in the cheeks."[1]

The usual dress of both sexes consists of a deer-skin garment reaching below the knees and girdled round the waist. The women are tattooed on the forearm, about the lips, and on the lower part of the cheeks, the pattern representing upturned mustachios of a light-blue shade, and producing an ugly effect. The practice has recently been prohibited by the Government.

[1] *Unbeaten Tracks*, ii. *passim*.

Yet their religious notions are of the very vaguest, and their gods mere wooden sticks and posts, whittled so as to let the shavings fall down in curls, and stuck about in their houses, on the hill-tops, by running waters, on precipices and mountain passes. But, as amongst some of the neighbouring continental tribes, "their chief divinity seems to be the bear, who, however, is eaten as well as worshipped. A young bear, captured in the early spring and confined in a cage, is kept in the chief's house, where it is suckled by a woman, and played with by the children until it gets strong and dangerous, when the great Bear Feast is celebrated.

"Truth is of value in their eyes; infanticide is unknown, and aged parents receive filial reverence, kindness, and support, while in their social and domestic relations there is much that is praiseworthy."[1]

The Ainos are entirely subject to Japan, and are now chiefly occupied with hunting and fishing. They are also employed to collect seaweed, and at one spot Blakiston even found them engaged in tillage, and raising a few crops of millet, potatoes, and turnips.

The Japanese.

During many ages of isolation in their insular homes the Japanese people have developed many characteristic physical and mental traits, by which they may easily be distinguished from the neighbouring continental races. Nevertheless their slightly oblique eyes, small nose, black lank hair, sparse beard, salient cheek-bones, and yellowish complexion, still attest their relationship with the great Mongolian family. But from what branch of that stock they more immediately descend, whether from the Chinese, Manchu, Korean, or Mongul proper, is a question which

[1] *Unbeaten Tracks*, ii. *passim.*

is still far from being determined. Its ultimate solution will probably depend upon a more exhaustive study of the Japanese language.

But the race has been so long separated from the parent stock that its linguistic affinities have become obscured almost beyond the hope of recovery. The Chinese elements in the national speech are all of comparatively recent date, and directly introduced since the dawn of the historic period. They lie entirely on the surface, and in no way affect the inner structure of the language, which has had time to become differentiated into a very distinct and at present completely isolated form of speech. It is an extremely soft and musical tongue, being in this respect fully on a level with the Italian, especially when spoken by ladies of the upper classes.

Both the pronunciation and the writing system present some very formidable obstacles to the student. Thus the Russian traveller Golovnin, who was for a long time a captive in Japan, tells us that, for instance, the word *fi* (= fire) seems to oscillate between the sounds of *fi, hi, psi, fsi.*

Although the standard speech is everywhere understood and spoken with considerable uniformity, every province has its peculiarities, increasing in the ratio of the distance from the capital. The best Japanese is perhaps spoken in Niigata, and by the very highest circles in Tokio. The men and women have also a different pronunciation, to which the ear takes some time to get accustomed. As in English, the absence of verbal, nominal, and other inflexions, gives at first sight a false impression of ease and simplicity. But as our studies progress, we come upon idiomatic forms, shades of expression, and modifications, which more than outweigh the apparent advantage secured by the absence of those grammatical elements.

Still more intricate is the writing system. The Japanese are the only people known to history, who have spontaneously and without outward pressure adopted their characters, literature, philosophy, and moral code from a foreign nation. All these things they have borrowed from the Chinese, retaining the cumbrous Chinese ideographic system, even after they had themselves developed a true phonetic syllabary. This "I-ro-ha" system, as it is called, from the names of the three first letters, consists of forty-seven syllables, of which there are two distinct forms, the Kata-kana and Hira-gana. Unfortunately the matter is still further complicated by the great variety of forms assumed by the Hira-gana according to the whim of the writer. Hence the student is compelled first to master the Chinese ideographic signs, then the two phonetic alphabets, and lastly a general mixture of all three—this last being at present the ordinary style. Of the Chinese hieroglyphics there are further three distinct forms—one restricted to printed works and poetry; the second to official documents, indentures, and records of all sorts; the third, or "grass" character, confined to correspondence and everyday use.

Although based on that of China, the Japanese culture has always presented many striking analogies to the European standard, to which it is now being altogether assimilated. This extraordinary and surprisingly rapid social revolution must be mainly attributed to their capacity for at least imitating the features of foreign institutions. Possessed of considerable mental endowments and quickness of apprehension, the Japanese are pre-eminently distinguished by their love of knowledge and their appreciation of the higher interests of humanity. Although on the whole of a kindly and lovable disposition, especially when compared with the Chinese and other branches of the Mongolian family, their character is the reverse of childlike or puerile. Beneath many genial and amiable qualities there is often betrayed a spirit of treachery, suspicion, and revenge, which will for years pursue its victim under the cloak of the most seemingly cordial friendship. A mercenary disposition and unbridled licentiousness are also amongst the darker shades of a picture which is nevertheless apt, by its cheerful and brighter aspects, to beguile the unwary stranger; for the Japanese of all classes are highly courteous and obliging, personally brave and proud of their forefathers' great deeds, altogether a warlike people, distinguished beyond all others for their contempt of death and for an almost morbid sense of honour. This latter sentiment, leading often to duels, has also given rise to the peculiar institution of the hara-kiri,[1] or "happy despatch," an abnormal and cruelly refined method of self-immolation indigenous in Japan from time immemorial, but now happily fallen into desuetude, and abolished as an official punishment.[2]

All travellers describe the first appearance of Japan as the embodiment of some Eastern legend, full of surprises and arousing a sense of bewilderment in the stranger. He rubs his eyes, asks himself—Can all this be actually true, and not rather some day-dream or tale out of *The Thousand and One Nights?* Here everything has a smiling outward appearance—the bright skies, the sparkling waters, the glorious vegetation, man himself with his parti-coloured dress and sprightly humour. The natives seem to be always gossiping, free of cares, almost frivolous, but still courteous, and this is as true of the

[1] From *hara*=stomach, and *kiru*=to cut; but the more usual native name is seppuku, which is in fact the Chinese reading of the same characters.

[2] A full and accurate description of a "suicide to order" of this description may be seen in Mitford's *Tales of Old Japan.*

sturdy porter as of the pretty waitress in the tea-house. Even the mendicant seeks rather to create interest by means of all manner of professional jokes and buffoonery. The whole family, down to the third generation, may be

JAPANESE WARRIOR IN FEUDAL TIMES.

seen solemnly amusing themselves flying fantastic kites or firing off rounds of squibs and crackers in broad daylight! A common species of fireworks consists of rockets and bombshells, which when shot up let off, not fire but smoke of the brightest and most varied tints, or else figures of paper and wire, which unfold in mid-air and flutter slowly to the ground.

In all cities there is the so-called *Joshiwara*, or quarter entirely set apart for pleasure, and enlivened by troops of clever acrobats, strolling players, jugglers, and clowns, all performing in the open streets. The numerous theatres may also be visited for a trifling fee, and here the audience remains squatted in family groups for hours together.

Wrestling is even a more popular entertainment than the playhouse, often exciting a degree of frantic enthusiasm resembling that of the Spaniards at a bull-fight. The Japanese are altogether a gay, pleasure-seeking people, usually devoting the whole evening to relaxation, which is always begun with the inevitable bath.

It may be a doubtful question whether the social vices are much more prevalent in Japan than in the West. The ideas associated with love and marriage may be of a simpler, less restrained character, but they are quite as rigidly adhered to. Marriage itself was formerly a mere formality—a mutual written engagement—the husband arranging matters with the parents and bringing his bride home without the intervention either of church or state. She then became the housewife, as in Europe, directing all domestic affairs and admitted to her husband's confidence. Her very first act was the sacrifice of her beauty. She made herself old and repulsive, shaved her eyebrows and applied a black enamel to her teeth. Although the Japanese are monogamists, separations were not unfrequent, the wife being put away after the Jewish fashion by a writing of divorcement. Infidelity was punished with great severity. But now all this is changed. Marriage is recognised as a civil contract.

In Japan the women are fond of dress. The hitamono, or silken under-garment, is generally of a bright-

JAPANESE TEA-HOUSE.

To face page 628.

red colour. Over it, according to the season and occasion, are worn two or three, and even as many as five or six flowing robes, the so-called "kimono," which fall down over the feet. These also are mainly of silk or crape, those underneath of a light, the others of a dark colour,

JAPANESE GIRLS.

and all are girdled round the waist by the "obi," six or eight feet long, a foot wide, generally of satin or some heavy silk material. The ends of this girdle are tied into a large square bow behind. Umbrellas and fans are worn by both sexes, but beyond this, about every third man you met formerly wore merely a wrap round the loins, and some elegant tattoo patterns in red and blue on various parts of his body. These were mostly porters, fishermen, "bettos" or grooms, and others of the lower classes, all of whom are now obliged to go decently clothed, while tattooing is forbidden. Men of patrician rank, princes or daimios, and high functionaries, wear silk, and formerly carried two swords in their girdle. They had also the exclusive right of riding, and letting the hair grow long; but this privilege has recently been extended to all classes, who have also largely taken to the European dress.

The houses are mostly one-storied, of bamboo, with projecting roofs and very large rooms, but little furniture to speak of—neither chairs, tables, nor bedsteads, articles which are all replaced by the indispensable mats and quilting. This is the case with all classes, the only perceptible difference consisting in the more or less costly material of the matting.

The Japanese bed consists of the so-called "futon," a capacious wadded garment with sleeves, into which the sleeper creeps and then draws another coverlet over him. The "makura," or pillow, consists of a little wooden box 8 inches long, on which is laid a paper cushion, renewed every night. The wealthy classes alone have fireproof stone houses, the outer wall in all the others consisting merely of a bamboo grating, so that the inner court is visible through four or five rooms from the street. Oiled paper is often substituted for bamboo, an arrangement which explains the fact that Tokio is burned down and renewed about every seven years. Paper is also universally used instead of glass for lamps and lanterns, and the risk of fire is increased by the recklessness with which the children play with the dangerous element. The towns are divided into five districts, regularly visited during the night by the "Kanobo," or firemen, and each district has a well-organised "shikashi," or fire-brigade, an observatory, guard-house, and the usual appliances for extinguishing fires.

A singular object, which never fails to rivet the attention of strangers on their first arrival, is the "jinrikisha," or hackney-coach, a sort of go-cart on two wheels, drawn by the "ninsoku," or native porters. The streets are crowded with these vehicles, which are by no

THE JINRIKISHA.

means uncomfortable, and move along with amazing rapidity. *Jinrikisha* is the Chinese, *kuruma*, or "wheel," the Japanese, name of this vehicle, which, although introduced only eight or ten years ago, has already become an institution in the country.

Of the many religious sects three only have acquired a special prominence. The oldest of these is the so-called Shinto religion, the cult of the "Kami-no-michi," or spirits, one of whom is supposed to be represented in the flesh by the reigning sovereign.[1] The followers of this creed talk vaguely about a sublime being diffused throughout the universe, far too holy and ethereal to be directly addressed in prayer. In its oldest and simplest form the Shinto religion identifies the "tenka," or heavens, with the deity, of which they are the abode. It seems also to have believed in the immortality of the soul, and in everlasting rewards and punishments in the after state. But these notions have become antiquated, or completely forgotten. Material objects of its worship are the heavenly bodies, the elements as well as the powers of nature, which had, at a very early period, been already conceived as self-existing personalities, and addressed in prayer as spirits, or "kami." Of the kami there are "eight millions"—that is to say, they are innumerable. But the whole system, which is exceedingly barren and empty, resolves itself into a vague reverence for ancestry, and especially for that of the Mikado, who reigns in virtue of his divine origin. For the present Mikado is the 121st lineal descendant of Jimmu Tenno (660 B.C.), fifth in descent from Amaterasu, the Sun-Goddess, and the great divinity of the Shinto religion.[2] Hence Shintoism has become a convenient instrument in the hands of the government for enforcing obedience to the reigning dynasty.

On this system was based the Shinto theocracy, which usurped the place of the military authority. But it was subsequently compelled to yield to Buddhism, which was introduced about 550 A.D., and struck deep root, although not without undergoing profound modifications. The

[1] Kami-no-michi is the Japanese reading of the two Chinese signs for Shin-to = the way of the gods.

[2] "The reigning house of Japan descends from the Sun-Goddess Amaterasu" (*Japan nach Reisen und Studien*, by J. J. Rein; Leipzig, 1881; i. p. 245).

Buddhist shrines and temples, which cover the land, are of unique structure and often of vast size, and are filled with all sorts of idols, the abortions of the most distorted

TEMPLE IN KAMAKURA.

fancy, and far more calculated to strike terror than to inspire confidence in their votaries. On all festive occasions the people troop in crowds to these places. To all the larger temples are attached regular spectacles, playhouses, panoramas, besides lotteries, games of various sorts, including the famous "fan throwing," and the shooting galleries, where the bow and arrow and the blowpipe take the place of the rifle.

But since the recent revolution, Buddhism, like all institutions formerly encouraged by the Shogun, has fallen into disfavour. The accumulated wealth of the priests has been confiscated, the temples spoiled of their artistic treasures, the monks driven from their monasteries, and many of these buildings converted into profane uses. Countless temple bells have already found their way to America, or been sold for old metal.

The third religious phase of thought is the so-called "Siza," a philosophic system, which is a feeble imitation of the ethical teachings of Confucius, but which has never found much favour with the ruling classes. It seems to be a sort of refined materialism, such as in fact underlies the whole religious thought of the nation. It relies altogether on moral truths and axioms. While outwardly conforming to the Kami religion, most of the *literati* and upper classes have hitherto been adherents of this school. But the Chinese is now being supplanted by what is called the "English Philosophy," represented in Japan chiefly by Buckle, Mill, Herbert Spencer, Darwin, Huxley, many of whose works have already been translated into Japanese.

Christianity was introduced by the Jesuit missionary Francis Xavier, about the middle of the sixteenth century. An impartial study of the early records shows that the Roman Catholic priests were sometimes preparing the minds of the people for the subversion of the native rule, and subjugation of the country to Spain or Portugal. Hence arose the persecution of 1596, which ended in the extirpation of Christianity and the expulsion of the Portuguese from Japan.

But since the restoration of the Mikado to supreme power, perfect liberty has been extended to the Christian religion. No obstacles are now thrown in the way of proselytism, and native preachers are allowed to proclaim their doctrines throughout the land, although foreign missionaries are still restricted to the Treaty ports.

Of the foreign missionaries the Russian "pope," Father Nikolai, attached to the Muscovite embassy in Tokio, seems to have hitherto had the greatest success. The Greek church is always crowded, and the people in the neighbourhood are selling off their shrines and idols, fearing they may become a drug in the market, so "fashionable" is the "orthodox" rite becoming. The Roman Catholic clergy are said to be very successful in the rural districts.

Meantime the Buddhists, alarmed at the progress of their rivals, have been stimulated to fresh exertions.

In Japan there exist properly speaking two nations side by side—the nobles and the commons. The nobles, who have been stripped of most of their prerogatives by the recent changes, were divided into two classes—the "Kuge" or "Court Nobility," and the "Buke" or "Aristocracy of the Sword"—of which there were again various degrees with six main divisions.

The whole people have from the earliest times been grouped into seven or eight classes according to their various social ranks and occupations:—1. The Daimios, the highest nobles of the state, originally powerful feudal lords, standing in somewhat the same relation to the Mikado that the imperial princes and counts did in mediæval Germany to the emperor. Their number corresponded at first to that of the "Koku," or provinces, so that there were originally eighteen "Kokushiu Daimios." 2. The hereditary nobles, from which class were chosen the provincial governors, the generals and

Government officials. In this class are included the "Hattamotto," or lower aristocracy. 3. The Sinto and Buddhist clergy. 4. The military class of the "Yakonins," and "Samurai." These were the four higher classes, who were privileged to carry two swords and to wear loose trousers. 5. The upper middle classes—physicians,

YAKONIN AND ATTENDANT.

officials, notaries, etc. 6. Merchants and traders, whatever their income. 7. Retail dealers, small shopkeepers, pedlars, hawkers, artists, painters, artisans. 8. Sailors, fishermen, peasants, day labourers.

There were also the so-called "yeta," or "yetori," a sort of pariah class numbering from 250,000 to 300,000, and scattered all over the empire. They were tanners, skinners, workers in leather, and public executioners and cremators. They were not permitted to dwell with the other classes in towns and villages, nor were they included in the census returns. They were also excluded from inns and tea-houses, and when travelling were compelled to eat from separate dishes in the open air.

But all these classifications have now been merged in the following three:—1. The Kwazoku, or nobles; 2. Shizoku, or old military class; 3. The Heimin, or commoners. All social privileges have been abolished, and every man is now equal before the law of the land.

8. *Topography: Chief Towns—Hakodate, Tokio, Yokohama, Niigata, Kioto, Osaka, Nagasaki.*

In Yesso the only places that can be called towns are Sapporo, the capital, on the west coast; Matsumai and Hakodate, both on Sugaru Strait over against Hon-do. Hakodate is probably the most important northern seaport, with a spacious and safe harbour and excellent anchorage. The town, in which is concentrated more than one-fourth of the whole population of the island, is a very flourishing place, with a large export trade in fish, skins, trepang, and other local produce. Here are several missionary stations, 17 schools, several large public buildings, and a British Consulate But the foreign trade has almost entirely ceased, foreign merchandise being now imported by native merchants in native vessels, and the exports sent also in native vessels to Hon-do and China.

In Hon-do there are scarcely any large towns north of Tokio, the capital, which lies about the middle of the east coast at the head of Yedo Bay, in 35° 50′ N. Although Yedo Bay, now renamed the Gulf of Tokio, is one of the largest and finest in the world, Tokio itself enjoys few of the advantages of a seaport, even the smallest vessels being obliged to lie some two or three miles off the shore. Like so many other Japanese ports, it has been partly obstructed by the alluvia of the River Tone, the southern branch of which enters the head of the gulf close to the capital.

The city is spread over an undulating plain, enclosed on the south by the bay, on the east and north by the broad and picturesque River Sumida, westwards partly by flat paddy fields, partly by a ridge of low hills planted with conifers and bamboo thickets. East of the Sumida River lies the large suburb of Honjo, continued southwards by the village of Shinagawa, which is itself an extension of the suburb of Takanawa.

The old divisions of Tokio were Shiro, Soto-shiro, Mitsi, and Honjo. On a wooded eminence in the centre of Shiro—that is, in the very heart of the capital—formerly stood the palace occupied till 1868 by the Shogun, but destroyed by fire in 1870. The park-like grounds were enclosed by lofty fortified walls and a broad moat flooded with running water. Round the Shiro spreads the Soto-shiro quarter, which is again surrounded by a wide ditch connected with the palace moat. This district is everywhere intersected by countless canals communicating both with the inner circular moats and with the sea. Here resides the commercial community, and here are situated the finest warehouses and public buildings.

Towards the north-east another wooded height was formerly covered with splendid Shinto and Buddhist temples, as well as with the tombs of several Shoguns of the last usurping family. But nearly all the temples were destroyed during the troubles of 1868-69, and this hill is now laid out as a public park, with a permanent exhibition building in the centre.

But the great feature of old Tokio were the numerous "yashiki," or palaces of the feudal lords. Some of these establishments occupied extensive wooded enclosures, and

A CANAL IN TOKIO.

consisted of clusters of houses surrounded by one-storied and whitewashed outhouses occupied by the Samurai, or "two-swordsmen," and other retainers of the Daimios. But since their owners have been stripped of their privileges, these strongholds have been dismantled and mostly destroyed.

The industries of Tokio are chiefly limited to artistic and fancy goods, such as bronzes and lacquer ware. The old law obliging the Daimios to reside for six months of the year in the capital, to which their families were confined all the year round, tended greatly to the development of the city. But since the repeal of that law in 1868 the population has fallen from about 1,500,000 to under 1,000,000 in 1880.

On the south side of the Gulf of Tokio lies the city of Yokohama, which, with the neighbouring port of Kanagawa, is open to Europeans. Thanks to its commodious harbour in the vicinity of the imperial capital, Yokohama has become the chief centre of the trade with the West, and the headquarters of the European establishments in the empire. It has been connected since 1872 with Tokio by a well-constructed railway 18 miles long, the first opened in the empire.

Niigata, capital of the province of Echigo, although its harbour is no longer accessible to large vessels, derives some importance from the fact that it is the only Treaty port on the west coast between Hakodate and Nagasaki, a distance of 1100 miles. It is a fine prosperous place of some 50,000 inhabitants, with large hospitals, barracks, normal schools, a geological museum, a missionary station, but no foreign trade. One of the streets, called the Teremachi, or Temple Street, is lined on one side for nearly its whole length by Buddhist temples, grounds, and priests' houses. The temples are mostly large handsome buildings, with magnificent high altars, candelabra, bronzes, and statues of Buddha, "with glories round their heads, in gorgeous shrines, looking like Madonnas" (*Miss Bird*).

The most interesting city in the interior is undoubtedly Miako, or Kioto, on a branch of the Yodo River about 8 miles west of Lake Biwa. Kioto is the ancient religious capital of the empire, and former residence of the Mikado. It is said to contain 93 Shinto and 935 Buddhist temples, amongst which are two conspicuous above all the rest for their size and splendour.

STATUE OF DAIBUTS—KAMAKURA.

According to the native accounts, one of these enshrines 333,333 idols, while the other contains the famous colossal figure of Buddha seated on the lotus. The Japanese look upon this city as a sort of Athens, from time immemorial the centre of learning and literature, where the most classic form of speech is current. It is still the seat of the book trade and a focus of native literature, while at the same time constantly crowded with Buddhist pilgrims from all quarters. Its principal industries are porcelain and bronzes. But although recently renamed Saikio, or the western capital, Miako is now in no sense a capital, the residence of the Mikado and the whole administration having been definitely removed to its great rival Tokio, the "eastern capital."

Kioto is connected by the River Yodo with Ohosaka or Osaka, the "Venice of Japan," and the queen of Japanese cities. It is intersected in all directions by innumerable canals which are crossed by no less than 3500 bridges, and the people live here altogether more on the water than on land. But since the opening of so many Treaty ports Osaka has lost much of its former wealth and importance, and its harbour is now so obstructed with sand-banks that the whole of its foreign trade and shipping have been gradually passing northwards to Yokohama and Tokio.

Facing Osaka, on the coast of the Inland Sea, are the seaports of Hiogo and Kobe, which, strictly speaking, form but one city, always spoken of by the natives under the latter, by Europeans under the former name. Hiogo, which is now connected by rail with Osaka and Otsu on Lake Biwa, commands the markets of Central Japan, and forms an important connecting link between the various ports thrown open to trade along the east coast.

The southernmost of these "Treaty Ports" is Nagasaki, on the west coast of Kiu-shiu. Nagasaki occupies a convenient position at the head of a romantic inlet which here forms a magnificent land-locked harbour about 500 miles from Shanghai, and 600, by the Inland Sea, from

Yokohama. Here is the rocky islet of Takoboko ("Papenberg"), from which native Christians are said to have been hurled into the sea at the close of the persecutions in the sixteenth century. Not far off is the artificial island of De-shima, occupied by the factory, where the Dutch monopolised the trade of Japan for two hundred years. But all restrictions are now removed, and the barriers between the station and the mainland have recently been cleared away. Thanks to its genial climate, Nagasaki has become a sort of sanitarium for the European traders settled on the Chinese seaboard.

9. *Highways of Communication—Railways.*

In Hon-do the three great historical highways are:—1. The Oshiu-Kaido, running from Tokio for 444 miles northwards to Awomori on Sugaru Strait; 2. The Tokaido, skirting the south-east coast for 307 miles between Tokio and Kioto; 3. The Nakasendo (Kishū-Kaido), running through the heart of the country for 323 miles, also between Tokio and Kioto. This alternative route is much more difficult than the coast road, and is often blocked by snow in winter.

Niigata on the west coast is connected with Tokio by two routes, 264 and 225 miles long respectively, both running for a part of the distance along the Nakasendo line. From Tokio the Koshiu-Kaido, 77 miles long, leads to Kofu in the province of Kai, whence it is continued for 32 miles to the Nakasendo road at Shimono-Suwa.

None of these highways are macadamised, and those of Shikoku and Kiu-shiu especially are said to be in a very bad state. But even in Nippon many of the roads in the interior are about as bad as they well can be.

"The road from Kurumatogé westwards is so infamous that the stages are sometimes little more than a mile. Yet it is by it, so far at least as the Tsugawa River, that the produce and manufactures of the rich plain of Aidzu with its numerous towns, and of a very large interior district, must find an outlet at Niigata. At present it is a perfect quagmire, into which great stones have been thrown, some of which have subsided edgewise, and others have disappeared altogether."[1]

Two short lines of railway have already been constructed, one 18 miles long, between Yokohama and Tokio, opened in 1872; the other 50 miles long, between Hiogo and Osaka, and since extended 16 miles farther to Kioto and Otsu on Lake Biwa, making a total distance of 66 miles. Both lines are substantially built, with neat stations in the English style. Lines of 500 miles have also been projected, but the country is everywhere so mountainous that it is doubtful whether railways will ever acquire any great development in Japan.

On the other hand, steam navigation is making rapid progress, and several lines of steamers now ply regularly between all the chief seaports of the empire. Several of these vessels have been built in Japan, and are worked entirely by native hands.

10. *Administration: The Mikado and Shogun — The Revolution — Army and Navy — Education — Art — The New Ideas.*

The peculiar political and social institutions of the Japanese people, which are now rapidly yielding to the influence of Western ideas, were slowly and independently developed during the long ages that they have

[1] *Unbeaten Tracks*, i. p. 186.

been in possession of their secluded island empire. How long they may have here been settled it is now hard to say, their historic records dating no further back than the third century of our era. These records embrace two clearly-defined epochs, the "oshei" and the "hashei," as they are called. The first, reaching to the year 1192 A.D., covers the period of the absolute authority of the Mikadoes, or hereditary emperors of divine origin, who were gradually transformed to spiritual sovereigns."[1]

The second period corresponds with the growth of the power of the Shoguns, or military rulers. This is the mediæval period, which was not brought to a close till 1868, when the Revolution restored the Mikadoes to the supreme authority. The power of the Shogun, the true significance of which has long been misunderstood, was not fully and legally recognised till the beginning of the sixteenth century. The popular idea is that there were two Emperors, the temporal Shogun residing in Yedo (Tokio), and the spiritual Mikado residing in Miako (Kioto). But in point of fact the Shogun was never anything more than the military vicegerent of the Mikado, without any legal claim to joint sovereign rights.[2] In course of time, however, he usurped the control of state affairs, retaining the Mikado in a sort of honorary imprisonment in Miako. Consequently the Revolution merely put an end to the Shogun's military despotism, and reinstated the Mikado in the full exercise of his sovereign rights.

[1] *Mikado* (from *mi*=sublime, and *kado*=gate) has etymologically the same meaning as the "Sublime Porte." The idea seems to be that "the Mikado is too sublime a being to be spoken of directly or otherwise than in a figurative sense" (Rein, *Japan*, i. 245).

[2] From *Sho*=general, and *gun*=army ; hence *Shogun*=commander-in-chief. The early European writers spoke of the Shogun as the Tykoon, from the Chinese *Tai-kun*="Great Lord."

This event was followed by perhaps a still more important event in 1873, when the Mikado voluntarily promised to his subjects a representative constitution. This promise was followed in 1875 by the appointment of a Senate, and of the "Dai-shin-in," consisting of the Supreme Council (Shoïn), and the College of Ministers (Uïn). The scheme of representation was further developed by the establishment in 1878 of the provincial and departmental Assemblies, and is to be completed by a full National Assembly, which, by the imperial decree of October 12, 1881, is to meet for the first time in the year 1890.

The original nine divisions of the empire, which had long ceased to answer to the requirements of the executive government, were replaced in 1871 by the three "fu," or urban districts, and the 66 "ken" or prefectures, which have since been reduced to 60. Each ken is governed by a "ken-rei," directly dependent on the Ministry of the Interior, and its name corresponds in almost every case with that of its chief town.

According to the law of 1873 regulating the conscription, all Japanese subjects are liable to serve from their twentieth year, and must pass three years either in the army or the navy. They then pass into the first reserve for two years, taking part in the annual exercises, after which they join the second reserve, which is only called out in case of a general levy. According to this new system the peace footing comprises 42 regiments of the line, 3 of cavalry, 16 batteries of artillery, and 10 battalions of engineers, or 31,680 men altogether, which in time of war is raised to 46,000 regulars of the front line of defence. The Japanese troops, who have at all times been distinguished by extreme bravery, are now disciplined after the French method, although no foreigner receives a permanent appointment in the army.

The navy consists of 21 ships (of which 3 are ironclads), with 60 guns and 3680 men. Most of the ships are built in the local dockyards, except the ironclads, which are commissioned in England. The crews are noted for their skill, courage, and coolness.

Special attention has of late years been paid to the question of public instruction. For educational purposes the country has been divided into seven circles, each with a "dai-gak'ko" or High School in the central town. Of middle schools ("chin-gak'ko") there are 224, and of elementary (sho-gak'ko") 47,000.

Japan has also joined the International Postal Union, and already possesses 17,000 miles of telegraphic lines.

The chief branches of local industry resemble those of China, from which country Japan has received its culture. The ancient and world-renowned bronze, lacquer, and paper wares still continue to be produced, as well as the porcelains and works in enamel, an art introduced some three centuries ago. The chief centres of the porcelain industry are Satsuma, Hizen, Kioto, and Kaga. In Hizen there are inexhaustible deposits of kaolin, or china clay, and here is produced the so-called "egg-shell" ware.

Extremely interesting are the paintings on paper or silk, and the books illustrated with woodcuts. Strips of tissues several yards long and about a yard wide are painted with figures in a somewhat conventional style, and with plants, in which a happy mean is observed between mannerism and the realistic style. Great skill and exuberance of fancy are displayed in the numerous illustrated works, which include many sketch-books with rich and varied landscapes, studies of men and animals, genre scenes and the most extravagant caricatures.

The Japanese are still mainly an agricultural people, who have brought certain branches of industry to great perfection, thanks to the excellence of the raw materials, combined with a patience and conscientious workmanship scarcely paralleled elsewhere.

Meantime the revolution of 1868 forms the starting point of a new era; and although it has been followed by many sweeping changes and reforms of a fundamental character, the country must still pass through a restless and anxious period of transition before the new ideas can become thoroughly assimilated. The nation has been suddenly awakened, as it were, from sleep, and seized all at once with a passionate desire to share with the peoples of the West in all the intellectual and material triumphs of a slowly-matured culture. This movement constitutes an almost unique feature in the history of Asiatic nations. But in the eagerness to appropriate the results of such a culture, mistakes must be expected. Fleets and armies have been improvised, and magnificent buildings, such as the Mint at Osaka, erected at a needless expenditure of vast sums that might have been better applied. Costly embassies are maintained in Washington and at the chief European Courts, while large sums are devoted to the education of several hundred students in Europe and America. Every inducement has been held out to foreign professors to settle in the country; orders of chivalry have been instituted on the European model; the natives have been forced to adopt the European dress; and the idea has even been broached of abolishing the national speech and replacing it by the English language and letters. According to the most judicious opinions, however, the crisis cannot be said to have been passed; errors springing mainly from excessive zeal have yet to be repaired. Many enthusiastic observers may hope that the collective wisdom of the nation will guide the Japanese race to a bright goal. Still, prudent judges will suspend their opinion regarding the moral and intellectual future of this interesting race.

11. *Statistics.*

Areas and Populations.

	Islands.	Area in square miles.	Pop. 1874.	Pop. 1880.
Nippon.	Hon-do	90,500	25,478,834	Details wanting.
	Kiu-shiu	14,200	4,986,613	
	Shi-koku	7,100	2,484,538	
	Sado	24	362,177	
	Tsushima	17		
	Awaji	12		
	Oki	8		
	Iki	3		
	Small Islands in Japan proper	?60	...	
	Yesso	35,700	144,690	
	Kuriles	4,900		
	Liu-kiu Islands	830	167,073	
	Bonin Islands	38	69	156
	Total Japanese Empire	153,384	33,623,373	35,925,313

Area and Population (Satow, Unpublished MS.)

Circuits.	Area in square miles.	Pop. 1876.
Gokinai	2,967	2,094,354
Tokaido	15,478	7,710,282
Tosando	40,478	7,279,408
Hokurokudo	6,785	3,448,199
Sanindo	6,490	1,660,755
Sanyodo	8,407	3,648,170
Nankaido	8,784	3,375,724
Saikado	16,795	5,280,740
Hok'kaido, Yesso, and Kuriles	36,270	149,554
	147,582	34,647,486

Population to the square mile, about 230.

Chief Towns.	Pop. 1884.
Tokio (Yedo)	902,837
Kioto (Miako)	255,403
Osaka	353,970

Trade Returns.

Year.	Imports.	Exports.	Total.
1873	£5,480,000	£4,130,000	£9,610,000
1878	6,660,000	5,250,000	11,910,000
1880	7,250,000	5,510,000	12,760,000
1884	5,764,000	6,603,000	12,367,000

Revenue and Expenditure, balanced (1884)	£16,621,000
National Debt (1885)	64,942,000

FOREIGN TRADE (1880).

	Imports.	Exports.
Great Britain	£4,645,000	£787,000
British Possessions	420,000	52,000
China	1,183,000	1,128,000
France	762,000	1,140,000
United States	594,000	2,240,000
Germany	335,000	6,600
Italy	31,000	100,000
Sundries	116,000	36,000

CHIEF STAPLES OF TRADE (1879).

Imports.		Exports.	
Yarns	£1,510,000	Silk and cocoons	£1,844,000
Cottons	1,035,000	Tea	882,000
Woollens	1,158,000	Rice	928,000
Raw cotton	614,000	Dried fish	206,000
Metals	377,000	Coal	171,000
		Porcelain (1880)	100,000

SHIPPING.

Year.	Tonnage.	Of which British.
1865	166,872	99,649
1869	1,069,398	410,105
1873	1,039,407	234,459
1878	749,529	417,691

Japanese steam tonnage (1880), 36,543 tons.
„ junk navy „ 468,750 „
„ fishing craft „ 433,000 boats.

TRADE OF THE CHIEF TREATY PORTS (1880).

	Imports.	Exports.
Yokohama and Kanagawa	£5,190,000	£3,710,000
Hiogo and Osaka	1,750,000	1,160,000
Nagasaki	280,000	487,000
Hakodate and Niigata	45,000	150,000

MINERAL RETURNS (1879).

Gold raised	£46,000
Silver „	90,000
Copper „	287,000
Lead „	5,000
Iron „	21,000
Coal „	342,000
Petroleum	63,500 gallons.
Coal-fields of Yesso	150,000,000,000 tons.
Total yield of minerals (1881)	£1,000,000

AGRICULTURAL RETURNS (1880).[1]

Land under forest	60,000 sq. miles.
Land capable of tillage	50,000,000 acres.

[1] From United States Consul-General's Trade Report for 1881.

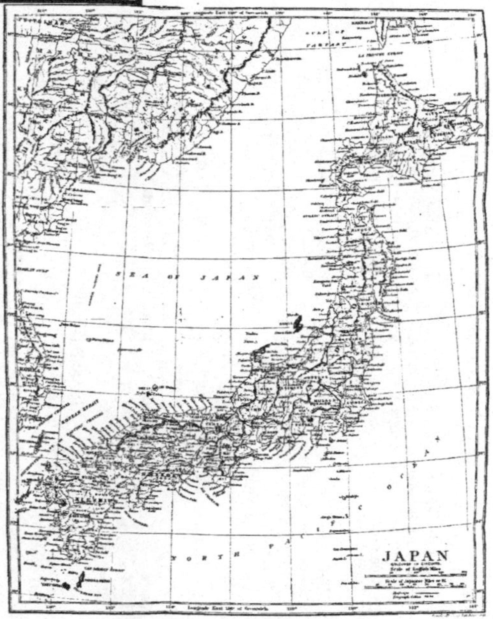

London: Edward Stanford, 26 & 27, Cockspur St. Charing Cross, S.W.

Land under cultivation . .	12,530,000 acres.
Land under rice . . .	6,800,000 ,,
Land under other crops . .	4,250,000 ,,
Rice crop . . .	205,000,000 bushels.
Barley crop . . .	60,000,000 ,,
Wheat crop . . .	38,000,000 ,,
Tea crop	90,000,000 lbs.
Sorghum sugar crop . .	65,000,000 ,,
Tobacco crop . . .	90,000,000 ,,
Pulse crop . . .	50,000,000 bushels
Silk, total yield . . .	£11,000,000
Silk and cocoons exported .	£2,310,000

Foreign Residents and Firms in the Open Ports.

	British.		Others, including Chinese.	
Years.	Residents.	Firms.	Residents.	Firms.
1874 . . .	1170	155	3961	310
1878 . . .	1067	92	4438	191

Finance (1880).

Revenue and Expenditure, balanced . . .	£11,130,000
National Debt	72,070,000

Chief Source of Revenue, the Land-tax, 87 per cent of the whole.

Postal Returns (1883).

Letters forwarded, 55,711,215.
Newspapers, packets, etc., 56,615,000.
Post-offices, 5315 ; mail routes, 36,052 miles.
Telegraph lines, 5000 miles.
Telegraphic messages, 2,732,810.
Submarine cable, connecting the various islands, completed 1880.

Railways open (1881), 76 miles.
Railways projected, 500 miles.
Coinage struck at the Osaka Mint (1870-80), £17,000,000.
Paper money in circulation (1880), £30,000,000.
Newspapers (1879), 211 ; circulation, 29,000,000.

Public Instruction (1883).

	Number.	Teachers.	Male.	Female.
Elementary Schools . . .	29,081	84,765	2,084,624	919,513
High ,, . . .	173	984	13,043	84
Normal ,, . . .	78	649	5,272	697
Other ,, . . .	1,225	1,927	45,599	10,829
Universities	2	165	1,897	...

Miscellaneous.

Army.—Peace footing, 35,560 ; War, 50,230 ; Reserve, 20,000.
Navy.—Ships, 23 ; Men, 4242 ; Guns, 149.
Police.—23,334.

CHAPTER XIV.

INDO-CHINA AND MALACCA.

1. *Boundaries—Extent—Area.*

THE south-eastern section of Asia, commonly spoken of collectively as Indo-China, Further India, Trans-Gangetic India, or the "Golden Peninsula," really consists of two distinct peninsular regions—Indo-China proper and Malacca—which differ profoundly from each other in their physical conditions no less than in the ethnical affinities, culture, and religion of their inhabitants. In all these respects the northern peninsular mass of Indo-China still belongs to the Asiatic mainland, whereas Malacca, projecting southwards parallel with the neighbouring island of Sumatra, forms, strictly speaking, an integral part of the great Oceanic world. Amongst the great schemes of canalisation projected since the piercing of the Isthmus of Suez, not the least ambitious is the recently-proposed connection of the Bay of Bengal with the China Sea by a canal across the narrow Isthmus of Krah, by which Malacca is at present connected with Indo-China.

Meantime, Further India, taken in its wider sense, lies almost exactly within the northern torrid zone, stretching from about 3 degrees beyond the tropic of cancer southwards to Cape Romania, which is the southernmost point of the continent, and which approaches almost to within 1 degree of the equator. This gives an extreme length of about 1800 miles, with a breadth varying from 700 miles in the north to a little over 60 miles in the Isthmus of Krah, a total area of over 870,000 square miles, and a population vaguely estimated at from 35 to 40 millions.

The coast-line of Indo-China is far more diversified by bays, bights, gulfs, islands, and headlands, than is the somewhat monotonous seaboard of British India. The west coast is watered by the Bay of Bengal, which here forms the Gulf of Martaban, while contracting southwards to the Malacca Strait, between Sumatra and the mainland. East of Malacca the coast is washed by the storm-swept China Sea, which here develops the great Gulfs of Siam and Tonking, between which the continent is rounded off by the graceful curve of the Cochin-Chinese seaboard.

2. *Relief of the Land: Mountain Systems—Cochin-Chinese Coast Range.*

The interior of Indo-China is one of the least-known regions in Asia, and here are concentrated some of the most interesting orographic and hydrographic problems that still await solution from modern research. The surface is covered with a number of parallel mountain ranges running mainly north and south, with intervening longitudinal river valleys broadening southwards to extensive alluvial plains, where are developed some of the largest deltas on the globe. We are not only still ignorant of the real character of these mountain ranges, but we do not even know where any of the large intervening rivers take their rise. Nor does the solution of these difficulties depend on a wider knowledge of the interior of the peninsula itself so much as on the further exploration of the Tibeto-Chinese frontier lands, whence the mountains radiate and whence the rivers flow southwards. Do these mountains form an independent highland system? or are they, as many suspect, simply the south-eastern continuation of the Tibetan plateau, cut up into so many separate ridges by fluvial action? In any case, it is clear that the sources of the great rivers must be sought for on the Tibetan plateau itself, and probably in the great lacustrine region where rise the Hoang-ho and Yang-tse, and where Nain-Singh speaks of chains of lakes draining eastwards. From the more recent explorations of A. K. (Pundit Krishna), it seems evident that here also the Irawady, Salwin, and Mekhong have their rise.

In the east of the peninsula the Cochin-Chinese coast range separates the Mekhong basin from the numerous short streams flowing to the China Sea. But the term "range" is somewhat inaptly applied to an intricate system of moderately elevated ridges crossing each other at all angles, and giving rise to a number of rivers, flowing some to the Mekhong, others to the coast. This upland system runs north and south, throwing off numerous spurs and offshoots which project seawards, breaking the Cochin-Chinese seaboard into a number of bays, bights, and inlets. It terminates in the extreme south with the headland of Cape St. James, at the entrance of the Saigon River.

The chain running parallel with the coast range between the Mekhong and Menam valleys merges southwards in low plateaux west of the Tonlé-sap, or Great Cambodian Lake. But it again acquires a considerable elevation at Shantabun, near the coast, whence it runs south-east along the Gulf of Siam, and then trends round to the north-east, here culminating with the Pursat or Krevanh ("Cardamom") hills south of the Great Lake.

3. *Hydrography: The Irawady, Salwin, Menam, Mekhong, and Song-ka Rivers—Lake Tonlé-sap.*

North of the Brahmaputra valley the eastern section of the Tibetan plateau seems certainly to be geologically continued eastwards far into Yunnan and Se-chuen. But in this little-known region, inhabited by the Mosso, Lolo, Si-fan, and other semi-independent aboriginal tribes, there are pressed together an extraordinary number of separate ridges, possibly produced by the action of running waters. Parallel with the Tant-la run several chains, mainly north and south, nearly at right angles with the Kuen-lun system, and these "Cross Ridges," as Blakiston calls them, penetrate far into Burma and Siam, where they form the Indo-Chinese mountain system. Their general direction is indicated by the course of the great rivers, some of which at all events take their rise on the Tibetan plateau, and which flow first north-east, parallel with the Tant-la and the other cross ridges on the Tibeto-Chinese frontier. All these rivers, amongst which must be included the Irawady, Salwin, and Mekhong, as well as the Yang-tse and Min, then trend gradually round to the south, flowing in this direction for hundreds of miles in the closest proximity. Nowhere else is there any instance of so many large streams flowing in independent parallel valleys, separated only by single ridges, without uniting into one general water system. All the large rivers which reach the coast between the Irawady and Yang-tse deltas, a distance of at least 5000 miles, are confined in their upper courses within the comparatively narrow tract which lies between the eastern tributaries of the Brahmaputra and the head-waters of the Hoang-ho.

Margary, Gill, M'Carthy, Soltau, Szechenyi, and others who have recently traversed the ground between Yunnan and Upper Burma, all speak of the numerous river valleys running north and south which they had to cross between Lake Ta-li-fu and Bhamo. Justus Perthes' map (1881) of Szechenyi's route from Sayang to Bhamo, a distance of about 160 miles as the bird flies, lays down, besides numerous tributaries, no less than four main streams identified by that explorer as the Mekhong, Salwin, and the two great forks of the Irawady. The same phenomenon is described by Desgodins, whose route lay far to the north between Se-chuen and Tibet, and by A. K., who in 1882 also traversed the remarkable region of contiguous river valleys between the Yang-tse and the Brahmaputra.

Going eastwards, and keeping within the limits of Indo-China, first comes the Irawady itself with its two great and still unexplored head-streams, and its vast delta, already described in Chapter VII. Here it will be sufficient to add that the Abbé Desgodins and A. K. (Pundit Krishna) have finally disposed of the claim of the Irawady to be regarded as the continuation of the Tibetan San-po. During a three years' residence in the valley of the Upper Salwin, at a convenient place for studying the question, Desgodins ascertained that the two rivers could not possibly be connected, and that consequently the San-po must flow to the Brahmaputra.

Parallel with the Irawady is the Salwin, whose upper course is the Lu-kiang, flowing from the Langtan or Gulong-sigong hills, on the Yunnan frontier. The Salwin flows thence first along the eastern frontier of Burma, and lower down between Pegu and Siam to its mouth in the Gulf of Martaban. It has a known but very imperfectly-explored course of at least 700 miles, during which it receives no large tributaries. Nor does it form a delta at its mouth, in this respect differing from most other great Asiatic rivers. It appears to be greatly obstructed by rapids, and is probably not navigable by large craft for more than 100 miles from its mouth.

Still farther east comes the Menam, the "Mother of Waters," the only large Indo-Chinese river whose course lies entirely within the geographical limits of the peninsula. It flows through the Laos States and Siam proper,

VALLEY OF THE IRAWADY.

mainly southwards, to the head of the Gulf of Siam, which it enters through three channels. Of these the easternmost is the most navigable, but even this is obstructed by a bar with scarcely 4 feet of water at ebb and 12 at flow. Hence large vessels proceeding to Bangkok, 38 miles from its mouth, discharge most of their cargoes in the roadstead. Throughout most of its course the Menam is fringed by forest trees, behind which the low-lying rice and sugar plains are regularly flooded during the inundations. A few miles above its mouth the Menam communicates with the Meklong, which is really an independent river, although from this circumstance often represented as a branch of the Menam.

Beyond the Menam follows the Lantsang or Kinlung-kiang, better known as the Mekhong, or great river of Cambodia.[1] Thanks to the famous French expedition of 1864 and several subsequent explorations, there no longer remains much doubt as to the true course of this, the longest of all the Indo-Chinese rivers, which rises in East Tibet and flows through Yunnan and between Siam and Cochin-China to its delta in Cambodia, at the south-eastern extremity of the continent. Its upper course is separated by a single narrow ridge from that of the Kinsha-kiang or Yangtse-kiang, the two streams here flowing for a long distance in parallel meridional valleys along the eastern range of the Tibetan plateau.

In its lower course the Mekhong is connected with the Tonlé-sap, or Great Lake of Cambodia, which drains to the river at low water, but which during the inundations receives a back current from the river. The Tonlé-sap, which is almost the only lake in Indo-China, has a mean area of about 1000 square miles. But during the summer floods its level is raised nearly 40 feet, which is sufficient to increase its size three or four times. It abounds to such an extent in fish of every sort, that their capture and cure for exportation forms one of the chief industries of the country.

"The entry of the great Cambodian lake is at once

[1] Mekhong is the Lao name of this river, which the Cambodians call the "Tonlé Thom"—*i.e.* the "Great River"—whence the European expression, the Great River of Cambodia, or simply the Cambodia River.

grand and beautiful. It presents the aspect of a vast inland strait, with its low banks covered with dense and half-submerged forest growths, but encircled in the distance by a vast mountain range, whose farthest crests merge in the azure sky or disappear in the hazy atmosphere" (*Mouhot*).

The Tonlé-sap may be regarded as a remnant of the marine inlet which formerly penetrated up the present Mekhong delta as far inland as the 13° 30′ parallel. This inlet has been filled in partly by slow upheaval of the land, partly by the alluvia of the great river, by which the delta is still steadily advancing seawards (*Aymonnier*).

The last river reaching the coast in an independent channel is the Song-ka (Sang-koi, Nhi-ha, Hong-kiang), or Red River of Tonking, which flows from the south Chinese highlands south-eastwards to the head of the Gulf of Tonking. This river, the navigation of which is open to the Chinese frontier town of Lao-kai, possesses considerable commercial importance, and the French have already made several attempts to open up a trade with the southern provinces of China through this channel. But a portion of its course is occupied by independent wild tribes, while it is obstructed at several points by rapids, which would require the formation of difficult and dangerous portages. Hence the Song-ka does not offer the great advantages which the French at first expected to derive from it.

Both the Mekhong and the Song-ka are joined by several important tributaries, many of which have recently been carefully explored by Aymonnier, Neis, Harmand, and other French naturalists. Dr. Neis completed in 1881 the survey of the Dong-nai, fixing the position of its source in the rugged highland region, whence it flows to the Mekhong delta. Here he determined two distinct lofty ranges, with an extensive intervening plateau, which it took seven days to cross.

In 1877 Dr. Harmand explored a considerable portion of the Si-bang-hieng, an important affluent of the Mekhong, famous for its magnificent tropical scenery

4. *Natural and Political Divisions: Burma—Siam—Annam—Cambodia—French Cochin-China—Malacca and Straits Settlements.*

The political condition of Indo-China has been largely determined by its prominent natural features. To the great river valleys of the Irawady, Menam, and Mekhong, correspond the ancient historical kingdoms of Burma, Siam, and Annam (Cochin-China), while the still more ancient empire of Cambodia, founded by the primitive Caucasic race of the peninsula, has been gradually restricted to the broad alluvial plains and delta of the lower Mekhong by the later Mongoloid intruders from the north. Malacca, also, almost physically detached from the mainland, has from prehistoric times been occupied by petty States, founded by peoples of Malay stock, either here indigenous or more probably intruders from the neighbouring Archipelago of Malaysia.

Burma.

At the beginning of the present century the Burmese empire was by far the largest and most powerful in Farther India. It occupied nearly the whole of the Irawady, Sitang, and Salwin basins, with a coast-line stretching for about 900 miles from the head of the Bay of Bengal to the Isthmus of Krah. Since then a series of disastrous wars with the English has caused the gradual loss of all the coast regions—Arakan, Pegu, and Tenasserim—which now constitute the flourishing province of British Burma. Thus entirely cut off from the sea, the country remained in a state of chronic trouble until the close of the year 1885, when King Thebaw, last of the native rulers, was dethroned, and the administration of Independent Burma taken over by the British authorities.

Within its present limits the late kingdom of Ava, as it was often called from one of its ephemeral capitals, is hemmed in on the north, west, and south-west by British Burma, on the north-east by the Chinese province of Yunnan, on the south-east by the kingdom of Siam. With an extreme length north and south of about 500 miles and a mean breadth of 300 miles, it has a total area of about 190,000 square miles and a population of perhaps 4,000,000. It is divided into three distinct sections—Burma proper, between 24° 30′ and 18° 50′ N. latitude, inhabited by the pure Burmese people; North Burma, occupied by the Sing-fu and other semi-independent hill tribes; and the tributary Shan States to the east. All the Shan or Laos States stretch eastwards to the Mekhong valley; but those subject to Burma lie mainly between the 24th and 20th parallels and between 97° to 101° E. long.

The Burmese rule, which was severely felt by the districts in the proximity of Mandalay, the present seat of government, became continually less oppressive as we proceed eastwards. In the north-east it was, so to say, overlapped by the Chinese authority, so that it was here often difficult to say where the one ceased and the other began. In some districts the triennial tribute due to the Burmese Court consisted of such trifles as gilded wax tapers, a little salt and tea, or perhaps a pair of embroidered shoes, a gold drinking-cup, a silver plume, or suchlike tinsel, and these presents were sent by several of the Shan districts both to China and Burmah. The dignity of "tsanwab" or "thabwa"—that is, feudal lord

—is hereditary in all the ruling families, but the Burmese Court conferred the investiture on each successive lord, and designated the next heir. In the principalities, ruled jointly by the Chinese and Burmese, both suzerains generally came to an understanding in the choice of the next heir; but, in case of disagreement, two chiefs were appointed, and fought it out.

As in other Indo-Chinese States, the white elephant ranked in Burma next to royalty itself. This elephant had a palace to himself, with a personal chamberlain and estates in the most fertile cotton districts, besides four gold umbrellas and a suite of thirty courtiers. At the same time the expression "white elephant" is extremely elastic, the colour being often of a dirty yellow, or even brown, if only a few light specks can be shown behind the ears, on the forehead or trunk.

Siam.

From Burma and the region of the Irawady we pass eastwards to the basin of the Menam, politically comprising the kingdom of Siam or Tai—that is, of the "Free." Siam occupies the heart of the Indo-Chinese peninsula between Burma and British Burma on the west, Yunnan on the north, Annam and Cambodia on the east. Southwards it includes the strip of territory between Tenasserim and the Gulf of Siam, as far as the Isthmus of Krah, in 10° N. latitude; and beyond this point all the northern section of Malacca nearly to Perak. Siam has thus an extreme length of at least 1000 miles, while its breadth varies from about 400 to 60 miles, with an area of nearly 320,000 square miles, and a total population of less than 6,500,000.

Siam proper consists mainly of the low-lying alluvial basin of the Menam and its numerous tributaries, branches,

and backwaters, which form an extensive and intricate delta, like that of Cambodia, continually advancing seawards. The northern background occupied by the Laos States is a more or less hilly country, stretching from the Mekhong to the Salwin, along the debatable ground between Burma, China, and Annam. Both sides of the Menam basin are skirted by densely-wooded terrace-like ranges, forming the water-partings towards the Salwin and Mekhong, but whose structure, form, and general elevation are almost entirely unknown. In fact, the greater part of the kingdom still remains an almost unexplored wilderness, mostly covered with dense tropical forests. The better-known parts of the country consist of extremely fertile alluvial plains intersected by numerous streams and canals, and producing magnificent crops of rice, sugar, cotton, maize, and indigo.

In Malacca the dependent States are Quedah, Kalantan, Patani, Ligor, Talung, Tringanu, whose Malay sultans pay merely a nominal tribute of a gold or silver tree or flower, sent every third year to Bangkok.

Annam and Cambodia.

East of Siam the remainder of the Indo-Chinese peninsula is occupied by the kingdom of Annam in the east, and the French possession of Cochin-China with the neighbouring vassal kingdom of Cambodia in the south.

The French, who have taken direct possession of the Mekhong delta, have during the past twenty years gradually extended their influence throughout the whole of this region. For Annam itself has by the recent treaties of 1885 become a French vassal State. It consists of a comparatively narrow strip of coast lands stretching nearly due north and south between the China Sea and the coast range skirting the left bank of the

Mekhong. It consists of three distinct sections—Tonking, or Dang-gnoi—that is, the "Eastern Land," watered by the Song-ka, or Hong-kiang; Cochin-China, or Dang-kong—that is, the "Interior Land;" and Champa, or Tsiampa, in the extreme south-east corner of the peninsula. To these must also be added the domain of the semi-independent wild tribes (Moi), and the section of the Lao nation settled on the left bank of the Mekhong, who are subject to Annam. Formerly Annam claimed jurisdiction over Cambodia and the Mekhong delta, but even within its present restricted limits it stretches north and south across thirteen degrees, between 23° and 10° N., with a total area of some 200,000 square miles, and a population vaguely estimated at from ten to twenty-one millions.

Annam proper, or Dang-kong, a narrow strip from 10 to 20 miles wide, extends from about 12° N. on the Champa frontier northwards to Tonking. It is enclosed on the west by bare hills covered with a very sparse vegetation. The domain of the Moi stretches west of this province from about 10° to 16° N. Under the general designation of Moi, the Annamese comprise all the numerous hill tribes known to the Siamese as Kha, and differing widely from each other in speech, type, and usages. The land of the Laos subject to Annam lies north of Cambodia and the Mekhong, and varies in breadth from 20 to 24 miles. In the south and west are several settled districts, but the east is an arid waste. The plains are enclosed on the north by two ranges, and the rivers are mere mountain torrents. Yet all the accounts of recent travellers represent the country as in a prosperous state, inhabited by a peaceful and industrious people living under the authority of patriarchal chiefs. They cultivate the land, and have some silk and earthenware manufactures.

Tonking is very mountainous in the north, where it presents the same general features as the neighbouring Chinese provinces of Yunnan and Kwang-si. The eastern districts are almost flat, merging seawards in an extensive alluvial plain. Most of the streams, which flow mainly in a south-easterly direction, contain large quantities of auriferous sands, the washing of which employs thousands of hands. An old wall running along the southern frontier towards Annam from the hills to the coast, has been rendered useless since the union of the two States under one sovereign. Next to Korea, Tonking had persisted most obstinately in the exclusive system as regards foreigners, and the policy even of the central government had been to keep Tonking in complete seclusion from outward influences. But it was at last thrown open to the trade of the world by the commercial treaty concluded between France and Annam in 1874. In 1885 the French protectorate was formally recognised by China.

Tonking is extremely fertile, and abounds in mineral wealth of all kinds. As many as 14 gold, 7 silver, 3 copper, 1 tin, 17 iron, and 3 salt mines have already been opened, while the extensive coal-fields, some reaching down to the coast, still remain almost untouched. But the vast resources of the country must remain undeveloped, while it continues a prey to anarchy and to the attacks of regularly-organised bands of marauders. A writer in a recent number of *Globus* tersely remarks that "before an active foreign trade can be developed the banditti must be exterminated."

The ancient kingdom of Cambodia, which formerly comprised a large portion of Indo-China, has long been restricted to the lower course of the Mekhong between Lake Tonlé-sap and the delta. Till recently it even stood for some time in the position of a vassal State to Siam. But since the French occupation of Lower Cochin-China the king of Cambodia has transferred his allegiance to France. His territory forms an extensive and exceedingly fertile alluvial plain watered by the Mekhong, the great Lake Tonlé-sap and their numerous affluents, branches, and connecting channels. The plain is diversified in the west by isolated hills and short ridges, and is confined in the north by the Phnom Dangrek range. The space between this range and the northern shores of the lake is strewn with the stupendous ruins of Angkor and many other remains, which still attest the former greatness of the Cambodian empire, when it formed one of the chief centres of Buddhist culture in the East. These monuments, the finest of which are now included within the limits of Siam, contain vast archæological and architectural treasures, which are still very imperfectly known. But Lieutenant L. Delaponte was entrusted at the end of the year 1881 with a mission for the systematic exploration of the interior of Cambodia, from which great results are anticipated.

Cambodia has an extreme length of 240 miles north and south, with a breath of 180 miles, a total area of 40,000 square miles, and a population of under 1,000,000.

French Cochin-China.

The French colony of Lower Cochin-China, which has become the centre of political power east of Siam, comprises the whole of the hot and marshy Mekhong delta. It consists of six provinces detached in 1863 from Annam, and now administered by the French marine department. It has an area of some 22,000 square miles of extremely fertile land, with a population of 1,600,000.

Malacca—The Straits Settlements.

The Malay peninsula stretches from 13° 45′ N. first

southwards to 8° 50′ N., and thence south-eastwards to Cape Romania in 1° 35′ N. The angle thus formed is marked by the Isthmus of Krah, which is about 3000 feet above sea-level. South of this point rise the Malacca mountains, whose irregular masses fill the whole of the peninsula to its southern extremity. It has an extreme length of about 900 miles, with an area of 83,000 and a population of about 1,000,000. The northern section, with an area of 14,000 square miles, and a population of 400,000, is held by petty Malay States tributary to Siam. All the rest is occupied partly by the British possessions, collectively known as the Straits Settlements, partly by a number of Malay States, which have accepted the British protectorate.

The Straits Settlements, which lie scattered along the west coast, include the province of Wellesley, Malacca, and the two important islands of Singapore and Pulo-Penang. Of the protected States the most important are Perak (Perah), Salangor, Rumbo, and Johor.

The interior, which consists of magnificent wooded ranges, intersected by numerous fertile river valleys, is still very little known. But a systematic survey of Perak and the neighbouring districts was begun in 1879 by H. S. Deane, who determined the height of the Shin (6000 to 7000 feet) and Titi Wangsu (7000) ranges, and ascended the navigable River Plus, a tributary of the Perak, for 50 miles to the Jeram Dina rapids. Here there are altogether over 1200 miles of water transport, and this region excels in the extent of its forest lands available for plantations. In a total of 5,000,000 acres, fully 200,000 are well suited for the cultivation of tea, coffee, cinchona, and indigo. The chief mineral is tin, of which there seem to be vast deposits. Tin also abounds in other parts of the peninsula, and the rich mines of the island of Thalang (Junkseylon) on the west coast, about 8° N., employed over 30,000 Chinese hands till 1872, when the works were partly abandoned.

But a much more important island is Pulo-Penang ("Prince of Wales"), which lies off the coast of the British province of Wellesley at the northern entrance of Malacca Strait. Although scarcely 15 miles long by 7 broad, such is the fertility of its soil that it produces large quantities of rice, pepper, cloves, nutmegs, betel, cotton, tobacco, tea, coffee, sugar, and cocoa-nuts.

At the southern entrance of Malacca Strait, and close to the mainland, lies the little island of Singapore, which, since its occupation by the English in 1818, has become one of the great centres of trade in the East. Formerly almost entirely uncultivated, it is now covered with pepper, sugar, rice, sago, and gambier plantations, on which a large number of Chinese coolies are employed. But the rapid progress of Singapore is due not so much to its agricultural produce as to its geographical position on the great trade route between India and China, combined with the enlightened policy and liberal institutions of its present rulers.

The Straits Settlements have a total area of about 1600 square miles, with a mixed Malay and Chinese population of over 200,000.

5. *Climate.*

Owing to its position between two oceans, and almost entirely within the tropics, the climate of Further India may be described as normally hot, moist, and relaxing, and about the large deltas distinctly enervating and malarious. There is scarcely any cold season except on the northern uplands towards the Tibeto-Chinese frontier, and the rainfall, due mainly to the south-west monsoons, from April to October, ranges from about 90 inches in Singapore to over 200 in parts of Burma. Owing to this abundant rainfall the Irawady is a copious river even before emerging from the last Tibetan highlands, and this, more than any other circumstance, has lent a colour to the theory of its connection with the San-po.

In the Lower Menam and Mekhong valleys, notwithstanding the excessive heat, the atmosphere remains charged with an unusual quantity of moisture both day and night throughout the greater part of the year. To this cause are due the many diseases, such as dysentery and typhoid fevers, which are here endemic. Exposure to the solar rays is also frequently attended by fatal sunstrokes, although the cholera, which is also endemic, commits less ravages than might be expected. The climate of French Cochin-China and Cambodia seems even to have become less fatal to Europeans of late years. Here the rainfall is scarcely more than 54 inches, while the temperature at Saigon averages about 80° F., the extremes being 75° and 85°, as results from the observations recorded between the years 1874 and 1881.

Notwithstanding its proximity to the equator, Malacca would seem to be, on the whole, rather cooler and drier than Indo-China proper. Thus the temperature of the Malacca district at the southern extremity of the peninsula is described as salubrious and equable, the glass ranging between 72° and 85° F. Even in Singapore and the province of Wellesley it is said seldom to rise above 87° or 88° F., and although very damp, the climate of Singapore agrees well with Europeans. On the other hand, that of Pulo-Penang is very oppressive and enervating, with a rainfall varying from 60 to 90 inches. The winter months are here the driest, and the northern winds the coolest and most invigorating.

6. *Flora and Fauna.*

Owing to its hot and moist climate and naturally fertile soil, the vegetation of the Indo-Chinese peninsula scarcely yields in exuberance and variety to that of the neighbouring archipelagoes. A great part of the surface is still everywhere covered with dense primeval forests, in which teak, eaglewood, gum-trees, the gutta-percha plant, bamboo, dye-woods, cardamum, vanilla, and many other useful tropical plants, are found in great abundance. The staple of agriculture is rice, of which vast quantities are produced, especially in the Irawady, Menam, and Mekhong valleys. Other cultivated plants are cotton, tobacco, indigo, the sugar-cane, cloves, cinnamon, coffee, tea, sago, pepper, ginger, besides maize, wheat, and tropical fruits in endless variety and abundance. In Cambodia and Cochin-China the hills are mostly overgrown with wild vanilla, various species of caoutchouc, many oil, resin, gum, and lacquer yielding plants, while the shores of the southern islands and many parts of Malacca are fringed with the cocoa-nut palm.

Of larger wild animals the most common are the elephant, tiger, leopard, wild boar, rhinoceros, and crocodile. The gibbon and other large species of apes, snakes, and birds, abound in all the wooded districts, while the rivers and especially Lake Tonlé-sap teem with every variety of fish. The chief domestic animals are the buffalo, ox, and horse, besides the tame elephant, which, contrary to the generally received opinion, breeds in confinement (*Dr. Harmand*).

7. *Inhabitants: The Kakhyens and Burmese—The Shans, Laos, and Siamese—The Annamese and Cambodians.*

Till recently Indo-China proper was supposed to be

exclusively occupied by peoples of Mongoloid stock, allied in speech to the Tibetan and Chinese branches of that family. But the systematic researches of the French naturalists carried on in the Mekhong valley during the last twenty years have determined the presence in that region of a second ethnical element apparently of Caucasic type, and speaking languages akin to those of the Malayo-Polynesian linguistic family. The Malay race itself has been settled probably from prehistoric times in Malacca, while the true aborigines of this peninsula seem to be a dark race akin to the Negritos of the Eastern Archipelago, and of which a few surviving tribes still linger in the interior of the country. All the inhabitants of Indo-China, taken in its widest sense, are thus reducible to four distinct tribes, as under:[1]—

(*a.*) MONGOLOID STOCK.

Tibeto-Burman Family.	Singfo or Kakhyen	Lenna Kara Lakone Kowrie Khalung Kunung	Kakhyen Hills, North Burma, Tapeng valley, and eastwards to Momein.
	Burmese	. . . .	Irawady basin.
Tai Family.	Shan	Tai-shan Tai-neua	Shan States, North Burma, North Siam, and Yunnan.
	Siamese	. . . .	Menam basin.
	Lao	Lau-pang-kah Lau-pang-dun	Middle course Mekhong River. North and East Siam.
	Tongkinese	. . . .	Tongkin.
	Annamese	. . . .	Cochin-China.
	Unclassified wild tribes	Moi[2] Kha[2] Phnom[2]	Siam, Cambodia, and Annam frontiers.

[1] For the races of British Burma see Chapter VIII. p. 289.

[2] *Moi* is the Annamese, *Kha* the Laotian and Siamese, *Phnom* or *Penom* the Cambodian collective name for all these aboriginal wild tribes, which seem to belong partly to the Mongol, partly to the Caucasic stock.

(*b.*) CAUCASIC STOCK.

Khmer (Cambodians)	Cambodia.
Khmer-dom (Kuy)	Cambodia and Siamese frontier.
Samré	South Lao, near Cambodian frontier.
Charay	Annamese and Cambodian frontier
Stieng	Annamese and Cambodian frontier
Cham	Southernmost districts of Annam.
Chong	South-east Siam, Gulf of Siam.

(*c.*) MALAY STOCK.

(See vol. *Australasia* of this series.)

(*d.*) NEGRITO STOCK.

(See vol. *Australasia* of this series.)

From the recent researches of Jenkins, Sladen, Forbes, and others, in North Burma, it appears that the Singfo and Kakhyens, hitherto regarded as two distinct races, are really one and the same people. Although split up into a great number of small tribes, they everywhere call themselves Singfo—that is, "men,"—and are always spoken of by the Burmese as Khyen or Kakhyen. This is again the same word as Karen, another form of Rakhaing, whence the province of Arakan takes its name. The Singfo claim to be the elder branch of the Burmese family, and although nominally subject to the "wun" or governor of Magong, they pay little heed to his mandates, and on all occasions show contempt and aversion for their "younger brothers," the civilised Burmese. They reach eastwards as far as Momein, and are generally regarded as a savage, unruly, and treacherous race. Major Sladen, however, found them friendly and intelligent, although extremely suspicious of strangers. They are active traders, and would willingly abandon their lawless and predatory habits, were regular commercial relations established across their country between Assam and China. Their religion consists mainly in the worship of good and evil spirits ("nats"), to whom they offer sacrifices. Mountains, valleys, trees, rivers, the sun and moon themselves, are under the influence of these nats, who seem to be sometimes confounded with the spirits of the departed.

On the other hand, the civilised Burmese all profess Buddhism, which in Burma seems to have preserved itself freer than elsewhere from intruding divinities. Here also the monastic vows are more faithfully observed than in other Buddhist lands, and the bonzes have generally promoted the education of the people. A complete national system of public instruction has been developed, all youths being obliged by law to reside for three years in a "khyung" or religious house, where they minister to the "phungys" or priests, and are by them instructed in reading, writing, the elements of arithmetic and of religion. Hence a knowledge of letters is universal in Burma; and here also the women enjoy a remarkable degree of freedom.

In their character the Burmese have much in common with the Chinese. They possess a considerable degree of intelligence and independence, and are shrewd and enterprising, although somewhat indolent. Free from the spirit of caste and national prejudices, they readily acknowledge the superiority of the Europeans, and are eager to learn from them. While extremely tolerant, or rather indifferent, to other religious sects, they remain steadfastly attached to their own tenets. Owing to local maladministration there was a constant migration from Independent to British Burma, until the events of 1885.

The Shans and Laos, who are essentially one race under two names, stand in much the same relation to the Siamese proper that the Talaings and Kakhyens do to the Burmese. Their domain occupies the whole of North Siam and a portion of East Burma, whence it stretches along the Salwin valley far into Yunnan, and down the Mekhong River to the frontier of Cambodia. Their allegiance is thus divided between Burma, China, and Siam. But their ethnical and linguistic affinities are entirely with the Siamese proper, all being so many closely related

BURMESE TYPES.

members of the Tai—that is, "Free" or "Noble" race—which seems to have occupied the Yang-tse basin before the arrival of the Chinese in that region. By the Chinese they were partly absorbed, partly driven southwards to their present homes in Yunnan and Further India. Here many, especially of the Laos tribes, have become intermingled with, and often assimilated to, the Kha, as they collectively call the aboriginal wild tribes of the peninsula.

But the pure Tai stock, which has almost universally adopted Buddhism, is everywhere distinguished by its low stature, light yellow complexion, black hair and eyes, small nose, dilated nostrils, and somewhat dull, unintelligent expression in the eye. The teeth are often dyed an ebony black, while the thick lips acquire a deep-red colour from the universal custom of chewing betel. The Siamese shave the hair of the head, leaving nothing but a tuft on the crown, which is always carefully dressed, especially by the women. The type is on the whole decidedly ugly, although the children are often pretty, and the women retain a certain comeliness till their twentieth year. In the *Land of the White Elephant*, Vincent remarks that outwardly there is little to distinguish the two sexes, both wearing the "languti," or loin cloth of coloured silk or cotton, with an upper garment varying with the season. The women often add a vest or strip of cloth folded across the breast. But rings, charms, earrings, and other jewellery, are reserved chiefly for the children of the upper classes, whose naked bodies are often profusely decked with gold and silver spangles and suchlike trinkets. Mouhot tells us that he saw in Bangkok a royal prince, some six or eight years old, so over-weighted with these objects that he was unable to stir.

Besides betel-chewing, tobacco-smoking is very general, and nearly every one has a cigarette stuck like a clerk's quill behind his ear. The staple food is rice and fish, varied with vegetables, fruits, and spiced soups.

The Lao domain, now subject to Siam, is divided into a great number of provinces, which are ruled partly by hereditary "Kiao," or princes, partly by governors appointed from Bangkok. The present Kiao of Bassac, on the Cambodian frontier, is the last survivor of the old Lao dynasty, which was deposed in 1828 by the Siamese. The national assemblies in the Lao States are usually conducted in the same way as in Siam and Cambodia. Under the Kiao are three dignitaries—the "Opalat," somewhat like the second king of Siam, the "Latsvong," and the "Latsbut." These positions are also held by members of princely families chosen by the Siamese Government, but all other officials are named by the ruling prince.

The development of the Lao States is much retarded by the practice of slave-hunting, which prevails to a far greater extent than is generally supposed. Regular expeditions are organised by the Lao rulers themselves, or by their immediate subordinates. Constant forays are made amongst the wild tribes, and often even amongst the half-caste Lao communities, especially along the banks of the Mekhong, and the captured victims either distributed among the wealthy Lao families, or forwarded in gangs to Korat, to Cambodia, and even to Bangkok, where they are publicly sold as slaves. These facts being apparently unknown in Europe, it may be well to quote the account of the traffic given by Dr. Harmand, an eye-witness of some of the scenes described:—"The brother of the prince of Bassac told me without any reserve that he was about taking a trip to the left bank of the Mekhong in order to hunt down the Khas. It seems that when times are bad the Lao mandarins organise these expeditions against the savages. Under some slight pretext a favourable camping-ground is selected, whence forays are made against the surrounding villages.

"When a sufficient number of all ages and both sexes have been captured, they are bound together and led to

Bassac, Stung-treng, and Attroppeu. Here they are purchased by native, Chinese, and especially Malay traders, who form them into gangs forwarded chiefly to Bangkok, Korat, and Phnom-penh, capital of Cambodia."[1]

The inhabitants of Annam, while mainly of Mongoloid stock, present great differences both physically and mentally. The highlanders are, as a rule, of taller stature, lighter complexion, and ruder habits, than the lowlanders,

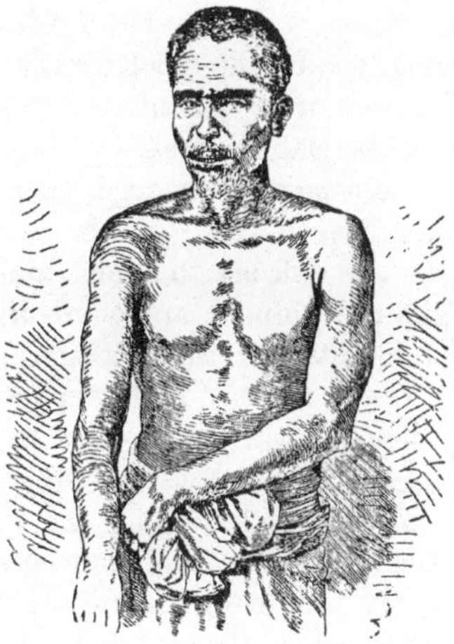

LAOS.

and many of these tribes still lead a nomad life. The civilised Annamese, while more advanced in social culture and even more industrious, are at the same time less truthful, more corrupt and treacherous, than the Moi, as they collectively name the hill tribes.

The Annamese are altogether a thoughtless, pliant, yet brave and lively race. Other accounts are less flat-

[1] "Le Laos et les Populations Sauvages de l'Indo-Chine," in *Tour du Monde*, July 5, 1879.

tering, and H. Bineteau describes them as weak, cowardly, and cruel, addicted to marauding on land and piracy on the high seas. The fairs and markets are very numerous, but their commercial relations are still mainly restricted to the interior, the king having till recently reserved to himself a monopoly of the export trade. And although by the treaty of 1874 with the French Annam is now thrown open to the trade of the world, the Chinese have hitherto derived the greatest benefit from the change. About 1000 Chinamen have settled in Haiphong, one of the chief Treaty ports, where they have monopolised every pursuit requiring skill, perseverance, and commercial acumen. "Lazy as an Annamese" has become a proverb amongst them, apparently with good reason.

Although rice and fish are the chief staples of food, the Annamese, like the Chinese, are omnivorous, greedily devouring snakes, locusts, certain earthworms, rats, dogs, cats, tigers, and other animal flesh repugnant to the European taste. On the other hand, they are remarkably temperate, drinking nothing but tea, or other infusions of aromatic herbs. Yet notwithstanding the great fertility of the soil, and their capacity of supporting existence on the scantiest food, the people of Annam are often visited by frightful famines, which leave whole districts for a time unpeopled.

The Cambodians, whose national name is Khmer, may be taken as the typical representatives of the Caucasic element in Further India, of which they are probably the aboriginal inhabitants. Although now restricted to the valley of the Lower Mekhong, their domain formerly occupied a large portion of the peninsula, where they are still represented in a fragmentary way by the Kuys, Charays, Chongs, Stiengs, and other tribes scattered over the hills and forests of East Siam and West Annam.

But next to the Cambodians proper, by far the most important section of the race are the Kuys, or Khmer-dom, —that is, "Primitive Khmers"—who are divided into several branches, occupying an extensive region round the northern and eastern shores of Lake Tonlé-sap. Here was the original centre of Cambodian culture, the magnificent remains of which now lie mostly buried amidst a rank vegetation of a thousand years' growth. The human figures sculptured on these Buddhist monuments, many of which must be some twenty centuries old, are not only of regular Caucasic type, but are often a faithful reflex of the features, dress, and ornaments of the present Khmer populations.

Some of these wild tribes are still distinguished by a gentle disposition, a certain innate politeness and courtesy, as well as a surprising artistic taste and skill lavished on their dress, ornaments, pipes, quivers, and other objects. These traits may well be the faint reflection of a now extinguished culture still cherished by these children of nature, lost for ages amid their dense woodlands, which they believe to be the centre of the universe, and which nothing can ever induce them to leave (*Mouhot*). But the Cambodians themselves seem to have retained little of their former greatness, except an overweening pride and arrogance. They are being gradually absorbed by the surrounding Annamese and Laos populations. A strange mystery hangs over this Cambodian race, which, fully 2000 years ago, built cities and raised monuments amid the swamps of Tonlé-sap, vying in size and grandeur with those of the Mesopotamian and Nile valleys. Their culture is certainly of Buddhist origin. But by what channel did Hindu influences penetrate to this remote corner of the continent? And whence came the race itself, with its European features and polysyllabic speech, totally distinct from that of the surrounding Mongoloid peoples, but showing marked affinities to the Oceanic linguistic groups? Lastly, by what barbaric hordes was their development arrested, their culture extinguished, their cities wasted, their stupendous shrines and monuments left to crumble in the midst of the exuberant tropical vegetation of Angkor and Battambang? These are questions which cannot yet be answered.

8. *Topography: Mandalay—Bhamo—Bangkok—Ayuthia—Luang-Prabang—Hué—Haiphong—Phnom-penh Saigon—Singapore.*

As might be expected, the few large towns of Burma all lie in the Irawady valley, which is distinguished both for its picturesque scenery and great fertility. Here is Mandalay, one of the many places to which the seat of government had in recent times been shifted, partly in consequence of revolutions or changes of dynasty, partly through superstitious motives or royal whims. Ava, which had been the capital since 1364, gave place about 1740 to Mutshob, which yielded in 1782 to Amarapura. In 1819 the Court returned to Ava, whence it again passed in 1837 to Amarapura, and in 1857 to Mandalay, where it has since remained. This place lies a few miles above Amarapura, on the same side of the Irawady. The houses on the river and in the suburbs are of the usual Burmese type—frail structures of bamboo and matting erected on piles sunk in the mud. They are generally small and packed closely together, and their materials are inflammable. Consequently fires once breaking out become extremely disastrous. The main streets are lined mostly with brick houses, but even here the bricks are merely attached to the wooden framework. A pretty effect is presented by the Chinese shops, which are often two stories high. In every direction the eye lights on gilded or painted pagodas, temples, and "Khyungs," or cloisters with schools attached. The city is encircled by a stout brick wall, with a ditch crossed by one or two rude bridges. Here was the scene of the horrible butcheries which accompanied the accession of the late King Thebaw in 1879. Mandalay was occupied by the British under General Prendergast on November 28, 1885.

On the Upper Irawady lies the important station of Bhamo, in 24° 15′ N. at the junction of the Tapeng. This is the starting point of caravans proceeding eastwards to Yunnan, and should a regular overland trade be established between British India and West China, Bhamo must from its position become one of the great emporiums of the East. At present it is a small stockaded town with a few hundred houses, occupied chiefly by Chinese and Shan traders, with a few Burmese officials.

Bangkok, the present capital of Siam, lies near the mouth of the Menam a few miles below Ayuthia, the former seat of government. Bangkok presents a pleasant prospect from the water, whence a panoramic view is commanded of its glittering palaces and temples surrounded by dense masses of a gorgeous tropical vegetation. But the charm is soon dispelled by a closer inspection of the place itself, which consists of a confused aggregate of narrow, muddy lanes, stagnant canals, and wretched hovels, occupied by a mixed Siamese and Chinese population, estimated at from 400,000 to 500,000. As in so many other parts of Further India, the Chinese have here almost monopolised the local trade, while the foreign exchanges are mostly in the hands of English merchants.

Beyond Ayuthia, the site of which is marked by some magnificent Buddhist ruins and a large elephant park, there are no towns of any consequence until we reach Luang-Prabang on the Mekhong. Even this place has a population of little more than 10,000 although credited by Bishop Pallegoix with 80,000. A simple monument marks the spot where the naturalist Henri Mouhot was carried off by fever in 1861.

Hué, capital of Annam, occupies an important strategical position at the mouth of the River Hué on

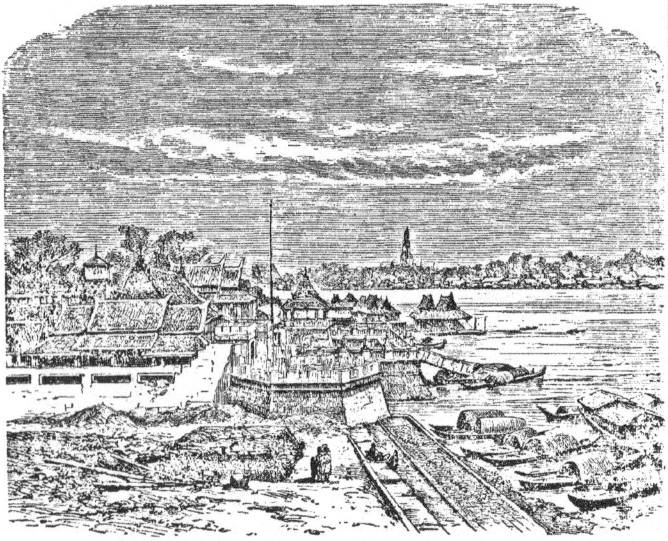

BANGKOK.

the Cochin-Chinese coast. It has been strongly fortified with skilfully-planned outer and inner lines by French engineers. With its extensive arsenal, magazines, walls 60 feet high, and moat 90 feet broad, Hué ranks as one of the strongest military positions in Asia. The king resides in a spacious and fortified palace; but all these defensive works have not prevented him from sinking to the position of a vassal of France, in whose favour he signed a treaty in 1874 practically accepting the protectorate of that country.

In Tongkin the chief Treaty port is Haiphong on the Lower Song-ka, which is navigable to this point by vessels drawing 8 to 10 feet. At Haiphong the French have a naval station for the ostensible purpose of suppressing piracy on the high seas, but which also serves to keep the country in subjection to the mistress of the Mekhong delta. Kesho, capital of Tongkin, which lies some miles above Haiphong, is accessible to large junks drawing from 4 to 6 feet.

Cambodia has recently shifted its capital from Udong to the neighbouring Phnom-penh, which is more conveniently situated at the junction of all the navigable waters in the kingdom.

The French have chosen for the capital of their Cochin-Chinese settlements the town of Gia-dinh, renamed Saigon, picturesquely and conveniently situated on the Dong-nai. Although lying at some distance from the sea, Saigon is accessible to large vessels, and is much frequented by English and Chinese traders engaged in the foreign and coast trade of Cambodia and Lower Cochin-China. The neighbouring island of Pulo Condor is well adapted to become a depot for the produce of the surrounding districts. The staples of the export trade are cotton, pepper, raw silk, tobacco, betel leaves, sugar, gums, hides, horns, and fish.

In Siamese Malacca the chief place is Toneah on the island of Junk Seylon, which is said to have a population of 30,000. Of the towns in British Malacca the largest are—Perak, capital of the protected State of that name; Malacca, on the south-west coast of the peninsula to which it gives its name; Penang, on the same coast;

and Singapore, unquestionably the most important place in the whole of Indo-China. Notwithstanding its proximity to the equator, the "Lion City," as its name means, enjoys a fairly healthy climate, and this British free port has in the course of sixty years been

FRUIT-SELLER, SINGAPORE.

transformed from an almost uninhabited rock into one of the great centres of Eastern trade. Very few places in the old or new world present such an example of rapid development as does the city of Singapore, which now counts its population by many tens of thousands and its exchanges by millions. It possesses a magnificent harbour, well sheltered and of easy access, lying at the converging point of the great trade routes between India, China, Japan, the Dutch East Indies, and Australia. Singapore, which is well laid out, with a Chinese, Malay, and European quarter, is the capital of all the Straits Settlements, with a governor's residence, Protestant cathedral, hospital, schools, and several benevolent institutions.

9. *Highways of Communication.*

In Further India there are nowhere any regular roads, and most of the trade routes follow the course of the great rivers and their affluents. Thus travellers and explorers wishing to penetrate from the west into China, ascend the Irawady to Bhamo, where they follow the caravan route up the Tapeng River valley through the Kakhyen highlands to Manwyne, and so on to Momien on the Yunnan frontier. This route has been frequently traversed in recent times, and is now reported by the Rev. Mr. Soltau (1881) to be safer than at any former period.

In the same way the chief highway to the interior of Siam follows the course of the Menam from Bangkok to Ayuthia and Prabat. Here it trends east over the Dong Phya Phai range to Korat, whence it runs due north through the Laos country to Luang-Prabang on the Mekhong. Bangkok is also connected with Cambodia and the Lower Mekhong by a track running eastwards through the Shong country, and over the Coast Range to Battambang and thence along the south side of the Great Lake to Udong and Phnom-penh. From Battambang a branch runs round the west end of the lake to Angkor, thus giving access to most of the magnificent ruins of the ancient Cambodian Empire.

In Annam all the chief towns are connected by a coast route running from Kesho southwards through

Thanh-hoa-noi, Koang-tri, Hué, Kwang-nai, Quin-hon, and Binh Thuan, to Saigon. From Kesho a track runs also west to the Mekhong at Kiang-kheng above Luang-Prabang, and thence across the Siamese Shan States, through Kiangmai to the Lower Salwin valley and Martaban.

10. *Administration.*

The Burmese Government was a pure despotism, his majesty of "the Golden Feet," ruling as an absolute monarch. Even the British Envoys were required to appear barefooted [1] in the presence of these despots, whose chief characteristics have too often been cruelty, licentiousness, and arrogance. Hence while court etiquette was rigidly maintained, the progress of the country was retarded by the policy of its rulers. Anarchy and decay were the prevailing features of the regions subject to the influence of the central government, while the more remote Shan States enjoyed a comparative degree of peace and immunity. The resources of the land were also further drained by the steady migration of the more intelligent and wealthier classes, who were glad to find a refuge in the neighbouring province of British Burma.

The Siamese and kindred peoples form the only nation that has two kings, although this old custom has at present lost all its significance. The "Wangna," or second king, usually chosen from amongst the nearest relatives of the first king, resides quietly in his palace without at all interfering in State affairs.

Siam itself is an electorate, the succession being determined by the nobles, who, however, are bound to elect a "Chao Fa"—that is, a prince of royal blood—on both

[1] The great "Shoe Question," which never was settled, is fully discussed in Col. Laurie's new work on *Ashé Pyee; or, The Superior Country.* London. 1882.

sides. Hence to secure a posterity capable of succeeding to the throne, the first king is bound to select a consort from amongst the daughters of the second reigning king, or of some former first or second king, or at least a princess from the ancient royal families of the Lao States. But in the latter case some of the public jurists doubt whether the issue of such alliances are entitled to the succession.

In theory all the inhabitants of the land stand in the relation of serfs or slaves to their sovereign. At the same time, there exist certain substantial class distinctions denoted by special tattoo marks on the wrist, all being so branded except the nobles and officials, with their families. In the artisan classes the son is compelled to follow his father's trade. During war military service is of universal obligation, but in peace the army is recruited from the working corporations under the control of the war office. But notwithstanding the ancient traditions of oriental despotism, Siam has made considerable social progress under the enlightened administration of the reigning sovereign. Free scope has been given to the action of European influences, and while Buddhism remains the State religion, Christianity is allowed to be preached and practised without any restrictions. To Bishop Pallegoix, for many years head of the French missions in Siam, science is indebted for some of the most valuable contributions to the history and ethnology of that State.

The Government of Annam is so far absolute that the whole authority is centred in the king or emperor, and emanates from him. The sovereign has his privy council, besides a ministry for the administration of justice, a war office, a ministry of public worship, a board of works, and a home and foreign office. But all this machinery of government has lost its significance since Annam has, by the treaties of 1874 and 1885 become a dependent State of France.

The ancient kingdom of Cambodia in 1864 accepted the French protectorate. The two provinces of Battambang and Angkor, seized by the Siamese at the end of the last century, but the occupation of which had never been acknowledged by Cambodia, were finally ceded to Siam by the treaty of 1867, when the new boundary-line was fixed between the two States. Thus were separated from Cambodia the two historical provinces, in which are situated the ancient capital, Angkor, and most of the grand monuments of former Cambodian culture.

The king of Cambodia is in theory absolute master of the life and property of his subjects, as well as of all the land. The only check to his despotism are certain long-established usages, combined with the fear of causing troubles which might require the intervention of the French. The nobles and officials are divided into four classes, at the head of which are the prime minister (*Shanfea*), the ministers of justice, finance, and trade. For administrative purposes the kingdom is divided into five *dey* or departments, and fifty-six *khet* or districts, the dey being governed by the ministers of state, the khet by inferior mandarins named by the king. Under the governor are a *balat* or lieutenant, two *snâng* or prefects, besides other smaller functionaries, and the *me srok* or village magistrate.

Formerly attached to British India, the Straits Settlements now form a Crown colony, with a separate administration vested in a Governor at Singapore, a Lieutenant-Governor at Penang, and a Resident Councillor at Malacca, appointed by the Queen. The Governor is assisted by Executive and Legislative Councils, of which the Lieutenant-Governor and the Resident Councillor, above mentioned, are members. The lieutenant-governors of the several settlements are also members of these bodies. Seat of government, Singapore.

INDO CHINA AND MALACCA.

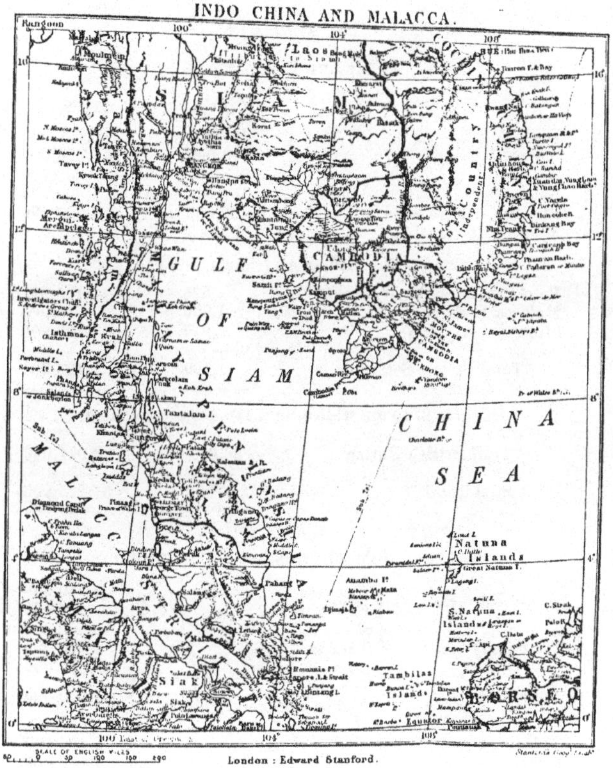

London : Edward Stanford.

11. *Statistics.*

Areas and Populations.

	Area in sq. miles.	Population.
Burma	190,000	?5,000,000
Siam	310,000	?6,000,000
Annam	200,000	?20,000,000
Cambodia	40,000	1,000,000
French Cochin-China . .	22,000	1,700,000
Malacca States	32,000	?250,000
Straits Settlements (census 1881).		
Pulo-Penang, with Wellesley .	325	190,597
Malacca District. . . .	1,000	78,000
Singapore	275	139,208
Total Straits Settlements .	1,600	423,384

Of these 2769 are whites, and 136,439 are coloured.

Add British Burma .	88,000	3,011,000
Grand Total . . .	883,000	37,275,000

Approximate Population according to Races.

Tibeto-Burmese	Burmese	4,750,000
	Kakhyens	250,000
Tai Family	Shans	1,500,000
	Siamese	2,000,000
	Laos	1,500,000
Annamese	Tongkinese . . . }	20,000,000
	Cochin-Chinese . . . }	
Moi, Kha, or Penom Wild Tribes		?500,000
Chinese		2,000,000
Malays		1,000,000
Negritos		?

Chief Towns.

Burma . . .	Mandalay	100,000
	Bhamo	5,000
Siam . . .	Bangkok	500,000
	Ayuthia	30,000
	Chantabun	30,000
	Luang-Prabang . . .	12,000
Annam . . .	Hué	100,000
	Haiphong	...
	Kesho (Ha-noi) . . .	100,000

French Cochin-China and Cambodia	Saigon	50,000
	Phnom-penh	...
	Udong	12,000
	Kamput	2,000
Malacca and Straits Settlements	Toneah	30,000
	Singapore	60,000
	Penang	...
	George Town	...
	Malacca	13,000

Siam.

Shipping of Bangkok (1884), 241,149 tons.
Mercantile Marine—58 vessels of 40,000 tons.
Navy—14 steamers of 5815 tons and 51 guns.
Revenue (1884), £2,000,000.

(Bangkok.)

Year.	Exports.	Imports.
1884	£2,052,348	£1,145,447

Cambodia.

Average Revenue	£130,000
Average Exports	200,000

French Cochin-China.

Average Imports, £2,750,000; Exports, £3,500,000.

Straits Settlements.

Revenue (1884)	£629,921
,, (1885, estimated)	653,612

Year.	Imports.	Exports.
1884	£20,043,903	£18,673,846

Shipping, 3,634,174 tons, besides 266,594 tons of native craft.

APPENDIX.

ETHNOLOGY AND PHILOLOGY OF THE ASIATIC RACES.

1. *Two Fundamental Physical Types—Mongolic and Caucasic.*

ASIA, the *officina gentium*, or mother of nations, as it has been called, claims this title rather on account of its teeming multitudes than on the ground of the number of distinct types to which it may have given birth. For if we exclude the dark aborigines of Malacca and the Deccan, two only of the five or six main varieties of the human family would appear to have originated in this continent. The inhabitants of Mongolia and Caucasia are commonly assumed to be typical specimens of these two varieties, whence the collective terms Mongolic and Caucasic, by which they are usually designated.[1] Many objections have been made to these, as there have against most other scientific names. But they are now so well established, and their meaning is so thoroughly understood, that the advantages to be derived from any change are not obvious. For Caucasic recent German writers frequently use the geographical expression "Mediterranean," so as to embrace the peoples settled round the shores of the great inland sea, all of whom belong to this variety. But although acceptable as an alternative, Mediterranean cannot be taken as a substitute for Caucasic, because it would exclude the many branches of the race which have from the remotest times occupied regions far removed from the Mediterranean. More convenient alternative terms are "Yellow" and "Fair," as indicating the prevailing colours of the two types. But these are by no means constant factors, the yellow Mongul often merging in light and brown shades scarcely to be distinguished from ordinary

1 The forms *Mongolic* and *Caucasic* are here adopted as the general names of the types in contradistinction to *Mongolian* and *Caucasian*, the ethnical names of two branches of those types.

European complexions, while the fair occasionally assumes very decidedly dark and even black tones.

2. *Diverse Interminglings.*

These modifications are due to many causes, prominent amongst which are climate, diet, social habits, and, above all, the intermingling of the two original stocks, which has been going on all over the continent for untold generations. South-Western Asia, including parts of India, the Iranian plateau, and the peninsulas of Anatolia and Arabia, are doubtless assumed to be the proper domain of the Caucasic, the rest of the continent that of the Mongolic type. But modern research has already exploded this idea, and the primordial diffusion of Caucasic elements throughout the remotest eastern and northern regions, notably Further India, Japan, Korea, Manchuria, and the Altai uplands, seems to be now fairly well established. All that can therefore be said with any confidence is that the east and the west were probably the respective original homes of the Mongolic and Caucasic stocks. If originally one, where they have become differentiated and whence dispersed it would be rash even to conjecture. We can but deal with facts, and it is a fair conclusion of modern science that the two primitive stocks have existed in more or less close proximity and contact with each other for many tens of thousands of years. History merely scratches the surface, and goes back through Akkadian, Semitic, Aryan, and Chinese echoes of the dim past, at most some 7000 or 8000 years. But even then Asia was already peopled from Arctic to Indian Ocean, from Sarmatia to the Pacific seaboard, much as it now is, not merely by two sharply-defined types, but by a hundred fully-developed modifications of those fundamental types.

Beyond these faint echoes history is absolutely silent. But all the more eloquent are the shell mounds, the rude palæolithic and other humble remains of primeval man, which in Asia as elsewhere associate him with the flora and fauna of the early quaternary epoch. During the long intervening ages till the dawn of history there has been ample time for the countless migrations, crossings, and overlappings, by which the fair and yellow elements have become endlessly modified and diffused throughout the whole continent, the adjacent Oceanic islands, the western peninsula of Europe, and the northern section of Africa. For all these regions belong from prehistoric times altogether or in part to the Asiatic ethnical domain. But while the Mongolic type has largely prevailed in Asia itself, the Caucasic has elsewhere been most widely diffused.

Beyond the limits of the continent the Mongol element is mainly represented by the Finno-Tatars of Europe and the Malays of the Eastern Archipelago and Madagascar, whereas the Caucasic stock is represented by the Aryans of the greater part of Europe, by the Hamites and Semites of North and North-East Africa, by the Indonesians of Malaysia, and by the large brown Polynesians of the Pacific Ocean.[1] But within the limits of the continent nearly the whole of the Chinese Empire, Indo-China, and Siberia, besides parts of India, Irania, and Turkestán, are occupied by various branches of the Mongol stock, while the Caucasic is restricted chiefly to the south-west, to parts of India, and to scattered enclaves in Further India and Japan.

3. *Their Distribution.*

At the same time, all these regions can only in a relative sense be spoken of as the respective domains of the rival stocks. It is correct to say that there are more Mongol elements in the east, more Caucasic in the west; but both are now found almost everywhere, interpenetrating each other to such an extent that few, if any, districts have remained in the exclusive occupation of either. Tolerably pure Caucasic blood may still survive in the Arabian peninsula and in parts of the Caucasus, while most of the Tibetan plateau probably still remains in the possession of its original Mongol inhabitants. But, speaking generally, the two types have in Asia become nearly everywhere more or less intermingled, and consequently modified, so that it often becomes difficult to say where one begins and the other ends. Owing to this confusion—the inevitable result of causes operating throughout long ages—some modern anthropologists have felt disposed to call in question the original distinction between the two fundamental types, and M. A. Hovelacque has even gone so far as to deny the reality of the Mongol type.[2] But Hovelacque merely succeeds in proving what is here insisted upon, that the Asiatic peoples are very much mixed, and that it is consequently difficult anywhere to find pure ideal specimens of the original stocks.

4. *Their Salient Features contrasted.*

The salient features of such ideal specimens are nevertheless fairly well established, and will be found contrasted in the subjoined comparative table.

1 See the Ethnological Appendices to the *Africa* and *Australasia* of this series.

2 "Le Type Mongolique," in *Revue Internationale des Sciences*, 1878.

	Ideal Mongolic Type.	Ideal Caucasic Type.[1]
Shape of Head	Normally brachycephalic, *i.e.* round horizontally.	Normally dolichocephalic, *i.e.* long horizontally.
Facial Angle	Prognathous, index Nos. 76 to 68.	Orthognathous, index Nos. 82 to 76.
Features	Square, angular, and flattened.	Rounded off and oval.
Cranial Capacity	1200 to 1300 cubic centimetres.	1300 to 1400 cubic centimetres.
Cheek-Bones	High and prominent.	Low and inconspicuous.
Ears	Large and standing out from the head.	Small, well formed, close to the head.
Mouth	Large with thick lips.	Small, with bright-red, moderately-thin lips.
Nose	Broad, flat, short, and somewhat concave.	Long, narrow, high, straight or somewhat convex—tip projecting beyond the nostrils.
Forehead	Low, receding, narrow.	Straight, broad below, fully developed.
Eye	Small, almond-shaped, oblique upwards and outwards, orbits wide apart, iris black.	Large, round, straight; orbits rather close set; iris normally blue or gray, but very variable.
Chin	Very small and receding.	Full and slightly projecting.
Neck	Short and thickset.	Long, slender, and shapely.
Figure	Squat, angular, heavy, muscular, inclining to obesity.	Symmetrical, slim, active, robust.
Hands and Feet	Disproportionately small.	Medium sized or large.
Stature	Below the average—5 ft. to 5 ft. 4 in.	Medium or above the average—5 ft. 4 in. to 5 ft. 9 in.
Complexion	Pale-yellowish, tawny or olive, inclining to a leathery-brown and cinnamon—no red or ruddy tinge.	Fair or white, inclining to brown and swarthy—normally with ruddy tinge.
Hair	Dull-black, long, coarse, stiff, and lank, cylindrical in section.	Long, wavy, and normally light brown, but very variable—glossy jet-black, flaxen, red, etc.; elliptical in section.

[1] Of the Caucasic there are no doubt two well-marked varieties—a dark brachycephalic and a fair dolichocephalic; but the latter is here taken as, perhaps, the most typical. The dark seems to be the result partly of climate partly of intermixture with prehistoric elements dating back to the Stone Age; but these elements, connected by some writers with the Lapp and Eskimo stock, have not been strong enough to destroy the fundamental unity of the Caucasic type. The men of Caucasic stock, whether Berbers of North Africa, Basques, Bretons, or Teutons of Europe, Arabian Semites, or Iranian Tajiks, can always be recognised and distinguished from any other by the possession of a greater average proportion of the characteristic qualities specified in this table, and race is on the whole a question of averages.

Beard . .	Very scanty or absent.	Full, bushy, often very long.
Eyebrows .	Straight and scanty.	Arched and full.
Expression .	Heavy, inanimate, monotonously uniform.	Bright, intelligent, infinitely varied.
Temperament	Dull, taciturn, morose, lethargic, but fitfully vehement.	Energetic, restless, fiery, and poetic.

These details can, however, be regarded as little more than broad generalisations approximately indicating either what the first Mongol and Caucasic men were like, or what the respective races tended to become. In reality they have long ceased to be pure races, and are now merely physical, political, and social amalgams properly called peoples and nations rather than races. The characteristics here specified may at most supply an anthropological basis for our classifications. But the classifications themselves must for the present be considered rather as ethnographic than strictly scientific.

5. *Numerous Fundamental Linguistic Types.*

Some help in determining the ethnical affinities of the Asiatic peoples is no doubt occasionally afforded by a consideration of the various languages current amongst them. But this feature cannot be relied upon to any great extent, and is in many cases a fruitful source of confusion and embarrassment. A very curious problem is suggested by the fact that to the two fundamentally distinct physical stocks correspond not two, but probably as many as thirty fundamentally distinct linguistic stocks. Thus the various branches of the Caucasic family, all comparatively slight varieties of one original type, are grouped in six or eight separate divisions, according to the six or eight radical forms of speech, such as the Aryan, Semitic, Georgian, etc., spoken by them. In the same way the Mongolic ethnical group is split up into the Ural-Altaic, Annamitico-Chinese, Tibeto-Burman, and other radical linguistic groups. It will be asked, how does it happen that there are so many more radical forms of speech than there are radical physical types? If the Caucasic peoples are genetically one—that is, had a common starting-point—how comes it that their Aryan, Semitic, and other tongues are not also genetically one—that is, reducible to a common stem?

There seem to be only two possible solutions to this problem. Either these radical forms of speech have been imposed upon the various members of the Mongolic and Caucasic stocks by other races that have long since perished, or else they have been independently evolved or developed during the course of ages by the several branches of the Mongolic and Caucasic races themselves. The first supposition must be rejected both as unnecessary and to the last degree improbable, and even absurd. For it seems preposterous that we should have to invent a fresh type of mankind for every fresh philological crux that may be discovered by modern linguistic science. To the second view it may be objected that the different forms of speech evolved by the various branches of the Mongolic and Caucasic families must still have had two common starting-points, just as the branches themselves had two common starting-points. Hence all should still possess some of the original elements of speech common to all.

Some etymologists pretend to discover these original elements in certain fanciful verbal resemblances which they detect in the vocabularies of the Georgian, Cambodian, Japanese, Nubian, Ashanti, Algonquin, Bribri, Peruvian, and other languages, separated from each other by vast spaces and countless ages. But the methods of these writers are unscientific, and their conclusions to the last degree inconsequent. Modern comparative philology has already given up the idle search for the traces of a primeval universal speech, and Frederic Müller endeavours to meet the difficulty by supposing that his twelve main divisions of mankind were differentiated prior to the development of their languages. But such a violent supposition could be accepted only on the ground that to the twelve ethnical correspond twelve linguistic divisions, which is far from being the case.

Nevertheless, the phenomenon of two or more radical forms of speech current within the same ethnical group seems to admit of an obvious explanation. It is due to the more or less evanescent nature of speech as compared with the relatively more persistent character of physical types. After dispersion from some common centre the various branches retain for an indefinite period the marks of their common ancestry. But the elements of their common speech, still in a somewhat infantile state, would in course of time gradually become obliterated, even developing new morphological forms until all traces of a common parentage are irrecoverably lost. No doubt some languages, like the Semitic, display an amazing vitality, resisting all serious change for ages, or departing but slightly from the original form, even after the race itself has developed several well-marked varieties. But in such cases the physical changes are due partly to great differences in the outward surroundings, partly to intermixture with other races, while the language undergoes little modification because already fully developed, and, so to say, fixed in a definite groove prior to the dispersion.

6. *Interchange of Physical and Linguistic Types.*

But it often happens that through conquest, migration, and other causes, whole peoples may be compelled to change their language in the course of ages several times over. Others, again, may have become profoundly modified in physique by interminglings, while still retaining their original speech in comparative purity. Notable instances are many of the Túrki peoples, originally of Mongol type, but by constant alliances now largely assimilated in appearance to the Caucasic type, while still speaking Túrki dialects. The reverse process is seen in the case of the Hazaras and Aymaks of North Afghanistán, who retain their original Mongol features although they have long since exchanged their Mongol speech for a Persian dialect. Hence neither speech nor appearance can always be regarded as a safe test of origin, and as it is now impossible to say how often such changes and shiftings may have taken place in prehistoric times, it is obvious that a complete genetical classification of the Asiatic peoples is no longer possible. In the subjoined scheme they are merely grouped tentatively on the basis of their most probable physical and linguistic affinities. Where language and features correspond in more or less isolated areas, as in the Arabian peninsula and the Tibetan plateau, we shall have probably a near approach to the truth; but the classification can otherwise lay claim to little scientific value. As already stated, it deals rather with peoples and nations than with races —that is, with groups clearly connected by consanguinity.

7. *Classification of the Asiatic Races.*

On the other hand, the affiliation of the various members of the linguistic groups is never doubtful. However mixed their vocabularies may be, their inner structure always resists the action of outward influences. Hence it is that we can speak with far greater confidence of linguistic than of ethnical families; and while the expression "Aryan race," for instance, will convey little meaning to the anthropologist, the expression "Aryan language" will be perfectly intelligible to the philologist. For this reason the organic Asiatic languages here occupy the first column to the left, and to them are referred the various races and branches of the second and third columns.

The grouping thus becomes largely linguistic, an arrangement which, if less satisfactory from the scientific point of view, has at least the advantage of stating definite facts. For the classification of languages stands on a totally different and far more solid basis than that of races. The organic languages of mankind may be regarded as practically so many distinct species, fixed though not immutable, absolutely independent of each other, and, like all true species in the animal and vegetable kingdoms, incapable of mutual combination to produce new species. But "with the varieties of man it is otherwise. They have never so far separated as to answer to the physiological definition of species. All races are fertile with one another, though perhaps in different degrees. Hence new varieties have constantly been formed, not only by segmentation, as it were, of a portion of one of the old stocks, but also by various combinations of those already established." [1] A striking illustration of this important truth is afforded by the subjoined scheme. Here none of the stock languages in the first column have any clear or essential relation to each other. They differ in their mechanism, and to a large extent even in their vocabularies, as profoundly as if all had sprung from independent centres. They are true species, which refuse to amalgamate, and thus form new species, so that fresh varieties are developed only within each separate stock language. But the races of the other columns are all mere physiological varieties of one common species, and are consequently constantly recombining, and thus giving rise to fresh modifications in endless variety. Many Mongolo-Caucasic mixed varieties, such as the Anatolian Turks, the Uzbegs, and Tajiks of Turkestán, have in this way been already developed. But no such phenomenon as, for instance, a Semitico-Aryan, or a Malayo-Taïc mixed language—mixed, that is, in its inner structure—is known to the science of comparative philology.[2]

8. *Ethnical and Linguistic Scheme.*

I. Mongolic or Yellow Type.

Stock Languages.	Races.	Main Divisions.
1. Tibeto-Burman	Tibetans	Bod-pa.
		Tangutan.
		Si-fan.
		Himalayan tribes.
		North Assamese tribes.
	Burmese	Burmese
		Kakhyen.
		Arakanese.
2. Khasi		South Assamese tribes.
3. Mon		Khasia tribes.
		Talaings of Pegu

[1] Prof. W. H. Flower's Address to the Department of Anthropology, British Association, September 1881.

[2] The Pehlevi, a curious mixture of Aryan and Semitic forms, might be supposed to be an exception. But Pehlevi was not a natural growth, but an artificial creation formed at the Court of the Sassanides, and entirely restricted to the learned classes. It was never a spoken language in the ordinary sense of the term.

Stock Languages.	Races.	Main Divisions.
4. Tai	Tai	Siamese. Shan. Lao. Ahom.
5. Sinico-Annamitic	Chinese Annamese	Chinese. Tongkinese. Cochin-Chinese.
6. Koreo-Japanese	Koreans. Japanese. Lu-Chu.	
7. Ural-Altaic	Finno-Tatars	Mongolian. Tungus and Manchu. Túrki. Samoyede. Ugrian.
8. Malayan	Malays	Malay. Formosan.

II. Caucasic or Fair Type.

Stock Languages.	Races.	Main Divisions.
9. Kartveli	Caucasians	Georgian and Mingrelian. Svan, Khevsur, Pshav, and Laz.
10. Cherkess	Caucasians	Circassian. Abkhasian. Kabard.
11. Chechenz	Caucasians	Chechenz.
12. Lesghian	Caucasians	Daghestán tribes.
13. Aryan	Iranians	Tajik Baluch. Kurd. Ossetian. Armenian. Afghan.
	Galchas	Zarafshán. Wakhi. Siah-Posh Kafir.
	Hindus	Panjábi. Hindi. Bengali. Assamese. Nepalese. Oriya. Marathi.
14. Semitic	Semites	Assyrian. Aramæan. Hebræo-Phœnician. Arab. Himyaritic and Abyssinian.

III. Races and Languages of Doubtful Affinities.

15. Brahui of Baluchistán.
16. Dravidian } of the Deccan. Subdivisions at p. 287.
17. Kolarian }
18. Sinhalese of Ceylon.
19. Khmer of Cambodia. Subdivisions at p. 672.
20. Aino of Yesso and Sakhalin.
21. Chukchis and Koriaks }
22. Yukaghirs } Hyperboreans of North-east Siberia.
23. Kamchadales }
24. Giliaks }
25. Aborigines of South-West China. Subdivisions at p. 572.
26. Negritos of Andaman Islands and Malacca.

9. *The Races of Isolating Speech.*

The first five languages in this table form the so-called "Monosyllabic," or "Isolating" family, which differs in some important respects from all other forms of speech. It is supposed to represent the most primitive type, consisting of crude monosyllabic roots, which are incapable of any modification of form to express the various grammatical relations. Each root is a term, which stands apart (whence the expression "Isolating"), and which acquires its definite meaning solely in virtue of its position in the sentence. Thus in Chinese *ngò tà ni* = I strike you : but *ni tà ngò* = you strike me. It is a mistake, however, to suppose that the roots are necessarily monosyllabic, for modern research has shown that most of them were originally in fact polysyllabic, and have been reduced to their present monosyllabic form by a slow process of phonetic decay.[1] Thus the Chinese *i* = to doubt, represents an original *tadaka*, a word of three syllables (*Terrien de la Couperie*).

From this it appears that, so far from representing a primitive condition of human speech, these languages are in an advanced state of phonetic corruption, by which their roots have been reduced to mere fragments of more ancient and fuller forms. The result has been an embarrassing number of homophones, the various meanings of which can be determined only by the tone with which they are uttered. Hence the introduction and gradual development of the regular tonic system, by which the isolating are distinguished from all other languages. It is as if the English

[1] The most recent school of philology seems to have abandoned the old idea that all roots are necessarily monosyllabic originally. The monosyllabic state is now, in fact, regarded as rather the result of phonetic decay than a primeval condition of speech.

words *bad*, *back*, *bag*, *ban*, *bar*, *bat*, were all reduced by phonetic decay to the common form *ba*, pronounced with six different tones corresponding to its six different original meanings. The best collective name of this group, spoken by fully one-fourth of mankind, would therefore seem to be the "Isolating Toned Family."

But although thus constituting a distinct morphological order of speech, it must not be supposed that these idioms are otherwise connected by the ties of descent from a common mother-tongue. At least all traces of such community of origin have long vanished, and within the group there are certainly five if not more that must be regarded as practically stock languages. No amount of ingenuity can now reduce the Tibetan, Khasi, Siamese, and Chinese [1] roots to one source. Tibetan differs even in structure from the other members of the family, and has already developed a number of grammatical forms, in virtue of which it occupies a peculiarly interesting position intermediate between the isolating and agglutinating states.

Nearly all the isolating tongues are written with syllabic letters derived either directly from the Devanágari, or from the Pali form of that system introduced into Indo-China by the Buddhist missionaries probably 2000 years ago. But the Sinico-Annamitic still adheres to the cumbrous ideographic writing supposed to be of Chinese invention. Although at present and originally consisting of pure ideographs, or signs representing not sound but ideas, a large number of true phonetic signs was developed at a very early date, and it is now ascertained that the Chinese was a mixed ideographic and phonetic system, like the Egyptian, down to the twelfth century B.C. But about that time began the tendency to discard the phonetic signs, and this tendency was carried to its logical conclusion in 820 B.C., when the whole system was recast on the present purely ideographic basis. This was undoubtedly a retrograde movement; but it was necessitated by the failure to work out a consistent alphabetic system applicable to the whole empire. Being unintelligible to the peoples unacquainted with the official language, the phonetic signs became a source of extreme embarrassment, and had consequently to be given up. The spread of empire thus brought about the backward step to ideographic writing, which probably more than any other cause has retarded the natural development of the Chinese intellect and the material progress of the race.

[1] Probably one-half of the Chinese roots are common to the Tai (Shan, Laos, Siamese, etc.); but these form no part of the original Chinese stock roots, having been borrowed from the aboriginal Shan populations of the Yang-tse-kiang basin.

10. *The Races of Agglutinating Speech.*

To the agglutinating order of speech belong all the other Asiatic stock languages except the Aryan and Semitic, besides those of the Caucasus (Nos. 10-13), which may be regarded as holding a position intermediate between agglutination and true inflection. In this system to the unmodified root are loosely attached the determining particles, either as prefixes, infixes, or suffixes, these particles having themselves been originally notional terms worn down by constant use. The word thus becomes an aggregate of elements resembling a rude piece of mosaic still showing all the joinings rather than a perfectly finished organism. It receives its highest development in the Ural-Altaic group, in which an almost indefinite number of suffixes may be agglutinated to the root, and in which greater fusion is effected by the principle of vocalic harmony causing all the vowels of the suffixes to be regulated by that of the root. Thus in Turkish the infinitive particle is *mak* or *mek* according to the vowel of the verbal root: *yazmak* = to write; *sevmek* = to love. Yet even here so slight is the agglutination that the suffixes readily shift their place like so many beads on a string, while the form and position of the primary root remain unchanged. In this way reflective, causal, reciprocal, negative, passive, and many other particles may be tacked on to an almost unlimited extent, each supplanting or yielding to the others according to certain definite laws of use and harmony. Thus the modal particle *mek* of *sevmek* will be thrust one place forward by the causal *dir*, which in its turn will have to make room for the reflective *in*, and so on, whence such combinations as:—

sevmek = to love.
sevdirmek = to cause to love.
sevinmek = to love one's self.
sevindirmek = to cause one's self to love.
sevishmek = to love one another.
sevishdirmek = to cause one another to love.
sevilmek = to be loved.
sevinilmek = to be happy.
sevdishilmek = to be impelled to love.
sevmemek = not to love.
sevishmemek = not to love one another, etc. etc.

The mutual relations of the peoples of Ural-altaic speech, often vaguely grouped together by the collective term "Turanian," pre sent almost endless difficulties to the anthropologist. Comparative philology has placed beyond reasonable doubt the fundamental unity of the Mongolian, Tungus, Túrki, Samoyede, and Finno-Ugrian, which are the five great branches of the linguistic family. But ethnologists are very far from being of accord as to the fundamental unity of the five corresponding racial branches. The reason is, because the languages have in all essentials remained faithful to the original type of the mother-tongue, whereas the race has, through conquest, migration, and other causes, largely departed from the type of the original Mongol stock. Intermixture with the Caucasic peoples in diverse degrees is betrayed in the blue eyes, light-brown and wavy hair, full beard, regular features, and fair complexion, prevalent especially amongst the Manchu, Finnic, and many branches of the Túrki race. The intermingling has been going on from the remotest times, for the old Chinese records speak of the "green eyes," "red hair," and tall figures of the Wu-sun, Ting-ling, and other South Siberian peoples, probably identical with the prehistoric civilised "Chudes" or Finns, traces of whose culture are found scattered over the mineral districts of the Altai and Ural highlands. Farther south and west "Iran" and "Turan" became at an early date so blended together that in many parts of Central and Western Asia we should scarcely suspect the presence of Mongolic blood but for the Jagatai, Turkoman, and other Altaic forms of speech current there.

Elsewhere occurs a phenomenon of a different but scarcely less perplexing order. For we find the inhabitants of Korea and Japan, as well as the Brahuis of Baluchistán, all of more or less mixed Mongolo-Caucasic type, speaking agglutinating languages, which can in no way be affiliated to the Ural-Altaic stock, or, in fact, to any other known forms of speech. W. G. Aston certainly shows some grounds for suspecting a common parentage of the Korean and Japanese, but all attempts to connect the Koreo-Japanese with the Ural-Altaic have hitherto failed. So also with the Brahui, whose assumed kinship with the Dravidian languages of Southern India rests on the flimsiest arguments. All these peoples have been so long separated from the parent stock that their common linguistic elements have become obscured, or rather effaced beyond the possibility of recovery.

The Dravidians themselves, as well as the Kolarians of the Indian peninsula, present difficulties of a still more formidable character. Here we have two agglutinating languages radically distinct from each other, spoken by the dark pre-Aryan aborigines, now of somewhat uniform type, but apparently descended from two distinct stocks. Too much importance has perhaps been given in this region to the element of colour, which may well be due to the slow influence of an extremely hot and moist climate. Some plausible reasons might none the less be advanced to support the theory of an Oceanic origin of the Kolarian-speaking hill tribes, who seem to constitute the autochthonous element in India, and whose affinities may possibly be with the Negritos of the Andaman Islands at one time spread over the Malay and Philippine Archipelagoes. On the other hand, the Dravidians should probably be traced to a continental source. Like the more recent Hindus, they seem to have arrived from the north-west; but they have become so intermingled both with the dark Kolarians and fair Hindus, that their original type can now no longer be determined. There are Dravidians, or, at least, Dravida-speaking hill tribes, such as the Todas and others in the Nilgiris, who might well be taken for Kolarians; and there are others, such as the cultured Telugu and Tamil peoples of the Deccan, who cannot now be distinguished from the Bengali, Orissa, Marathi, and other peoples of Aryan speech. Here again, but for the still surviving fundamentally distinct Kolarian and Dravidian tongues, the presence in India of such diverse ethnical elements might not have been easily suspected, and all the inhabitants of the peninsula might well have been referred to two distinct types—the Aryan and the Kolarian.

Pickering, Logan, and Latham have endeavoured to detect resemblances between the Dravidian and Australian tongues, and out of these attempts has been evolved Huxley's "Australoid Family." But sounder methods of inquiry have shown that the supposed similarity of speech rests merely on a few accidental verbal coincidences, while the physical differences are such as to preclude any direct relationship. The two races have little in common beyond what they have in common with the whole human species.

The other races of agglutinating speech (Malays and Negritos) are treated in the Ethnological Appendix to the *Australasia* of this series.

11. *The Races of Inflecting Speech.*

Regarding the dark, yellow, and fair physical types as three phases in the upward development of the human race, and the isolating, agglutinating, and inflecting linguistic types as three phases in the upward development of human speech, it is significant to find that the Aryan and Semitic—that is, the two most highly-developed inflecting languages—are the exclusive inheritance of the two most highly-developed branches of the fair type. For whatever view may be taken as to the relative excellence of the two orders, there can be no doubt that inflection grows out of agglutination, of which it is the natural outcome. They merge so imperceptibly one into the other that it is not always easy to draw the line between them. But wherever the affix becomes so fused with the root as to be no longer separable from it, we have true inflection. A direct and obvious consequence of this intimate fusion is the modification of both, whence it follows that a modified root is an unerring test of inflected speech. Such forms as *foot*, *feet*, *sing*, *sang*, due to the influence of now vanished suffixes, or to reduplication, establish the claim of English to be regarded as an inflecting language, notwithstanding its extreme poverty of grammatical forms. These very examples show that, instead of being an original condition of speech, inflection is the result of phonetic decay acting simultaneously on the root and its formal elements, and producing a sort of chemical union of the several parts, which can only be resolved by scientific analysis. We have now, so to say, a perfect organism, a complete synthesis, which, if not the best instrument of thought, is certainly the highest morphological evolution of which language is capable.

The Aryans.

The tendency of all speech seems undoubtedly to be from isolation through agglutination to inflection, and from inflection back again to isolation, or from analysis to perfect synthesis, and from synthesis back to analysis.[1] The Latin form *amatur* was built up during its prehistoric life on the stem *am-a*, the third personal pronoun *t*, the connecting or euphonic vowel *u*, and the reflective *se*, with the usual change from *s* to *r*: amatuse=amature=amatur. But this synthetic form has been again, during its historic life, broken up by the Romance tongues into the modern analytical forms *ama-si*, or *si-ama*, *s'aime*, etc. By an exactly similar process have been developed from the synthetical Sanskrit, through the Prakrits, the Panjábi, Bengali, Marathi, and all the other so-called Gaurian or Neo-Sanskritic languages of Northern India. In the same way

[1] Pott, Rénan, and others, have questioned this view, but their arguments seem to be somewhat metaphysical. Speech as an organism, a living growth, must conform to the fundamental laws of evolution from the simple to the complex, from the homogeneous to the heterogeneous—that is, from isolation to inflection of some kind. We cannot take fully-developed languages any more than we can take fully-developed mammoths as starting-points of inquiry. But unfortunately all the known languages of the world are already, and have long been, in a more or less developed state. Here we have, so to say, no protoplasm, no raw material, not even any monads to work upon. But the protoplasm and the monads must be postulated, or else we must assume that fully-formed languages—that is, complex organisms—have always existed as such; in other words, that languages are created, not evolved. There is no alternative.

the modern Afghan, Persian, Kurdish, Baluch, Armenian, are all found to be in an advanced state of decomposition or analysis when compared with their prototypes, the ancient Baktrian, Persian of the Achemenides, and Haik or old Armenian.

All these languages of India and Irania are commonly grouped in two great divisions, forming the Indic and Iranic branches of the Aryan linguistic family. But more recent research has tended to show that the Afghan and Armenian ought to be regarded as indedependent members of the family, while room must now be made for the Galcha, another independent member, of which science has hitherto taken no account, but which seems, like the Afghan, to occupy an intermediate position between the Indic and Iranic branches.[1] The Asiatic division of the Aryan family will accordingly consist, not of two, but of five independent branches, as under :—

Indic Branch :—Sanskrit; The Prakrits . . .	Kashmiri. Nepali. Panjábi. Sindhi. Gujaráti. Maráthi. Hindustáni. Bengali. Assamese. Oriya. Sinhalese (?)
Galcha Branch . . .	Karateghin. Darwazi. Wakhi. Siah-Posh Kafir. Chignani.
East Iranic Branch :—Baktrian (Zend); Pushtu (Afghan).	
West Iranic Branch :— Achemenidian . .	Pehlevi; Parsi; Neo-Persian. Baluch. Kurdish. Ossetian.
Haïk Branch :—Ancient and Modern Armenian.	

None of the Galcha dialects have yet been reduced to writing. The Armenians employ an alphabet based on the Greek, and intro-

[1] The Galcha tribes occupying the upland valleys on both sides of the Hindu-Kush have been recently studied especially by Major Biddulph (*Tribes of the Hindu-Kush*), Ch. de Ujfalvy, and Prof. Tomaschek, of Graz. The type is described as that of a primitive Caucasic race of Aryan speech, with blue eyes, fair complexion, light hair, and brachycephalous or round head, consequently quite different from that of the Iranian Tajiks, who are mostly dark, with dolichocephalic or long heads. The languages seem to incline more towards the Iranic than the Indic branch, while presenting many marked peculiarities entitling them to be regarded as independent members of the organic Aryan tongue. See also Van den Gheyn's *Les Dialectes du Pamir*, 1881.

duced by their Apostle Mesrob in the fourth century. The old Zendic and the cuneiform letters of the Achemenidian epoch have been supplanted by the modern Persian, an elegant form of the Arabic alphabet current amongst all the Muhammadan peoples of Irania and India. But the Sanskritic languages are properly written in various forms of the Devanágari, "the most perfect system of alphabetical notation in existence" (*J. Dowson*). The origin of this syllabic alphabet has long been a subject of much controversy, and although the balance of opinion seems to favour a Semitic descent, the arguments advanced by Dr. Burnell and others in support of this view are extremely weak and unsatisfactory. The question cannot here be discussed ; but it may be stated that further inquiry will probably show that the Indian alphabet is a local invention. The very possibility of such a feat seems to be doubted by those who forget that a similar feat has in recent times been performed by a Cherokee Indian, and that fully as great a miracle was wrought over 2000 years ago by the Hindus themselves, when they invented the so-called "Arabic" ciphers, a numerical system not merely unapproached by any other, but absolutely incapable of improvement.[1]

The Semites.

While presenting some slight analogies to the Aryan, the Semitic form of inflection is marked by several distinctive features of a very remarkable character. Foremost amongst these are its triliteral roots and the astonishing number of grammatical forms produced by mere modification of the root-vowels. So strong is the singular love of triliteration, that even borrowed words are made, by reduplication or otherwise, to conform to the principle, as if the evolution of the language were directed or controlled by some invisible academy of pedantic grammarians. The triliteral root *qtl* yields in Arabic such forms as *qatala*=he killed, *qutila*=he was killed, *uqtul* = to kill, *qatil* = killing, *qatl* = murder, *qitl* = enemy, *qutl*=murderous, and many others, in which the three consonants are always invariable factors, while the vowels are subject to almost endless change. On the other hand, the Semitic tongues have never developed the faculty of word-building, or derivation by composition of two or more roots, or of roots and suffixes, which plays such a wide part in the Aryan group. Altogether the two

[1] The Devanágari system would also appear to be more ancient than is commonly supposed, for M. Terrien de la Couperie and M. de Rosny have recently shown that even before the introduction of writing from China, the Japanese and Koreans possessed an alphabet of Indian origin.

systems have little in common beyond the broad principle of inflection, and whatever germs they may have once jointly possessed have almost entirely disappeared during the course of an independent evolution dating back many thousand years before the historic epoch.

During these long ages the Aryan organic speech has been split up into several distinct groups, still flourishing both in Asia and Europe, whereas the Semitic never developed more than four or five closely-related varieties, none of which have survived except the Arabic and Abyssinian. Arabic is supposed by Dr. Hommel[1] and others to represent the purest type of Semitic speech; but the recently-discovered remains of the long-extinct Assyrian have revealed more archaic grammatical forms, although the Arabic phonetic system seems undoubtedly to have remained most faithful to that of the organic tongue. The Assyrian records as preserved in the brick tablets of Nineveh are written in a peculiar variety of cuneiform characters intermediate between the Akkadian of Babylonia and the Achemenidian of West Persia. All the other Semitic tongues employ different forms of the so-called Phœnician alphabet, which is of hieroglyphic origin, and which is the source of all the European writing systems.

The Semitic domain is confined in Asia to the south-western extremity of the continent, mainly comprising the three regions of Mesopotamia, Syria, and Arabia. Here the race and its peculiar speech are supposed to have become differentiated from the other branches of the Caucasic stock, by which they have at all times been hemmed in and prevented from expanding eastwards. But in a paper recently read before the Anthropological Institute of Great Britain and Ireland, M. Bertin has attempted to show that North-East Africa, or the Nile valley, was the original home of the Semites, whence they separated at a remote period from the Egyptian Hamites. Many striking affinities have certainly been traced between the Hamitic and Semitic tongues, pointing at a common origin of both. But such a possible common Hamitico-Semitic speech carries us back to such a remote epoch that it would be idle to speculate on the probable centre of dispersion. Abundant proofs of a stone age in Egypt have lately been brought forward by General Pitt Rivers and Professor Haynes of Boston.[2] Similar proofs of a stone age in Mesopotamia are not wanting; but there is no evidence to connect either the Hamites or Semites with the races flourishing

[1] *Die Semiten und ihre Bedeutung für die Kulturgeschichte.* Leipzig 1881.

[2] See *Popular Science Monthly* for April 1879, and the *Memoirs of the American Academy of Arts and Sciences*, 1881.

during those ages. They may be their lineal descendants, or they may be more recent intruders, and in either case the migrations of the assumed Hamitico-Semitic people may have been just as well, if not rather more probably, from the Mesopotamian valley westwards, than from the Nile valley eastwards.

The Caucasians.

Although presenting here and there some marked varieties, all the aborigines of the Caucasus seem to belong to the Caucasic stock. There is a general uniformity of type and even of usages and traditions pointing at a common descent. The innumerable dialects spoken in the "Mountain of Languages," as the Caucasus has been called, have also at least a common phonetic system, whose chief characteristic is extreme harshness. In this respect they far exceed all other languages in the Old World, and are rivalled in the New only by the Thlinkeet, Chinook, Apache, and some others on the north-west coast. The climate of Caucasia, in some places excessively hot and moist, elsewhere of an almost Arctic rigour, is inadequate to explain this extraordinary and universal roughness of articulation, which almost exceeds the limits of credibility. Thus the old capital of the kingdom of Georgia is spelt *Mtzkhet*, though foreigners well may ask how such an accumulation of hard initial consonants is to be pronounced. Any effort of an Englishman to do so would probably sound more like *Manx Cat* than anything else. Yet we are told that Georgian is musical compared with Cherkess and Abkhasian, which are themselves quite soft compared with Chechenz and some of the Daghestán dialects.

Many of these dialects, although grouped together in our scheme under the collective name of Lesghian, present profound structural differences, almost sufficient to entitle them to be classed as fundamental stock languages. In fact the actual number of stock languages spoken in Caucasia is still far from being determined. No doubt the Georgian, Svanian, Mingrelian, Laz, and all the other southern idioms, form a single linguistic family commonly known as the Kartvelian, while the Abkhasian, Cherkess, Kabard, and the other western and west central, are also derived from a common source. But in the east the Chechenz is certainly distinct from the Avar, and the affinities of the Dargo, Dido, Ude, Duodez, and many others current in Daghestán, are so uncertain that all attempts have hitherto failed to reduce them to a common origin. Some differ even in their morphology—those of South Daghestán having apparently already arrived at the inflecting state (*M. Smirnov*), while the Georgian and Cherkess are regarded as still agglutinating, although here it often seems hard to draw the line between the two orders.

The surprising number of stock languages thus developed within the narrow area of the Caucasus, and which have here been current as far back as historic record goes, finds elsewhere no parallel, except, perhaps, in the Sudan and some parts of North and South America. Neither climate nor the physical features of these secluded upland valleys are at all sufficient to account for the phenomenon. A plausible explanation has been suggested by the geographical position of this region—lying on the highway of migration between two continents, and serving as a refuge for numerous races of the surrounding plains, which may have escaped to these mountain fastnesses from the sword of devastating hordes. All these causes may doubtless have contributed to the result, which, however, seems mainly due to the vast period of time during which the Caucasian highlands have been peopled. The favoured and isolated valleys of the Rion, Ingur, Upper Kuban, and Terek rivers—centres of evolution for several animal and vegetable species—have been probably inhabited long enough to have also become centres of evolution for several linguistic species. There was a time when Europe and Asia were separated by the broad Ponto-Caspian Strait, preventing any spread of the human race northwards, protecting Caucasia from hostile inroads in that direction, and thus promoting the early and rapid peopling of its fertile upland and lowland tracts. Here the very Caucasic type itself may well have become differentiated from the Mongolic, and it is a remarkable fact that some Suabian immigrants settled in the Kura valley have in a few generations become assimilated to, without intermarrying with, the surrounding Caucasian peoples.[1] Certainly this is a region of intense physical energy, where the vital forces have been in operation for an incalculable period of time in a number of secluded valleys cut off from intercourse with the outer world, and thus affording a most favourable field for independent local development. The multiplicity of the Caucasian languages may therefore be accounted for without supposing them to be surviving fragments of linguistic families formerly spread over wider areas. Of such families not a trace is elsewhere to be found, and the attempts made to connect, for instance, the Cherkess with the Basque of the Pyrenees have now been abandoned as hopeless.

1 "Il paraît que dans l'espace de deux générations les colons souabes ont changé physiquement d'une manière remarquable sous l'influence du milieu. Quoiqu'il n'y ait point eu de croisement entre eux et leurs voisins, la plupart ont maintenant la chevelure foncée, les yeux noirs, la figure ovale et régulière, la taille élégante et souple. Ils ne ressemblant plus á leurs cousins restés dans la mère-patrie" (*Reclus*, vi. p. 225).

12. *Races and Languages of Doubtful Affinities.*

A recent writer in the *Atlantic Monthly* well observes that "there never was a primitive Mongolian [Mongolic] mother-tongue in the sense in which there was a primitive Aryan mother-tongue. The common ancestors of Japanese, Chinese, Tungusian, and Mongol never at any time lived together in one great society, welded into a unit of community of language, traditions, and customs, as was the case with the common ancestors of Roman, Teuton, and Hindu. On the contrary, the aboriginal yellow men must have roamed about in detached tribes, like the blacks of Australia or the red men of America, with half-formed languages fluctuating from generation to generation, diverging with great rapidity, and speedily losing all traces of their origin. . . . With such divergent careers as these, we need not expect to find evidence of linguistic community among the different branches of the yellow race. If we find one set of linguistic phenomena in China, and a totally different set in Japan, and yet another set among the barbarous Mongols and Tunguses, this is no more than we might have expected. We find just such a state of things as would follow from the isolated independent development of a number of languages either without any original kinship, or with the original kinship blurred and destroyed almost from the very beginning."

These remarks are quite as applicable to peoples of the Caucasic and other stocks as they are to those of the Mongolic race. Hence it is not surprising to find in almost every part of Asia sporadic groups of languages, which can no longer be affiliated to any recognised linguistic families. The Brahui, Dravidian, and Kolarian have already been referred to (p. 703). The Sinhalese of Ceylon seems to occupy a somewhat doubtful position between the Aryan and Dravidian groups. The race itself, of which the Weddhas are probably the original stock, is extremely mixed, betraying obvious traces of Aryan, Dravidian, Kolarian, and other elements. But the language, notwithstanding a large admixture of Sanskrit words, due to Buddhist influences, appears to be fundamentally of Dravidian type.

The Khmers of Cambodia may be taken as the most typical representatives of the Caucasic stock in Indo-China, whence they sent off branches at a vastly remote period into the Eastern Archipelago and the Pacific Ocean. Hence the affinities of the Khmer language are with the so-called Malayo-Polynesian group, as elsewhere shown by the writer.[1]

[1] *On the Relations of the Indo-Chinese and Oceanic Races*, by A. H. Keane. Trübner, 1880.

A very distinct and important branch of the Khmers are the Chams (Ciam or Tsiam), also an historical people, who are supposed to have been the dominant race in Further India before the rise of the Cambodian power. They are still found in scattered communities spread over a wide area, although their chief home is the Champa country, in the south-east corner of the peninsula. In a paper on the Cham dialect of Dambang-dek recently read before the Académie des Inscriptions, M. Aymonier states that they have three distinct languages—the *Dalil* or sacred language, the *Cham* or vernacular, and the *Bani*, a Muhammadan idiom which has now supplanted the others. Several Sanskrit inscriptions of great antiquity have been discovered in the country, and the people themselves have a tradition that their culture was received from India some 2000 years ago. Since then they have been affected by Malay influences to such an extent that Crawford wrongly supposed they were a branch of the Malay race. The migrations were from Indo-China to the Eastern Archipelago, with an occasional reflux back to the mainland, as in Malacca and Champa.

The Ainos, another offshoot of the Caucasic stock, seem to have been the earliest inhabitants of the Japanese Archipelago, where they still survive in the northern island of Yesso and in Sakhalin. Their speech is entirely distinct, having lost all traces of any former connection with the other Caucasic tongues of the mainland. But their type, as graphically described by Miss Bird in *Unbeaten Tracks*, is obviously a mere variety of the Caucasic, although a recent writer has attempted to affiliate them to the Mongolic.[1] If the Ainos, with their full bushy beards, hairy bodies, straight eyes, and light complexion, are to be regarded as Mongols, then Havelacque will be fully justified in rejecting the term Mongolic altogether as a distinct variety of mankind. The Ainos are either Caucasic or else the last survivors of some third extinct stock, of which science has no further knowledge.

The scattered north-east Siberian tribes, commonly grouped as "Hyperboreans," all seem to be physically of Mongolic origin. But their languages show no perceptible relationship with any others in the Old or New World. Whatever affinities they may have among themselves are of an extremely vague character, except in the case of the Chukch and Koriak, evidently dialects of a common mother-tongue. The Chukchis themselves, regarding whom so many extravagant theories have recently been advanced, seem to be merely a branch of the Koriak race, who have at one or two points of the coast come in contact with Eskimo tribes from America.

[1] J. J. Rein in *Japan nach Reisen und Studien*. Liepzig, 1881, vol. i. p. 443.

So little is known respecting the Solos, Man-tze, and other aboriginal tribes of South-West China, that any theorising on their possible origin and affinities would be premature. All that can be stated with any certainty is that they differ greatly in physical appearance, some approaching the Mongolic others the Caucasic type, while their languages seem to belong mainly to the isolating order. Yet Gill speaks of the Sumu, a large Man-tze nation in Se-chuen, whose speech is of Aryan type. Should this statement be verified, it will rank as perhaps the most surprising discovery made in recent times in the field of ethnology.

INDEX.

THE END.

Printed by R. & R. Clark, *Edinburgh.*

Spotted something that isn't quite right?

Let us know and we'll fix it for future reprints. And as a thank you for your diligence, we'll send you a free Sothis book of your choice.

publisher@sothis.press

This page intentionally left blank

Made in the USA
Monee, IL
08 November 2026

68028671R00454